Superconductors

Superconductors

A.V. Narlikar

OXFORD
UNIVERSITY PRESS

OXFORD
UNIVERSITY PRESS

Great Clarendon Street, Oxford, OX2 6DP,
United Kingdom

Oxford University Press is a department of the University of Oxford.
It furthers the University's objective of excellence in research, scholarship,
and education by publishing worldwide. Oxford is a registered trade mark of
Oxford University Press in the UK and in certain other countries

© A.V. Narlikar 2014

The moral rights of the author have been asserted

First Edition published in 2014

Impression: 2

Published in the United States of America by Oxford University Press
198 Madison Avenue, New York, NY 10016, United States of America

British Library Cataloguing in Publication Data
Data available

Library of Congress Control Number: 2014933417

ISBN 978–0–19–958411–6

Printed in Great Britain by
Clays Ltd, St Ives plc

Acknowledgements

This book is about superconducting materials. It focuses on super-conductors that stand out for three reasons: (1) they are scientifically exciting, with novel properties, and indicate new physics and unusual superconductivity mechanisms, or (2) they are technologically already established as first-generation superconductors for important practical applications, or (3) they constitute newly discovered systems, possessing exciting features to serve as future generations of practical supercon-ductors for applications. The book contributes to an already rich schol-arship in the field, and brings unique value-added in at least two ways. First, most book-length accounts have focused on the phenomenon of superconductivity, whereas in this book, it is the materials themselves—with all their distinctive properties and applications—that form the focus of the study. Second, while there are already several useful accounts that concentrate on a specific class of superconducting materials, this book provides an analysis of a wide range of materials that transcends type and class. The subject matter should be of interest to graduates and final-year undergraduates reading Natural Sciences and Engineering. It is also likely to be directly relevant to specialists working on partic-ular superconductors and trained engineers interested in the possible new-generation superconductors for applications.

I am grateful to the Indian National Science Academy, New Delhi, and its Science Promotion programme for supporting my supercon-ductor research and the preparation of this book. Thanks are extended to Dr Praveen Chaddah, the Director of UDCSR, for making available various facilities and resources of the Consortium. Helpful assistance provided by Arjun Sanap, Madhulika Sinha, U. P. Deshpande, and Suresh Bharadwaj is acknowledged. I am grateful also to my former colleagues at NPL, New Delhi, in particular Drs Hari Kishan, Anurag Gupta, and V. P. S. Awana for many useful discussions. A substantial portion of the writing of the book was carried out during my various visits to Cambridge, where I hold a stimulating affiliation with the Applied Superconductivity and Cryoscience Group at the Department of Materials Science and Metallurgy, University of Cambridge. I am thank-ful to Professor Glowacki for this and for many valuable discussions in his group. I remain thankful to my wife Dr Aruna Narlikar for her invaluable help, patience, and constant support throughout, and for her useful suggestions on many occasions during the preparation of the book. For her technical help, I would also like to thank our daughter, Dr Amrita Narlikar, who in fact also first persuaded me to take up this project. Finally, I would like to extend my grateful thanks to Dr Sönke Adlung for his sustained patience and for being most cooperative and considerate. It was a pleasure working with him, and his various suggestions have been extremely useful for the project.

A.V. Narlikar

June 2013

Contents

CHAPTER I

Onnes' discovery and one hundred years of superconductors

The field of superconductors began more than a century ago, but it remains a rare example among the numerous scientific developments of the twentieth century where the original excitement of discovery has not just been retained but indeed has continued to grow so long after the original event. Indeed, one seldom finds any new type of material or phenomenon that remains a front runner in both basic and applied research so many years after its discovery. Generally, a time span of a few decades is sufficient for a scientific discovery to move from the journals reporting the research frontiers to become a topic in undergraduate textbooks. The fact that the field has already been the subject of six Nobel Prizes to as many as 12 researchers provides but one piece of evidence of its richness, and it is not surprising that superconductors are among the most promising technologically relevant materials of the present century.

Interestingly, the phenomenon is exhibited by an incredible diversity of materials that include pure elements, alloys, intermetallic compounds, ceramics, inorganic polymers, and others. Even some organic salts, doped fullerides, complex oxides, pnictides, and novel heavy fermion compounds are found to be superconductors with intriguing properties. Moreover, superconductors can be single-crystalline, polycrystalline, thin films, highly disordered, or even in the amorphous state. In several instances, the phenomenon manifests itself in intriguing ways in various forms, and under different conditions such as *re-entrant superconductivity*, *proximity-effect-induced transient superconductivity*, *pressure-induced superconductivity*, *magnetic-field-induced superconductivity*, and even *photo-induced metastable superconductivity*.

The field that was originally a mere academic curiosity for a handful of experimental and theoretical physicists has now transformed into an area of intense and highly competitive research in diverse disciplines including structural chemistry, ceramic engineering, materials science, metallurgy, solid state electronics, electrical and electronic engineering, and cryogenics. Today, superconductivity is found to occur at temperatures as high as 164 K (Gao et al., 1994) for a mercury-based cuprate under a pressure of 30 GPa, and there seems every possibility that novel superconductors may be discovered at even higher temperatures.

1.1 Onnes' discovery

In 1911, Heike Kamerlingh Onnes at the University of Leiden in the Netherlands discovered the pure element mercury as the first superconductor (Onnes, 1911). The discovery took place about three years after his remarkable achievement, on 10 July 1908 (Onnes, 1908), in liquefying helium gas at a temperature of 4.2 K at ambient pressure. Subsequently, on pumping down the vapour pressure of the liquid, he could reach a temperature of 0.9 K, which was then the lowest temperature attained anywhere in the world. The behaviour of electrical resistance of metals at low temperatures was among the first problems investigated by Onnes. For gold, of varying purity, he found the resistivity to gradually decrease with temperature, but to become saturated before reaching liquid helium temperature as a result of impurity scattering as per Matthiessen's rule. He therefore chose mercury, which could be readily purified by multiple distillation, and studies of its low-temperature resistivity yielded the most unexpected result (inset in Figure 1.1). As with gold, its resistivity initially decreased (Figure 1.1) gradually with temperature, but a little below 4.2 K it dropped abruptly to a value indistinguishable from zero. These measurements were made by Onnes' co-workers, Gilles Holst, a graduate student, and Cornelius Dorsman, a post-doctoral researcher, while the cryogenic part of the procedure was monitored by Onnes and his chief technician, Gerrit Flim (de Bruyn, 1987). The measurements were made on 8 April 1911, and Onnes reported his surprising result at the meeting of the KNAW (the Dutch Royal Society) on 28 April 1911.

Initially, Onnes thought (de Bruyn, 1987) that the resistivity to have fallen to some low value, but soon he realised this to be a novel zero-resistance state, which he called 'supraconductivity' or *superconductivity*, as we know it today, and the temperature at which it occurred the *critical temperature* T_c. Onnes' subsequently found tin and lead also to be superconductors, with T_c values of 3.72 and 7.20 K, respectively. In 1913, Onnes received the Nobel Prize in Physics for liquefying helium gas and discovering the new phenomenon of superconductivity.

It is worth mentioning that even with the much improved precision of measurements available today, it is seemingly impossible to assert that the resistance has indeed fallen to zero—only an upper limit to the value can be set. Interestingly, even today, the method used to estimate the resistivity of superconductors is the same as the one that Onnes followed a century ago. An electric current is induced in a closed ring of superconductor by applying a changing magnetic field. If the superconducting ring of inductance L has a resistance R, the trapped current $I(0)$ at time

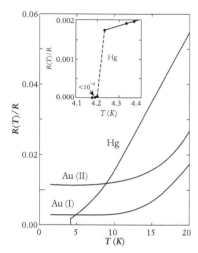

Figure 1.1 Onnes' discovery of superconductivity in mercury (see also the inset). Also shown is the resistance behaviour of gold samples of higher purity, Au (I), and of lower purity, Au (II).

$t = 0$ will decay with time in accordance with $I(t) = I(0) \exp(-Rt/L)$. For a ring of normal metal such as copper, or for a superconductor in the *normal state* (i.e. the state commencing just above T_c), the induced current decays within a fraction of a second, while in a superconducting loop it takes many years of observations to detect any such decay. This is conveniently studied in terms of the loss in the magnetic field of the decaying current. A long series of such measurements, performed by Onnes over several months, and much later by many others using more sensitive techniques such as Hall probes and NMR on different superconductors (Roberts, 1964; Skolnick et al., 1987), have helped to set $10^{-23} - 10^{-18}$ Ωm as the upper limit to the resistivity of the superconducting state. The larger value (Skolnick et al., 1987) is usually for high-T_c cuprate superconductors at the higher temperature of liquid nitrogen, where there is an added dissipation of current due to flux creep, to be described later. These figures, when compared with the value of 10^{-8} Ωm for the low-temperature resistivity of purest copper, are taken as zero. Owing to their very long decay time, these currents are termed *supercurrents* or *persistent currents*. Their prominent applications are in *high-field superconducting electromagnets* and *bulk permanent superconducting magnets*. In the former, by shorting the two current terminals of the superconducting solenoid with a superconducting wire after energising, the current flowing in the solenoid becomes trapped in the persistent state and the magnet behaves as a permanent magnet, operating without a power supply (as long as it is kept cooled below T_c). Similarly, a bulk superconductor, magnetised in a high magnetic field, will develop a large trapped magnetic flux, maintained by induced supercurrents. This also allows such superconductors to serve as permanent magnets (again as long as their temperature is below T_c).

1.2 One hundred years of superconductors

During the hundred years following Onnes' discovery, several thousand superconductors have been found, with more than half of the periodic table to date exhibiting superconductivity. Over these years, the field of superconductors has proved to be extraordinarily rich and dynamic, frequently manifesting some unexpected turn in the form of a novel property or phenomenon. Table 1.1 summarises the prominent milestones achieved along this hundred years of progress. Since various developments listed in the table will be discussed separately in the forthcoming chapters, here we only briefly examine some of them, with a historical perspective.

It is worth recalling that one of the first ideas pursued by Onnes, immediately after his discovery, was to wind a superconducting electromagnet to generate a high magnetic field of about 10 T, free from Joule dissipation, but he soon discovered that even small magnetic fields

Table 1.1 Summary of major developments occurring in the field of superconductors during the last hundred years

No.	Years	Discovery
1	1911	Onnes' discovery, with Hg as the first superconductor
2	1911–1930	Various elemental superconductors, with Nb having the highest T_c of 9.2 K Existence of critical current and critical field and Silsbee's rule
3	1930–1940	Meissner effect London phenomenology High-field behaviour of alloys, Superconductivity in the amorphous state
4	1940–1950	NbN ($T_c = 16$ K) Pippard's non-local modification of London theory and range of coherence and formation of intermediate state
5	1950–1960	Ginzburg–Landau phenomenology Abrikosov's theory of type II superconductors Isotope effect and Fröhlich's suggestion of the role of phonons Formulation of BCS theory Discovery of various A-15 superconductors, with $T_c = 18$ K for Nb$_3$Sn
6	1960–1970	Discovery of Giaever and Josephson tunnelling Commercial development of Nb–Zr and Nb–Ti conductors for superconducting electromagnets producing 6–10 T magnetic fields
7	1970–1980	Nb$_3$Ge as highest-T_c superconductor Large magnets for MHD, bubble chambers, etc. Bronze process for Nb$_3$Sn conductor Ternary stannides, borides, and chalcogenides manifesting interplay of superconductivity with magnetic order Re-entrant behaviour Transition metal dichalcogenides and heavy fermion superconductors
8	1980–1990	Superconducting organic charge-transfer salts Discovery of high-T_c cuprate superconductors with $T_c > 130$ K Superconducting bismuthates
9	1990–2000	$T_c = 164$ K of Hg-1223 under pressure Quaternary borocarbides Doped fullerides Ruthenates and ruthenocuprates Metallonitride halides
10	2000–2012	Magnesium diboride Non-oxide perovskites Pyrochlore oxides Sodium cobaltate hydrate Ferro-pnictides with $T_c \approx 55$ K BiS$_2$-based superconductors

below 0.03 T, called the *threshold* or *critical field* H_c, abruptly destroyed the phenomenon. A similar effect resulted on passing a small electric current through the superconducting wire when this current exceeded a certain critical value I_c. Silsbee (1916) pointed out (*Silsbee's rule*) that the two effects are essentially the same, with the critical current I_c being that which produces the critical magnetic field H_c on the wire's surface. This was a major setback in terms of practical applications

of the phenomenon. It should be mentioned that subsequent work at Leiden had found a lead–bismuth alloy possessing a larger critical field of about 2.5 T. However, a solenoid wound with this material failed to sustain a high current density, and superconductivity was lost at disappointingly low transport currents. It became clear that the major constraints on the use of superconductors for any viable application were (a) low transport current densities in high magnetic fields and (b) the necessity for liquid helium temperatures. Clearly, breakthroughs in these areas were needed, the former calling for a high-field/high-current superconductor and the latter a high-temperature superconductor. The first breakthrough came in the early 1960s, with the universal recognition of the existence of the so-called type II superconductors, followed by the discovery of high-field/high-current properties of niobium-based alloys (Hulm and Blaugher, 1961) and intermetallic A-15 compounds (Kunzler et al., 1961). The second breakthrough, in 1986, was the discovery of high-T_c ceramic cuprates (Bednorz and Müller, 1986), that is, the so-called HTS (high-temperature superconducting) cuprates.

Until the 1930s, superconductors were characterised simply by the vanishing of their electrical resistance. Then, Meissner and Ochsenfeld (1933) found that in the superconducting state, the material, besides becoming perfectly conducting, also becomes perfectly diamagnetic. In the *Meissner effect*, a field-cooled sample expels the magnetic field from its bulk when its temperature falls just below T_c and it becomes superconducting. The phenomenon is reversible in that with a rise in temperature to just above T_c, the expelled field re-enters the sample, making it normal (i.e. non-superconducting), as abruptly as it had been expelled, just below T_c, to make it superconducting. The Meissner effect is an independent property of a superconductor that does not follow from zero resistivity. Realisation of this led F. and H. London (1935) to propose their phenomenological explanation of the perfect diamagnetism in conjunction with infinite electrical conductivity. The superconductor excludes all the magnetic field from its bulk, leaving a thin surface layer of thickness λ_L, called the *London penetration depth*, over which persistent currents flow so as to balance the external magnetic field and make the bulk of the superconductor field-free. The reversible nature of the superconducting transition in a magnetic field led Gorter and Casimir (1934) to apply thermodynamic considerations to the phenomenon.

Penetration depth measurements in the mid-1940s revealed shortcomings of the London phenomenology, which, for instance, was not able to account for the positive surface energy that must exist between the normal and superconducting phases. This, along with other experimental observations, led Pippard (1950) to introduce *coherence* as a basic property of a superconductor that makes any perturbing influence on the superconducting order parameter *non-local*. The order parameter is affected over a distance from the location of the disturbance equal to the range of coherence, ξ. Ginzburg and Landau (1950) subsequently developed a comprehensive phenomenological theory, taking account of the non-local character of the superconducting state, which proved to be a distinct improvement over the London phenomenology.

Abrikosov (1957) extended the Ginzburg–Landau theory to the seemingly improbable situation of negative surface energy between normal and superconducting phases, and in this way he predicted a whole new class called *type II superconductors* that included alloys and compounds. In contrast to superconducting pure elements, which were termed *type I superconductors*, in which superconductivity was abruptly destroyed by a small critical magnetic field H_c, in type II superconductors the phenomenon persisted up to a much higher critical field, called the *upper critical field H_{c2}*, and the destruction of superconductivity by the field was gradual and not abrupt. Interestingly, these two distinct types of behaviour had already been experimentally observed by Shubnikov et al. (1937), but owing to the outbreak of the Second World War had failed to attract much attention. Abrikosov's work established that a type II superconductor enters what is known as a mixed state (also called the *Shubnikov phase*) that is composed of a lattice-like arrangement of quantised magnetic flux vortices, and it turns out that in order for such a material to carry large transport current densities without resistance in high magnetic fields, the movement of these flux vortices has to be anchored, or pinned down. Interestingly, various microstructural features in the form of extended crystal defects and inhomogeneities in the material can act as pinning centres for flux vortices. Thus, by introducing appropriate types of microstructures through metallurgical treatments, type II superconductors can be made to sustain large critical currents in high magnetic fields to meet the technological requirements for various applications. This was a major breakthrough during the early 1960s that involved materials scientists, metallurgists, and electrical engineers putting together R & D efforts in the field of superconductors, which had hitherto been the province of a small group of academic physicists. Around the same time, Matthias (1957), who had long been exploring superconductivity in a wide spectrum of alloys and compounds, proposed empirical rules that led to the discovery of numerous superconductors with relatively high T_c values, prominent among which were binary and pseudobinary A-15 compounds such as V_3Si ($T_c = 17.1$ K), Nb_3Sn ($T_c = 18.5$ K), Nb_3Al ($T_c = 18.9$ K), and Nb_3AlGe ($T_c = 21$ K). By the early 1970s, Nb_3Ge could be synthesised in thin-film form with a T_c of 23.2 K, which, until the discovery of the high-T_c cuprates in 1986, remained the highest known critical temperature (Testardi et al., 1974).

In parallel with the above developments, attempts were continually being made to find a microscopic explanation of superconductivity, but no significant clues were in sight. An important development was the discovery of the *isotope effect* (Maxwell, 1950), in which the T_c of pure elements varies inversely as the square root of the isotopic mass. This was a convincing indication of the involvement of phonons. Interestingly, Fröhlich (1950) had independently linked superconductivity with the electron–lattice interaction, suggesting that this leads to an effective attraction between the electrons. In addition, several studies of thermal properties carried out in early 1950s were indicative of the formation of an energy gap in superconductors.

These developments served as crucial clues for Bardeen, Cooper, and Schrieffer (1957) in formulating their well-known BCS theory, according to which superconductivity occurs when the repulsive interaction between two electrons at the Fermi level is overcome by an attractive interaction resulting from a mechanism involving phonon exchange. This results in two electrons of opposite spin and momentum forming a bound state, called a *Cooper pair*, which moves without resistance. The triumph of the BCS theory rested on its convincing explanations for a number of features, such as the zero resistance, the Meissner effect, the isotope effect, and the energy gap. Careful measurements (Deaver and Fairbank, 1961; Doll and Nähbauer, 1961) showed the magnetic flux trapped inside a superconducting ring to be quantised in units of $h/2e$, rather than h/e, which is evidence of the presence of electron pairs. The theory was supported further by the phenomena of tunnelling of electrons through a thin insulating barrier between a superconductor and a normal metal, known as *Giaever tunnelling* (Giaever, 1960), and between two superconductors, known as *Josephson tunnelling* (Josephson, 1962). These studies corroborated the existence of both the energy gap and electron pairing. Apart from providing a firm footing for the BCS theory, Josephson tunnelling has over the years become an experimental tool in the form of superconducting quantum interference devices (SQUIDs) (Clarke, 1982) of unsurpassed sensitivity for detecting very small magnetic fields with noise levels as low as $1\,\mathrm{fT}/\mathrm{Hz}^{1/2}$.

By the mid-1970s, commendable progress had been made in superconducting magnet technology, with small superconducting magnets finding increasing use in low-temperature laboratories and large ones being successfully tested for use in big installations such as prototype magnetohydrodynamic (MHD) devices, nuclear bubble chambers, and particle accelerators (Iwasa and Montgomery, 1975). The so-called *bronze route*, based on solid state diffusion, was successfully developed into a variety of commercially viable processes for fabricating fully stabilised multifilamentary long conductors of the brittle compound Nb_3Sn that today find extensive use in winding superconducting magnets for fields of $10-20$ T.

The search for new superconducting systems during the 1970s led to the discovery of Chevrel-phase superconductors, which are ternary molybdenum chalcogenides (Chevrel et al., 1971), some possessing the highest critical fields (>50 T) known at that time. In contrast to most other superconductors, where the magnetic elements had a deleterious effect on superconductivity, their presence in the Chevrel phases, surprisingly, supported it. Studies of these novel materials, along with others such as rare-earth-based ternary borides and stannides, which had been discovered around the same time, led to the start of an exciting era of research into the mutual competition between magnetism and superconductivity (Subba Rao and Shenoy, 1981). This led to numerous interesting developments, such as the discovery of *re-entrant behaviour*, the *coexistence of superconductivity and magnetism*, and the advent of *magneto-superconductors*. The interplay has since been extensively studied also in the rare-earth-based superconducting *quaternary borocarbides* (Mazumdar et al., 1993; Nagarajan et al., 1994), for example RNi_2B_2C, where R is a rare-earth (RE) ion. The interesting feature of borocarbides is that T_c and the magnetic ordering temperature T_M are not too small and are close to each other, which has proved to be a considerable asset in their investigation.

The 1970s witnessed the advent of another fascinating class of superconductors, the *heavy fermion* (HF) *superconductors*, which contain regular arrays of either lanthanide or actinide ions, notably cerium and uranium (Steglich et al., 1979). Although T_c in most cases is small, about 1 K or even lower, these materials exhibit unusual normal-state and superconducting behaviour due to their effective electron mass being one or two orders of magnitude larger than the bare electron mass. This results in unusual features such as the coexistence of superconductivity and magnetic order, significant T_c depression with non-magnetic doping, and strong coupling of the heavy quasiparticles to the cell volume, each pointing to an unconventional mechanism of Cooper-pair formation.

The year 1980 began with the discovery of superconductivity in an organic salt $(TMTSF)_2PF_6$ (Jerome et al., 1980), with a T_c of about 1 K under pressure and with spin density waves (SDWs) present at temperatures a little above T_c. This was followed shortly by the synthesis of a higher-T_c organic system $(ET)_2X$, with $T_c = 2.5$ K for X = ReO_4 (Perkin et al., 1983) and $T_c = 12$ K for X = $Cu[N(CN)_2]Br$ (Williams, 1990). With the advent of fullerides during the 1990s (Haddon et al., 1991), it was found that thin films of these substances, when suitably doped with alkali metals (M_3C_{60}, where M is an alkali metal), showed superconductivity with T_c of 18 and 29 K, respectively, for M = K (Hebard et al., 1991) and M = Rb (Rosseinsky et al., 1991). On further doping of the latter with Tl, the T_c was increased to about 40 K.

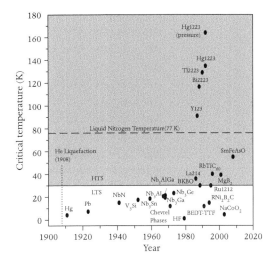

Figure 1.2 Years of discovery of some prominent superconductors along with their critical temperatures. LTS, low-temperature superconductors; HTS, high-temperature superconductors.

Although ever since Onnes' discovery, the maximum known value of T_c had been rising continuously, the average rate at which it advanced was around 1 K every three years. Figure 1.2 depicts the critical temperatures of some of the prominent superconductors, along with the year of their discovery. Until 1986, the highest T_c recorded was 23.2 K for a Nb_3Ge thin film, and this value had remained stagnant for about 12 years. Bednorz and Müller's (1986) exciting announcement of a T_c of around 30 K for a ceramic sample of $La_{1.85}Ba_{0.15}CuO_4$ was most unexpected. The excitement was not due to the high T_c alone, although in itself this was a remarkable feat. More than that, this was also a breakthrough in the type of materials that were being investigated. The ceramic oxides are generally insulators, and no one had ever thought of cuprates as serious candidates for superconductivity, let alone for high T_c. Within months of Bednorz and Müller's discovery, $YBa_2Cu_3O_{7-\delta}$ was found (Wu et al., 1987) as the first superconductor with $T_c \approx 90$ K, well above the boiling point of liquid nitrogen at about 77 K. Clearly, a new family of high-temperature superconductors, commonly termed HTS, in the form of cuprates was emerging. Those materials with T_c in excess of 30 K are generally considered as HTS, while lower-T_c materials are termed LTS (i.e. low-temperature superconductors). In 1986, the maximum known T_c started rising rapidly (Figure 1.2), and finally, in 1994, the highest T_c of 164 K was reached for the cuprate $HgBa_2Ca_2Cu_3O_{8+y}$, under pressure (Gao et al., 1994).

During the mid-1990s, the ruthenocuprates were discovered as variants of the high-T_c cuprates, with two distinct compositions: $RuSr_2RECu_2O_8$ (Ru1212) (Bauernfeind et al., 1995) and $RuSr_2(RE, Ce)_2Cu_2O_{10}$ (Ru1222) (Felner et al., 1997). Much of the interest in these two systems has been due to the exciting interplay between their high-temperature superconductivity and the complex magnetic orders that emerge at different temperatures. The search for non-copper-based HTS has led to superconducting bismuthates, with $Ba_{0.6}K_{0.4}BiO_3$ (Cava et al., 1988) having the highest $T_c \approx 35$ K in this class.

The news in early 2001 of superconductivity at 39 K in MgB_2 (Nagamatsu et al., 2001) was received with much surprise, since this binary material had been in commercial production in quantities of tons per annum since the 1950s. It was astonishing that its relatively high T_c had remained unnoticed for so long. Interestingly, Swift and White (1957) had failed to notice the specific heat anomaly at about 40 K present in their heat capacity data on MgB_2. With its impressive T_c of 39 K, MgB_2 holds exciting potential to replace Nb—Ti ($T_c = 10$ K) and Nb_3Sn ($T_c = 18.5$ K), which are presently used in practically all high-field (10–20 T) applications.

From 2001 to date, besides MgB_2, many novel superconductors have been announced almost every year, the latest being those based on BiS_2, discovered in 2012. These, together with sodium–cobalt oxide hydrate (Takada et al., 2003) and various iron-based superconductors containing phosphorus (Kimihara et al. 2006) and arsenic (Takahashi et al. 2008), are the prominent developments of the present century. The oxy-ferro-pnictides $RFeAsO_{1-x}F_x$ (where R is a rare-earth ion) are found to have T_c values up to 55 K, with critical fields of 50–100 T, which places them next to the HTS cuprates.

Technologically, the HTS cuprates, owing to their higher T_c, are of particular interest for practical applications. However, they are inherently brittle and are highly anisotropic systems, both of which are serious handicaps. Also, with their very small range of coherence, their grain boundaries manifest a pronounced *weak-link effect* that drastically lowers their current densities. In addition, their functioning at liquid nitrogen temperatures is seriously impaired by flux creep effects. All of these factors have constituted major challenges for the development of these materials as conductors, in the form of wires and tapes, for high-field electromagnets and transmission cables functioning at liquid nitrogen temperatures. These challenges are being effectively met, however, and the overall progress that has been achieved with these materials has been truly impressive.

Despite numerous attempts, the mechanism behind the high values of T_c has to date remained elusive. This is mostly because the cuprates and pnictides have turned out to be most unusual materials with intriguing properties even above T_c; that is, even their *normal state* remains to be fully understood. Besides HTS cuprates and pnictides, there are several low-T_c systems such as organic superconductors, heavy fermion materials, ruthenates, and cobaltates, among others, whose superconductivity is unconventional, with both their mechanism of electron pairing and the symmetry of their order parameter being different from what was originally envisaged in the conventional BCS theory.

1.3 Progress with LTS and HTS applications

Although this book is not about superconductor applications, we briefly provide in this section some highlights of the progress that has been made. Since the mid-1960s, LTS have revolutionised high-field electromagnet technology on a global scale. In particular, Nb–Ti alloys ($T_c \approx 10$ K) and the intermetallic compound Nb_3Sn ($T_c = 18$ K) have proved to be the backbone of present-day superconducting magnet technology for fields up to 20 T. Today, these two superconductors alone are responsible for a billion-dollar superconducting wire industry. For instance, they are ubiquitous in magnetic resonance imaging (MRI), in magnetic confinement of high-temperature plasma in fusion reactors, in high-speed maglev trains, and in beam steering and focusing magnets in large particle accelerators, such as the Large Hadron Collider (LHC) at CERN, Geneva. In fact, the feasibility of gigantic superconducting magnets several metres in diameter and producing magnetic fields of about 5 T at the liquid helium temperature of 4.2 K was well established way back in the late 1960s for applications such as MHD power generation, liquid hydrogen bubble chambers, and superconducting magnetic energy storage (SMES).

At present, the most common medical application of LTS is in MRI, for the production of large-volume stable magnetic fields for whole-body imaging. This application alone represents a multibillion-dollar global market. Experimental high-energy physics research hinges on being able to accelerate subatomic particles to nearly the speed of light using superconducting magnets. The LHC could not have been conceived without the use of superconductors. It has over 1200 large superconducting dipole magnets aligned over a circumference of 27 km, with an additional few hundred quadrupole magnets being used to keep the beams focused. Small magnets producing fields of 5–20 T find extensive use in a variety of research apparatus all over the world. Before LTS-based magnets became available, the equipment required for producing high magnetic fields was extremely expensive and such facilities were available at only a few places.

Turning to HTS, the field has come a long way since their discovery in 1986. The technology of HTS has progressed from exciting basic research to viable applications, and in many cases working prototypes have been produced. However, the performance of HTS at liquid nitrogen temperature continues to pose a significant challenge. These materials are brittle ceramics that are expensive to manufacture in a form suitable for coil winding. However, novel processing techniques have been developed with promising performances that will be described later in this book.

Today, the applications of HTS are in areas where they have specific advantages. For example, they are used as liquid-nitrogen-cooled, low-thermal-loss current leads for high-current LTS devices, in radiofrequency and microwave filters for mobile phone base stations, in ultrasensitive selective receivers for military use, and in fast fault current limiters. These current limiters provide power grid operators with unsurpassed reliability in handling the hazardous power surges caused by short circuits and lightning strikes and in maintaining an undisrupted power supply. Magnetic billet heating is another important area, with HTS being used with much better results than conventional induction and gas heating. Over the years, the transport current density of HTS has improved to a level suitable for a variety of applications. These include rotating electrical machines, synchronous AC and DC homopolar motors and drives, generators and condensers, magnetic ore separators, and extra-high-field inserts for high-field LTS magnets to boost the field to 25–30 T, albeit at liquid helium temperature.

Superconducting underground electric power transmission cables of both LTS and HTS provide a viable means for transmitting large quantities of power in densely populated cities where health, safety, and environment are the prime concerns. Such HTS cables are being widely developed and tested in China, Japan, Korea, Russia, Western Europe, and the USA.

An equally spectacular success of LTS has been their use as SQUID elements for commercially available ultrasensitive magnetometers, which have become common tools of materials research in laboratories all over the world. SQUIDs using $Nb/AlOx/Nb$ Josephson junctions today at liquid helium temperature achieve a magnetic noise floor of 1 $fT/Hz^{1/2}$, which allows diagnostically relevant magnetic detection of human brain signals in magnetoencephalography (MEG). HTS SQUIDs functioning at liquid nitrogen temperature have approached this magnetic sensitivity to within an order of magnitude (Narlikar, 2004) and are available commercially for non-destructive evaluation of faults in computer chips and aircraft.

The US navy has been using HTS tapes for reducing the length of very low-frequency antennas employed on submarines, thus markedly lowering antenna losses. Another noteworthy military application of HTS is in 'E-bombs', which make use of strong magnetic fields to create a fast high-intensity electromagnetic pulse to disable an enemy's sensitive electronic equipment.

It is likely that future exciting developments in applications of superconductivity will be based primarily on the discovery of new superconductors, and their performance will play a key role. We should expect further significant breakthroughs and applications as new superconducting systems are found with novel properties or with even higher critical temperatures than have been achieved to date.

1.4 This book

This book is neither about basic aspects of superconductivity nor about its applications, but rather its main focus is on superconducting materials themselves. It is not meant to serve as an encyclopaedia, describing each and every superconductor that exists, but instead is about those that constitute some of the important milestones described in Section 1.2. Our focus is on superconductors that have emerged as scientifically exciting, that have already been established in technological applications, or that pose a variety of challenges. This book contains a further 20 chapters. The next five chapters present the basic aspects of superconductivity. The subject matter included in them is the bare minimum needed to understand all the subsequent chapters. To ensure that the presentation is readily accessible to readers from diverse scientific and technical disciplines, the use of complicated mathematics has been kept to a minimum in the main text, with essential mathematical derivations being confined to short appendices to the relevant chapters.

The existing literature concerning superconductors is huge and exhaustive, and our aim is not to provide a reference to each and every published work. We have cited only those references that we consider to be necessary to satisfy the interests of our proposed broad readership and to help provide them with further reading.

1.5 Summary

In this chapter, we have noted superconductivity as an extraordinary scientific discipline where the curiosity and excitement about the challenges posed are as great, if not greater, a century after Kamerlingh Onnes' discovery. We have taken a quick look at the last 100 years of the path-breaking advances in the field of superconductors and have briefly sketched the astonishing progress made in their technological applications.

CHAPTER 2

The superconducting state

Although superconductivity was discovered in 1911, it took more than four decades before Bardeen, Cooper, and Schrieffer (1957) could advance a satisfactory microscopic explanation of the superconducting state, characterised by perfect conductivity and perfect diamagnetism. Much of the initial understanding of the phenomenon came through experimental studies of different physical properties of superconductors and their phenomenological explanations. While these form the subject matter of Chapters 3–6, in this chapter we choose to deviate from the historical sequence of development and briefly outline the essential features of the microscopic theory so that we can understand the forthcoming chapters in a better perspective. We start with the origin of resistance in normal metals and try to understand in simple terms the microscopic nature of superconducting state as envisaged by Bardeen, Cooper, and Schrieffer (1957).

2.1 Electrical conduction in metals and the origin of resistance

In metals and metallic materials, the motion of the outer or valence electrons of the atoms is responsible for electrical conduction. These electrons have to satisfy Pauli's exclusion principle, which implies that each energy state can be occupied by at most two electrons, and that these must have opposite spins. At 0 K, starting from the lowest-energy state, the energy states are filled in order of increasing energy state until all the available electrons are accommodated; the highest filled energy state is the Fermi level E_F. The density of states per unit energy range per unit volume, $N(E)$, varies as $E^{1/2}$:

$$N(E) = \frac{1}{2\pi^2}\left(\frac{2m}{\hbar^2}\right)^{3/2} E^{1/2}, \tag{2.1}$$

where m is the electron mass and $\hbar$ is Planck's constant divided by 2π.

The parabolic variation of $N(E)$ with E is shown in Figure 2.1. At 0 K, the Fermi level is very sharp, but for $T > 0$, it becomes diffuse. As the temperature rises, there is an increase in the amplitude of the lattice vibrations, which continually scatter conduction electrons

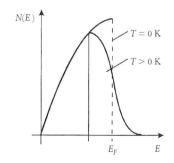

Figure 2.1 Variation of density of states $N(E)$ with E.

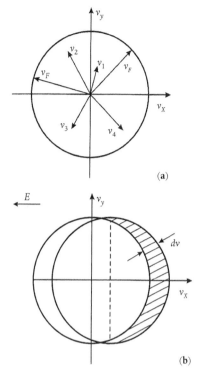

Figure 2.2 Electrons in a conductor: (a) in the absence of an electric field; (b) when an electric field E is applied.

from occupied to unoccupied states. Since the lattice vibrations have energy k_BT, where k_B is Boltzmann's constant, the only electrons that are affected by scattering are those situated within the energy range $\Delta E = k_BT$ near the Fermi energy. A dynamic equilibrium is soon established in which the energy lost and the energy gained in the scattering process are equal.

We may now understand the conduction process. If the electrons move randomly in all possible directions, there is no net velocity. The situation changes when an electric field is applied, and it can be conveniently visualised by plotting the velocities of the electrons in velocity space. The sphere constructed in Figure 2.2(a)—the Fermi sphere—is of radius equal to the Fermi velocity v_F. The points inside the spherical Fermi surface correspond to the endpoints of velocity vectors drawn from the centre. As all the points inside the sphere are occupied, the velocity vectors cancel each other pairwise and thus the resultant net velocity of the electrons is zero. With the application of an electric field E, the Fermi sphere is displaced in the direction opposite to that of the field, that is, towards the positive electrode (Figure 2.2(b)). Although the majority of electron velocities still cancel each other, some, as shown by the shaded area in the figure, remain uncompensated and therefore there is a net flow of electrons accelerated under a force $F = eE$ towards the positive electrode, and it is this that results in conduction. However, both the steadiness of the current flow under a fixed electric field and the fact that when the field is switched off, the electrons do not continue to drift with constant velocity suggest that the moving electrons experience resistance in their passage through the metal.

The electrons moving inside a solid under an electric field are continually being scattered, resulting in dissipation of their energy. If the crystal lattice is perfect, the individual electron waves scattered in the forward direction are in phase and thus they interfere constructively without losing any energy. If, however, the periodicity is disturbed by the presence of lattice imperfections such as vacancies, interstitial and impurity atoms, dislocations, and grain boundaries, or by thermal vibrations of the atoms, the scattering becomes incoherent. Now, there are more electrons that lose their energy to the lattice than those that gain energy. An equilibrium is reached in which the energy transferred to the lattice equals the energy gained from the electric field. Scattering due to imperfections is temperature-independent and gives rise to what is known as the residual resistance of metals at low temperatures. At higher temperatures, the lattice vibrations become more important, and since their amplitude is larger at elevated temperatures, the transfer of energy to the lattice becomes faster. To maintain the steady-state equilibrium, the electric field has to be increased, which amounts to an increase in resistance. In the classical free-electron theory, the motion of the electrons accelerated by an electric field E and simultaneously decelerated by a frictional force is described by the equation

$$m\frac{dv}{dt} + \eta v = eE.$$

Here the second term, where η is a constant, represents the damping or frictional force responsible for resistance. In the steady state, when the electrostatic force eE is balanced by the frictional force, the electrons reach a steady velocity, the Fermi velocity v_F, such that $\eta v_F = eE$, giving $\eta = eE/v_F$, and the above equation becomes

$$m\frac{dv}{dt} + \frac{eE}{v_F}v = eE,$$

the solution of which is given by

$$v = v_F\left[1 - \exp\left(-\frac{eE}{mv_F}t\right)\right]. \tag{2.2}$$

The factor mv_F/eE, which has units of time, is called the relaxation time τ and measures the average time interval between the two successive scattering collisions. The current density constituted by n conduction electrons per unit volume, moving with a velocity v_F, is given by

$$j = nev_F = \sigma_n E,$$

and the electrical conductivity and resistivity are given respectively by

$$\sigma_n = \frac{nev_F}{E} = \frac{ne^2t}{m},$$

$$\rho_n = \frac{mv_F}{ne^2l_e}, \tag{2.3}$$

where $l_e = v_F\tau$ is the electron mean free path between two consecutive collisions. At temperatures above the Debye temperature, the energy of ionic vibrations and the collision cross-section of the ion become proportional to temperature, and hence the resistivity varies linearly with temperature. As the temperature is lowered, departures from linearity are seen, and at very low temperatures, the resistivity due to ionic vibration varies as T^5. With decreasing temperature, the electron mean free path increases, and for pure homogeneous metals it tends to infinity as the temperature approaches 0 K. However, ρ_n can never become zero in this way—first owing to the unattainability of the absolute zero of temperature, and second because of the presence of crystal imperfections in real materials that give rise to the residual resistance at low temperatures. *The reason for the occurrence of zero resistance in superconductors is quite different.*

2.2 Microscopic nature of superconducting state

2.2.1 Difficulties in theoretical formulation

A prominent hurdle to the understanding of superconductivity was the fact that the interaction energy responsible for it is very small. As we will see in Chapter 4, purely from thermodynamic considerations, the superconducting state represents a more ordered state of lower energy than the state above T_c, that is, the *normal state*. The difference in energy between the two states, known as the *condensation energy* $\frac{1}{2}\mu_0 H_c^2$ is, however, extremely small. Here μ_0 is the permeability of the free space and H_c is the critical magnetic field (to be described in Chapter 3) above which superconductivity is destroyed. At 0 K, the condensation energy is of the order of 10^{-8} eV/atom and is very small in comparison with the Fermi energy, which is of the order of a few eV. This was at the root of the problem in formulating a microscopic theory, which necessitated identifying and picking out just one crucial interaction between electrons required for superconductivity and leaving out all the remaining interactions that were the same in the two states. No theory for any phenomenon that had hitherto been developed permitted such a calculation of electronic energies to one part in 10^8. This difficulty was further aggravated by the fact that the phenomenon was occurring in too wide a variety of materials with differing crystal structures to invite any special explanations involving crystal lattice structures. Furthermore, material features such as crystal imperfections and impurities did not seem to have much effect on the basic phenomenon, although in some instances they altered details such as transition width and temperature. As a result, there was no microscopic explanation of superconductivity for 40 years after Onnes' discovery.

Valuable clues to the nature of the superconducting state and of the relevant interaction came with the experimental work in the 1950s, which eventually led to the successful formulation of a microscopic theory. Two types of experimental observations proved to be particularly rewarding in the stimulus they gave to theoretical development, namely (i) the isotope effect and (ii) measurements of thermal properties.

2.2.2 Isotope effect

In 1950, Maxwell (1950) and Reynolds et al. (1950) independently observed that the T_c of Hg isotopes varied as the inverse square root of their isotopic mass M, that is, $T_c \propto M^{-\alpha}$, where the exponent $\alpha = 0.5$. The isotopes of a few other superconducting elements such as Zn, Cd, Sn, Pb, and Tl also showed similar behaviour, which is termed the *isotope effect* (Table 2.1).

Since the Debye temperature θ_D also varies as $M^{-0.5}$, these observations strongly suggest that the lattice vibrations and phonons are

Table 2.1 The exponent α of the isotope effect observed for different elements

	Hg	Zn	Cd	Sn	Pb	Tl
T_c (K)	4.1	0.85	0.51	3.7	7.2	2.38
α	0.50	0.45	0.50	0.47	0.48	0.50
$\Delta\alpha/\alpha$	$\pm$ 0.03	$\pm$ 0.01	$\pm$ 0.10	$\pm$ 0.02	$\pm$ 0.01	$\pm$ 0.10

From Parks, R. D. (ed.). *Superconductivity*, Vol. I. Marcel Dekker, New York (1969), p. 125.

important in determining T_c and are likely to be involved in the mechanism of superconductivity. The isotope effect provided clear evidence that superconductivity somehow arose from an interaction between conduction electrons and lattice vibrations. Interestingly, quite unaware of this, Fröhlich (1950, 1954) had visualised an attractive interaction between electrons developing through their coupling with lattice vibrations and had ascribed superconductivity to an altered distribution of conduction electrons at E_F resulting from such an interaction. Cooper (1956) soon pointed out that if there was a net attraction, however weak, between a pair of electrons at the Fermi level, these electrons would form a bound state. Extending this further, Bardeen, Cooper, and Schrieffer (1957) formulated a comprehensive microscopic theory, called the *BCS theory of superconductivity*.

Before proceeding further, it is worth pointing out that although the isotope effect provided the clue for finding a theoretical explanation of superconductivity, similar studies subsequently made on a few other superconducting elements have revealed considerable deviations from the isotope exponent value of $\alpha = 0.5$. For example, the transition metal elements Mo ($T_c = 0.91$ K), Re ($T_c = 1.7$ K), and Os ($T_c = 0.66$ K) show much smaller exponent values of 0.35, 0.39, and 0.23, respectively, while no isotope effect has been observed at all for Zr ($T_c = 0.61$ K) and Ru ($T_c = 0.50$ K). Furthermore, α-U ($T_c = 2.4$ K) is found to exhibit a negative isotope effect, with $\alpha = -2.2$. Similarly, measurements of the oxygen isotope effect in high-T_c cuprate superconductors (HTS) have yielded very small exponent values that increase with decreasing T_c, and the phononic contribution in these materials is believed to be slight. The existence of these anomalies has led to considerable discussion, and in some cases the possibility of alternative or *unconventional* mechanisms of superconductivity has also been seriously proposed.

2.2.3 Heat capacity and thermal conductivity measurements

As will be described in Chapter 4, the specific heat and thermal conductivity of a metal at low temperatures have contributions from both the conduction electrons and the thermal vibrations of the crystal lattice (i.e. phonons). The latter contribution remains essentially unaffected through the superconducting transition. Interestingly, however, at temperatures sufficiently below T_c, the contributions to both heat capacity and thermal conductivity associated with the conduction electrons exhibit an exponential temperature dependence of the form $e^{-bT_c/T}$, where b is a constant. This could arise (Mendelssohn, 1945; Goodman, 1953) from electrons becoming thermally excited across a possible gap in the energy spectrum of the superconductor, with a width $b = 2\Delta$ of around $3.5 k_B T_c$. This was indeed a very significant result for theoretical developments and was later corroborated by tunnelling and optical measurements.

2.2.4 Cooper pairs

Bardeen, Cooper, and Schrieffer (1957) showed that the interaction of conduction electrons with the crystal lattice could lead to an attractive electron–electron interaction in the following manner. At 0 K, an electron moving through the lattice with a momentum $\mathbf{k}$ distorts the lattice and gets scattered. In the language of field theory, this means the creation of a *virtual phonon*. This phonon of momentum $\mathbf{q}$ is absorbed by a second electron with momentum $\mathbf{k}'$, which gets scattered with a momentum $\mathbf{k}' + \mathbf{q}$ (Figure 2.3). It is important that the available phonons are restricted to a small energy range $\hbar\omega_D \approx k_B\theta_D$, where ω_D is the Debye frequency and θ_D is the Debye temperature. If $e_{\mathbf{k}}$ and $e_{\mathbf{k}'}$ are the energies of the two electrons, the theory shows that they will attract each other if $e_{\mathbf{k}} - e_{\mathbf{k}'} = \hbar\omega_D$. As already mentioned, Cooper (1956) had earlier shown that if an electron system develops a net attractive interaction, the system will condense into a bound state of *Cooper pairs*.

The pairs of two electrons with momenta $(\mathbf{k}_1, \mathbf{k}_2)$ must be chosen in such a way as to maximise the possibility of transition from any one paired state to all other paired states $(\mathbf{k}'_1, \mathbf{k}'_2)$ without any change in the net momentum:

$$\mathbf{k}_1 + \mathbf{k}_2 = \mathbf{k}'_1 + \mathbf{k}'_2 = \mathbf{K},$$

where $\mathbf{K}$ is the constant momentum of the pairs. Since the whole interaction is mediated by phonons available in the small energy range of $\hbar\omega_D$, the pair condensation occurs in a small shell of width $\hbar\omega_D$ at E_F (Figure 2.4). It is clear that momentum conservation limits the interaction to the phase-space area of the shaded region, where the two shells intersect each other. The ground state, at 0 K and with zero current density, with the greatest overlap of the phase-space domains where the transitions are allowed, is achieved for $\mathbf{K} = 0$. This means that the electrons are associated in pairs with zero total momentum, and the electrons forming a pair have equal and opposite momenta $(\mathbf{k}, -\mathbf{k})$. Further consideration, including exchange terms, shows that in addition

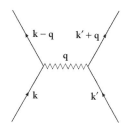

Figure 2.3 The BCS phonon mechanism.

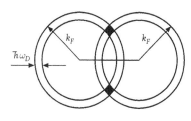

Figure 2.4 Pair formation in a shell of width $\hbar\omega_D$ at E_F.

to the momentum correlation, the electrons in a pair must also have a spin correlation, which means that energetically it is most favourable to restrict the pairing to electrons with opposite spins ($\mathbf{k}\uparrow$, $-\mathbf{k}\downarrow$), indicated by oppositely directed arrows. The latter correspond to $\pm^1/_2$ spins for the pairing electrons. In the state of current flow, there is a displacement of the ground state in momentum space by $m\mathbf{v}_S$, where $\mathbf{v}_S$ is the velocity of the paired electrons. In this situation, an electron with momentum $\mathbf{k} + m\mathbf{v}_S$, with spin up, is paired with an electron of momentum $-\mathbf{k} + m\mathbf{v}_S$, with spin down. The total momentum is identical for all pairs, and equal to $2m\mathbf{v}_S$. The current carriers in the superconducting state (which until the BCS theory came along were referred to as super-electrons) are now identified as the electron pairs (Cooper pairs). The electrical charge carried by a Cooper pair is double the electronic charge, that is, $2e$, which can be negative or positive, depending on whether the pertinent charge carriers are electrons or holes. In many cuprate super-conductors, for example, where the positively charged holes are the current carriers in the normal state, the supercurrent below T_c is carried by the hole pairs, each pair having a positive charge of $2e$. The fact that the current carriers in the superconducting state do indeed possess an electric charge $2e$, negative or positive, was corroborated soon after the advent of the BCS theory through flux quantisation experiments (Deaver and Fairbank, 1961) as well as by tunnelling studies (Nicol et al., 1960).

In contrast to the ferromagnetic state, where the spins are aligned in a parallel configuration, the paired electrons in a superconductor have opposite spins. The two phenomena, superconductivity and fer-romagnetism, have therefore been considered mutually incompatible. Ferromagnetic impurities, with their strong exchange forces, in general, are found deleterious to superconductivity, and their presence even in small concentrations may cause a drastic reduction in T_c. Similarly, an applied magnetic field, if it is sufficiently large, can destroy supercon-ductivity in two ways. First, it tends to align the electron spins along the field direction (Zeeman effect), and thereby destroys their antipar-allel spin correlation. Second, the magnetic field gives rise to a Lorentz force, and, since the two electrons forming a Cooper pair have opposite momenta, it acts in opposite directions on the two electrons, breaking the pairing (orbital effect).

The electron pairs in the superconducting state possess a lower energy than the single electrons of the normal state, although the energy difference (i.e. the condensation energy), as mentioned previ-ously, is very small. As a result, there is a very small energy gap 2Δ above them at E_F that inhibits the kind of collisional interactions responsible for the resistive state. At $0 < T < T_c$, the electron pairs are continually being scattered by lattice vibrations, but since the pair assembly involves a high degree of correlation in their motion, the net momentum is always conserved. There is no transfer of momentum from the moving pairs to the lattice, and thus there is no resistance. For a transition to the resistive state, the pairs have to be broken into single electrons or *quasiparticles*, as they called, by the supply of energy equal to the energy gap. As the temperature is raised, the lattice vibrations acquire sufficient energy to break the pairs, and at T_c all the pairs are broken. At intermediate temperatures, $0 < T < T_c$, both pairs and single electrons are present, but the latter are effectively 'shorted-out' by electron pairs that continue to move without resistance.

2.2.5 Energy gap

The detailed derivation of the BCS theory is too technical for this book, although in Appendix 2A, we present some elementary ideas and a simplified formulation for the interested reader. According to the BCS theory, at 0 K, the energy gap 2Δ is given by

$$[2\Delta]_{0K} = 4\hbar\omega_D \exp\left[-\frac{1}{N(0)V}\right], \tag{2.4}$$

where $N(0)$ is the electron density of states (DOS) at E_F, ω_D is the Debye frequency, and V is the electron–phonon interaction parameter. For $k_B T_c << \hbar\omega_D$ (or $T_c << \theta_D$), the temperature above which all the pairs are broken, T_c, is given by the well-known *BCS equation*,

$$k_B T_c = 1.14\hbar\omega_D \exp\left[-\frac{1}{N(0)V}\right], \tag{2.5}$$

or

$$T_c = \theta_D \exp\left[-\frac{1}{N(0)V}\right] = \theta_D \exp\left(-\frac{1}{\lambda}\right), \tag{2.6}$$

where $\lambda = N(0)V$ is the electron–phonon pair potential, which measures the strength of the electron–phonon interaction. Clearly, to realise a higher T_c, the Debye temperature θ_D, the DOS $N(0)$ at E_F, and the electron–phonon coupling constant V should be large. For metals and alloys, $N(0)V$ is generally small (<0.5) and the Debye temperature is around 300 K. Increasing θ_D generally lowers the coupling constant V, which sets an upper limit on T_c of around 30 K for the electron–phonon-mediated coupling in metallic superconductors. Until the discovery of high-T_c cuprates in 1986, the highest T_c recorded was 23.2 K for the A-15 superconductor Nb_3Ge, which was well within the realm of conventional phononic superconductivity.

As θ_D is proportional to $M^{-1/2}$, equation (2.6) leads to the isotope effect. Any significant deviations experimentally observed in the isotope effect are therefore important and necessitate modifications to the theory or a search for possible alternative or unconventional mechanisms of electron pairing.

Combining equations (2.4) and (2.5) yields the bandwidth of the energy gap at 0 K,

$$[2\Delta]_{0K} = 3.52 k_B T_c. \tag{2.7}$$

The numerical factor 3.52 is known as the *gap coefficient*.

2.2.6 Weak and strong coupling

The value of 3.52 for the gap coefficient is for the situation when $k_B T_c << \hbar \omega_D$, or, more conveniently, when $T_c / \theta_D << 0.1$. This situation is referred to as *weak electron–phonon coupling* and is found in superconducting elements such as Al, In, and Sn. On the other hand, Pb and Hg, for example, possess a larger value of T_c / θ_D, which makes them *strongly coupled*, and for such materials the equations in the preceding subsection do not strictly hold. The gap coefficient of strongly coupled systems, which include many superconducting alloys and compounds, and also the high-T_c oxide superconductors, is larger than 3.52 and can reach as high as 10. Interestingly, the high-T_c cuprates, which are highly anisotropic, exhibit strong-coupling behaviour in the basal plane but weak-coupling behaviour in the perpendicular direction. The mechanism responsible for the high T_c of the cuprates, however, remains to be fully settled.

The theory was extended to the strong-coupling regime using Eliashberg's (1962) coupled integral equations by considering the frequency dependence of $\Delta(\omega)$, which is small and therefore had earlier been neglected in the weak-coupling case. It was shown by Schrieffer et al. (1963) that the electron DOS in superconductors could be measured directly through tunnelling experiments with the help of conductance versus voltage plots. McMillan and Rowell (1965) gave an inversion program to yield the phonon spectrum from tunnelling measurements. McMillan (1968) extended the BCS theory to strongly coupled superconductors, giving

$$T_c = \frac{\theta_D}{1.45} \exp\left[\frac{-1.04(1 + \lambda_e)}{\lambda_e - \mu(1 + 0.62\lambda_e)}\right], \tag{2.8}$$

where the dimensionless parameter λ_e, which for the strongly coupled superconductors is greater than unity, is given by

$$\lambda_e = 2 \int_0^\infty \frac{\alpha^2 F(\omega)}{\omega} d\omega, \tag{2.9}$$

where $F(\omega)$ is the phonon DOS at the Fermi surface and α is a measure of the electron–phonon interaction strength. The parameter λ_e is related to the electron DOS through the relation

$$\lambda_e = \frac{N(0)\langle I^2 \rangle}{M^* \langle \omega^2 \rangle}, \tag{2.10}$$

where $\langle I^2 \rangle$ is the electronic matrix element, M^* is the effective atomic mass, and $\langle \omega^2 \rangle$ is the mean-square phonon frequency. Theoretical estimations of T_c values for both strong- and weak-coupling superconductors have met only with limited success. Using the model described here, McMillan (1968) predicted T_c values for a number of strongly coupled systems. Although the agreement was good for pure Pb, the values predicted for Nb and different Nb and V alloys were about 100% too high.

2.2.7 Conventional and unconventional electron pairs: unconventional superconductors

Unlike a single electron (or hole) carrier, which has half-integer spin and is a *fermion*, obeying *Fermi–Dirac statistics*, an electron pair has integer spin (in the case described so far, the net spin $S = 0$) and an electron pair therefore behaves like a *boson*, obeying *Bose–Einstein statistics*. At low temperatures, bosons behave differently from fermions, and, since they are not constrained by the Pauli exclusion principle, free from any restriction they can condense into the same energy state. The spins of the electrons forming a pair can be antiparallel ($S = 0$), or parallel ($S = \pm 1$). The symmetry of the superconducting order parameter (i.e. the pairing state) is classified by the orbital quantum number L, which for singlet pairing (i.e. $S = 0$) takes even integer values 0, 2, 4, . . . (*even-parity pairing*) and the pair states are respectively labelled by s, d, g, etc., corresponding to s-wave, d-wave and g-wave pairing, etc. In the case of triplet pairing ($S = \pm 1$), the orbital quantum number L is an odd integer, that is, 1 (*p-wave* pairing), 3 (*f-wave* pairing), etc., leading to a symmetry that corresponds to an odd-parity pairing. The pair assembly in the BCS theory corresponds to isotropic *spin singlet s-wave* pairing. This is essentially the outcome of the simplifying approximation (2A.2) in Appendix 2A, which assumes the electron–electron interaction potential to be spherically symmetric. This leads to a spherical and isotropic gap function of the s-wave state that is free from nodes and that characterises *conventional pairing*. All others pairings—p-wave, d-wave, f-wave, etc.—are referred to as *unconventional pairing*. There also exist unconventional forms of s-wave pairing, including *anisotropic* and *extended s-wave pairing*. The pairing state of a superconductor can be a mixture of these or a combination of two different pairings, such as $s + d$ or $s+g$, and in general it is experimentally a very challenging and difficult task to identify or separate out two essentially similar gap states of an unconventional superconductor. Some of the gap states of even and odd parity are depicted in Figure 2.5 As well as the differences in the gap function, whether superconductivity is conventional or not is also decided by the way in which the pairs are formed. In the original BCS framework, the formation of Cooper pairs is mediated by electron–phonon interaction. In *unconventional superconductivity*, the

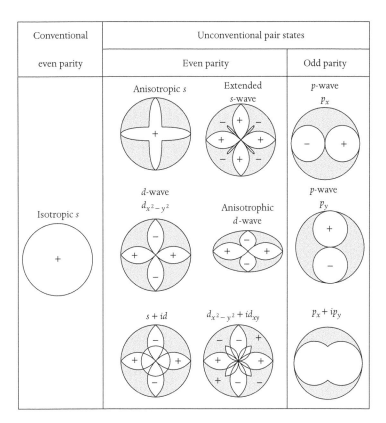

Figure 2.5 Conventional and unconventional pairing states of even and odd parity.

pairing interaction is non-phononic, involving spin- or charge-density waves, magnons, excitons, bipolarons, spin bags, etc. The *generalised BCS theory* is claimed to account for unconventional superconductivity (Anderson and Morel, 1961; Balian and Werthamer 1963)—a claim that emphasizes the universality and strength of the underlying concepts of the BCS theory.

As described in Section 2.2.5, from equation (2.6) in the conventional BCS framework, there is an upper limit to the superconducting critical temperature of about 30 K. Consequently, for materials with T_c in excess of 30 K, it is necessary to consider alternative or unconventional superconductivity mechanisms. Since 1986, a wide variety of high-T_c cuprates have been found with T_c values considerably exceeding 30 K, and reaching 135 K or higher, which, barring a few exceptions such as MgB_2, BKBO, and doped fullerides, cannot be accounted by the conventional electron–phonon mechanism. For instance, in the case of the Fe-based superconductor (FBS) $LaFeAsO_{1-x}F_x$ (Kamihara et al., 2008), calculations predict (Boeri et al., 2008) that if only the conventional electron–phonon coupling were responsible for superconductivity, its T_c would be less than 1 K, as against the observed value of 26 K. To achieve this latter value, the electron–phonon coupling would have to be five times as strong. Muon spin rotation experiments (Drew et al., 2008) carried out on a still higher-T_c (> 50 K) compound of the same family (Gao et al., 2008) found a rapid increase in magnetic fluctuations as T_c was approached, indicating their prominent involvement in superconductivity.

In contrast with most low-temperature superconductors, where the conventional phonon-mediated BCS pairing mechanism works, to date there seems no such unique mechanism universally accepted for high-temperature superconductors. Several model mechanisms and theories have been proposed for high-temperature superconductivity. Broadly, these may be grouped in two types. On the one hand, there is a generalised BCS framework in which the origin of superconductivity is non-phononic and the Cooper pairing of charge carriers results through the exchange of different electronic excitations, such as charge transfer, excitons, and plasmons, arising from charge fluctuations in the system (Griffin, 1987; Jagadish and Sinha, 1987; Kresin, 1987; Ruwalds, 1987; Varma et al., 1987; Ashkenazi and Kuper, 1989). Electronic excitations, because they have much a higher energy than phonons, favour a much larger T_c than that arising from a phononic mechanism. The other type of mechanisms proposed consider the unusual features of high-temperature superconductors in both normal and superconducting states, to be described later in this book, and have little to do with the BCS pairing. These mechanisms are all magnetic in origin and include the *antiferromagnetic spin exchange model* (Miyake et al., 1986), the *resonating valence bond* (RVB) model, originally due to Anderson (Anderson, 1987, 1997; Anderson et al., 1987; Baskaran et al., 1987; Kivelson et al., 1987), and the *spin-bag model* (Schrieffer, 1988). In the RVB model, there exist two types of quasiparticles: *holons*, which are positively charged bosons that condense in the superconducting state, and spinons, the neutral fermions that represent spin excitations. It is *spinon–holon scattering* that accounts for the observed data. All these models are, however, theoretically too involved for the present book and are not discussed here. The interested reader may find the original references cited above to be useful, and may also benefit from some concise review articles written about these models (Baskaran, 1989; Jha, 1989; Sinha, 1990).

One potential candidate mechanism for many high-temperature superconductors is d-wave pairing, and unconventional pairing is also being seriously considered as a mechanism in organic superconductors, ruthenocuprates, and various Ce- and U-based heavy fermion superconductors (although many of these are low-temperature superconductors with a very low T_c). These heavy fermion superconductors include the so-called *ferromagnetic superconductors*, in which there is strong evidence for the coexistence of what might be considered incompatible phenomena, namely superconductivity and ferromagnetism, the former being conventionally characterised by antiparallel and the latter by parallel

Table 2.2 Possible unconventional superconductors

- Ce-based heavy fermion superconductors: $CeCu_2Si_2$, $CeCoIn_5$, $CeIrIn_5$, $CePt_3Si$
- U-based heavy fermion superconductors: UPt_3, UBe_{13}, URu_2Si_2, UNi_2Al_3, UPd_2Al_3
- Metal oxide superconductors: high-T_c cuprates, oxypnictides, Sr_2RuO_4, ruthenocuprates, cobalt oxide hydrate, β-pyrochlore, KOs_2O_6
- Organic superconductors: $(TMTSF)_2X$, $(ET)_2X$

spin correlation. In these systems, instead of phonons, the pairing of electrons seems to be mediated by the exchange of spin fluctuations and, to favour ferromagnetic order, the formation of a spin-parallel coupled superconducting state is envisaged. The spin triplet state of odd parity ($S = 1$) with p- or f-wave pairing seems more relevant here than the singlet state of even parity ($S = 0$) with s- or d-wave pairing. Table 2.2 gives a list of some of the potential candidates for unconventional electron pairing.

There are many other systems showing both conventional and unconventional superconducting traits. All these superconductors will be discussed later in this book.

2.3 Summary

In this chapter, we have briefly described the origin of resistance in normal metals and the microscopic basis of the superconducting state. We have introduced various terms, such as phonon mediation, Cooper pairing, and energy gap, that arise in the BCS theory and that will be used in forthcoming chapters dealing with the properties of super-conductors. The detailed microscopic theory is too complicated and beyond the scope of this book, although some preliminary aspects of the theory are provided in Appendix 2A. Finally, some of the important unconventional superconductors that seem to call for alternative pairing mechanisms have been mentioned.

Appendix 2A: BCS ground state and the energy gap

It is of interest to determine the difference in the ground-state energy of a superconducting state with respect to that of the normal state. Here we follow a rather simplified approach to describe some elementary features of the BCS ground state; for a detailed derivation, the reader may refer to the original paper of Bardeen, Cooper, and Schrieffer (1957).

(i) $T = 0K$

The difference in ground-state energy between the superconducting and normal states at 0 K may be written as

$$W(0) = \Delta E_{\text{kin}} + \Delta E_{\text{cor}},$$

that is, the difference in kinetic energy ΔE_{kin} plus the correlation energy ΔE_{cor} for all possible transitions $(k, -k)$ to $(k', -k')$, spin suffixes being dropped for convenience. With occupation probability h_k of the pair $(k, -k)$, the difference in electron kinetic energy is

$$\Delta E_{\text{kin}} = 2 \sum_k e_k h_k,$$

where e_k is the energy of electron state k, and the numerical factor 2 comes in because there are two electrons in a pair. The correlation energy is obtained by summing over all transitions from k to k'. One starts with the pair occupying the state $(k, -k)$ with probability h_k. The transition to the state k', as per Pauli's exclusion principle, is possible only if it is empty, and this will happen with a probability $1 - h_{k'}$. In the final state, k' is occupied and k is empty, the probabilities of which are $h_{k'}$ and $1 - h_k$, respectively. The correlation energy is obtained by multiplying the transition probabilities by a suitable matrix element $V_{kk'}$ having dimensions of energy. The correlation will be given by

$$\Delta E_{\text{cor}} = \sum_{k,k'} \left\{ -V_{kk'} [h_k (1 - h_{k'}) h_{k'} (1 - h_k)]^{1/2} \right\}.$$

The negative sign makes the interaction attractive. The total energy of the state must be minimised with respect to the probability h_k, giving

$$0 = \frac{\partial W(0)}{\partial h_k} = 2e_k - \frac{1 - 2h_k}{[h_k (1 - h_k)]^{1/2}} \Delta_k, \tag{2A.1a}$$

where

$$\Delta_k = \sum_{k'} V_{kk'} [h_{k'} - (1 - h_{k'})]^{1/2}, \tag{2A.1b}$$

which is an important parameter whose significance will be seen in the following. Equation (2A.1a) has solutions for h_k and $h_{k'}$, given by

$$h_k = \frac{1}{2} \left(1 - \frac{e_k}{E_k} \right), \qquad h_{k'} = \frac{1}{2} \left(1 - \frac{e_{k'}}{E_{k'}} \right),$$

where

$$E_k = \left(\Delta_k^2 + e_k^2 \right)^{1/2}, \qquad E_{k'} = \left(\Delta_{k'}^2 + e_{k'}^2 \right)^{1/2}.$$

Substituting these expressions into (2A.1b) yields

$$\Delta k = \frac{1}{2} \sum_{k'} V_{kk'} \left[1 - \left(\frac{e_{k'}}{E_{k'}} \right)^2 \right]^{1/2} = \frac{1}{2} \sum_{k'} V_{kk'} \frac{\Delta_{k'}}{E_{k'}}.$$

The matrix element $V_{kk'}$ is given by

$$V_{kk'} = \begin{cases} V = \text{constant} & \text{if } e_k, \, e_{k'} < \hbar\omega_D, \\ 0 & \text{otherwise.} \end{cases} \tag{2A.2}$$

Thus, for $|E_k| < \hbar\omega_D$, Δ_k is replaced by $\Delta(0)$, which is independent of k and is a parameter characterising the energy gap. Changing the summation to integration gives

$$\Delta = \frac{V}{2} \int\limits_{-\hbar\omega_D}^{\hbar\omega_D} \frac{\Delta(0)}{E_{k'}} \frac{\partial k'}{\partial e} \, de.$$

$\partial k'/\partial e = N(E)$, the density of states, and since the phonon energy is small compared with the usual electron energies in the metal, $N(E)$ may be assumed constant and equal to $N(0)$, the value at E_F. Thus,

$$\frac{1}{N(0)V} = \int\limits_0^{\hbar\omega_D} \frac{1}{[\Delta^2(0) + e^2]^{1/2}} \, de = \ln\left[\frac{e}{\Delta(0)} + \left(\frac{e^2}{[\Delta(0)]^2} + 1\right)^{1/2}\right].$$

Here V is called the pair potential and the composite parameter $N(0)V$ is referred to as the coupling constant.

For $N(0)V \ll 1$,

$$\left(\frac{e^2}{[\Delta(0)]^2} + 1\right)^{1/2} = \left|\frac{e}{\Delta(0)}\right|,$$

and consequently

$$\frac{1}{N(0)V} = \ln\left[\frac{2\hbar\omega_D}{\Delta(0)}\right],$$

giving for the gap the well-known relation of BCS theory

$$[2\Delta]_{0K} = 4\hbar\omega_D \exp\left[-\frac{1}{N(0)V}\right].$$

One may now consider the simplest excitation of the superconducting ground state, namely the breaking up of a single electron pair, leading to two single (or quasi-) particles and one empty pair state. The sum of the kinetic energy and the correlation energy gives the energy of the first excited state above the ground state as $2E_k$. Since $E_k = \{[\Delta(0)]^2 + e_k^2\}^{1/2}$, the smallest value of E_k is when $e_k = 0$. The minimum excitation energy for a pair is $2\Delta(0)$, which represents a gap between the ground state and the first excited state, depicted in Figure 2.6

(ii) $0 < T < T_c$

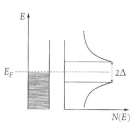

Figure 2.6 Energy gap 2Δ in the superconducting state.

The situation for $0 < T < T_c$ requires the use of statistical mechanics and leads to an integral equation containing a temperature-dependent gap function $\Delta(T)$:

$$\frac{1}{N(0)V} = \int_0^{\hbar\omega_D} \frac{de}{\left([\Delta(T)]^2 + e^2\right)^{1/2}} \tanh\left[\frac{\left([\Delta(T)]^2 + e^2\right)^{1/2}}{2k_BT}\right]. \quad (2A.3)$$

Numerical solution of this equation yields the temperature dependence of the gap, as shown in Figure 2.7. At 0 K, the gap is maximal and it vanishes at T_c. At $0 < T < T_c$, the electron pairs are continually scattered by lattice vibrations, but since the motion of the pairs involves a high degree of correlation, the net momentum is always conserved. Thus, under an electric field, there is no loss of momentum from pairs to the lattice, and so in the superconducting state there is no resistance, until T_c is reached, when all the pairs are broken. At intermediate temperatures, the material contains both paired electrons and single electrons, but the latter still cannot carry any current because they are 'shorted-out' by the former.

For $k_BT_c < \hbar\omega_D$, equation (2A.3) yields

$$T_c = 1.14\frac{\hbar\omega_D}{k_B} \exp\left[-\frac{1}{N(0)V}\right],$$

or

$$T_c = \theta_D \exp\left[-\frac{1}{N(0)V}\right].$$

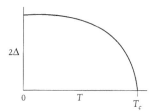

Figure 2.7 Variation of the energy gap 2Δ with temperature.

CHAPTER 3

The superconducting transition and its basic phenomenology

3.1 Fundamental characteristics of the superconducting transition

A superconductor is characterised by two fundamental and unique properties at $T \leq T_c$:

1. The DC electrical resistance vanishes and the electric field within the material is zero. This corresponds to *perfect conductivity* and is termed the *resistive transition*.
2. The magnetic permeability vanishes, in what is known as the *Meissner–Ochsenfeld effect* (or just the *Meissner effect*). When a superconductor is placed in a magnetic field, the field is abruptly expelled from the bulk of the material, that is, the magnetic induction $B = 0$ and the material becomes *perfectly diamagnetic* in what is termed the *magnetic transition*.

Both transitions are fully reversible and, upon warming the sample to T_c, the original resistivity is sharply restored and the expelled magnetic field re-enters the sample as abruptly as it had been expelled when the sample became superconducting. The two transitions are mutually independent; that is, one does not follow from the other.

Just as the critical temperature T_c represents a thermal threshold above which superconductivity is destroyed, there is a similar magnetic threshold, called the *thermodynamic* or *threshold critical field* H_c, above which the superconducting properties (or the *superconducting state*) suddenly revert to normal properties (or the *normal state*). Similar to the transition at T_c, the transition at H_c is reversible; that is, on lowering the field to $H < H_c$, superconductivity is restored. Thermodynamic considerations show (see Appendix 4A to Chapter 4) that the free energy per unit volume of the superconducting state is lower than that of the normal state by $\frac{1}{2}\mu_0 H_c^2$, which is known as the *condensation energy*. Further, the transition at T_c, in the absence of a magnetic field, is of *second order* (i.e. there is no latent heat involved in the transition), while the transition at H_c is of the *first order*. Here, we point out that a superconductor responds differently to an applied magnetic field depending on whether it is a type I superconductor (such as pure elements), or a type II (alloys and compounds). We are presently ignoring this difference, and, unless otherwise mentioned, the description in this chapter relates to elemental superconductors.

Besides T_c and H_c, there is one more threshold if the superconductor (at $T < T_c$) is carrying an electric current I (or current density J), which is expressed by the *critical current* I_c or the *critical current density* J_c. If $I > I_c$ (or equivalently $J > J_c$), a voltage appears across the sample, that is, the current is no longer lossless, and superconductivity is destroyed. As with the other two thresholds, the process is reversible, that is, on lowering the current below the critical value, superconductivity is restored.

3.2   The critical field H_c

The magnetic threshold, that is, the thermodynamic critical field H_c, decreases with increasing temperature, roughly following the parabolic relation $H_c(T) = H_c(0)(1 - t^2)$, where $t = T/T_c$ is the *reduced temperature*. $H_c(0)$ is the value of the critical field at $T = 0$ K and, at T_c, the critical field is zero. From the measured $H_c(T)$ values, one can estimate $H_c(0)$ using the above relation. Table 3.1 gives the $\mu_0 H_c(0)$ values of some of the elemental superconductors, the largest value being approximately 0.2 T for pure Nb. Most of the other elements possess H_c values that are 10–100 times lower. Figure 3.1 shows the characteristic parabolic behaviour of some typical superconductors. The parabolic threshold curve represents the phase boundary separating the superconducting phase (on its left) from the normal phase (on the right).

We will see in Chapter 5 that, in contrast to pure elements, in general, superconducting alloys and compounds remain superconducting up to much a higher magnetic field, called the *upper critical field* H_{c2}. For a number of superconducting systems, such as Chevrel phases, HTS cuprates, and the recently discovered ferro-pnictide superconductors, $\mu_0 H_{c2}$ ranges from 40 to 400 T. Despite very large values of the upper critical field, their thermodynamic critical field is a few hundred times smaller. The critical temperature is highest when $H = 0$ and the critical field is highest when $T = 0$.

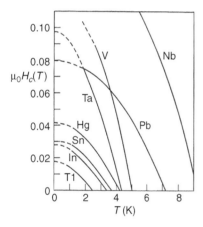

Figure 3.1   Parabolic temperature dependence of the thermodynamic critical field H_c.

Table 3.1 $H_c(0)$ of some superconducting elements

Element	Nb	V	Ta	Mo	Ti	Pb	Sn	Ga	Cd
$\mu_0 H_c(0)$ (T)	0.206	0.141	0.097	0.0096	0.0056	0.08	0.031	0.0059	0.0028

From Ekin, J.W. *Materials at Low Temperatures* (eds R. P. Reed and A. F. Clark). American Society for Metals, Materials Park, OH (1983), p. 465.

3.3   The critical current

The third threshold above which superconductivity is destroyed, I_c, is closely related to the critical field. An electrical transport current I (in A) flowing through a cylindrical superconducting ($T < T_c$) wire of radius R (in cm) produces a circumferential magnetic field (i.e. the self-field) $\mathbf{H}_I$ on its surface, with magnitude (in Oe) given by $H_I = I/5R$. Clearly, at a certain critical value I_c of the transport current, when H just exceeds H_c, superconductivity will be destroyed. This is known as *Silsbee's rule* (Silsbee, 1916). The critical current above which superconductivity is lost is $I_c = 5H_cR$. When the transport current flows in the presence of an external magnetic field $\mathbf{H}_e$, the superconductivity will be destroyed when the magnitude of the vector sum of $\mathbf{H}_e$ and $\mathbf{H}_I$ just exceeds H_c. In the situation where $\mathbf{H}_e$ is transverse to the transport current, there is a demagnetisation factor $D = {}^1\!/_2$, and the criterion for destruction of superconductivity is $2H_e + I/5R > H_c$. On the other hand, when the external field is parallel to I, the two fields $\mathbf{H}_e$ and $\mathbf{H}_I$ are mutually perpendicular and the above criterion instead becomes $H_e{}^2 + (I/5R)^2 > H_c{}^2$. In the above situations, the destruction of superconductivity is gradual, owing to the formation of an *intermediate state* to be described later. Nevertheless, for *pure elemental superconductors*, Silsbee's rule works, and because their critical fields are small, it presents a practical limitation on the use of superconducting pure elements for electrical circuits, in particular for winding high-field superconducting electromagnets that need several hundred amperes of current flowing without dissipation. This limitation, as we will see later, can be circumvented by the use of superconducting alloys and compounds, which can carry very large lossless currents in high magnetic fields. Instead of Silsbee's rule, their critical currents are determined by the presence of various kinds of crystal lattice defects and inhomogeneities in the material that effectively pin down the movement of magnetic flux vortices produced by the magnetic field. Flux pinning allows superconductors to sustain large transport currents without resistance in a high magnetic field, and it forms the core of superconductor magnet technology for a variety of cutting-edge applications mentioned in Chapter 1.

3.4   Resistive transition

Onnes' original discovery of superconductivity in mercury was by observing its DC resistive transition where the resistance–temperature (R–T) curve manifested a sharp drop (Figure 1.1) to zero resistance at T_c. In R–T measurements, one must take the obvious precaution that the direct current passed through the sample is sufficiently low (around a few mA or less) that its self-field does not drastically lower the value of T_c, let alone destroy the superconductivity altogether.

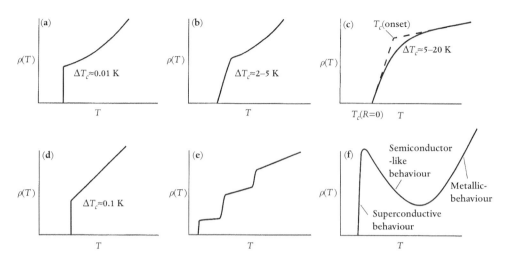

Figure 3.2 Schematic illustration of different forms of resistivity behaviour observed in superconducting materials.

Different forms of superconducting transitions commonly seen in R–T measurements of superconductors are shown schematically in Figure 3.2. The width of the transition can in general vary from 0.01 K, or even less, for homogeneous and pure metals (Figure 3.2(a)) to a few kelvin for inhomogeneous alloys and compounds (Figure 3.2(b)). For non-stoichiometric ceramic samples of high-T_c cuprates (HTS cuprates), for example, the transition width can be as large as 20 K or even more (Figure 3.2(c)), while their textured or epitaxially grown stoichiometric thin films exhibit single sharp transitions, as narrow as 0.1 K or less (Figure 3.2(d)). When the transition is broad, it is customary to mention the value of T_c(onset) as well as $T_c(R = 0)$, the former referring to the temperature at which the drop in the resistance commences. Its unambiguous estimation is in general difficult. It is common to linearly extrapolate the upper and lower parts of the knee of the curve, as shown in Figure 3.2(c), and their point of interception is taken as T_c(onset) (or T_{co}). Alternatively, the onset of transition is also taken as the temperature point on the knee of the curve where the resistance has dropped by 10% of its value where the rounding-off has been detected. If a sample is inhomogeneous and contains multiple phases with different T_c values intercepting the path of the current flow, then multiple superconducting drops are observed (Figure 3.2(e)) at different temperatures. It is worth mentioning that the resistivity behaviour in the normal state (above T_c) will also vary at low temperatures, and also differently for different superconductors. While the metallic elements and most of the alloys and compounds, above T_c, tend to show the expected T^5 behaviour of resistance, the HTS cuprates, for example, under optimal conditions generally exhibit linear behaviour. This is one of the attributes of an *unconventional normal state*. When their carrier concentration is low, the observed temperature dependence is activated or semiconductor-like, showing a negative dR/dT, prior to the superconducting drop (Figure 3.2(f)). The resistivity behaviour in both normal and superconducting states is sensitive to the sample processing and is influenced by the metallurgical state and microstructure.

3.5 Implications of perfect conductivity

If we simply regard a superconductor as a limiting case of a normal conductor with electrical conductivity σ tending to infinity, then it follows from Ohm's law $\mathbf{J} = \sigma\mathbf{E}$ that the electric field $\mathbf{E}$ within the material tends to 0. Since from Maxwell's equation curl $\mathbf{E} = -\dot{\mathbf{B}}$, for a perfect conductor, the magnetic induction $\mathbf{B}$ is invariant in time. Such a description, where a superconductor is regarded simply as a perfect conductor, is inadequate, as can be seen by considering the following situations.

Case (i): Zero-field-cooled sample

Let us start with a macroscopic body, such as a long cylindrical rod in an external magnetic field H parallel to the sample axis (so that the demagnetisation factor $D = 0$). Initially, while maintaining $H = 0$, the sample is cooled to a temperature $T < T_c$ so as to become perfectly conducting. Subsequently, we switch on the external magnetic field $H < H_c$. Persistent screening (eddy) currents are induced on the sample surface, preventing the field from penetrating the sample, so the field remains excluded from the sample (Figure 3.3(a)).

Case (ii): Field-cooled sample

We now start with $H_c > H > 0$ and the sample temperature $T > T_c$. The magnetic lines of force penetrate the sample, and B is therefore not equal to zero. We now cool down to $T < T_c$ so that the final physical conditions of the sample in respect of T and H are the same as in case (i). However, in contrast to Figure 3.3(a), the sample contains a magnetic induction B, and moreover, changing the external field has no effect on B, as the induced screening currents make B invariant and it remains trapped in the sample (Figure 3.3(b)).

The two cases just described imply that all changes in H are screened out in a perfect conductor, and any change in the external magnetic field induces currents on the surface of the sample such that the magnetic field of these currents just compensates the change in the external field to keep B constant.

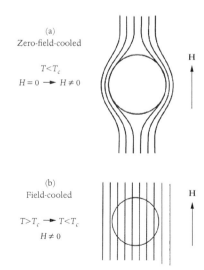

(a)
Zero-field-cooled

$T < T_c$
$H = 0 \longrightarrow H \neq 0$

H

(b)
Field-cooled

$T > T_c \longrightarrow T < T_c$
$H \neq 0$

H

Figure 3.3 Behaviour of a sample of material exhibiting perfect conductivity but not superconductivity: (a) zero-field-cooled; (b) field-cooled.

For a perfect conductor, by applying simple electrodynamics, one finds (see Appendix 3A)

$$\dot{B}(x) = \dot{B}(0)\exp(-x/\lambda). \tag{3.1}$$

where λ represents the depth, or the sample thickness below the surface, over which the induced screening currents flow to keep B invariant and is given by

$$\lambda = \left(\frac{m}{ne^2\mu_0}\right)^{1/2}. \tag{3.2}$$

Clearly, for $x \gg \lambda$, $\dot{B}(x) = 0$, and equation (3.1) implies that in the interior of a perfect conductor, below a depth λ, the magnetic field cannot change in time from the initial value it had when the sample became perfectly conducting.

However, as described below, Meissner and Ochsenfeld (1933) found that a superconductor behaved quite differently from what was expected for a perfect conductor.

3.6   Meissner–Ochsenfeld effect

Meissner and Ochsenfeld (1933) placed a single crystal of tin in a magnetic field $H < H_c$ at room temperature and then gradually lowered the temperature, as followed in Case (ii) above (Figure 3.3(b)). Above T_c, in the normal state, the external field penetrated the sample (Figure 3.4(a)). However, just below T_c, in sharp contrast to the behaviour expected for a perfect conductor (Figure 3.3(b)), they found that the field outside the sample changed abruptly in accordance with the sample completely expelling the interior magnetic induction, leading to a state where $B = 0$ (Figure 3.4(b)). This is known as the *Meissner–Ochsenfeld effect*, or just the *Meissner effect*, and the state corresponding to flux exclusion, that is, $B = 0$, is known as the *Meissner state*. When H exceeded H_c, the field abruptly penetrated the sample, destroying the superconducting

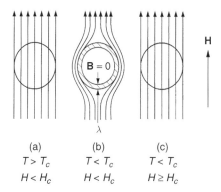

Figure 3.4   The Meissner effect.

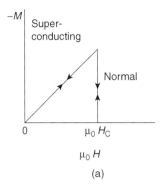

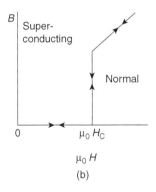

Figure 3.5 (a) Magnetisation and (b) B–H curves in the Meissner state.

state and returning the sample to its normal state (Figure 3.4(c)). The process was reversible; that is, on reducing the field below H_c, the field was suddenly expelled, making the sample once again superconducting. Clearly, the Meissner effect showed that superconductivity was not just a state of perfect conduction, but was also one of perfect diamagnetism.

Thus, since $\mathbf{B} = \mu_0(\mathbf{H} + \mathbf{M})$, for a *bulk superconductor* with $H < H_c$, since $B = 0, M = -H$ and the magnetic susceptibility $\chi = M/H = -1$, where M is the magnetisation per unit volume. A superconductor in a field $H < H_c$ behaves like a perfectly diamagnetic substance with permeability $\mu = 0$. However, as will be seen in the next section, the magnetic field does penetrate the sample over a small distance λ, called the penetration depth, over which the shielding supercurrents flow. Figure 3.5 shows the reversible M–H and B–H curves associated with the Meissner effect.

Samples with thickness $d \gg \lambda$ are considered as thick, and in many instances their properties can be better understood by assuming $\lambda = 0$. A specimen with a thickness $d < \lambda$ is considered as a thin superconductor, in which field penetration becomes important. Clearly, for a thin superconductor in the form of a film or a fibre, the Meissner effect is only partial, although it does manifest the $R = 0$ state for $T \leq T_c$. This, as we will soon find, has an interesting consequence in enhancing the critical field.

3.7 London phenomenology

In order to account for the Meissner effect, F. and H. London (1935) considered (Appendix 3A) superconductivity to arise from what they called *superelectrons*, different from normal electrons, and added to Maxwell's equations two more equations, namely $\mathbf{E} = \mu_0\lambda^2\mathbf{J}$ and $\mu_0\lambda^2\text{curl}\,\mathbf{J} + \mathbf{B} = 0$, with the first being the perfect-conductivity equation while the second was purely phenomenological. Instead of equation (3.1), this led (Appendix 3A) to

$$\mathbf{B}(x) = \mathbf{B}(0)\exp(-x/\lambda). \qquad (3.3)$$

Clearly, for $x \gg \lambda$, $\mathbf{B}(x) = 0$, which is in accordance with the Meissner effect. The London equations thus allow for finite penetration of the applied field and predict the field to decay to $1/e$ of its value over a distance λ beneath the sample surface (Figure 3.6). The parameter λ is called the *penetration depth*, or the *London penetration depth* and is given by

$$\lambda = \left(\frac{m^*}{n_s e^{*2}\mu_0}\right)^{1/2}. \qquad (3.4)$$

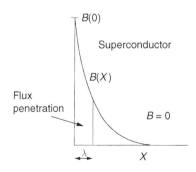

Figure 3.6 Flux penetration over the distance λ, the penetration depth.

It represents the distance over which supercurrents flow to counter the effect of external field. Note that equation (3.4) is the same as equation (3.2) but with the pertinent parameters corresponding to those of super-electrons instead of normal electrons: the effective mass $m^\star$, charge $e^\star$, and number density n_s of the superelectrons involved in superconductivity. The flux quantisation and tunnelling experiments of the early 1960s revealed that $e^* = 2e$. This is entirely consistent with the microscopic BCS theory of superconductivity (Bardeen et al., 1957), as described in Chapter 2, where the charge carriers in the condensed state are electron pairs (*Cooper pairs*) rather than single electrons.

3.8   Penetration depth

The existence of the penetration depth in superconductors has been confirmed experimentally, and it varies from 50 to 100 nm for pure metals and from 500 to 1000 nm for alloys and compounds. Several different ways to determine the penetration depth of superconductors experimentally are described by Shoenberg (1952). These measurements are generally made on samples in the form of either thin films or fine colloidal powders of dimensions comparable to the penetration depth. Magnetisation measurements, for example, involve the use of a mutual inductance method, where the change in the flux linkage between coils wound on a superconducting sample is studied as a function of temperature. As the penetration depth varies with temperature, the flux linkage between the two coils changes accordingly, and can be readily measured. Besides magnetisation experiments, microwave surface impedance measurements have also proved fruitful for studies of the penetration depth (Waldram, 1964). At high frequencies, the surface impedance of a superconductor is determined by the penetration depth λ, while in a normal metal it is determined by the skin depth δ. The resonant frequency of a cavity containing the superconductor is measured for both the normal and superconducting states, and from the observed shift in the resonant frequency and by using the Kramers–Kronig relation, the zero-frequency value of λ can be obtained. The observed temperature variation of the penetration depth, to a good approximation, is represented by the relation

$$\lambda(T) = \lambda(0)\left[1 - \left(\frac{T}{T_c}\right)^4\right]^{-1/2}, \tag{3.5-I}$$

where $\lambda(0)$ is the penetration depth at 0 K. A better fit than this, however, is given by the BCS theory (Bardeen et al., 1957):

$$\lambda(T) = \lambda(0)\left\{\frac{\Delta(T)\tanh\left[\frac{\Delta(T)}{2k_B T}\right]}{\Delta_0}\right\}^{1/3},$$

where $2\Delta(T)$ is the superconducting energy gap at temperature $T < T_c$ and $2\Delta_0$ is that at 0 K. Near T_c, however, both equations reduce to

$$\lambda(T) = \lambda(0) \left[2 \left(1 - \frac{T}{T_c} \right) \right]^{-1/2}. \qquad (3.5\text{-II})$$

The manner in which the penetration depth changes with temperature provides a useful way to infer the nature of the superconducting order parameter, namely whether it is *s*-wave, *p*-wave, or *d*-wave, etc.

3.9   Depairing current density

The current density associated with supercurrents flowing in the penetration depth λ can be determined by differentiating equation (3.3), which yields

$$J(x) = \frac{dB}{dx} = -\frac{B(0)}{\lambda} \exp\left(-\frac{x}{\lambda}\right). \qquad (3.6)$$

It is worth noting that this current has a maximum value of H_c/λ at the surface, for $x = 0$, when the field at the sample surface, $B(0)$, is equal to the thermodynamic critical field H_c. This is the maximum current density that a superconductor can tolerate before reverting to its normal state. It is termed the *depairing* (or *pair-breaking*) *current density*. As the name suggests, at this current density, all the Cooper pairs responsible for a lossless current flow are broken. Its value, which depends upon H_c and λ, ranges around 10^{12} A/m². In practice, the optimum values of the current density realised so far are generally two orders of magnitude smaller than the depairing current density.

3.10   Shortcomings of the London phenomenology

Experimental values of the penetration depth in pure metals are four to five times larger than the London value given by equation (3.4). They show a marked increase when impurities are present in the sample. Pure samples exhibit only a weak field dependence of the penetration depth, with its value changing by a mere 2% when the field is raised from zero to H_c. Impure samples, on the other hand, show a greater field dependence. None of these observations, however, seemed to follow from the London theory. Furthermore, the existence of the penetration depth λ itself posed a problem that was originally perceived by London (1950). He realised that the field penetration over a small but finite depth inside a superconductor should make it energetically favourable to split into a fine mixture of superconducting and normal domains, with the bulk of the sample remaining superconducting. The surface energy between such domains, on the basis of the London theory, is $-\lambda(\frac{1}{2}\mu_0 H_c^2)$. Here the term in parentheses is the condensation energy mentioned in Section 4.1, which represents the energy difference per unit volume between the normal and superconducting states. The negative sign implies that the creation of domains is favoured, and in such a situation no Meissner effect is to be expected. Clearly, it is somehow energetically unprofitable for a superconductor to undergo such a splitting—it must cost extra energy to create the interfaces separating normal and superconducting domains. In other words, there must be a larger positive contribution to the surface energy that would make the net sign of the overall interface energy positive. Studies of superconductors driven into the intermediate state were also in accord with the idea of positive boundary energy.

3.11   Intermediate state

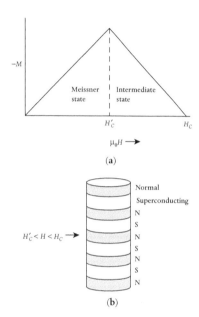

Figure 3.7 (a) For a sample of arbitrary shape, when the demagnetisation factor is not zero, field penetration occurs in the range $H'_c < H < H_c$, resulting in an intermediate state, shown for a cylindrical sample in (b).

The magnetisation behaviour of Figure 3.5 is for long cylindrical or flat samples parallel to the magnetic field such that the demagnetisation factor $D_m = 0$. But, for an arbitrarily shaped sample where $0 < D_m < 1$, some parts of the sample surface experience a larger field than the rest. The field penetration now begins at a field H lower than H_c, given by $H'_c = H_c(1 - D_m)$, and when it rises to H_c, the superconductivity is completely destroyed (Figure 3.7(a)). Between H'_c and H_c, the superconductor is in the intermediate state, comprising alternating superconducting and normal regions (Figure 3.7(b)), the volume proportion of the latter increasing with the applied field. The structure of the intermediate state at the specimen surface has been investigated using numerous experimental tools, such as local field probes, the Faraday magneto-optic technique, the magneto-resistive technique, high-resolution Bitter patterns, and the electron-beam shadow technique, as reviewed by DeSorbo and Healy (1964) and Livingston and DeSorbo (1969). The existence of the intermediate state structure immediately at the sample surface confirms the positive sign of the boundary energy between the normal and superconducting regions. If this were not so, then the normal regions would branch out into finer parts near the surface so that there would be no observable structure at the sample surface. The observation of the intermediate state was therefore an important milestone in theoretical development.

3.12   Filamentary superconductors
and Mendelssohn's sponge

An interesting consequence of the existence of the penetration depth is that a thin sample of thickness $d < \lambda$ possesses a higher critical field than the bulk material. As mentioned in Section 3.6, for a thin film or a fibre, the Meissner effect is incomplete and the increase in energy due to expulsion of the magnetic field is therefore less than for the bulk sample. Since the condensation energy per unit volume is the same for both, namely $\frac{1}{2}\mu_0 H_c^2$, field expulsion from a relatively smaller volume would lead to a higher value of H_c for the low-dimensional material. This was confirmed experimentally by several workers. Pontius (1937) found that the H_c of a lead wire was markedly increased when the wire diameter was reduced. Bean et al. (1962) forced several soft superconducting metallic elements such as Hg, In, and Pb etc. into leached Vycor glass with pore diameters of about 3–5 nm, so that the material now consisted of interconnected networks of ultrafine metallic filaments (Figure 3.8). The critical fields of some of these filamentary materials were found to exceed 5 T, whereas their bulk values were 100 times lower. Further, in zero field and below 4 K, they carried current densities of 10^8 A/m^2, although both these effects lasted only for a short time, since the forced pressed metals oozed out from the pores soon after they were introduced. Nevertheless, this is the first, and so far the only, example of a synthetic material made using soft elemental superconductors that could sustain large magnetic fields and electric currents.

Superconducting alloys in general possess a much higher critical field than pure elements. Before the advent of Abrikosov's type II superconductivity (Abrikosov, 1957), this was explained in terms of *Mendelssohn's sponge model* (Mendelssohn and More, 1935). The model envisaged the alloys as being inhomogeneous internally and forming interconnected fine filaments arranged in the form of a sponge. For the reason already mentioned, the filaments remained superconducting at a much higher field, long after the external field had destroyed superconductivity in the bulk material. However, the exact nature of these filaments was not clear. Some, authors (Hauser and Buehler, 1962; Hauser and Helfand 1962) suggested that crystal dislocations played this role. However, subsequent experimental results raised serious doubts about the validity of the sponge model. Instead, the behaviour of superconducting alloys, as we will see, is adequately explained in terms of Abrikosov's mixed state of type II superconductors, as discussed in Chapter 5.

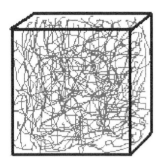

Figure 3.8 Schematic illustration of a synthetic high-field/high-current superconductor formed by forcing a soft superconducting metallic element such as Hg, In, or Pb through the interconnected ultrafine pores of Vycor glass. Such a composite containing a very fine mesh of metallic filaments of 3–5 nm diameter showed superconductivity up to magnetic fields exceeding 5 T; that is, more than about 100 times greater than the bulk critical fields of these metals (Bean et al., 1962).

3.13 Range of coherence and non-local theory

Against the background of a wide variety of facts and experimental observations, Pippard (1950) proposed *spatial coherence* as a fundamental property of a superconductor. Studies of elemental superconductors and their dilute alloys had revealed the following significant features:

1. The extreme sharpness of the superconducting transition in zero magnetic field suggested that cooperative behaviour of a large number of electrons was involved, although the transition became broader in the presence of impurities.

2. The very weak field dependence of the penetration depth indicated that the effect of a magnetic field was not confined simply to the penetration layer of thickness about 10^{-7} m, but was spread over a greater distance of the order of 10^{-6} m.

3. The estimated interfacial boundary energy between superconducting and normal domains was $\frac{1}{2}\mu_0\delta H_c{}^2$, where the length δ was of the order of 10^{-6} m, which was too large to arise merely from the mismatch between the two regions, although it could be accounted for if the boundary was spread out.

Pippard realised that if the local electronic state in a superconductor is characterised by an *order parameter* ψ, then any perturbation in ψ will be spread over a distance ξ (of the order of 10^{-6} m) from the centre of the disturbance; that is, the perturbing effect in a superconductor is primarily *non-local*. This concept is known as the *coherence of the superconducting state* and the distance ξ is known as the *range of coherence*. As a result, the current density $\mathbf{J}_s$ is not simply related to the value of the vector potential $\mathbf{A}$ at the pertinent local point in space, according to the relation (3A.6c) assumed in the London phenomenology, namely $\mu_0\lambda^2\mathbf{J} + \mathbf{A} = 0$, but depends on $\mathbf{A}$ in a surrounding region of volume approximately ξ^3. In the superconducting state, a very large number of electrons in this volume act together coherently, so that the transition is extremely sharp; otherwise, statistical fluctuations would have considerably broadened the transition. Electrons, however, do not behave so coherently over a distance larger than their mean free path l_F. The presence of impurities decreases l_F and can also lower the coherence length, and hence the coherence volume. This accounts for the broadening of the transition due to impurities or with alloying. To include the above effects, Pippard (1953) modified the London equation (3A.6c) and proposed

$$\mu_0\lambda^2\mathbf{J}_s + \frac{\xi}{\xi_0}\mathbf{A} = 0, \tag{3.7}$$

where ξ_0 is the range of coherence in the ideal pure-metal limit. In this phenomenology, for normal metals, the penetration depth λ and the coherence length ξ depend on the electron mean free path l_F through the relations (at 0 K)

$$\lambda = \lambda_0\left(\frac{\xi}{\xi_0}\right)^{1/2}, \tag{3.8}$$

$$\frac{1}{\xi} = \frac{1}{\xi_0} + \frac{1}{\alpha l_F}, \tag{3.9}$$

where $\xi \to \xi_0$ as $l_F \to \infty$. Here α is a constant of the order of unity. With a decrease in l_F due to the presence of impurities ξ is reduced, which contributes to an increase in λ.

3.14   Interface boundary energy

The existence of the parameter ξ, which represents a spatial variation of the order parameter ψ, makes the boundary between superconducting and normal domains broad and diffuse. It yields a positive contribution $\frac{1}{2}\mu_0 H_c^2 \xi$ to the interfacial boundary energy, due to loss of order. The flux penetration over a depth λ, as explained earlier, results in a decrease in the boundary energy by $\frac{1}{2}\mu_0 H_c^2 \lambda$. The net boundary energy is therefore

$$E_B = \frac{1}{2}\mu_0 H_c^2 (\xi - \lambda) = \frac{1}{2}\mu_0 H_c^2 \delta, \tag{3.10}$$

where $\delta = \xi - \lambda \sim 10^{-6}$ m for superconducting pure metals, which have a positive surface energy. As a result, it becomes energetically unfavourable for a superconductor in a field $H < H_c$ to form interfaces by splitting into normal and superconducting domains, and the Meissner effect is observed.

3.15   Summary

In this chapter, we have described the two fundamental and unique properties of superconductors, namely the zero electrical resistance and the Meissner effect, which are mutually independent. The basic phenomenology of the transition in the form of the London theory and its subsequent non-local modification have been briefly presented. Phenomenological considerations have led to the emergence of two very important characteristic lengths of a superconductor, namely the penetration depth λ and the range of coherence ξ. The relative size of these two lengths is crucial for the superconducting transition, particularly in determining how a superconductor will respond to an external magnetic field.

Appendix 3A: Electrodynamics of a perfect conductor and London phenomenology

The magnetic implications of perfect conductivity described in Section 3.2 follow from simple electrodynamics of a perfect conductor.

For such a conductor, in an electric field $\mathbf{E}$, the equation of motion of an electron of mass m, velocity $\mathbf{v}$, and charge e does not have a retarding term and is therefore given by $m\dot{\mathbf{v}} = e\mathbf{E}$. If n is the number density of electrons, then the current density $\mathbf{J} = ne\mathbf{v}$, and one can write

$$\mathbf{E} = \frac{m}{ne^2}\dot{\mathbf{J}} = \mu_0\lambda^2\dot{\mathbf{J}}, \tag{3A.1}$$

where

$$\lambda^2 = \frac{m}{ne^2\mu_0}. \tag{3A.2}$$

λ has dimension of length. Using Maxwell's equation curl $\mathbf{E} = -\dot{\mathbf{B}}$ gives

$$\mu_0\lambda^2\text{curl}\,\dot{\mathbf{J}} + \dot{\mathbf{B}} = 0. \tag{3A.3}$$

Applying another of Maxwell's equations, curl $\mathbf{B} = \mu_0\mathbf{J}$, one finds for a perfect conductor

$$\nabla^2\dot{\mathbf{B}} = \frac{1}{\lambda^2}\dot{\mathbf{B}}. \tag{3A.4}$$

The above equation has a solution (Von Laue 1932) showing that $\dot{\mathbf{B}}$ decreases exponentially with depth x inside the sample, and for a semi-infinite slab the solution is

$$\dot{\mathbf{B}}(x) = \dot{\mathbf{B}}(0)\exp\left(-\frac{x}{\lambda}\right). \tag{3A.5}$$

Clearly, for $x > \lambda$, $\dot{\mathbf{B}}(x) = 0$, showing that in the interior of a perfect conductor, the magnetic induction cannot change from the value it had when the sample became perfectly conducting. The induced supercurrents, which do not decay with time, are confined within a depth of λ from the sample surface, and it is they that are responsible for the observed shielding.

However, as described in Section 3.3, the observed response of superconductors to an external magnetic field, as manifested by the Meissner effect, is unexpectedly different. Instead of $\dot{\mathbf{B}}(x) = 0$, it corresponds to $\mathbf{B}(x) = 0$.

To explain the Meissner effect, F and H. London (1935) considered superconductivity to arise from superelectrons, different from normal electrons, and added to Maxwell's equations the following two equations:

$$\mathbf{E} = \mu_0\lambda^2\dot{\mathbf{J}} \tag{3A.6a}$$

and

$$\mu_0\lambda^2\text{curl}\,\mathbf{J} + \mathbf{B} = 0. \tag{3A.6b}$$

The second of these can be alternatively written as

$$\mu_0\lambda^2\mathbf{J} + \mathbf{A} = 0, \tag{3A.6c}$$

where $\mathbf{A}$ is a vector potential such that curl $\mathbf{A} = \mathbf{B}$, and the gauge chosen is such that div $\mathbf{A} = 0$.

Equation (3A.6a) is same as (3A.1), the equation for perfect conductivity, while (3A.6b) is analogous to (3A.3) but with the current density and the magnetic induction rather than their time derivatives. The justification for the introduction of these new equations was merely to explain the observed phenomenon, and thus the London approach was *phenomenological*. Following the same procedure as described in the case of (3A.3), instead of (3A.4) one obtains

$$\nabla^2 \mathbf{B} = \frac{1}{\lambda^2} \mathbf{B}, \tag{3A.7}$$

which, for a semi-infinite slab, has a solution given by

$$\mathbf{B}(x) = \mathbf{B}(0) \exp\left(-\frac{x}{\lambda}\right). \tag{3A.8}$$

Clearly, for $x >> \lambda$, $\mathbf{B}(x) = 0$, in accordance with the Meissner effect. The London equations thus allow for finite penetration of the applied field and predict the field to decay to 1/e of its value at the sample surface over a distance λ beneath the surface.

CHAPTER 4

Thermodynamics and general properties

4.1 Thermodynamic aspects of the transition

The discovery of the Meissner effect and the reversible nature of the superconducting transition led Gorter and Casimir (1934) to consider the transition as a thermodynamic phase change. They showed that the transition from the normal to the superconducting state lowered the free energy of the sample, and this negative contribution per unit volume, called the *condensation energy*, is given by (Appendix 4A)

$$\Delta F = F_n - F_s = \frac{1}{2}\mu_0 H_c^2. \tag{4.1}$$

In the Meissner state, the exclusion of the magnetic flux H raises the magnetic free energy of the sample by $\frac{1}{2}\mu_0 H^2$. As the external magnetic field is increased, the superconducting state gains energy until it fully compensates the negative condensation energy, which occurs at $H = H_c$, when superconductivity is destroyed. Physically, $-\frac{1}{2}H_c^2$ represents the energy that can be drawn upon to start the London supercurrents. It clearly suggests that the 'superelectrons' (Cooper pairs) possess a lower energy than the normal electrons. This produces an energy gap at the Fermi energy: the electronic states below the Fermi energy are pushed down, while those above are pushed up (Figure 4.1).

As mentioned in Section 2.2.1, the magnitude of the condensation energy is extremely small, of the order of 10^{-8} eV/atom, which had posed a major problem in identifying the interaction responsible for superconductivity.

Differentiating equation (4.1) with respect to temperature yields the entropy difference between the normal and superconducting states, and, as may be seen from Appendix 4A, the former has a higher entropy than the latter. Thus, the superconducting state is more ordered than the normal state. Further, in the absence of a magnetic field, the transition at T_c is of second order, while that at H_c is of first order.

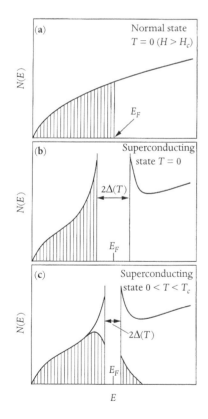

Figure 4.1 Density-of-states curve for (a) normal state, (b) superconducting state at $T = 0$, and (c) superconducting state at $0 < T < T_c$. (b) and (c) depict the superconducting energy gap 2Δ at E_F. In (c), there are quasiparticles present above the gap.

4.2 Thermal properties

4.2.1 Heat capacity

Heat capacity measurements on superconductors reveal (Figure 4.2) an abrupt increase in the electronic specific heat at T_c, which is one of the characteristics of a second-order phase transition. The results on superconducting Al (Phillips 1959) with $T_c \approx 1.2$ K are depicted in Figure 4.2. The heat capacity can in general be measured in both superconducting and normal states at $T < T_c$ and may be compared. For the measurement in the normal state at $T < T_c$, the superconducting state is first quenched by the application of a magnetic field $H > H_c(T)$. The normal-state data for Al at 0.03 T are shown in the figure. Modest magnetic fields do not have a detectable effect on electronic or lattice contributions. However, for the Chevrel phases, pnictides, and high-T_c cuprates, owing to their huge critical fields of 50–400 T, which are difficult to realise, it has been a formidable problem to investigate their normal-state behaviour by quenching the superconducting state at $T < T_c$.

At low temperatures, the specific heat C of a normal metal is expressed as $C = C_{en} + C_L$, where $C_{en} = \gamma T$ and $C_L = \alpha T^3$ are respectively the electronic and lattice contributions. The Sommerfeld constant γ is given by

$$\gamma = \frac{2}{3}\pi^2 k_B^2 N(0),$$

where $N(0)$ is the bare density of states at the Fermi level at 0 K and k_B is the Boltzmann constant. The coefficient α is related to the Debye temperature θ_D through

$$\alpha = \frac{12}{5}\pi^4 N k_B \frac{1}{\theta_D^3}.$$

The BCS theory gives the specific heat jump $\Delta C = 1.43\gamma T_c$, which for most superconductors holds true.

In the superconducting state, C_L essentially remains invariant. This is suggested by the absence of any observable change in the lattice parameters and by the detection of only minimal changes in the elastic properties (Ahlers and Walldorg, 1961). Consequently, the difference between the specific heat values in the normal and superconducting states, C_n and C_s respectively, arises only from a change in the electronic specific heat in the two states, $C_n - C_s = C_{en} - C_{es}$. A typical superconducting behaviour across the transition is shown in Figure 4.3. At $T \ll T_c$, $C_{en} > C_{es}$, but near T_c the latter rises rapidly, manifesting a sharp discontinuity at T_c. The behaviour can be best described in terms of the way the entropy changes during the transition.

The specific heat C of a material is expressed by the thermodynamic relation $C = V_T\, dS/dT$, where V_T is the volume per unit mass and S is

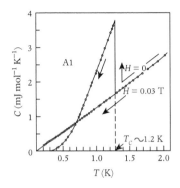

Figure 4.2 Electronic specific heat in superconducting and normal states of Al (Phillips, 1959).

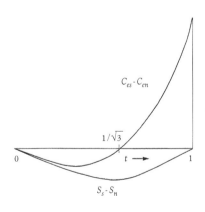

Figure 4.3 Temperature variation (schematic) of the differences in the electronic specific heat and the entropy of superconducting and normal states.

the entropy. The value of C can be at constant volume (C_v) or at constant pressure (C_p). With calorimetry, the latter is routinely measured, but for theoretical purpose the former is preferred. At low temperatures, below 25 K (i.e. for LTS), the two specific heats may be nearly equal, but their difference may rise to 1–2% at $T = 100$ K, which could be relevant for HTS. The sharp increase in the heat capacity at T_c is attributed to a rapid rate of decrease in the entropy resulting from the formation of the new ordered state of 'superelectrons'. The proportion of Cooper pairs continually increases, at the expense of the normal electrons, as the temperature is lowered, and at 0 K it becomes 100%. At $T < 0.3T_c$, in the superconducting state, the decrease in the specific heat with T for a conventional (BCS) superconductor is exponential: $C_{es} = A \exp(-B/k_B T)$, where A and B are constants. The observation of this exponential behaviour served as one of the vital clues for the microscopic theory. It indicated the presence of a gap in the electronic energy spectrum of superconductors below T_c (Corak et al., 1956), which was subsequently predicted by the BCS theory (Bardeen et al., 1957). The constant B is related directly to the energy gap 2Δ. Near 0 K, the electron pairs can only be excited to a higher state by receiving an energy equal to 2Δ, which explains why C_{en} is larger than C_{es} at very low temperatures (Figure 4.3). Any striking deviation from the above exponential behaviour below T_c is generally taken as an indication of unconventional superconductivity.

4.2.2   Thermal conductivity

Analogous to the specific heat, the thermal conductivity of normal metals is composed of electronic and lattice parts: $\kappa = \kappa_e + \kappa_L$. In contrast to the heat capacity, the thermal conductivity manifests the anisotropies of the crystalline sample. In normal metals, the transport of heat is due entirely to electronic heat conduction, as the lattice term is negligibly small. However, in many alloys, including HTS, κ_L may be larger than κ_e. The electronic component is primarily affected by scattering by impurities or defects, and also by phonons, the former becoming predominant at low temperatures. In the superconducting state, the Cooper pairs cannot carry the thermal energy, nor are they scattered by phonons. The electronic heat conduction therefore rapidly diminishes in the superconducting state, but at the same time conduction by phonons is enhanced as they are no longer scattered. At low temperatures, the phonon mean free path increases so much that the heat conduction by phonons may exceed that by electrons. These processes compete with each other and the dominant of the two determines the increase or decrease of thermal conductivity below T_c, quite often giving rise to a peak in the κ–T plot. In pure superconductors, the decrease in κ_e overrides increase in κ_L, and thus the overall thermal

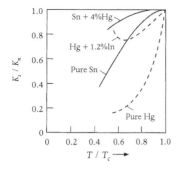

Figure 4.4 Temperature variation of normalised electronic thermal conductivity of pure and alloyed superconductors (after Hulm et al., 1950).

conductivity in the superconducting state is reduced. Figure 4.4 shows some typical variations of κ_{es}/κ_{en} for pure metals and alloys (Hulm, 1950). In impure metals and alloys, including HTS for example, the phonon component dominates, with the result that thermal conductivity increases after the superconducting transition. Clearly, it depends on the sample purity and the ratio κ_e/κ_L whether κ is limited by impurity, defect, or phonon scattering, which in turn decides the final response of the thermal conductivity. If the superconductor is driven normal by applying an external magnetic field, the thermal conductivity is restored to the larger value of the normal state. In the situation where the electronic thermal conduction in the superconducting state is limited by impurity scattering only (i.e. below liquid helium temperature), Bardeen et al. (1959) found κ_{es}/κ_{en} to vary exponentially with temperature. This had, in fact, been observed a few years earlier by Goodman (1953), which suggested the opening of an energy gap in superconductors below T_c. A deviation in the exponential behaviour of the thermal conductivity at low temperatures, below T_c, is a signature of unconventional superconductivity.

4.2.3 Thermal expansion

Thermal expansion of a material is of basic interest because it is part of the equation of state, which provides the inter-relation between pressure P, volume V, and temperature T. The *coefficient of volumetric thermal expansion* α is given by $\alpha = [dV/dT]_P/V$, which can alternatively be expressed in terms of the isothermal compressibility, free energy F, pressure P, and entropy S:

$$\alpha = \left(\frac{\partial \ln V}{\partial T}\right)_P = -K_T\left(\frac{\partial^2 F}{\partial V \partial T}\right) = K_T\left(\frac{\partial P}{\partial T}\right)_V = K_T\left(\frac{\partial S}{\partial V}\right)_T.$$

The ratio α/C_p can be normalised to obtain the dimensionless and nearly temperature-independent *Grüneisen parameter* γ_G, which is a measure of the anharmonicity of the material, given by (White, 1992)

$$\gamma_G = \frac{\alpha V}{C_p K_S} = \frac{\alpha V}{C_v K_T} = \text{constant},$$

where K_S and K_T are respectively the adiabatic and isothermal compressibilities and V is the crystalline molar volume. In contrast to the heat capacity, the thermal expansion displays the crystalline anisotropies of the sample. Analogous to the heat capacity and thermal conductivity, $\alpha(T)$ is composed of both electronic and phononic contributions at low temperatures, that is, $\alpha(T) = \alpha_e + \alpha_L = aT + bT^3$, and at sufficiently low temperatures < 10 K the linear electronic contribution predominates over the phononic T^3 part. The ratios of the respective electronic and phononic terms of $C_p(T)$ and $\alpha(T)$ give the corresponding

Grüneisen parameters. The value of the electronic Grüneisen para-meter $\gamma_G^{el} = \alpha_e V / K_S C_e$, is a measure of the volume dependence of the electronic density of states $N(E_F)$ at the Fermi surface:

$$\gamma_G^{el} = \frac{d \ln[N(E_F)]}{d(\ln V)}.$$

The superconducting transition affects the entropy and free energy. As stated earlier, the condensation energy per unit volume, $\frac{1}{2}\mu_0 H_c^2$, is extremely small. The differences in volume, length, energy, and elastic moduli between normal and superconducting states are usually not more than a few parts in 10^7. These differences are less of engineering than of fundamental interest, as they relate thermodynamically to the critical field H_c and critical temperature T_c (Shoenberg, 1952), given by

$$\Delta V = V_s(H_c) - V_s(0) = V_s \mu_0 H_c \frac{dH_c}{dP} + \frac{1}{2}\mu_0 H_c^2 \frac{dV_s}{dP} \tag{4.2}$$

and

$$\Delta \alpha = \alpha_n - \alpha_s = \mu_0 \left(\frac{dH_c}{dT} \right) \left(\frac{dH_c}{dP} \right). \tag{4.3}$$

This leads to the Ehrenfest relation for the pressure dependence of T_c:

$$\frac{dT_c}{dP} = VT_c \frac{\Delta \alpha}{\Delta C}, \tag{4.4}$$

where $\Delta C = C_n - C_s$ is the difference in the heat capacities of the nor-mal and superconducting states. For most superconductors, dT_c/dP is negative, since $C_n > C_s$, which makes $\alpha_n > \alpha_s$.

The thermal expansion is difficult to measure experimentally. For lead, the linear coefficient of expansion due to electrons was found to be proportional to T^3 between 1.3 K and 12 K when the sample was in the superconducting state. The value of α in the superconducting state was found to be less than in the normal state (White, 1962), as expected. Similar results have been noted also for V, Nb, and Ta from extrapol-ations of the thermal expansion versus temperature curves above and below T_c. Direct observations of the length changes δ_l at the super-conducting to normal transition were mostly prevented by hysteresis effects which made it impossible (White, 1962) to draw unambiguous conclusions about δ_l. However, more recently thermal expansion data on different HTS systems, such as YBCO, LSCO, and BSCCO, are avail-able (White, 1992) to serve technical purposes of designing compatible substrates, but more and better single-crystal data (such as those of Meingast et al., 1990) are needed to understand the lattice dynamics involved.

4.2.4 Thermoelectric power

The *thermoelectric power* (or *thermopower*, or *Seebeck coefficient*) S, generally expressed in $\mu V/K$, measures the magnitude of an induced thermoelectric voltage produced as a result of the temperature difference across the material, and is given by $S = \Delta V/\Delta T$. It describes the interaction between electric and thermal currents and displays the anisotropic features of the crystalline sample. The temperature difference causes the charge carriers in the material to diffuse from the hot side to the cold side, thus giving rise to a thermoelectric voltage. Measurements are generally made on a flat plate-shaped sample sandwiched between two flat copper plates, across which a temperature gradient is established, and the resulting voltage difference between them is measured. The value of S thus measured also includes a thermoelectric contribution from the copper, which has to be subtracted. For a superconductor below T_c, S vanishes similarly to the electrical resistance, although thermoelectric effects are still present. Measurements of S are among the most frequently pursued transport investigations (Kaiser and Uher, 1991) to gain fundamental information about charge carriers, whether holes or electrons.

As with other thermal properties mentioned above, there are two distinct contributions to the thermopower, namely S_d and S_g arising respectively from charge carriers and phonons, giving $S = S_d + S_g$. The diffusion thermopower S_d is due to the difference in the diffusion rates of charge carriers up and down the temperature gradient. S_g represents the phonon drag contribution, which arises because, under a thermal gradient, phonons are not in thermal equilibrium and while they move they lose momentum by interacting with the charge carriers. If phonon–electron interaction is predominant, the phonons tend to sweep or drag electrons to one end of the material, which contributes to the already present thermoelectric field. This phonon drag contribution is most important in the temperature region where phonon–electron scattering is dominant, and this happens for $T \approx \theta_D/5$, where θ_D is the Debye temperature. At lower T, there are too few phonons to drag the electrons, while at higher T, or when the carrier density is low, phonon–phonon scattering dominates over phonon–electron scattering. The phonon drag typically causes a peak at a temperature near $\theta_D/5$ in the S–T curve, which may be countered by strong electron–phonon interaction. As may be seen in Figure 4.5, showing the S–T curve (Putti et al., 2002) for MgB_2 ($\theta_D \approx 1000$ K), which has a strong electron–phonon interaction, the peak is not reached at $T = \theta_D/4$. For isotropic metals, the two contributions are given by $S_d = C_e/ne$ and $S_g = C_L/3ne$, where C_e and C_L are respectively the electronic and lattice specific heats (see Section 4.7.1) and n is the number density of carriers with charge e. For $T \ll \theta_D$, the thermopower S of metals thus assumes the simple form

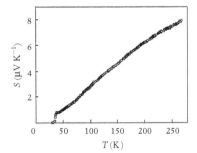

Figure 4.5 Thermoelectric power versus temperature for MgB_2 (after Putti et al., 2002).

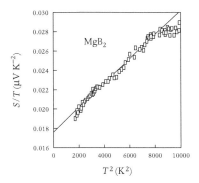

Figure 4.6 Thermopower measurements (after Putti et al., 2002) on MgB$_2$ ($T_c = 39$ K), showing the linear behaviour of S/T with T^2 in the normal state between 40 K and 90 K.

$$S = S_d + S_g = \frac{C_e}{ne} + \frac{C_L}{3ne} = AT + BT^3, \qquad (4.5)$$

where $A = \gamma/ne$ and $B = \alpha/3ne$, with γ and α as described earlier. Low-temperature thermopower data are found to be in fair agreement with the above equation, although deviation from the expected linear temperature behaviour of the diffusion thermopower has also been reported (Kaiser and Uher, 1991). Figure 4.6 shows the ratio S/T as a function of T^2 for MgB$_2$ ($T_c = 39$ K), which exhibits linear behaviour in accordance with equation (4.5).

Diffusion thermopower and Hall effect should in general give the same signs for the type of carrier. A disagreement between the signs can be commonly understood in multicarrier/multiband cases, where significant cancellation between electron and hole contributions, as found in high-temperature superconductors, may occur.

In the region just above T_c (where superconducting fluctuations occur), the thermopower may rise in some high-temperature superconductors (Kaiser and Uher, 1991), owing to the presence of Cooper pairs, which increase the mean free path of phonons. The phonon drag is greater owing to the increased phonon disequilibrium and the decrease in the density of carriers interacting with the phonons. The same effect accounts for the peak in thermal conductivity below T_c observed in high-temperature superconductors.

As mentioned earlier, although the thermopower in the superconducting state is zero, thermoelectric effects still persist (Ginzburg and Zharkov, 1978; Ginzburg, 1989). The diffusion of normal-state carriers down the temperature gradient is balanced by an opposing supercurrent that flows without the need for the usual thermoelectric voltage. A circulating current is thus established in the superconducting sample whereby the normal current is transformed into the supercurrent, and vice versa at the ends of the sample (Kaiser and Uher, 1991). The circulating currents contribute an additional term κ_{add} to the thermal conductivity κ_e of superconductors. This excess contribution is due to an extra heat current occurring as electrons diffuse down the temperature gradient owing to the thermoelectric effect when the electric field $E = 0$ In low-temperature superconductors, the ratio $k_{add}/k_e < 0.1\%$ ($S \approx 5\mu V/K$) near 300 K and much smaller below T_c.

4.3 Ultrasonic behaviour

Attenuation of ultrasonic waves is the result of scattering by the lattice, static imperfections and, conduction electrons. The scattering by the lattice and imperfections at low temperatures is essentially temperature-independent, and therefore the conduction electrons have an important role in the attenuation of ultrasonic waves and the temperature variation is determined by the excitation of thermally excited electrons.

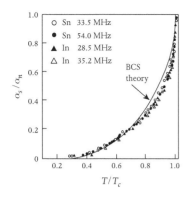

Figure 4.7 Ultrasonic attenuation of Sn and In (after Morse and Bohm, 1957).

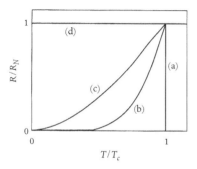

Figure 4.8 Schematic illustration of the behaviour of reduced AC resistance, measured with increasing frequencies, from (a) to (d), as a function of reduced temperature; R_N is the normal-state resistance. (a) corresponds to DC resistance behaviour and (d) to optical frequencies in the visible range, when the behaviour is indistinguishable from that in the normal state. Curves (b) and (c) correspond to the intermediate frequency range, where the transition width gradually increases with frequency.

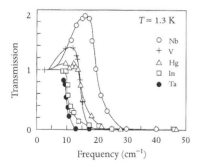

Figure 4.9 Optical measurements on various elemental superconductors, showing transmission losses resulting at high frequencies (after Richards and Tinkham, 1960).

Below T_c, the electron pairs can no longer interact with the lattice and consequently the attenuation is diminished. The ratio of the attenuation coefficients in the superconducting and normal states α_s/α_n at $T < T_c$ gives the relative proportion of unpaired electrons. The BCS theory gives

$$\frac{\alpha_s}{\alpha_n} = \frac{2}{\exp(2\Delta/k_BT) + 1}. \tag{4.6}$$

Ultrasonic attenuation measurements on several superconducting elements (Figure 4.7) are in accord with this equation (Morse and Bohm, 1957). The technique has been used to measure the energy gap and its anisotropy in superconducting single crystals.

4.4 AC and optical properties

The alternating current (AC) behaviour of superconducting alloys (type II superconductors), to be described later, differs significantly from that of metallic pure elements (type I superconductors), considered now. A bulk superconductor continues to be diamagnetic and resistance-less at low frequencies ($< 10^9$ Hz). Although no AC losses are expected at low frequencies, some losses are nevertheless observed, which are attributed to field penetration effects at the sample surface. Due to surface defects, hysteresis results when an AC field is applied. The resistive transition gradually becomes broader when the frequency of the electric field is raised, and in the infrared and visible range of the spectrum there is no superconducting state detected at all. This is illustrated schematically in Figure 4.8. When the frequency f of the electric field is below the value that corresponds with the superconducting energy gap, that is, $f < 2\Delta/h$, the energy is insufficient to break the Cooper pairs and the resistance continues to be zero. Thus, electromagnetic radiation in this frequency range will be transmitted through the superconductor (Figure 4.8). In the higher-frequency range, however, since the energy is larger, the Cooper pairs are broken, with absorption of the electromagnetic energy, and no superconductivity is observed. Since $2\Delta \approx 3.5k_BT_c$, for superconductors with T_c of about 1 K, f is in the microwave region, for T_c of about 10 K, it is in the far-infrared region, while for high-T_c cuprate superconductors with T_c of around 100 K, it is still higher. In this way, Biondi and Garfunkel (1959) were able to measure the energy gap of Al, with $T_c = 1.08$ K, by studying the absorption of microwaves. Richards and Tinkham (1960) studied the energy gap of several pure elements such as Nb, V, Ta, Hg, and In, with T_c in the range of 3.5–9.3 K, using a far-infrared monochromator, and showed that the frequencies in the range $f < 2\Delta/h$ were transmitted (Figure 4.9) and that there was an onset of an absorption edge at the gap frequency.

4.5 Tunnelling in the superconducting state

When two metals are kept apart, separated by an insulator, forming what is known as a *junction*, there is a finite probability of electron transport taking place from one to the other, provided the insulating layer is sufficiently thin—of the order of, or less than, the electron mean free path. This is a quantum mechanical phenomenon, which is termed *tunnelling*. However, tunnelling will take place, say from the metal on the right to the metal on the left, only if the latter has empty energy states to accept the incoming electron. The tunnelling phenomenon in superconductors is considered with respect to two situations:

(i) where the junction is between a normal metal and a superconductor, with a thin insulating barrier, termed an *N–I–S junction*;

(ii) where the junction is between two superconductors, termed an *S–I–S junction* (in place of the insulating layer I, one may have a thicker normal metallic layer N, resulting in an *S–N–S junction*).

The electron or single-particle tunnelling that can occur in both N–I–S and S–I–S junctions is generally referred to as *Giaever tunnelling* (Giaever, 1960). In the case of S–I–S junctions, where the electrodes on either side of the barrier are superconducting, under certain conditions, Cooper pairs can tunnel between them. This was predicted by Josephson (1962) and is called *Josephson tunnelling*.

4.5.1 N–I–S junctions

The energy diagram of an N–I–S junction is shown in Figure 4.10(a). At $T = 0$ K, all the states are filled up to the Fermi level E_F in the normal metal on the left side of the barrier, while they are filled up to $E_F - \Delta$ in the superconductor on the other side, where 2Δ is the superconducting energy gap. Above the superconducting gap of width 2Δ, the energy states are empty. In the absence of a bias voltage, the Fermi levels on either side are equalised, and clearly there will be no tunnelling of electrons across the insulating barrier. Electrons from the metal cannot tunnel into the superconductor, because of the superconducting gap, and also the electrons in the superconductor, which is filled to just below the gap, cannot flow the other way, because the energy states in the metal are occupied. The situation will remain unchanged in the presence of a bias voltage $V < \Delta/e$. At $V = \Delta/e$, there will be a sudden onset of current due to tunnelling of electrons from left to the right to fill up the empty states, where there exists a large density of states. With increasing bias voltage (Figure 4.10(b)), further empty states above the gap are filled, resulting in the *I–V* behaviour shown in Figure 4.10(c).

Figure 4.11(a) depicts the situation for $0 < T < T_c$. Some of the electrons in the normal metal have energies in excess of E_F, while

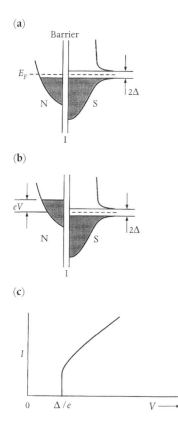

Figure 4.10 Energy diagrams of an N–I–S junction at $T = 0$ for (a) $V < \Delta/e$, (b) $V \geq \Delta/e$. (c) The resulting *I–V* behaviour.

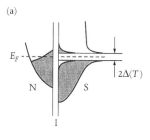

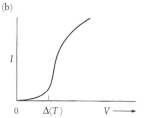

Figure 4.11 (a) Energy diagram of an N–I–S junction at $0 < T < T_c$. (b) The resulting I–V behaviour.

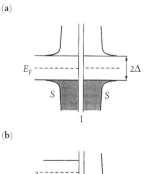

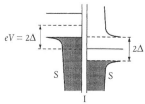

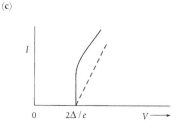

Figure 4.12 S–I–S tunnelling between two superconductors with identical energy gaps. (a, b) Energy diagrams. (c) I–V behaviour. The dashed line (linear behaviour) corresponds to the situation where both electrodes are normal.

there are electrons present above the superconducting gap, owing to the thermal breaking of Cooper pairs. At a relatively small bias voltage, tunnelling can now commence, although a significant rise will again occur when $V = \Delta(T)/e$ (Figure 4.11(b)). The nonlinear I–V behaviour observed in Giaever tunnelling provides a convenient way to measure the superconducting energy gap.

4.5.2 S–I–S junctions

4.5.2.1 *Junction electrodes with similar energy gaps*

At $T = 0$, in the absence of a voltage bias, the Fermi levels and the energy gaps in the superconductors on either side of the barrier are equalised (Figure 4.12(a)). Also, on both sides, all the energy states up to $E_F - \Delta$ are filled, and under these conditions no tunnelling is to be expected in the equilibrium condition. This situation will remain unchanged so long as the bias voltage $V < 2\Delta/e$. However, at $V = 2\Delta/e$ (Figure 4.12(b)), when the bottom of the gap on the left side matches with the top of the gap on the right side, there will be a sudden onset of tunnelling current, since the electrons on the left side of the barrier can now tunnel to the right side, to fill the vacant states above the gap. With a further increase in the bias, more vacant states get filled. This is manifested in the I–V curve of Figure 4.12(c). At a finite temperature $T < T_c$, the sharp features tend to get smeared out and, with increasing T, the knee of the curve moves towards the lower-voltage side. At T_c, when both electrodes become normal, the I–V behaviour becomes linear (dashed line in Figure 4.12(c)).

4.5.2.2 *Junction electrodes with different energy gaps*

If $2\Delta_1$ and $2\Delta_2$ are the energy gaps of the left and right electrodes, respectively, with $\Delta_1 < \Delta_2$, for $T = 0$ K, the tunnelling current starts only when the bias voltage reaches $V = (\Delta_1 + \Delta_2)/e$ (Figure 4.13(a, b)). The I–V behaviour is essentially very similar to that described in Section 4.5.2.1. However, at a temperature $0 < T < T_c$, electrons can be excited across the smaller gap while the energy states above the larger gap may be essentially empty (Figure 4.13(a)). With the application of even a small bias voltage, the excited electrons may tunnel to the unoccupied states (Figure 4.13(b)) of the other superconductor until $V = [\Delta_2(T) - \Delta_1(T)]/e$. With further increase in voltage, the number of tunnelling electrons is unaffected, but the pertinent density of states is now smaller, which results in a decrease in current. The current continues to decrease until the bottom of the smaller gap matches with the top of the larger gap, which happens at $V = [\Delta_1(T) + \Delta_2(T)]/e$. At this point, the current starts rising again with V, as was observed for the case $T = 0$. The I–V behaviour is as shown in Figure 4.13(c). As may be seen, between $[\Delta_2(T) - \Delta_1(T)]/e$ and $[\Delta_1(T) + \Delta_2(T)]/e$, there is a

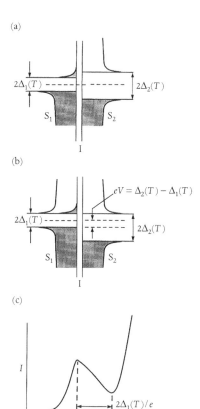

(a)

$2\Delta_1(T)$ ----- ----- $2\Delta_2(T)$

S_1 S_2

I

(b)

$eV = \Delta_2(T) - \Delta_1(T)$

$2\Delta_1(T)$ ----- ----- $2\Delta_2(T)$

S_1 S_2

I

(c)

I

0

$2\Delta_1(T)/e$

$V \longrightarrow$

$\Delta_2(T) - \Delta_1(T)/e$

$\longleftarrow \Delta_2(T) + \Delta_1(T)/e$

Figure 4.13 (a, b) S–I–S tunnelling between two superconductors with different energy gaps. (c) The I–V behaviour shows a negative-resistance region (see the text).

negative-resistance region, which is characteristic behaviour when both electrodes of the tunnel junction are in the superconducting state with different T_c values. Clearly, from the observed I–V behaviour, the gap energies of both superconductors can be readily obtained.

4.5.3 The Josephson effect

In Section 4.5.2, we have described the tunnelling phenomenon in superconductors involving single particles, which in 1960 had provided striking confirmation of the existence of the BCS energy gap. Also, it emerged as an accurate technique for determination of the gap width 2Δ as a function of temperature. It was Josephson (1962) who first realised that besides the single-particle tunnelling, one can have Cooper pairs tunnelling between two superconducting electrodes separated by a sufficiently thin barrier in the form of an insulating oxide layer of about 1 nm thickness. He further pointed out that the tunnelling of Cooper pairs would give rise to a direct current (DC) with zero bias voltage and an alternating current (AC) when the bias voltage had a finite value. These two distinct behaviours are respectively referred to as the *DC and AC Josephson effects*. As well as in the S–I–S-type tunnel junctions having a thin insulating barrier, these effects can be readily seen in a host of superconducting structures called *weak links*, where the superconducting electrodes are mutually connected through a fine constriction in the form of a narrow microbridge, a point contact, a thin disordered region, a narrow crack, etc. Different weak link structures are depicted in Figure 4.18 later in the chapter.

4.5.3.1 The DC Josephson effect

Consider two separate superconductors each described by a mutually independent wavefunction. If the two superconductors are, however, coupled together through a thin barrier, forming a tunnel junction, they may become mutually interlocked, with their phases extending across the barrier. For sufficiently weak supercurrents, the barrier behaves like a superconductor with no voltage difference across the junction. The applied current I controls the phases δ_1 and δ_2 of the two superconductors according to the following current–phase relation, known as the *first Josephson equation*:

$$I = I_0 \sin \delta, \tag{4.7}$$

where $\delta = \delta_1 - \delta_2$ represents the phase difference between the two superconductors and I_0 is the critical current, that is, the maximum DC supercurrent the junction can sustain with $V = 0$, and it corresponds to a phase difference $\delta = \pi/2$. This is known as the *DC Josephson effect*. For $I > I_0$, the phase difference can no longer adjust and the junction switches along the load line to the single-electron tunnelling region, as

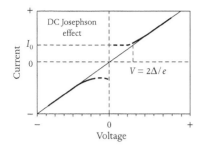

Figure 4.14 Schematic *I–V* characteristic in the DC Josephson effect.

shown in Figure 4.14. In the DC Josephson effect, a magnetic field causes modulation of the critical current, with the modulated current I_M being given by

$$I_M = I_0 \frac{\sin(\pi\phi/\phi_0)}{\pi\phi/\phi_0},\qquad(4.8)$$

where ϕ is the total magnetic flux in the plane of the junction and ϕ_0 is the flux quantum mentioned above. Thus, in the situation where the total flux through the junction is an integral multiple of ϕ_0, the net supercurrent that can flow through the junction becomes zero. Equation (4.8) has the $(\sin\theta)/\theta$ form familiar from the optical diffraction pattern due to a single slit, and accordingly the critical current is found to be modulated with magnetic flux as shown in Figure 4.15. For two such junctions connected in parallel in a superconducting ring, equation (4.8) is modified and one observes interesting quantum interference effects in the presence of magnetic flux (Figure 4.16) that resemble the optical diffraction that occurs with a double slit and the system behaves like a very sensitive fluxmeter. The critical current oscillates in the presence of external magnetic flux passing through the ring, with a period of one flux quantum. Such a device is known as a DC SQUID (*superconducting quantum interference device*). If the device is biased with a constant current, the voltage across the SQUID is also periodic in ϕ_0 and the device behaves like a transducer. Similar to a DC SQUID, there is also an RF SQUID, which comprises only a single junction in a superconducting ring. The ring is coupled to a tank circuit and the voltage across the circuit is periodic in ϕ_0.

4.5.3.2 The AC Josephson effect

When the current is increased from zero, initially there is no voltage across the junction, but for $I > I_0$, a voltage V appears and δ evolves with time according to the voltage--frequency relation

$$\delta = \frac{2eV}{h} = \frac{2\pi V}{\phi_0},\qquad(4.9)$$

Figure 4.15 Modulation (schematic) of the critical current of a single Josephson junction in a magnetic field. The behaviour is analogous to optical diffraction due to a single slit.

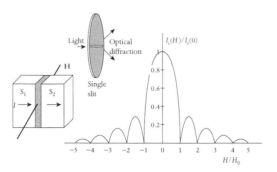

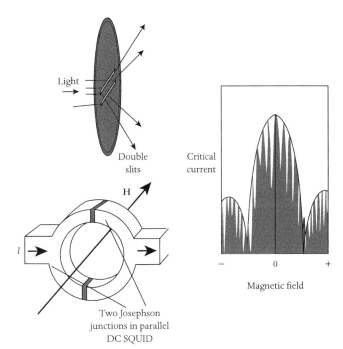

Figure 4.16 Behaviour of two Josephson junctions in parallel forming a DC SQUID. The behaviour is analogous to that of optical diffraction due to two slits.

where ϕ_0 ($= 2.067833636 \times 10^{-15}$ Wb) is the quantum of magnetic flux, given by $h/2e$, where h is Planck's constant. In this situation, the supercurrent oscillates between the two superconductors with a frequency $\nu = \dot{\delta} = 2e\mathrm{V}/h$ that corresponds to emission of electromagnetic radiation of frequency ν. The ratio $\nu/V = 2e/h$ is the ratio of two fundamental constants, with a value 484 MHz/V, and is termed the *Josephson constant*. Equation (4.9) is known as the *second Josephson equation* and the effect is known as the *AC Josephson effect*. Alternatively, if an RF field of frequency ν is imposed on the junction, a direct current will flow across the junction if the relation $2eV = nh\nu$ is satisfied, n being an integer. This is known as the *inverse Josephson effect*. In the presence of imposed RF radiation, the *I–V* behaviour of a Josephson junction exhibits steps at constant voltages (Figure 4.17), called *Shapiro steps*, which are related to the fundamental constants in the form $nh\nu/2e$. In the field of metrology, the AC Josephson effect thus provides a quantum standard for the electrical volt expressed in terms of the fundamental constants e and h. The *Josephson voltage standard* is used in metrology laboratories all over the world.

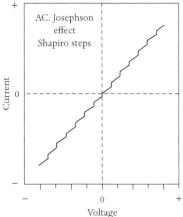

Figure 4.17 Schematic illustration of the Shapiro steps (AC Josephson effect) in the *I–V* behaviour when a Josephson junction is irradiated with microwave radiation.

4.5.3.3 Inverse AC Josephson effect

If an unbiased Josephson junction is subjected to RF radiation of frequency ν, a DC voltage V appears across the junction; this effect is known as the *inverse AC Josephson effect*. The voltage generated is sensitive to external magnetic fields and also varies with RF power and frequency.

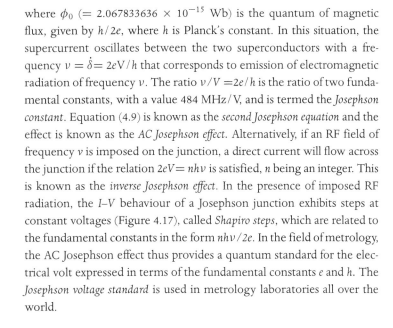

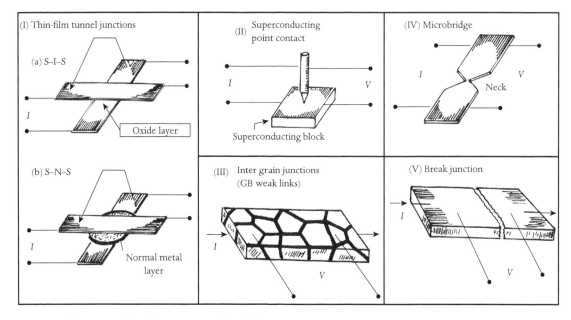

Figure 4.18 Different types of Josephson junctions and weak links.

The inverse Josephson effect has been used to detect the presence of high-temperature superconductivity in multiphase HTS samples.

Although both DC and AC Josephson effects were initially predicted for tunnel junctions, they have been manifested in different situations where the two superconductors are coupled through a variety of weak links (Figure 4.18), such as point contacts (a sharp point of a rod or a wire of superconductor pressed against a flat surface of another superconductor), a microbridge (i.e. a constriction formed on a superconducting thin film), a grain boundary separating two neighbouring grains in a superconducting material, and a break or a crack junction produced by simply breaking a superconductor at low temperatures. The above types of weak links, beside the Josephson tunnel junction, are successfully being used in Josephson devices. SQUID devices made using LTS, functioning at liquid helium temperature, have been commercially available for the last few decades, while those made using HTS, functioning at liquid nitrogen temperature, are in the development stage, awaiting an improved and more reproducible performance.

4.6 Summary

We started this chapter with the essential thermodynamic aspects of superconductors, leaving details to Appendix 4A for the interested reader. Four prominent thermal properties of superconductors discussed are (i) heat capacity, (ii) thermal conductivity, (iii) thermoelectric power, and (iv) thermal expansion, all of them being relevant to engineering applications. They are also fundamentally important in materials characterization, because they yield valuable microscopic parameters of superconductors such as the nature of the charge carriers, the electronic density of states, the Debye temperature, and the energy gap. As mentioned in the chapter, the first indication of the presence of an energy gap in superconductors came through heat capacity and thermal conductivity studies, which paved the way to a successful formulation of the microscopic theory of superconductivity. Ultrasonic, optical, and AC properties of superconductors have also been briefly presented. Different features of tunnelling phenomena, including Josephson effects in superconductors, have been described.

Appendix 4A

4A.1 Condensation energy

The condition of equilibrium between normal and superconducting phases is that their free energies be equal. At the critical field, $F_s(H_c) = F_n(H_c)$, where subscripts s and n refer to superconducting and normal states, respectively. As the normal state is generally non-magnetic or only weakly magnetic, $F_n(H_c) = F_n(0)$.

In the presence of $H < H_c$, owing to the Meissner effect, $B = 0$, and therefore the work done per unit volume in magnetization from 0 to H is

$$F_s(H) = F_s(0) + \mu_0 \int_0^H H dH = F_s(0) + \frac{1}{2}\mu_0 H^2.$$

The free-energy difference per unit volume between the normal and superconducting states is therefore

$$\Delta F = F_s(H_c) - F_s(0) = F_n - F_s = \frac{1}{2}\mu_0 H_c^2. \tag{4A.1}$$

The free-energy density of the superconducting phase in zero field is thus lower than that of the normal phase by $\frac{1}{2}\mu_0 H_c^2$, which is known as the *condensation energy* of the superconducting phase. At $T = 0$, the condensation energy is equal to $\frac{1}{2}\mu_0[H_c(0)]^2$. This is typically 10^{-8} eV per atom. Compared with the Fermi energy E_F of about 1 eV, the condensation energy is about eight orders of magnitude smaller. This was one of the major problems in formulating the microscopic theory.

4A.2 Entropy

Differentiating (4A.1) with respect to T gives the entropy difference between the two phases of volume V:

$$\Delta S = S_n - S_s = -V\mu_0 H_c(T)\frac{dH_c(T)}{dT}. \tag{4A.2}$$

As described in Section 3.1.1, H_c is a decreasing function of T and vanishes at $T = T_c$ without any discontinuity. The right-hand side of equation (4A.2) is therefore positive, confirming that the entropy of the superconducting state is lower than that of the normal state; that is, the former is more ordered. Further, the heat energy Q associated with the transition is

$$Q = T(S_n - S_s) = -TV\mu_0 H_c(T)\frac{dH_c(T)}{dT}. \tag{4A.3}$$

In the absence of a magnetic field, that is, when $H_c(T) = 0$ (which occurs at the transition for $T = T_c$ and when $\Delta S = 0$), $Q = 0$; that is, no heat energy is involved. Thus, in the absence of a magnetic field, the transition at T_c is of second order. On the other hand, when $H_c(T) > 0$, equation (4A.3) shows that the transition is of first order. There is absorption of heat in an isothermal superconducting-to-normal state transition, and a corresponding cooling of the specimen when this occurs adiabatically.

4A.3 Heat capacity

Since the specific heat $C = T\,dS/dT$, differentiating equation (4A.2) yields

$$C_s - C_n = TV\mu_0\left\{\left[\frac{dH_c(T)}{dT}\right]^2 + H_c\frac{d^2H_c(T)}{dT^2}\right\}. \tag{4A.4}$$

At $T = T_c$, $H_c = 0$, and equation (4A.4) reduces to

$$C_s - C_n = TV\mu_0\left[\frac{dH_c(T)}{dT}\right]^2\Bigg|_{T=T_c>0}. \tag{4A.5}$$

This is *Rutger's formula*, which shows that, at T_c, the specific heat of the superconducting state exceeds that of the normal state, which is manifested as a discontinuity. Further, the formula (4A.5) relates this discontinuity to the slope of the threshold curve near T_c. Its validity has been successfully tested for many elements.

The low-temperature heat capacity comprises two components, the lattice part C_L and the electronic part C_e. The lattice contribution is essentially the same in the superconducting and normal states, that is, there are negligible changes in the elastic constants, and thus $C_{nL} - C_{sL} = 0$. Consequently, $C_n - C_s = C_{en} - C_{es}$, with the electronic specific heat in the normal state $C_{en} = \gamma T$ and $C_{es} = \alpha T^3$. Now, from the fact that the entropies of the two phases must be equal at T_c and at $T = 0$ (because of the third law of thermodynamics: $S_n = S_s = 0$), we get $\alpha = 3\gamma/T_c^2$ and $C_{es} = 3\gamma T^3/T_c^2$. Thus, at $T = T_c$, $C_{es} = 3\gamma T_c = 3C_{en}$, which is corroborated experimentally. At $T = 0$, the free-energy difference per unit volume between the normal and superconducting states is

$$[\Delta F]_{T=0} = \frac{1}{4}\gamma T_c^2 = \frac{1}{2}\mu_0[H_c(0)]^2,$$

or

$$[H_c(0)]^2 = \frac{\gamma T_c^2}{2\mu_0}. \tag{4A.6}$$

Interestingly, this is essentially similar to the prediction of the microscopic theory, which gives

$$[H_c(0)]^2 = \frac{\gamma T_c^2}{0.17\mu_0}. \tag{4A.7}$$

CHAPTER 5

Advent of type II superconductors

5.1 Ginzburg–Landau phenomenology

The phenomenologies of London and Pippard were extended further by Ginzburg and Landau (1950), who introduced a theory of great power and versatility in treating spatial variations in the superconducting order parameter across the normal–superconducting interface boundary. Unlike the London theory, which is purely classical, the Ginzburg–Landau theory uses quantum mechanics in making its various predictions. Further, although it is still phenomenological, Gor'kov (1959) showed the scope of the theory to be really quite broad and, in fact, to represent the limiting form of the microscopic theory. The superconducting wave function ψ, which represents the order parameter in the Ginzburg–Landau theory, depicts the motion of Cooper pairs and is directly proportional to the gap parameter Δ of the BCS theory. Also, as will be described shortly, in the hands of Abrikosov (1957), the theory proved very successful in predicting a new class of superconductors, called *type II superconductors*. The combined theoretical work of Ginzburg, Landau, Abrikosov, and Gor'kov is commonly referred to as the *GLAG theory*.

A full description of the Ginzburg–Landau theory is too involved for this book. However, as per practice we are following, some relevant features of the theory are briefly explained for the interested reader in Appendix 5A. For more details, the reader may refer to the original paper of Ginzburg and Landau (1950), and to various textbooks on superconductivity (Lynton, 1962; Saint-James et al., 1969; Buckel, 1991).

Like Pippard, Ginzburg and Landau began by noticing the previously mentioned drawbacks of the London theory (see Chapter 3), especially the enormous positive interfacial energy between the normal and superconducting regions, for which the London theory did not find any justification. They further found that the critical fields of thin films were correctly predicted only by assuming that λ varied with the electron mean free path—this did not follow from the London theory. It is, however, surprising that Ginzburg and Landau failed to notice, among various shortcomings of the London phenomenology, the remarkable observation of Shubnikov et al. (1937) that the magnetisation of single crystals of Pb-In and Pb-Tl alloys showed what is now recognised as ideal type II behaviour (to be described later), in contradiction to the London theory.

The Ginzburg–Landau theory is essentially a two-fluid model of superconductivity in which, at $0 < T < T_c$, both normal electrons and superelectrons are simultaneously present. The superelectrons are described by an order parameter in the form of a wavefunction ψ, which is normalised such that $|\psi|^2 = n_s$, the number density of superelectrons. In order to include the effect of spatial variation in n_s across the boundary between normal and superconducting phases, Ginzburg and Landau added a kinetic energy term to the free-energy expression (4A.1) (see Chapter 4, Appendix 4A) of the superconductor in an external magnetic field H. Further, following the generalised theory of second-order phase transitions (Landau and Lifshitz, 1958), they expressed the free-energy difference between normal and superconducting states, in the absence of H, by a power series in $|\psi|^2$ containing temperature-dependent coefficients α and β of the form $\alpha(T)|\psi|^2 + \frac{1}{2}\beta(T)|\psi|^4 + \cdots$ in which the higher order terms were neglected. The modified Ginzburg–Landau expression for the free energy (per unit volume) of the superconductors in the presence of H thus obtained (equation (5A.1) in Appendix 5A), when minimised respectively with respect to the order parameter ψ and to $\mathbf{A}$, yielded two self-consistent nonlinear equations, known as the Ginzburg–Landau equations G-L(I) and G-L(II) (equations (5A.2) and (5A.3) of Appendix 5A). The former describes the spatial variation of the order parameter, leading to the range of coherence ξ, while the latter describes the equilibrium current distribution, which yields the penetration depth λ.

5.1.1 Two characteristic lengths (ξ and λ) and the Ginzburg–Landau parameter

The range of coherence x follows from the gradient term in the equation G-L(I), which makes any rapid variation of ψ energetically unfavourable. The distance over which this variation $\delta\psi$ can occur is the range of coherence. For a semi-infinite slab, the solutions for $\delta\psi$ are of the form $\exp(-x/\xi)$, where ξ is given by

$$\xi^2 = \frac{\hbar^2}{4\psi_0^2 m^* \beta} = -\frac{\hbar^2}{4m^*\alpha}, \tag{5.1}$$

where ψ_0 is the equilibrium value in zero field. With $\alpha = -(e^{*2}/m^*)\mu_0^2 H_c^2 \lambda^2$, this gives

$$\xi^2 = \frac{\hbar^2}{4e^{*2}\mu_0^2 H_c^2 \lambda^2}, \tag{5.2}$$

or

$$\xi = \frac{\phi_0}{4\pi \mu_0 H_c \lambda},$$ (5.3)

where ϕ_0 is the flux quantum given by $h/e^* = h/2e = 2 \times 10^{-15}$ Wb. The temperature dependence of α is given by $(T_c - T)/T_c$, that is, $1 - t$, where $t = T/T_c$ is the reduced temperature, and the temperature dependence of ξ is therefore given by $(1 - t)^{-1/2}$. Further, the coefficients α and β are related to the microscopic parameters such as $N(0)$, the density of states at E_F, and the electron mean free path l_F. The latter depends on the sample purity, and ξ can be expressed in the general form (Goodman 1966),

$$\xi(T) = \frac{0.522\xi_0}{[(1 - t)(1 + 0.852\rho)]^{1/2}},$$ (5.4)

where ξ_0 is the range of coherence for the pure limit $l_F \to \infty$ and ρ measures the purity of the sample. For an impure metal ($l_F << \xi_0$), $\xi(T)$ takes the form (Goodman, 1966)

$$\xi(T) = \frac{0.60(\xi_0 l_F)^{1/2}}{(1 - t)^{1/2}}.$$ (5.5)

The microscopic theory (Gor'kov, 1959, 1960) gives a similar expression, but with a numerical factor of 0.85 instead of 0.60. For pure materials ($l_F > \xi_0$), the equation reduces to the form given by Pippard (see equation (3.9) in Chapter 3).

In the BCS theory (Bardeen et al., 1957), the size of the Cooper pairs is given by ξ, and its value in the pure limit is

$$\xi_0 = \frac{\hbar v_F}{\pi \Delta_0} = \frac{0.18\hbar v_F}{k_B T_c},$$ (5.6)

where v_F is the Fermi velocity and $2\Delta_0$ is the superconducting energy gap at 0 K.

The second Ginzburg–Landau equation, G-L(II), describing the supercurrent distribution, yields the penetration depth λ. We consider G-L(II) in a low field, $H \approx 0$. In this situation, the order parameter ψ essentially remains constant in space, that is, grad $\psi \approx 0$ and $\psi(r) = \psi_0 = $ constant, and G-L(II) gives

$$\mathbf{J} + \frac{e^{*2}}{m^*}\mathbf{A}|\psi_0|^2 = 0.$$ (5.7)

The curl of the above equation has the same form as the London equation $\mu_0\lambda^2$ curl $\mathbf{J} + \mathbf{B} = 0$, and on comparison we find that

$$\lambda^2 = \frac{m^*}{|\psi_0|^2 e^{*2}\mu_0},$$ (5.8)

which is really the same as equation (3A.2) since in the Ginzburg–Landau theory $|\psi_0|^2 = n_s$. Substituting $|\psi_0|^2 = -\alpha/\beta$, and recalling that these coefficients depend on the temperature and the electron mean free path, gives (Goodman, 1966)

$$\lambda(T) = \lambda_0 \left[\frac{1 + 0.852\rho}{2(1-t)} \right]^{1/2}, \tag{5.9}$$

where λ_0 is the weak-field penetration depth in pure metal at 0 K. In the impure limit ($l_F << \xi_0$), from the microscopic theory (Gor'kov, 1960) the behaviour is given by

$$\lambda(T) = 0.62\lambda_0 \left[\frac{\xi_0}{l_F(1-t)} \right]^{1/2}. \tag{5.10}$$

We now define an important dimensionless parameter κ, called the Ginzburg–Landau parameter, whose value for a superconductor determines its behaviour in a magnetic field. At T near T_c, κ is given by

$$\kappa = \sqrt{2} \frac{e^*}{\hbar} \mu_0 H_c \lambda^2. \tag{5.11}$$

Since the temperature dependences of H_c and λ^2 tend to cancel each other, this parameter is essentially temperature-independent near T_c. The Ginzburg–Landau parameter is related directly to the two characteristic lengths λ and ξ by

$$\kappa = \frac{1}{\sqrt{2}} \frac{\lambda}{\xi}, \tag{5.12}$$

where ξ and λ are as given in equations (5.3) and (5.8) and, in their temperature dependences given by equations (5.4) and 5.9), both diverge as $(1-t)^{-1/2}$, which accounts for the temperature independence of κ. Both λ and ξ depend on the electron mean free path l_F, with the former increasing and the latter decreasing with decreasing l_F, which makes κ sensitive to the presence of impurities. All kinds of inhomogeneities, defects, disorder, impurities, alloying, etc. therefore tend to enhance this parameter. Following Goodman (1966),

$$\kappa \approx \kappa_p + 7.5 \times 10^3 \gamma^{1/2} \rho_n, \tag{5.13}$$

where $\kappa_p = 0.96\lambda_0/\xi_0$ is the value of the parameter in the pure limit of $l_F \to \infty$, γ is the coefficient of the electronic specific heat of the normal state in erg cm^{-3} K^{-2}, and ρ_n is the normal-state resistivity in $\mu\Omega$ cm.

5.1.2 Implications of κ: boundary energy

The Ginzburg–Landau description helps more realistically in evaluating the surface energy E_B between normal and superconducting regions. Clearly, E_B contains two terms: (1) a negative contribution due to flux penetration in the Meissner region and (2) a positive contribution arising out of the spatial variation in the order parameter. The equation for the free energy in an applied magnetic field H_c is of the form volume energy (bulk) + surface energy E_B = total free energy integrated over the whole space. For this one, needs the solutions of the Ginzburg–Landau equations for $\mathbf{A}$ and ψ to be inserted in the evaluation of the free energy. In general, the problem cannot be handled analytically, but it may be shown that $E_B = 0$ for $\kappa = 1/\sqrt{2}$. With numerical analysis, one finds that $E_B > 0$ when $\kappa < 1/\sqrt{2}$, and $E_B < 0$ when $\kappa > 1/\sqrt{2}$. Following equation (5.12), the former corresponds to the situation $\lambda < \xi$ and the latter to $\lambda > \xi$.

5.2 Sign of the surface energy and superconductor types

The sign of the surface energy between the superconducting and normal phases determines whether a sample exposed to a magnetic field $H < H_c$ will have normal regions in the superconducting matrix. As already mentioned, the boundary energy depends upon the relative size of λ and ξ, or, equivalently, the magnitude of the Ginzburg–Landau parameter κ.

(a) $\lambda < \xi$, or $\kappa < 1/\sqrt{2}$

The sign of the boundary energy is positive. The increase in energy due to a loss in order over a distance ξ compensates for the decrease in energy due to flux penetration over a distance λ (Figure 5.1(a)). Except for the field penetration over the depth λ, the bulk of the material remains completely diamagnetic for $H < H_c$, and beyond $H \geq H_c$ the superconductivity is completely destroyed.

Figure 5.2(a) and (b) respectively show schematic magnetisation and B–H curves of a superconducting sample free from any demagnetising effect, namely a sample with the shape of a long cylinder. Below H_c, there is Meissner effect with $B = 0$ and $M = -H$; above H_c, there is abrupt flux penetration, reverting the material to the normal state. The curves are reversible. This kind of behaviour is typical of pure elemental superconductors such as Al, In, Sn, and Zn, and characterises the magnetisation behaviour of the so-called *soft or type I superconductors* with a positive surface energy between the normal and superconducting phases.

(b) $\lambda > \xi$, or $\kappa > 1/\sqrt{2}$

This corresponds to the situation where the sign of the boundary energy becomes negative. Ginzburg and Landau (1950) had noted this situation, but surprisingly they did not pursue it further. It was Abrikosov (1957) who extended the Ginzburg–Landau theory to the case of $\kappa > 1/\sqrt{2}$ and thereby was able to predict a completely new class of superconductor of negative surface energy, called a *type II superconductor*.

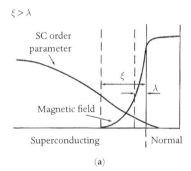

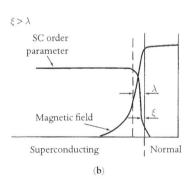

Figure 5.1. Penetration depth λ and range of coherence ξ for (a) a type I superconductor with a positive surface energy and (b) a type II superconductor with a negative surface energy.

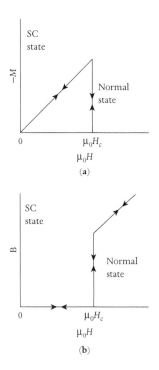

Figure 5.2, (a) Schematic magnetisation curve and (b) schematic *B–H* curve of an ideal type I superconductor.

In a type II superconductor, at a field $H < H_c$, the decrease in energy due to field penetration dominates over the increase in energy caused by the decrease in superconducting order, occurring over a much smaller length ξ than in case (a). The situation is depicted in Figure 5.1(b). To minimise its energy, the magnetic flux starts entering the sample at a field $H_{c1} << H_c$ in the form of quantised vortex lines that give rise to normal regions as they enter the superconductor. The diamagnetic state characteristic of the Meissner effect is thus observed only up to the field H_{c1}, known as the *lower critical field*. Although the ideal diamagnetic state is destroyed, the zero-resistance state remains unaffected. With increasing field, beyond H_{c1}, the field continues to gradually penetrate the sample until H_{c2} is reached, when the bulk superconductivity is destroyed. Between H_{c1} and H_{c2}, the material is in a *mixed state* where superconductivity exists in the presence of a magnetic field. Typically, most superconducting alloys and compounds, including high-T_c superconductors, exhibit the mixed state characteristic of type II behaviour. The κ values of superconducting transition metal alloys and compounds range from 10 to 50, while for high-T_c cuprate superconductors κ may exceed 100. Among pure elemental superconductors, only Nb and V have been conclusively identified as type II superconductors, with κ values between 1 and 2, while all the rest are type I, with $\kappa < 1/\sqrt{2}$. The magnetisation and $B–H$ curves are schematically illustrated in Figure 5.3(a) and (b). Interestingly, as mentioned previously, even about two decades before the advent of Abrikosov's theory (1957), Shubnikov et al. (1937) had experimentally observed this kind of magnetisation behaviour for Pb–In and Pb–Tl single crystals, but for a long time this had remained otherwise unnoticed. In recognition of that work, the mixed state is also referred to as the *Shubnikov phase*.

The curves in Figure 5.3 are reversible for a sample that is completely homogeneous and defect-free. When the sample contains inhomogeneities in the form of crystal defects and disorder, the magnetisation becomes hysteretic, as shown by the curve drawn with a broken line in Figure 5.3(a). Such a superconductor exhibiting an *irreversible type II behaviour* is termed a *hard superconductor* (irreversible type II superconductors were formerly termed *type III superconductors*).

In the case of type I superconductor, the field at which the magnetic flux abruptly enters the sample to make it normal defines the thermodynamic, threshold, or bulk critical field H_c. Determination of the H_c of a type II superconductor from the magnetisation behaviour is not so straightforward. However, if the magnetisation curve is fully reversible, simple thermodynamic considerations show that

$$-\int_0^{H_{c2}} M\, dH = \frac{1}{2}\mu_0 H_c^2,$$

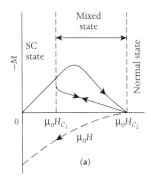

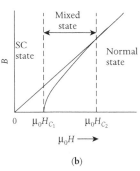

Figure 5.3. (a) Schematic magnetisation curve and (b) schematic *B*–*H* curve of an ideal type II superconductor. In (a), the outer curve is the irreversible magnetisation curve when the sample contains defects, and the broken line is the curve in a decreasing field.

and thus, from the area of the reversible magnetisation curve, one can estimate H_c of a type II superconductor. It lies between H_{c1} and H_{c2}; that is, $H_{c1} < H_c < H_{c2}$.

5.3   Mixed state and other characteristics

5.3.1   Mixed state

An understanding of the detailed structure of the mixed state was first provided by Abrikosov (1957a, b) by solving the Ginzburg–Landau equations for $\kappa > 1/\sqrt{2}$. He showed that field penetration in the mixed state occurs in the form of a triangular lattice (although originally a square lattice was considered) of *vortex lines* (also commonly called flux lines or flux vortices), as shown in Figure 5.4(a); this vortex lattice is commonly referred to as the *flux line lattice* (FLL). The existence of a triangular FLL has been confirmed by various experimental techniques such as neutron diffraction (Cribier et al. 1966), electron microscopy of Bitter patterns (Essmann and Träuble, 1967), NMR studies (Fite and Redfield, 1966), and, more recently, scanning probe microscopy (Troyanovski et al., 2001). Figure 5.4(b) depicts the triangular FLL in a superconducting amorphous thin film imaged using a scanning Hall microscope (Nishio, 2010). Observational studies have revealed the FLL to contain a wide variety of lattice defects, including vacancies, interstitials, dislocations, low- and high-angle grain boundaries, and stacking faults (Träuble and Essman, 1968, 1969). Dislocations in the FLL (called *flux lattice dislocations*) have been considered to be important (Dew-Hughes, 1971) in the phenomenon of *vortex flow*, also called *flux flow*.

A vortex line has a complex structure. Each vortex line (Figure 5.4(c)), which runs parallel to the direction of the magnetic field, has a normal cylindrical core carrying a magnetic field that is a single quantum of magnetic flux $\phi_0 = h/2e$, and consequently the mixed-state structure represents the state of highest subdivision. At the centre of the core, where the field is maximum, the order parameter $\psi = 0$. The order parameter rises to its maximum value over the length ξ, while the magnetic field drops to $1/e$ of its maximum value over the distance λ. The former implies that the cylindrical normal core of the vortex line has radius ξ. The normal core is surrounded by surfaces of constant ψ that correspond to vortices of supercurrents flowing to screen out the field over the distance λ.

5.3.2   Lower and upper critical fields

The lower critical field H_{c1} represents the field at which the first vortex line is nucleated in the sample (Figure 5.5: (I)) to form the mixed

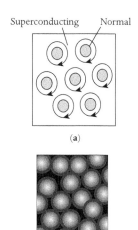

(a)

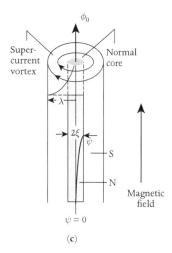

20 nm (b)

(c)

Figure 5.4 (a) Schematic of a triangular vortex lattice. (b) Vortex lattice in a superconducting amorphous thin film imaged using a scanning Hall microscope (courtesy of Nishio). (c) Structure of a vortex line.

Figure 5.5. (I) At H_{c1}, the first flux line is nucleated and flux penetration is very rapid. (II) Subsequently, when $H > H_{c1}$, the flux lines are a distance λ apart when their supercurrent vortices overlap and the penetration is slowed down. (III) At $H = H_{c2}$, the normal cores overlap and (IV) beyond that, the vortex structure loses its identity.

state. This holds for an ideal type II superconductor that is homogeneous and defect-free and does not exhibit hysteresis. As flux penetration continues, the mutual interaction of vortex lines becomes relevant. The interaction falls off exponentially according to $\exp(-x/\lambda)$, and thus at H_{c1} the penetration is fastest and the lines very quickly move to within a distance of λ of each other without much cost of energy. This is manifested by the near-infinite value of the slope of the magnetisation curve at H_{c1}. Once the mutual separation of the vortex lines becomes close to λ (Figure 5.5: (II)), such that supercurrent vortices of neighbouring lines overlap, the flux penetration becomes slower, and it continues until the vortex lines are ξ apart, when their normal cores start overlapping each other. This occurs (Figure 5.5 (III)) at the upper critical field H_{c2}, when the bulk of the material ceases to be superconducting. Beyond H_{c2} (Figure 5.5: (IV)), the vortex structure no longer exists. Both H_{c2} and H_{c1} can thus be related to ξ and λ by

$$\mu_0 H_{c2} = \frac{\phi_0}{2\pi\xi^2}, \tag{5.14}$$

$$\mu_0 H_{c1} = \frac{\phi_0}{4\pi\lambda^2}, \tag{5.15}$$

so that at H_{c2} and H_{c1} the vortices are respectively about ξ and λ apart.

From the Ginzburg–Landau theory, near T_c, the upper and lower critical fields H_{c2} and H_{c1}, respectively, are related to the thermodynamic critical field H_c through the Ginzburg–Landau parameter κ by

$$H_{c2} = \sqrt{2}\kappa H_c, \tag{5.16}$$

$$H_{c1} \approx \frac{H_c(\ln\kappa + 0.08)}{\sqrt{2}\kappa}, \tag{5.17}$$

while the slope of the M–H curve near H_{c2} is given by

$$-\left[\frac{dM}{dH}\right]_{H_{c2}} = \frac{1}{1.16(2\kappa^2 - 1)}. \tag{5.18}$$

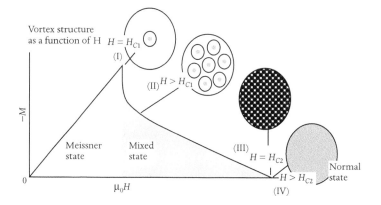

Although the three values of κ in the above three equations are same when measured near T_c, at lower temperatures they deviate from each other, as they have different temperature dependences (Maki, 1964a, b). Using equations (4A.7), (5.13), and (5.16) for $\kappa \gg \kappa_p$ and for $\xi_0 \gg \lambda_F$ yields

$$\mu_0 H_{c2}(0) = 3.09\gamma\rho_n T_c \tag{5.19}$$

(in T), where γ is in J m^{-3} K^{-2} and ρ_n is in Ω m. The upper critical field can also be determined from the initial slope of the $H_{c2} - T$ curve at $T = T_c$, given by (Werthamer et al., 1966)

$$H_{c2}(0) = 0.69 T_c \left[\frac{dH_{c2}}{dT} \right]_{T_c}. \tag{5.20}$$

The transition at H_{c2} is of second order, while that at H_{c1} is of first order if κ is small but otherwise is again of second order. In the case of Nb, the transition at H_{c1} was suggested (Serin, 1965) to be of the so-called lambda-type, with the magnetisation near H_{c1} varying logarithmically with the field.

It is clearly from equation (5.19) that to enhance the upper critical field, the parameters γ, ρ_n, and T_c should be increased. However, such an increase has a limit called the *Clogston–Chandrasekhar limit* (Clogston, 1962; Chandrasekhar, 1962). Accordingly, if the normal state, instead of being completely non-magnetic, is even weakly Pauli spin-paramagnetic, it will be energetically favourable (Zeeman effect) for the superconductor to become normal in a sufficiently high magnetic field independent of H_{c2}. By equating the condensation energy with the Pauli paramagnetic energy, in the BCS framework, this limit, which is termed the *Pauli-paramagnetically limited critical field*, is given by

$$H_p(0) = 1.84 T_c \tag{5.21}$$

(in T). This is the upper limit of the applied magnetic field at which superconductivity may exist, which becomes dominant for materials possessing very large spin susceptibility. While for most superconductors the Clogston–Chandrasekhar limit holds, there are exceptions where the above limit is surpassed. Both Maki (1966) and Werthamer et al. (1966) independently realised that spin–orbit interaction counteracts the Pauli paramagnetism, and the former is strengthened in alloys formed with the elements of higher atomic number in which the limit is exceeded. More recent work (Powell et al., 2003; Shimahara, 2004; Gabovich et al., 2004) has shown that the above limit does not hold for materials such as those of the heavy fermion category, various Chevrel phases, and organic superconducting systems where antiferromagnetism coexists with superconductivity, some of which exhibit field-induced superconductivity. The presence of charge density wave also enhances the paramagnetic limit (Gabovich et al., 2004; Ekino et al., 2005). All such materials are generally found to be *unconventional superconductors*, different from the BCS type.

It is worth pointing out that in the framework of the BCS model, orbital effects limit superconductivity and at $H_{c2}(0)$ the kinetic energy of the charges circulating around flux lines just exceeds the condensation energy of the Cooper pairs. $H_{c2}(0)$ is therefore often referred to as the *orbital critical field*. On the other hand, when the paramagnetic effect is dominant, superconductivity is destroyed at $H_p(0)$, given by equation (5.21), at which the polarisation energy just exceeds the energy of antiparallel spin alignment. It is the competition between the orbital and paramagnetic effects that decides the suppression of superconductivity by a magnetic field. When Pauli paramagnetism dominates, the transition to the normal state at very low temperatures changes from second order to first order. This is a very interesting situation, and gives rise to a novel type of superconducting state, called the *Fulde–Ferrell–Larkin–Ovchinnikov (FFLO) state* (Fulde and Ferrell, 1964; Larkin and Ovchinnikov, 1964) in some of the heavy fermion and organic superconductors, to be described in Chapters 13 and 14. The ratio $H_{c2}(0)/H_p(0)$ is called the Maki parameter α and indicates whether a material is orbitally or paramagnetically limited.

5.3.3   Sheath critical field H_{c3}

A type II superconductor possesses also a fourth critical field H_{c3} associated with surface superconductivity. Saint-James and de Gennes (1963) showed that if the applied field was parallel to the sample surface of a type II superconductor, the solution of the Ginzburg–Landau equation was quite different at the surface than in the interior. Their result showed that at H_{c2} the order parameter was zero everywhere except over a thin surface sheath of thickness ξ where superconductivity still persisted up to a field $H_{c3} > H_{c2}$, given by

$$H_{c3} = 1.695 H_{c2} = 2.392 \kappa H_c. \tag{5.22}$$

It follows from this equation that if $\kappa > 0.419$, then $H_{c3} > H_c$. Thus, a type I superconductor with $\kappa > 0.419$ will carry the superconducting surface sheath even after the bulk superconductivity is quenched at H_c. The phenomenon shows an angular dependence and, when the field is perpendicular to the sample surface, $H_{c3} = H_{c2}$. Equation (5.22) is valid only for the case where the boundary separates the superconductor from vacuum or an insulator. If the boundary conditions are different, the resulting H_{c3} is modified. A normal metal coating, if it is non-magnetic, will destroy the surface sheath if its resistivity is lower than the normal-state resistivity of the superconductor (Harault, 1966; Hauser, 1966). For low-κ type II superconductors, the presence of a surface sheath (Bertman et al., 1966; Barnes and Fink, 1966) promotes magnetic

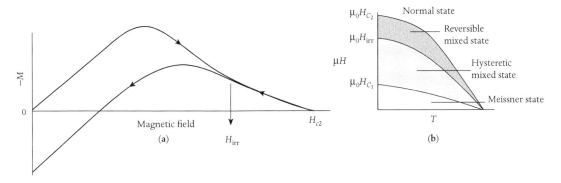

Figure 5.6. Schematic illustration of (a) the irreversibility field H_{irr} above which magnetisation behaviour is reversible. (b) Temperature dependence of H_{c1}, H_{irr}, and H_{c2}. The curve of H_{irr} versus temperature forms the irreversibility line, separating the reversible mixed state above it from the irreversible state below.

irreversibility, but in large-κ materials, the presence of inhomogeneities and defect structure is essential to make the magnetisation hysteretic.

5.3.4 Irreversibility field H_{irr}

The existence of the *irreversibility field* H_{irr} was discovered only after the advent of high-T_c cuprates when measurements of their magnetisation in the reverse cycle of a decreasing field showed magnetic reversibility until a field H_{irr} was reached. Below this field, the magnetisation curve began deviating from the curve obtained in an increasing field like any hysteretic sample. Above H_{irr} the magnetisation curve was that of a reversible sample. The field H_{irr} is smaller than H_{c2} but still far above H_{c1}. In high-temperature superconductors, thermal activation is dominant, leading to easier vortex line motion near H_{c2}. When H is reduced, the vortex lines at first move easily and M follows the same path as when H was increasing; that is, the superconductor is reversible. However, below H_{irr}, it changes to the original irreversible type. The values of H_{irr} at different temperatures form the *irreversibility line*, which separates the reversible state above from the irreversible state below (Figure 5.6). Clearly, for high-field applications, not only should H_{irr} be large, but, as we will see in Chapter 6, the material should also carry a large resistance-less current density J_c in a high magnetic field. A high H_{c2} by itself does not confer such an ability, and, as we will find in Chapter 6, to achieve a high J_c, a type II superconductor is required to have microstructures that will interact with the vortex lines so as to effectively inhibit or pin down their mobility.

5.4 Summary

In this chapter, we have seen how the phenomenological understanding of superconductors, described in Chapter 3, was vastly enhanced by the Ginzburg–Landau theory. Its success rested on the fact that it could quantitatively yield the two characteristic lengths λ and ξ and address the question of boundary energy between the superconducting and normal phases in terms of the Ginzburg–Landau parameter κ. Accordingly, for $\kappa < 1/\sqrt{2}$, the boundary energy is positive and these are type I superconductors, which include pure elements, where superconductivity is abruptly lost at H_c. When $\kappa \geq 1/\sqrt{2}$, a situation that was examined theoretically by Abrikosov, the boundary energy is negative, and such materials, which include impure metals and alloys, are type II superconductors. The field penetration now begins at a magnetic field H_{c1} ($<H_c$) and continues until H_{c2} ($>H_c$), when the bulk of the superconductivity is lost; between the two fields, the sample is in a mixed state. The mixed state comprises a triangular lattice-like arrangement of vortex lines, where an individual vortex line possess a complex structure described in terms of λ and ξ. The field H_{c2} can be as large as 100 T or more. However, for high-field applications, both J_c and H_{irr} are also required to be large.

Appendix 5A: Ginzburg–Landau equations

As with the non-local theory, Ginzburg and Landau (1950) chose to remove the absolute rigidity of the superconducting order parameter that was implicit in the London model, and considered the phenomenon, at $T < T_c$, in terms of two fluids, namely normal electrons and 'superelectrons'. All the superelectrons, responsible for superconducting order, represented in the form of an order parameter, are described by one and the same wavefunction ψ, which is normalised so that $|\psi|^2 = n_s$, the number density of superelectrons. Accordingly, $|\psi|^2 = 0$ at $T \geq T_c$ (in the normal phase), and $|\psi|^2 > 0$ for $0 < T < T_c$ (in the superconducting state). Following the original Landau–Lifshitz theory (1958), they expressed the difference in the free energy of superconducting and normal states, in the absence of magnetic field, in the form of a power series in $|\psi|^2$, that is, $\alpha(T)|\psi|^2 + \frac{1}{2}\beta(T)|\psi|^4 + \cdots$, where α and β are temperature-dependent coefficients, with $\alpha(T) < 0$ for $T < T_c$, $\alpha(T_c) = 0$, and $d\alpha/dT \neq 0$; $\beta > 0$ for $T \leq T_c$. As $|\psi|^2 = n_s$ is small close to T_c, terms in the series of higher order in $|\psi|^2$ were neglected. α and β were assumed to be continuous functions of T:

$$\alpha(T) = \text{constant} \times \frac{T_c - T}{T_c},$$

$$\beta(T) = \beta(T_c) = \beta_c.$$

For $T < T_c$, the free-energy minimum will occur for $\psi = \psi_0$, where $|\psi_0|^2 = -\alpha/\beta \propto T_c - T$, and the free-energy difference per unit volume (Section 5.1) is $F_s - F_n = -\alpha^2/2\beta = \frac{1}{2}\mu_0 H_c^2$.

The superconducting order parameter ψ is assumed to change from a temperature-dependent equilibrium value on the superconducting side of the boundary to zero on the normal side over a distance of the order of boundary width, which corresponds to a kinetic-energy term E. Consequently, in the presence of an external magnetic field $H < H_c$, constant in time, besides the increase in energy due to magnetic-flux expulsion given by $\frac{1}{2}\mu_0 H^2$, one must take into account the kinetic energy E, which corresponds to the spatial variation of the order parameter across the boundary between the superconducting and the normal phase. The Ginzburg–Landau free-energy expression in the presence of an external field therefore takes the form

$$F_s(T, H) = F_n(T, 0) + \alpha(T)|\psi|^2 + \frac{1}{2}\beta(T)|\psi|^4$$

$$+ \frac{1}{2}\mu_0 H^2 + \frac{1}{2m^*}\left|\left(\frac{\hbar}{i}\mathrm{grad} - e^*\mathbf{A}\right)\psi\right|^2, \qquad (5\mathrm{A}.1)$$

where the last term on the right-hand side represents the kinetic energy associated with the spatial variation in the order parameter ψ. Here $\mathbf{A}$ is the vector potential of the applied field, $\mathbf{H} = \mathrm{curl}\,\mathbf{A}$, and $e^*\,(= 2e)$ and m^* are respectively the effective charge and mass of the superelectrons, that is, the Cooper pairs. Minimising this free energy with respect to ψ and $\mathbf{A}$ respectively leads to two equilibrium equations, known as the Ginzburg–Landau equations, G-L(I) and G-L(II):

$$\mathrm{G\text{-}L\,(I)}: \quad \alpha\psi + \beta|\psi|^2\psi + \frac{1}{2m^*}\left(\frac{\hbar}{i}\mathrm{grad} - e^*\mathbf{A}\right)^2\psi = 0, \quad (5\mathrm{A}.2)$$

$$\mathrm{G\text{-}L\,(II)}: \quad \frac{1}{\mu_0}\mathrm{curl}\,\mathrm{curl}\mathbf{A} + \frac{e^{*2}}{m^*}\mathbf{A}|\psi|^2$$

$$- \frac{e^*\hbar}{2m^*i}(\psi^*\mathrm{grad}\psi - \psi\,\mathrm{grad}\psi^*) = 0. \qquad (5\mathrm{A}.3)$$

In G-L(II), as per Maxwell's equation, the first term $(1/\mu_0)$ curl curl $\mathbf{A}$ represents the current density and the equation describes the supercurrent distribution, while G-L(I) describes the equilibrium spatial variation of ψ. G-L(I) and (II) are coupled nonlinear equations in ψ and $\mathbf{A}$ that can be solved subject to various boundary conditions to yield insights into the superconducting characteristics linked directly with the parameter ψ and the vector potential $\mathbf{A}$.

CHAPTER 6

Critical current and flux pinning

6.1 Transport current in the mixed state

As pointed out in Chapter 5, besides T_c, the critical current, which represents the maximum electrical current that the material can sustain without resistance, is the most important parameter from the viewpoint of practical applications. High critical current density implies a considerable saving in the form of the size of the installation as well as the quantity of the materials and refrigeration costs involved.

As discussed in Section 3.1.1, from Silsbee's rule, the critical current for a type I superconductor is that which produces the critical field H_c at the surface of the superconductor. For a type II superconductor, this would correspond to a current that would produce a surface field $H_{c1} < H_c$, which represents the limiting field for the Meissner state. Since H_{c1} is small, the surface field reaches this value at a low current and the material subsequently enters the mixed state in the form magnetically quantised vortex lines. As described in Chapter 5, a vortex line comprises a thread of normal core of radius ξ carrying a quantum of magnetic flux $\boldsymbol{\phi}_0$, surrounded by vortices of supercurrent of radius λ. When the sample carries current in the mixed state it contains both the transport current (of density $\mathbf{J}$) and the magnetic flux $\mathbf{B}$ threading through the bulk of the material. Their coexistence gives rise to the Lorentz driving force $\mathbf{B} \times \mathbf{J}$, which acts on vortex lines, tending to move them in a direction perpendicular to $\mathbf{B}$ and $\mathbf{J}$. This force is maximum when the magnetic flux and the current are transverse to each other. If the material is completely homogeneous, there is no counteracting force preventing the vortex lines from moving and an unstable situation develops in which the vortices move along the direction of the force. Since each vortex line carries a quantum of magnetic flux $\boldsymbol{\phi}_0$, n such vortices, as they move with a velocity $\mathbf{v}_L$, induce an electric field $\mathbf{E}$, given by Faraday's law of induction

$$\mathbf{E} = n\mathbf{v}_L \times \boldsymbol{\phi}_0 = \mathbf{v}_L \times \mathbf{B}, \tag{6.1}$$

where $\mathbf{B} = n\boldsymbol{\phi}_0$ is the induction associated with n flux vortices. Since $\mathbf{E}$ is no longer zero, the superconductor now exhibits energy dissipation and shows a resistance, induced rather than ohmic, which can be an appreciable fraction of the normal-state resistance. In a type II superconductor, carrying a transport current, the mixed state is formed as soon as the self-field of the current (plus the external magnetic field, if any) exceeds H_{c1}, and the critical current in the mixed state is defined as that which produces a detectable voltage across the specimen as induced by moving vortex lines. For an ideal and magnetically reversible type II superconductor, where there is no hindrance to the motion of vortex lines, the lossless current above H_{c1} is zero. The mixed state, if the vortices are allowed to move, cannot carry any resistance-less current at all (Heaton and Rose-Innes, 1964). Clearly, such a material is of no interest for technical applications.

To carry a high current without resistance in a high magnetic field, the vortex lines must be pinned against a strong Lorentz driving force to keep their motion strictly inhibited. This is realised by introducing in the material various types of inhomogeneities, such as dislocations, precipitates, impurity particles, and grain boundaries. These microstructural features serve as pinning entities that interact with vortex lines and produce a pinning force that counteracts the Lorentz driving force and thereby promotes a stable non-equilibrium vortex distribution. Flux gradients resulting from non-equilibrium distributions of vortices can be equated with critical currents. The mixed state can now sustain a lossless transport current that corresponds to the driving force being just equal to the pinning force. As inhomogeneities are progressively added, the current-carrying capacity of the material markedly increases (Figure 6.1(a)) and simultaneously the magnetisation curve becomes more hysteretic (Figure 6.1(b)). In an increasing magnetic field, the flux enters the sample at a field H_p, several times larger than H_{c1}, which corresponds to the peak in magnetisation. Flux penetration, which is much slower than for the reversible magnetisation curve, is complete at H_{c2}. On reducing the external field below H_{c2}, the flux is prevented from escaping, which results in a trapped flux when the external field is reduced to zero. The magnetisation curve in both increasing and decreasing fields, for $H \gg H_p$, deviates symmetrically from the equilibrium or reversible magnetisation curve, and this deviation is related directly to the critical current of the material.

It should be remembered that the critical current I_c (or the critical current density J_c) does not represent a fundamental physical property of the superconductor, since it is defined by the force balance equation mentioned above. It corresponds to the onset of flux flow that leads to the appearance of a *detectable voltage* of $\sim 1\ \mu\text{V cm}^{-1}$ across the sample when the transport current has reached the critical value. An alternative criterion defines I_c or J_c at that point where the sample resistivity reaches $10^{-14}\ \Omega$ cm, a value that represents the quench condition for a superconducting magnet. This criterion is, however, less sensitive and yields I_c or J_c values that are 10–50% larger.

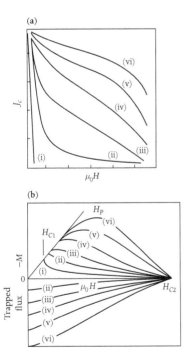

Figure 6.1 As inhomogeneities and defects are progressively added, from (i) to (vi), the critical current density of a type II superconductor in a magnetic field increases (a) and simultaneously the magnetisation becomes more hysteretic (b), showing a rise in the trapped flux when the magnetic field, after being increased to H_{c2}, is lowered to zero.

6.2   Driving force and the critical state

Under isothermal conditions, for a general non-uniform three-dimensional distribution of curved vortex lines, the driving force per unit length of the vortex line is given by (Campbell and Evetts 1972)

$$\mathbf{f}_L = \boldsymbol{\phi}_0 \times \mathrm{curl}\mathbf{H}(\mathbf{B}), \tag{6.2}$$

where $\mathbf{H}(\mathbf{B})$ is the external field that would be in equilibrium with the local induction $\mathbf{B}$. For $H >> H_{c1}$, however, the difference between $\mathbf{H}(\mathbf{B})$ and $\mathbf{B}$ is small: $\mathbf{H}(\mathbf{B}) \approx \mathbf{B}$. As curl $\mathbf{H} = \mathbf{J}$, equation (6.2) is reduced to

$$\mathbf{f}_L = \boldsymbol{\phi}_0 \times \mathbf{J}, \tag{6.3}$$

while the driving force per unit volume, that is, the *driving force density*, is given by

$$\mathbf{F}_L = \mathbf{B} \times \mathrm{curl}\mathbf{H}(\mathbf{B}), \tag{6.4}$$

or

$$\mathbf{F}_L = \mathbf{B} \times \mathbf{J}. \tag{6.5}$$

Considering for simplicity a two-dimensional system of straight vortex lines, equation (6.2) can be written as (Friedel et al., 1963) $f_L = \phi_0 \, dH(B)/dx = \phi_0 \, J$. The field gradient is equivalent to a macroscopic current normal to the direction of the gradient, which may be viewed as resulting from superposition of microscopic supercurrents flowing around each vortex. The vortex line is subjected to a macroscopic Lorentz driving force proportional to $\phi_0 dH(B)/dx = \phi_0 J$ tending to make the gradient zero. In a completely homogeneous type II superconductor, a non-uniform vortex distribution is therefore unstable and a static equilibrium is attained only when $\mathbf{f}_L = 0$, that is, $J = 0$. When the material contains inhomogeneities inhibiting vortex motion, a static non-uniform vortex distribution becomes stable, provided that $|\mathbf{f}_L| < |\mathbf{f}_p|$, where $\mathbf{f}_p$ is the maximum pinning force acting per unit length of the vortex line. The criterion $|\mathbf{f}_L| < |\mathbf{f}_p|$ implies that the pinning force is too large to be overcome by the Lorentz driving force, and the vortices therefore remain stationary in a static equilibrium. The force balance equation

$$-\mathbf{f}_L = -\boldsymbol{\phi}_0 \times \mathrm{curl}\mathbf{H}(\mathbf{B}) = \mathbf{f}_p \tag{6.6}$$

defines the maximum or the *critical* field gradient, or equivalently the critical current density J_c, that can exist in the sample. If $|\mathbf{f}_L| > |\mathbf{f}_p|$, or alternatively if $J > J_c$, the vortices move so as to make the gradient zero, and thus dissipation occurs. Equation (6.6) may alternatively be written as

$$-\mathbf{F}_L = -\mathbf{B} \times \mathrm{curl}\mathbf{H}(\mathbf{B}) = \mathbf{F}_p, \tag{6.7}$$

where $\mathbf{F}_p$ is the pinning force per unit volume, that is, the pinning force density. Thus, technologically it is important to have defects and inhomogeneities in the material that will give rise to strong pinning forces to withstand large flux gradients and stop vortices from moving.

The force balance equation (6.6) or (6.7) defines the *critical state* when it is satisfied over the entire sample. It follows that the situation is one in which the flux gradient is everywhere optimum or critical, or, alternatively, every macroscopic region of the sample carries a critical current density J_c that is a function only of the material and the local value of the flux density B. The existence of the parameter J_c is the basis of the *critical state concept* first conceived by Bean (1962, 1964). Its advantage is that if one knows how the critical current varies with the field, that is, the function $J_c(B)$, one can determine the field profile within the superconductor and work out the magnetisation curve. Alternatively, from the magnetisation behaviour, one can determine $J_c(B)$. Different forms of $J_c(B)$ have been successfully used by various workers to find an acceptable fit to their magnetisation data. These include $J_c(B)$ varying as B^{-1} (Anderson, 1962; Friedel et al., 1963; Silcox and Rollins 1963), $B^{-1/2}$ (Yasukochi et al. 1964, 1966), $(B + B_0)^{-1}$ (Kim et al., 1963), B^{-q} for $0 < q < 1$ (Irie and Yamafuji, 1967), $B^{-1.9}$ (Witcomb et al., 1968), $H_{c2} - B$ (Goedemoed et al., 1967), $B^{1/2}(H_{c2} - B)$ (Alden and Livingston, 1967; Campbell et al., 1968; Coffey, 1968), $B^{-1}[H_{c2}(T)]^{5/2}$ (Fietz and Webb, 1969), and an exponential form $\exp(-B/B_0)$ (Fietz et al., 1964; Chaddah, 1994). However, $J_c(B)$ is more microstructure-dependent than composition-dependent, and therefore samples with the same composition may possess different $J_c(B)$ forms depending on their metallurgical state.

In the critical state model, the maximum value of F_L measures the maximum pinning force, and therefore the pinning force per unit volume F_p at any induction B is equal to $BJ_c(B)$, where $J_c(B)$ is the current density at induction B. If J_c is in A m^{-2} (A cm^{-2}), and B in T (G), the pinning force $F_p(B)$ in N m^{-3} (dyn cm^{-3}) is given by

$$F_p(B) = \frac{BJ_c}{10}.$$

The plot of F_p versus the reduced magnetic field $h = H/H_{c2}$ or reduced induction $b = B/B_{c2}$ is termed the *pinning curve*. The position of its peak is dependent on the microstructure and the metallurgical history of the sample. The peak corresponds to the maximum F_p, which is generally believed to occur when all the pinning sites are occupied by vortex lines. The subsequent decrease is ascribed to the presence of unpinned vortices weakening the pinning force. The peak position accordingly should be invariant with sample temperature, but this is not found to be the case. Logarithmic plots of $BJ_c(T)$ versus $H_{c2}(T)$, as shown in Figure 6.2,

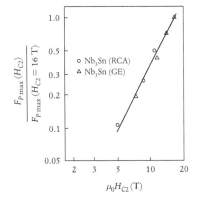

Figure 6.2 Normalised pinning force as a function of upper critical field for Nb$_3$Sn samples (after Kramer, 1973).

are found to be linear (Kramer 1973), suggesting that the pinning force can be represented by

$$F_p(T) = gp(h)H_{c2}^m(T),$$

where the prefactor g depends on the microstructure, $p(h)$ represents the field dependence of the critical current density J_c, and $H_{c2}^m(T)$ describes the temperature dependence of H_{c2}. The function $p(h)$ can be expressed in the general form $p(h) = h^p(1-h)^q$. The numerical exponents m, p, and q depend on the microstructure and are determined by suitably fitting the experimental data. If p, q, and m are known, one gets a scaling law for flux pinning. By measuring F_p and $p(h)$ at one temperature, it is easy to predict these parameters at any other temperature simply by scaling the results by H_{c2}^m (Kramer, 1973). In Nb$_3$Sn, in which the dominant pinning is by grain boundaries, the experimental data are generally found to fit with $p = 0.5$ and $q = 2.0$ and the pinning curve depicts a peak at $h = 0.3$ (Figure 6.3) (Kramer, 1973). These parameters and the peak position change as the dominant pinning centres vary. An important implication of this model (Kramer, 1973) is that it is inconceivable to increase $J_c(H)$ significantly simply by raising the number density of pinning centres. Instead, a more effective approach would be to raise the thermodynamic critical field H_c by increasing the critical temperature T_c and the coefficient of electronic specific heat γ. If a material, despite possessing a very high H_{c2} and a high density of effective flux pinning centres, has a very low thermodynamic critical field H_c with a large penetration depth, then its depairing current density is intrinsically low. Several extreme type II superconductors, such as the Chevrel phases, for example, seem to suffer from this inherent limitation.

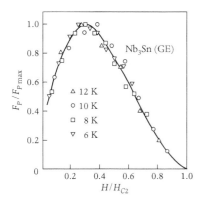

Figure 6.3 Pinning force curve for Nb$_3$Sn, where pinning is by grain boundaries; the peak is observed in the data obtained at different temperatures at $h = H/H_{c2} = 0.3$ (after Kramer 1973).

6.3 Vortex motion

Type II superconductors exhibit three types of vortex motion, described as (i) flux creep, (ii) flux flow, and (iii) flux jump, all leading to dissipation and restoration of resistance. Flux creep occurs under thermal activation when $f_L < f_p$, while flux flow takes place when $f_L > f_p$. When the vortices move abruptly, transporting a large quantity of magnetic flux, the motion is called a flux jump.

6.3.1 Flux creep

Strictly, there should be no vortex motion when $f_L < f_p$. Anderson (1962), however, pointed out that this would be true only at 0 K. At any finite temperature $T < T_c$, there will always be some thermal activation aiding the Lorentz force f_L to cause the pinned vortex lines to jump over the pinning barriers and move, despite $f_L < f_p$. Thermally activated movement of flux vortices past pinning centres is called *flux creep* in analogy

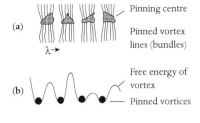

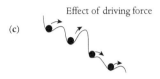

(a) Pinning centre

Pinned vortex lines (bundles)

$\lambda \rightarrow$

(b) Free energy of vortex

Pinned vortices

(c) Effect of driving force

Figure 6.4 Anderson's flux creep model. (a) Inhomogeneous regions in the sample trap or pin small bundles of flux lines. (b) Pinning occurs through a variation in the free energy of vortices due to defects, and vortices are pinned at the low-energy sites. (c) The effect of the driving force or the field gradient is essentially to tilt the free-energy wells, which effectively lowers the depth of the wells, allowing the vortices to jump off the barriers by thermal activation.

with mechanical creep, where the pinned dislocations move under a relatively low load due to thermal activation. Anderson considered small packets or bundles of vortex lines of radius λ getting pinned by inhomogeneous regions of the materials (Figure 6.4(a)). The inhomogeneities in a type II superconductor give rise to a local variation in the free energy of the vortex lines, with the result that the latter get pinned in potential wells at the low-energy sites (Figure 6.4(b)). The probability of a pinned vortex getting out by jumping over a barrier of height E_0 to get depinned is proportional to $\exp(-E_0/k_B T)$. In the absence of a Lorentz force, if the temperature $T < T_c$ is sufficiently low such that $E_0 >> k_B T$, then the probability of depinning is, in general, too small and essentially there will be no flux motion. In the presence of a field gradient or transport current J that effectively tilts the barriers (Figure 6.4(c)), the probability of pinned vortices jumping over the barriers is markedly enhanced. This will be proportional to $\exp[-(E_0 - F_L VS)/k_B T]$, where F_L is the Lorentz force per unit volume acting on a group of pinned vortex lines, anchored at a pinning site, of volume V. S is a characteristic length introduced to convert the force into energy, which may be related to the width of the pinning barriers. In the presence of F_L, the thermal activation now allows the vortices to jump across the barriers, with a hopping rate

$$R = R_0 \exp\left(-\frac{E_0 - F_L VS}{k_B T}\right),\tag{6.8}$$

where $R_0 \approx 10^5 – 10^{10} \text{ s}^{-1}$ is the vibrational frequency of the vortices. Clearly, because of flux creep, even a small F_L at a finite temperature will promote jumping of the pinned vortex lines over the barrier to start dissipation. This seems to contradict the critical state concept, where the pinned flux moves only when $F_L > F_p$. However, experimentally, when T is small, the flux creep becomes detectable only when F_L approaches $F_p \approx E_0/VS$. To accommodate the effect of flux creep, the critical state may be defined as that state in which, at a given temperature, the creep rate has fallen just below the detectable limit $\approx R_c$. In this situation, both the field gradient and the current are said to have reached their critical values. Accordingly, equation (6.8) can be expressed as

$$k_B T \ln\left(\frac{R_c}{R_0}\right) = [-(E_0 - F_L VS)]_{\text{critical}},\tag{6.9}$$

giving the critical state as

$$(F_L)_{\text{critical}} = |\mathbf{B} \times \mathbf{J}_c| = \frac{E_0}{VS}\left[1 - \frac{k_B T}{E_0}\ln\left(\frac{R_0}{R_c}\right)\right]$$

$$= \alpha_c(T).\tag{6.10}$$

The factor $\alpha_c(T)$ carries the pinning energy E_0 and is controlled by the microstructural features of the material and the temperature T. At

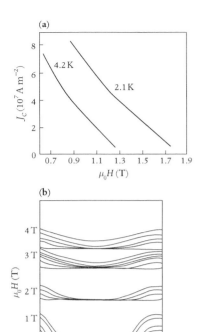

Figure 6.5 (a) J_c varying as $1/B$ for Pb–Bi alloy (after Campbell et al., 1968). The plot also shows that, for all magnetic fields, J_c(4.2 K) < J_c(2.1 K). (b) Flux profile, measured using a Hall probe, in a Nb–Ti sample, showing that the critical flux gradient, or J_c, decreases with increasing field (after Coffey, 1967).

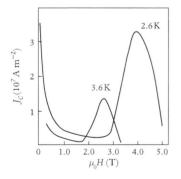

Figure 6.6 For Nb–80%Ti alloy, containing weakly superconducting omega-precipitates, over a rage of temperature and magnetic field, the critical current density J_c increases with rising temperature and magnetic field (after Baker and Sutton, 1969).

$T << T_c$, the pinning energy E_0 is expected to be temperature-independent, which makes $\alpha_c(T)$ decrease with increasing T. This is consistent with the general observation that at a constant B the critical current density J_c diminishes as the temperature is raised. If E_0 is field-independent, then $J_c \propto 1/B$, which again generally holds true. The results for Pb–Bi alloy (Campbell et al., 1968), showing an inverse dependence of J_c on B, at 2.1 and 4.2 K, are depicted in Figure 6.5(a). Perhaps the most convincing confirmation of this may be found in the results of Coffey (1967) in Figure 6.5(b). The field distribution in a Nb–Ti sample in increasing field is measured by a Hall probe placed in a transverse gap in the sample. As illustrated in the figure, the critical flux gradient, and therefore J_c, decreases with increasing magnetic field. Comparing the two curves of Figure 6.5(a), one can see that for the whole range of magnetic field, J_c at a higher temperature is always smaller than at a lower temperature. On the other hand, if E_0 is strongly field- and temperature-dependent, the resulting J_c behaviour is interestingly different. An example of this is when the pinning entities themselves are weakly superconducting, such as fine particles of omega phase formed in Nb–80%Ti alloy (Baker and Sutton, 1969). In this situation, over a small range, J_c increases with increasing temperature and magnetic field (Figure 6.6), which is contrary to the generally observed behaviour (Figure 6.5a) described above.

In low-temperature superconductors, at $T << T_c$, the dissipation resulting from flux creep is in general low and does not seriously lead to any significant rise in the decay of persistent current. However, flux creep is still very important, since it is the origin of flux jumps, which, for example, significantly limit the performance of superconducting magnets. Further, as to be expected, in high-temperature superconductors, their performance near liquid nitrogen temperature is particularly marred by pronounced flux creep occurring at elevated temperatures. In both LTS and HTS, the direct approach to reducing flux creep and enhancing J_c is to make the vortex pinning substantially stronger.

6.3.2 Flux flow

When $f_L > f_p$ (or, equivalently, current density $J >$ critical current density J_c), flux creep changes into a highly dissipative form of flux motion called *flux flow* in which the vortices move transverse to the transport current. The flux flow state is commonly studied by measuring the DC conductivity of a flat strip of the material (at $T < T_c$) in the mixed state produced by the application of a magnetic field transverse to the flat surface. The current I is passed along the length of the flat surface and the longitudinal voltage V generated is measured as a function of current. Until $I < I_c$, $V = 0$, showing that a static mixed state offers no electrical resistance. The flux flow state starts when I just exceeds

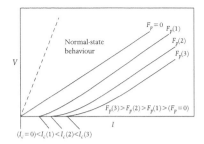

Figure 6.7 Flux flow behaviour.

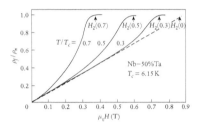

Figure 6.8 Variation of ρ_f/ρ_n with temperature for Nb–Ta alloy (after Usui et al., 1968).

I_c and is characterised by appearance of a voltage drop that increases linearly with I (Figure 6.7). The *flux flow resistance* R_f, given by the derivative dV/dI, is a significant fraction of the normal-state resistance of the material at the pertinent temperature. While I_c depends on the pinning force, R_f does not. The latter originates solely from moving vortices and is therefore characteristic of a homogeneous sample, unperturbed by the microstructure. In an external magnetic field H, the flux flow resistivity ρ_f approaches the normal-state resistivity ρ_n as $H \rightarrow H_{c2}$, and the ratio $\rho_f/\rho_n = H/H_{c2} = \xi^2/d^2$, where d is the vortex spacing in the field H and ξ is the range of coherence, which represents the spacing of the vortex lines at $H = H_{c2}$. The results of Usui et al. (1968) on Nb–Ta alloy are depicted in Figure 6.8. Since the normal vortex core has a radius ξ, the above expression for ρ_f/ρ_n suggests that the flow resistivity is caused by normal cores. Accordingly, in the flow state, the current must pass through the normal cores of moving vortex lines and thereby cause dissipation.

The flux flow voltage is found to be independent of time, which indicates that vortices move with a constant velocity. Thus, the metal behaves like a viscous medium and prevents vortices from being accelerated to infinite velocity by the Lorentz force. If $\mathbf{v}_L$ is the vortex velocity, the viscous force $\mathbf{f}_V$ is given by

$$\mathbf{f}_V = -\eta \mathbf{v}_L, \tag{6.11}$$

where η is the *coefficient of viscosity*, or the *damping coefficient*. Its origin is linked with the normal electrons comprising the vortex core (Bardeen and Stephen, 1965; Nozières and Vinen, 1966; van Vijfeijken and Niessen, 1965).

As mentioned in Section 5.8, the flux line lattice (FLL, vortex lattice) typically contains a variety of lattice defects. The close similarity of the vortex lattice with the conventional real crystal lattice has led to the possibility of flux flow taking place through the movement of flux lattice dislocations (FLDs) in an FLL. For example, an array of edge dislocations, all of the same sign, can give rise to a flux gradient commonly resulting from the passage of electric current. But if the current is too large and the flow is too rapid, the vortex lattice may break down, giving way to an amorphous structure (Dew-Hughes, 1971).

6.3.3 Flux jumps

Flux jumps are essentially instabilities manifested by erratic movements of vortex lines that lead to sudden penetration of a large magnetic flux into the sample. This behaviour is again analogous to the discontinuous mechanical yielding of metals in which pinned dislocations are suddenly released under mechanical stress. Flux jumps are technologically important because they can drastically reduce J_c and cause unexpected premature quenching of superconducting magnets. Such quenching is most hazardous since it can lead to permanent damage from the resulting large electric fields or temperature rise.

Flux jumps occur due to local breakdown of the critical state represented by equation (6.6). If for some reason the equilibrium is disturbed locally, leading to $f_L > f_p$, then vortex motion will occur, leading to dissipation and a local temperature rise, deleterious for superconductivity. The latter generally causes a decrease in the pinning force whereby J_c is locally lowered, and may lead to nucleation of local normal regions that are propagated in the rest of the conductor. As an increase in temperature will cause further flux motion, this builds up into a cascade process that results in a large *flux avalanche* causing rapid quenching.

The origin of flux jumps may be related to flux creep. At a finite temperature $T < T_c$, owing to flux creep, the vortex lines are continually moving under thermal activation. This may be harmless initially, but can gradually build up into increased vortex motion, leading to large flux jumps and an avalanche as described above. The situation is markedly aggravated by mechanical movement of the conductors, vibrations, mechanical shock, or sudden and rapid increase in transport current. In technological applications, it is essential to protect all coil windings of superconducting magnets against movement and vibration resulting from magneto-mechanical forces. Care must also be taken to avoid rapid charging of the coils, particularly in high fields; otherwise the magnets run the risk of unexpectedly becoming quenched at low transport currents.

An interesting situation giving rise to large flux jumps was pointed out by Evetts et al. (1964) and Silcox and Rollins (1964). They found that the magnetisation curves in the region around $-H_{c1}$ were highly prone to large flux jumps (Figure 6.9), which seemed independent of the material or its microstructure. In this region, negative vortex lines encounter previously trapped positive ones, and the energies released in their mutual annihilation are sufficient to heat the sample above T_c. If this heat is not conducted away sufficiently fast, a local temperature rise will occur at the flux front, driving huge instabilities near $-H_{c1}$.

6.4 Stabilisation of superconductors

Flux instabilities indeed presented a tall hurdle in the initial stages of magnet development during the 1960s when Nb–Zr magnets had just arrived. Broadly, there exist four approaches to stabilise superconductors: (1) through microstructures, (2) cryostatic stabilisation, (3) dynamic stabilisation, and (4) adiabatic stabilisation.

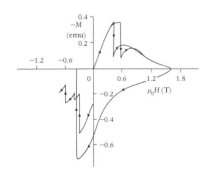

Figure 6.9 Pronounced flux jumps at $-H_{c1}$ in a sample of Pb alloy (after Evetts et al., 1964).

6.4.1 Stabilisation through microstructures

The temperature rise resulting from flux jumps can be curtailed by increasing the specific heat of the material by suitably alloying it, taking care that this does not too adversely affect superconductivity. Second, one may introduce pinning centres having a high specific heat, which will be less susceptible to temperature rises. Alternatively, as mentioned previously, pinning centres that are weakly superconducting give rise to a positive dJ_c/dT, which should combat the adverse effect of the temperature rise occurring in flux instability. Two commonly cited examples illustrating the role of weakly superconducting additives are the alloys of PbInSn (Livingston, 1966) and Nb–80%Ti (Baker and Sutton, 1969). Later, we will come across further instances in HTS cuprates and other systems. However, this approach still remains to be exploited for developing superconductors for practical applications.

6.4.2 Cryostatic stabilisation

In this approach, the superconductor is bonded to a highly conducting sheath of a metal such as copper. The large thermal conductivity of copper ensures better cryogenic cooling of the superconductor, and its large electrical conductivity allows the sheath to serve as an effective shunt for current sharing in the event of accidental quenching. However, a disadvantage of cryostatic stabilisation is that a very large volume fraction, more than 70% of the composite, is taken up by copper, which markedly lowers the overall current density and significantly adds to the weight and volume. This is not a serious problem with large installations, but for smaller magnets cryostatic stabilisation is not the answer, and instead one has to resort to adiabatic stabilisation.

6.4.3 Dynamic stabilisation

This, as the name suggests, does not represent a steady-state condition, and it requires that the rate of release of magnetic energy as heat during the instability, given by the magnetic diffusivity D_M, must be smaller than the rate at which this heat is dissipated, given by the thermal diffusivity D_T, i.e., $D_T/D_M > 1$. For most superconductors, however, $D_T/D_M < 10^{-3}$, and therefore the situation is unfavourable for dynamic stabilisation. However, $D_T/D_M > 1$ can be achieved if the superconductor is bonded to a highly conducting metal such as copper, and consequently the approach followed for cryogenic stabilisation also works for dynamic stabilisation.

6.4.4 Adiabatic stabilisation

Flux jumps are inhibited if the thickness of the superconducting strip is reduced (Chester, 1967; Wilson et al., 1970). Superconductors can be fabricated in a multifilamentary form in which fine superconducting filaments are embedded in a high-conductivity metal such as copper to simultaneously obtain the advantages of both cryogenic and dynamic stabilisation. At the same time, adiabatic stabilisation requires a much smaller amount of normal metal, which helps reduce weight and means that the overall, or engineering, current density is not drastically lowered. Wilson et al. (1970) were the first to consider the response of such a multifilamentary composite to an applied magnetic field. If the resistivity of the matrix in which the superconducting filaments are embedded is large, then the application of a magnetic field generates loops of supercurrent that remain strictly confined to individual filaments, which thereby retain their individual identity (Figure 6.10(a)). But, if the matrix resistivity is small, voltages generated by field sweep are sufficiently large to drive the current across the matrix to form a single large loop occupying the entire width of the composite (Figure 6.10(b)). The individual identity of fine filaments is lost as the composite now behaves like a thick conductor, losing the advantages of stability. Wilson et al. (1970) were able to identify a characteristic length l_c that measures the distance over which the transverse currents of the loop are spread. This length depends on the sweep rate of the magnetic field, the resistivity of the matrix, and the filament size. Clearly, if the length of the conductor is less than l_c, only a fraction of the supercurrent loops will traverse the matrix, with the remaining loops being confined to the individual filaments (Figure 6.10(c)), each retaining its identity. In a long conductor, this is simply achieved by twisting the conductor with a pitch $<< l_c$. In general, for slow sweep rates of about 0.1 T s^{-1}, twisting is usually not necessary, but for more rapid rates of field change, such as 5–6 T s^{-1}, twisting of the conductor is needed, or alternatively the low-resistivity copper matrix should be replaced by a more resistive cupronickel matrix. Such multifilamentary superconductors formed with a resistive matrix are the appropriate choice for AC applications.

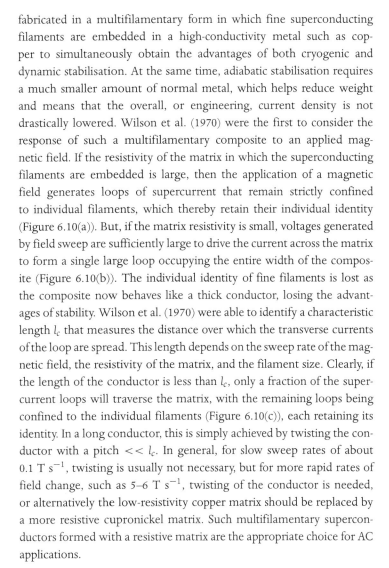

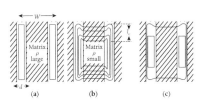

Figure 6.10 Current distribution in a multifilamentary superconductor in a magnetic field: (a) highly resistive matrix; (b) conductive matrix; (c) conductor length reduced by twisting with a switch $< l_c$ (after Wilson et al., 1970).

6.5 Pinning centres

Various kinds of extended lattice defects, such as clusters of point defects, dislocations, particles of a second phase, grain boundaries, subgrain boundaries, and twin boundaries, give rise to local variations in the electron mean free path in which superconducting properties are locally altered in relation to the surrounding matrix. In the defect regions, potential wells are created (Figure 6.4), in which vortex lines are trapped. The greater the difference between the superconducting properties of the defect region and the surrounding matrix, the deeper are the potential wells and the stronger is the pinning. Broadly speaking, the pertinent region causing flux pinning may be (1) more strongly superconducting than the matrix (i.e. possessing higher T_c and H_{c2}), (2) less superconducting, or (3) non-superconducting (i.e. normal or even magnetic). Alternatively, this situation is more conveniently described in terms of the pinning entity and the matrix possessing different Ginzburg–Landau parameters κ, and the pinning is then referred to as κ-pinning or $\Delta\kappa$-pinning (Narlikar and Dew-Hughes, 1964, 1966; Dew-Hughes, 1971; Hampshire and Taylor, 1972; Dew-Hughes and Witcomb, 1972). The larger the difference $|\Delta\kappa|$, the stronger is the pinning. The precise nature and appearance of such regions or inhomogeneities serving as pinning entities may vary from material to material. Nevertheless, they are all essentially defects in the ideal periodicity of the crystal lattice and can be described as *microstructures*, a term that broadly refers to the size and distribution of lattice defects.

As we have seen in Chapter 5, a vortex line has a normal core of radius ξ, the range of coherence, which may vary from 0.2 nm to 50 nm, depending on the material, whether it is an HTS cuprate or an LTS metallic alloy. Since pinning is essentially controlled by a spatial variation in superconducting properties over a distance ξ, a relatively much finer defect structure is called for to achieve optimum pinning in HTS in comparison with LTS. Because of the very short coherence length of HTS cuprates, coupled with their pronounced anisotropic nature, to date it has not been possible to create the ideal configuration of pinning centres needed for optimum pinning in these materials. Furthermore, for the same reason, the defect structures, instead of contributing to flux pinning, give rise to weak-link effects that, for strongly anisotropic superconductors such as HTS cuprates become particularly intense along certain orientations and degrade J_c. A pronounced grain alignment thus becomes a mandatory requirement with HTS cuprates.

Microstructural features in which all three dimensions are much smaller than ξ are not expected to serve as effective pinning centres, since superconducting properties are not able to change significantly over lengths shorter than ξ. Although this holds true for individual point defects, such as vacancies or interstitial atoms, which may not contribute significantly to flux pinning, aggregates of such defects can assist in flux pinning. Clusters of vacancies resulting from neutron irradiation may collapse into a dense distribution of small dislocation loops that provide flux pinning (Good and Kramer, 1970).

Turning now to line defects, interaction of vortex lines with individual dislocations is believed to be weak and the pinning effects due to a uniform distribution of dislocations are experimentally found to be feeble (Narlikar and Dew-Hughes, 1966; Nembach, 1966). Figure 6.11(a) is a TEM image of a highly cold-worked Nb–Ta alloy showing a very large density of uniformly distributed dislocations that fail to produce strong flux pinning (Narlikar and Dew-Hughes, 1966). On the

(a)

(b)

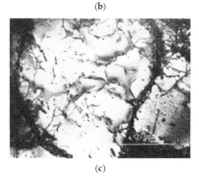

(c)

Figure 6.11 (a) Heavily cold-worked Nb–Ta alloy showing a uniform high density of dislocations. (b) After annealing, when the dislocation structure becomes polygonised (non-uniform), showing subgrain boundaries, the ensuing flux pinning exhibits a dramatic increase. (c) The pinning is enhanced yet more when the polygonised sample is further cold-worked. The subgrain boundaries now become denser and better defined, which promotes greater flux pinning. (After Narlikar and Dew-Hughes (1966).)

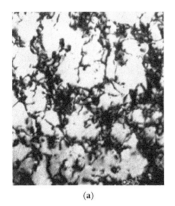

(a)

(b)

(c)

Figure 6.12 (a) Typical dislocation cell structure in deformed Mo–34%Re alloy. (b) Polygonised structure formed in the annealed compound. (c) Particles of sigma-phase formed in Mo–Re alloy serve as additional pinning entities besides the dislocation cell walls. (After Witcomb et al. (1968).)

other hand, in deformed transition metals and alloys of Nb, V, Nb–Zr, Nb–Ti, Mo–Re, etc., there is a great deal of evidence (Narlikar and Dew-Hughes 1964, 1966; Witcomb et al., 1968; Baker, 1970; Neal et al., 1971; Dew-Hughes and Witcomb, 1972; Hampshire and Taylor, 1972; Witcomb and Narlikar, 1972) that strong flux pinning occurs only when dislocations are arranged into dense tangles, forming the walls of a three-dimensional cell structure. A typical example of a cellular dislocation structure (Witcomb et al., 1968) is seen in the TEM image of cold-worked Mo–34at%Re alloy in Figure 6.12(a). The dislocation structure is markedly inhomogeneous in the sense that individual cells have low dislocation density while their walls are a thousand times denser. Pinning increases as the cell size decreases and the cell walls become denser and better defined. Although in single-phase materials this configuration of dislocations realised by heavy cold working is found to yield the strongest pinning, an enhanced pinning ability has been noted in the polygonised dislocation structures formed after low-temperature (500–900°C) annealing of the above materials, again representing a non-uniform distribution of dislocations (Narlikar and Dew-Hughes, 1966; Witcomb et al., 1968; Baker 1970; Dew-Hughes, 1971). The polygonised dislocation structure formed in Mo–Re alloy and producing strong pinning is shown in Figure 6.12(b). Besides dislocations, the Mo–Re alloy has particles of non-superconducting sigma-phase (Figure 6.12(c)), which also contribute to flux pinning. Similarly, the above-mentioned Nb–Ta alloy, which when cold-worked reveals a uniform dislocation distribution and poor flux pinning, after dislocation polygonisation exhibits (Figure 6.11(b)) a dramatic increase in its flux pinning ability (Narlikar and Dew-Hughes, 1966). The pinning is further enhanced when the polygonised sample is cold-worked. The dislocation subgrain boundaries now become denser and better defined, which promotes flux pinning (Narlikar and Dew-Hughes, 1966). On the other hand, the bulk twin boundaries, which again are surface defects, do not seem very effective in promoting pinning (Witcomb et al., 1968), unless they are formed on a very fine scale, such as those formed in a martensitically transformed phase.

In the case of important polycrystalline intermetallics and ordered compounds, such as Nb_3Sn, V_3Ga, and Nb_3Ge, belonging to the A-15 structural class, and in MgB_2, another potential intermetallic that was discovered at the beginning of the present century (Nagamatsu et al., 2001), the high-angle grain boundaries are natural crystal defects in abundance, and are responsible for strong pinning, although the broad grain boundary width in the latter poses a serious connectivity problem for high-current applications. As with dislocation cell structure, the pinning increases as $1/D$, where D is the grain diameter, and consequently, to achieve a higher J_c, sample processing must aim to realise a small D and slow grain growth.

Normal particles, formed either as precipitates of a second phase or introduced artificially during sample processing, are three-dimensional defects capable of strong flux pinning. The J_c of sintered MgB_2 (Dou et al., 2005; Gupta and Narlikar, 2009) and both bulk samples (Gupta and Narlikar, 2007) and thin films (Huhtinen et al., 2010) of HTS cuprates is found to be significantly enhanced by artificially adding nanosized normal particles of $BaZrO_3$, SiC, or SiO_2 or of various forms of carbon, including synthetic nanosized diamonds and carbon nanotubes, etc. Enhancing flux pinning by artificially incorporating novel particles with different properties and in varying sizes and concentrations has emerged as a powerful approach in HTS cuprates and other potential systems. *Artificial pinning centres* (APCs) in the form of nanodiamonds, carbon nanotubes, nanorods, etc., have been extensively used to enhance the J_c of HTS cuprates, pnictides, and MgB_2.

6.6   Pinning interactions

The vortex energy comprises two relevant components associated with its unique structure: (1) the condensation energy associated with the normal core of radius ξ, and (2) the magnetic energy of the surrounding supercurrents, spread over a distance λ. When the vortex lines interact with the microstructure through their normal cores, this gives rise to what is termed as *core pinning*. On the other hand, when the interaction takes place via circulating supercurrents, the pertinent interaction is *magnetic pinning*.

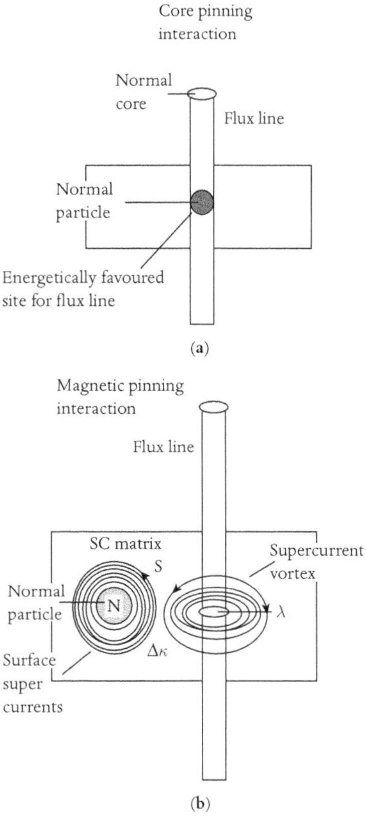

6.6.1   Core pinning

If the superconducting matrix contains a distribution of normal particles ($\kappa = 0$), the latter become energetically preferred sites for the normal cores of the vortex lines to nucleate (Figure 6.13(a)). Clearly, it would cost extra energy to nucleate them elsewhere in the superconducting matrix, since this would necessarily demand the creation of normal regions to accommodate the normal cores. Consequently, a flux core will have a free energy in a normal particle that is $\frac{1}{2}\mu_0 H_c^2$ per unit volume smaller than it would be in the matrix, and this is what creates the potential well responsible for pinning in the core interaction. Detailed considerations show (Dew-Hughes, 1971) that in the situation where the pinning takes place at the walls of dislocation tangles forming a cell structure of diameter a, the critical current density is given by

$$J_c = \frac{\mu_0 H_{c2} h(1-h)\Delta\kappa}{4.64a\kappa^3}, \tag{6.12}$$

where $h = H/H_{c2}$ is the reduced upper critical field. As can be seen, J_c increases with $\Delta\kappa$ and is higher for low-κ materials, which is found experimentally to be the case.

Figure 6.13 Pinning interaction: (a) core pinning; (b) magnetic pinning.

6.6.2 Magnetic pinning

The magnetic interaction for pinning basically arises from the difference in the magnetisation (or the magnetic induction) between the pinning entity and the matrix. Because of the difference $|\Delta\kappa|$ between the matrix and the pinning entity, the magnetisation of the pertinent pinning entity differs from that of the matrix in which it is embedded. In this situation, there is always a circulating supercurrent around the pinning entity at its interface with the superconducting matrix. It is this surface current that interacts with the circulating supercurrent of vortices to cause the magnetic pinning (Figure 6.13(b)). Bean and Livingston (1964) pointed out that the surface current resulting from a variation in magnetic induction across the interface creates an irreversible surface barrier responsible for flux pinning. The ensuing pinning force is given by (Campbell et al., 1968)

$$F_p(B) = \frac{S_v \Delta M(B) \phi_0^{3/2}}{\lambda B^{1/2}}, \tag{6.13}$$

where S_v is the surface area of the pinning interface per unit volume normal to the Lorentz force and $\Delta M(B)$ is the difference in the reversible magnetisation between the matrix and the pinning centre. If the pinning centres are normal non-magnetic entities (their Ginzburg–Landau parameter being zero and $\Delta\kappa = \kappa$), then $\Delta M(B)$ represents the magnetisation of the matrix at induction B. Campbell et al. (1968) found excellent agreement between theory and experiment for a PbBi eutectic where pinning is caused by normal particles of Bi. Strong pinning by dislocation tangles in cold-worked transition metal alloys has also been explained in a similar way. The κ value at the walls of dense dislocation tangles is higher than within cells having much lower dislocation density, and the resulting current density J_c is given by (Dew-Hughes and Witcomb, 1972)

$$J_c \approx \frac{\mu_0^{1/2} S_v \phi_0^{1/2} H_{c2}^{1/2} h^{-1/2} |1 - 2h| \Delta\kappa}{2.48 \lambda \kappa^3}. \tag{6.14}$$

The critical current density thus calculated agrees well with the estimated value from magnetisation data on MoRe alloys (Dew-Hughes and Witcomb, 1972). This approach has been successfully applied to explain flux pinning by various microstructural features such as clusters of point defects, dislocation tangles and networks, normal particles, and grain boundaries (Dew-Hughes, 1971; Campbell and Evetts, 1972).

Pinning, in general, may occur owing to both core and magnetic interactions. The ratio of magnetic to core interactions approaches 1 for large-κ materials, suggesting that the two interactions are equally favoured. The dominant pinning mechanism is therefore determined by

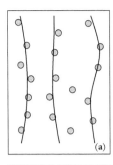

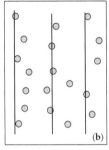

Figure 6.14 An elastically more relaxed flux line lattice (a) can accommodate more pinning centres and cause stronger pinning in comparison with a rigid lattice (b).

microstructural details, namely the size and distribution of pinning entities. If the distribution of pinning centres has a wavelength larger than λ, the pinning is expected to be magnetic. Low-κ materials therefore tend to exhibit magnetic pinning. If the pinning centres are more closely spaced than λ, then this should favour core pinning. The magnetic interaction is expected to decrease as the vortex density rises, whereas the core interaction should remain relatively invariant until near H_{c2}, where the vortex cores start overlapping. Core pinning is generally manifested by the critical current falling steeply to zero as H_{c2} is approached. On the other hand, this approach to zero J_c is more gradual if the pinning interaction is magnetic.

In the discussion so far, we have considered the interaction of individual vortex lines with the pinning centres, while their mutual interaction has been ignored; that is, the flux line lattice (FLL) has been mostly disregarded. This is, in general, justified if the FLL is sufficiently relaxed to allow pinned vortex lines to take up the minimum-energy positions without any restraint, as shown in Figure 6.14(a). With a relatively stiff lattice, vortex lines (Figure 6.14(b)) are unable to accommodate most of the pinning centres around them, and the pinning therefore is weak. Clearly, a completely rigid FLL cannot be pinned, since its displacement through one lattice spacing does not lead to any change in energy.

6.6.3 Intrinsic pinning

Besides core and magnetic pinning interactions, there is yet one more possibility, namely *intrinsic pinning* (Narlikar and Agarwal, 1988), which may be important in highly anisotropic materials such as HTS cuprates. We have previously mentioned the interesting possibility that, under the influence of F_L, vortex motion in the FLL may take place through the movement of flux lattice dislocations (FLDs). In high-temperature superconductors possessing a very short range of coherence ξ along the c-direction, the FLL may experience a force equivalent to the Peierls–Nabarro force that opposes dislocation motion in conventional crystals. Such a force will serve as an obstacle to FLDs moving along the c-direction and will thereby cause intrinsic pinning. In HTS cuprates, the intrinsic pinning will therefore become relevant when, for instance, both the applied field and the transport current are confined to the basal plane of the crystal and the resulting F_L is along the c-direction. Indeed, in this configuration of transport current and applied magnetic field, $J_c(B)$ is optimal for HTS cuprates.

6.7 AC losses

Because of the presence of the mixed state, the problem of AC losses is more complicated in superconducting alloys and compounds than in

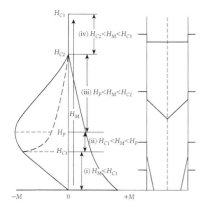

Figure 6.15 Magnetisation curve (left) and magnetic flux profile (right) under an AC field.

pure metals, discussed in Section 4.4. Vortex motion under an AC field gives rise to a rapid dissipation and the resulting temperature rise turns the sample normal. It is immaterial whether the AC magnetic field is applied externally or produced by transport of an alternating current. The losses increase with the strength of the peak AC field H_M on the sample surface and are found (Beall and Meyerhoff, 1969) to vary as $(H_M/H_{c1})^3$. The behaviour may be summarised as follows with respect to four regions of the magnetisation curve along with their flux profiles as illustrated in Figure 6.15.

When the peak AC field $H_M < H_{c1}$, the situation is no different from that in pure elemental type I superconductors (Section 4.4), and essentially no losses are to be expected at low frequencies. However, small losses do occur due to the penetration depth and hysteresis resulting from the presence of surface defects. When $H_M > H_{c1}$, the sample enters a mixed state and the losses markedly grow. However, if strong pinning centres are present in the sample, this situation is significantly countered and the losses remain mostly confined to the sample surface. Owing to pinning (Figure 6.1), flux penetration is considerably retarded until a magnetic field $H_p > H_{c1}$ is reached, where H_p, which can be several tens of times larger than H_{c1}, corresponds to the field at which the magnetisation of the sample is maximum. For $H_p < H_M < H_{c2}$, the entire volume of the sample is exposed to the AC field, and the losses become volume-dependent and very large. In the region $H_{c2} < H_M < H_{c3}$, only the thin surface sheath continues to be superconducting, and this is not of practical use. Nevertheless, losses are to be expected through surface defects.

As per the above discussion, materials such as Nb possessing either inherently large H_{c1} of about 0.1 T and various alloys and compounds such as A-15 superconductors noted for an appreciably large $H_p > 0.5$ T resulting from their strong flux pinning are to be preferred for AC applications. Among recent materials, MgB_2 and RE-123 of the HTS family could become natural choices for AC applications if their pinning behaviour can be enhanced.

6.8 Summary

A large upper critical field H_{c2} is a basic requirement of a type II superconductor for high-field applications, but that alone does not confer upon the superconductor an ability to carry a high electrical current density without resistance in a large magnetic field. In the mixed state, the transport of electrical current in the presence of a magnetic field results in a driving force that causes vortex lines to move, and moving vortices give rise to dissipation. The presence of various kinds of crystal defects and inhomogeneities is mandatory for high J_c. Various interesting approaches to stabilise conductors for applications against the adverse effects of flux instabilities have been described. The relevant basic aspects of vortex motion and its hindrance by pinning centres have been discussed in terms of the established concepts of critical state, flux creep, flux flow, and pinning interactions.

CHAPTER 7

Superconductors in abundance

Superconductors today present an enormous variety of materials, including pure elements, metallic alloys and intermetallic compounds, semimetals, ceramic materials, inorganic and organic polymers, fullerides, and others. Some of these, discovered since 1986, with T_c exceeding 30 K, are referred to as high-temperature superconductors (HTS), while all the others, having lower T_c, comprise the low-temperature superconductors (LTS). In general, superconductivity occurs in both bulk materials and thin films, in single crystals and in polycrystals, and even in the amorphous state realised in quench-condensed thin films or metallic glasses formed by rapid quenching of bulk materials from melt. Superconductivity may occur in magnetic materials under applied pressure or, in some unusual situations, the phenomenon may be observed under an imposed magnetic field, which otherwise is commonly known to destroy superconductivity. In both these situations, the effect is metastable or transient in that the superconductivity tends to disappear when the pressure is removed or the imposed magnetic field is switched off. Similarly, optical excitation can help to induce metastable superconductivity in oxygen-deficient copper oxide samples by enhancing their carrier density and charge mobility. Another fascinating situation inducting superconductivity is the *proximity effect*—in many instances, a normal material manifests signatures of weak transient superconductivity when placed in close proximity to a superconductor.

Clearly, as superconductivity occurs in so many different categories of substances, and under different physical situations and metallurgical conditions, the possibility of there being any direct correlation between the occurrence of superconductivity and crystallographic structure seems remote. This chapter presents a swift overview of the occurrence of superconductivity in different materials. Since a majority of the materials mentioned here will be discussed in detail in later chapters, we confine ourselves here to only brief descriptions.

7.1 Low-temperature superconductors (LTS)

7.1.1 Superconductors in the periodic table

Presently there are 54 elements known to exhibit superconductivity, which are indicated in the periodic table of Figure 7.1, together with their T_c values. Of these, 30 are found to be superconducting in their stable bulk phase at ambient pressure. There are also several elements that otherwise are non-superconductors but that exhibit superconductivity only under an externally applied pressure or only when they are in thin-film form, as fine particles, or in the amorphous state. However, under whatever physical conditions, the superconducting pure elements belong to the LTS category, with $T_c < 10$ K.

The alkali metals Na, K, and Rb, for example, of group IA, possessing good electrical conductivity, do not become superconducting down to the lowest temperature studied. This also holds true for Ag, Au, and Cu of group IB, noted for their high electrical conductivity. Broadly, this is consistent with the BCS theory, where the electron–phonon interaction, responsible for the Cooper pair formation, is weak in these high-conductivity metals. Nevertheless, it is suggested that Au should become superconducting at about $100\,\mu$K and Ag and Cu at about $10\,\mu$K

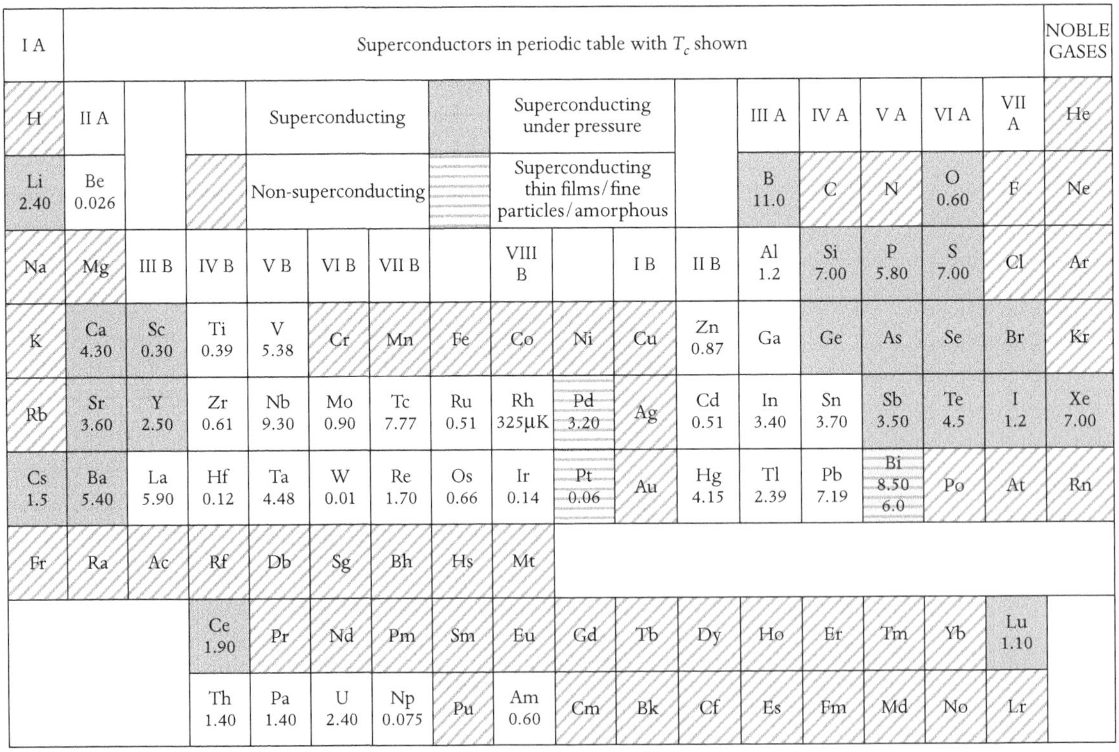

I A					Superconductors in periodic table with T_c shown												NOBLE GASES
H	II A			Superconducting				Superconducting under pressure				III A	IV A	V A	VI A	VII A	He
Li 2.40	Be 0.026			Non-superconducting				Superconducting thin films/fine particles/amorphous				B 11.0	C	N	O 0.60	F	Ne
Na	Mg	III B	IV B	V B	VI B	VII B	VIII B		I B	II B		Al 1.2	Si 7.00	P 5.80	S 7.00	Cl	Ar
K	Ca 4.30	Sc 0.30	Ti 0.39	V 5.38	Cr	Mn	Fe	Co	Ni	Cu	Zn 0.87	Ga	Ge	As	Se	Br	Kr
Rb	Sr 3.60	Y 2.50	Zr 0.61	Nb 9.30	Mo 0.90	Tc 7.77	Ru 0.51	Rh 325μK	Pd 3.20	Ag	Cd 0.51	In 3.40	Sn 3.70	Sb 3.50	Te 4.5	I 1.2	Xe 7.00
Cs 1.5	Ba 5.40	La 5.90	Hf 0.12	Ta 4.48	W 0.01	Re 1.70	Os 0.66	Ir 0.14	Pt 0.06	Au	Hg 4.15	Tl 2.39	Pb 7.19	Bi 8.50 6.0	Po	At	Rn
Fr	Ra	Ac	Rf	Db	Sg	Bh	Hs	Mt									
			Ce 1.90	Pr	Nd	Pm	Sm	Eu	Gd	Tb	Dy	Ho	Er	Tm	Yb	Lu 1.10	
			Th 1.40	Pa 1.40	U 2.40	Np 0.075	Pu	Am 0.60	Cm	Bk	Cf	Es	Fm	Md	No	Lr	

Figure 7.1 Periodic table showing the occurrence of superconductivity in pure elements.

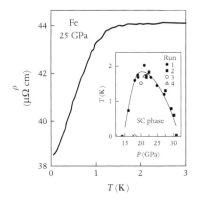

Figure 7.2 Hexagonal and non-magnetic Fe formed under a pressure above 15 GPa manifests superconductivity with the onset-T_c showing a peak near 2 K for a pressure of about 25 GPa (after Shimizu et al. 2001). No superconductivity is observed for P above about 30 GPa. The resistive transition does not drop down to zero, owing to contact resistance due to connecting wires of Au/or Pt.

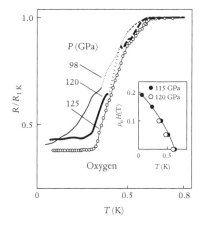

Figure 7.3 Solid oxygen exhibits superconductivity under an imposed pressure of about 100 GPa and the optimum onset-$T_c \approx 0.6$ K for $P = 125$ GPa. The residual resistivity is believed to be due to non-superconducting solid oxygen resulting from an inhomogeneous pressure distribution. The inset shows the extrapolated critical field of 0.2 T at 0 K. (After Shimizu et al. (1998).)

(Herrmannsdörfer and Pobell, 2005), although these predictions remain to be experimentally proven. The alkali metals Li and Cs and the alkaline earths Ca, Sr, and Ba exhibit superconductivity only under imposed pressure. Only Be of the alkaline-earth group becomes superconducting at ambient pressure, albeit at a very low temperature of 0.026 K. Among the non-transition-metal elements, superconductivity occurs in Zn, Cd, and Hg of group IIB; Al, Ga, In, and Tl of group IIIA; and Sn and Pb of group IVA.

Among elements, Nb (of group VB) has the highest T_c of 9.3 K, while Rh (group VIIB) becomes superconducting at the lowest temperature of 325 µK. The other transition elements such as Ti, Zr, Hf, V, Ta, Mo, W, Tc, Re, Ru, Os, and Ir show superconductivity at intermediate temperatures. However, the magnetic $3d$ elements Cr, Mn, Fe, Co, and Ni and the lanthanides, from Ce to Yb, owing to their strong exchange and Coulomb forces, are non-superconductors. Interestingly, the non-magnetic form of Fe (the hexagonal epsilon phase), produced by pressurising magnetic Fe at 15–30 GPa, exhibits superconductivity (Shimizu et al., 2001) at 1.5–2 K (Figure 7.2 and its inset). On the other hand, owing to suppression of spin density waves, Cr under pressure does not become superconducting (Jaramillo et al., 2008, 2009), although this may happen at very high pressures and lower temperatures. In this context, oxygen (group VIA) presents an interesting case. Frozen oxygen in solid form is weakly ferromagnetic, but when it is subjected to a pressure of about 100 GPa, which suppresses magnetism, superconductivity (Figure 7.3) is observed and, at 125 GPa, $T_c = 0.6$ K, with a critical field (0 K) of about 0.2 T (Shimizu et al., 1998). The last of the lanthanides, Lu, has a nearly filled $4f$ shell and is superconducting at about 0.1 K. Divalent Eu possesses a strong local magnetic moment that suppresses superconductivity, but it has recently been shown to exhibit superconductivity at 1.8 K under a pressure of >70 GPa, and at 2.75 K under 142 GPa (Debessai et al. 2009) (Figure 7.4). Among the actinides, Th, Pa, and Am exhibit superconductivity below 2 K.

Pure carbon (group IVA) exhibits superconductivity in none of its allotropic forms. However, diamond formed by doping with 2.8% B (Ekimov et al., 2004), graphite intercalated with Na (Belash et al., 1987), and C_{60} solid doped with alkali metals (Hebard et al., 1991; Tanigaki et al., 1991) do show superconductivity at about 4 K, 5 K, and above 17 K, respectively. The results of Ekimov et al. on the B-doped polycrystalline diamond (Figure 7.5) show that its superconductivity is suppressed with pressure. Interestingly, there are unconfirmed reports of carbon nanotubes exhibiting superconductivity above 15 K.

Turning to other elements that show superconductivity only under imposed pressure or when formed by thin-film processing, the examples include the semimetals Si, Ge, As, Sb, Bi, Se, and Te. Under pressure, they become more metal-like and exhibit superconductivity. Also in this

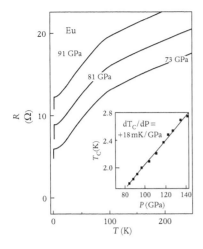

Figure 7.4 Above a pressure of 70 GPa, the pure lanthanide metal Eu exhibits superconductivity below 2 K. T_c rises to about 2.8 K (inset) at $P \approx 140$ GPa. (After Debessai et al. (2009).)

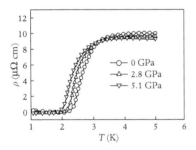

Figure 7.5 Polycrystalline diamond synthesised with 2.8% boron shows superconductivity with a broad transition and an onset-T_c of about 4 K, which decreases with imposed pressure. (After Ekimov et al. (2004).)

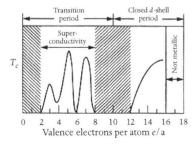

Figure 7.6 Matthias' rules.

category are P and S, which show superconductivity under pressure. Interesting pressure effects leading to superconducting behaviour are manifested by Sc, Y, various lanthanides such as Ce, Pr, Nd, Eu, and Yb, and the actinide U.

In bulk form, Bi is crystalline, semimetallic, and non-superconducting. However, its quench-condensed film in the amorphous state is metallic and superconducting below 6 K, but above 20 K turns crystalline and semimetallic (Buckel and Hilsch, 1954). The behaviour of bulk Bi and its alloys is in general unusual. For instance, the bulk alloy Bi_3Au shows superconductivity at 1.8 K (Shoenberg, 1938), although neither Bi nor gold, in themselves, exhibit superconductivity. Similarly, bulk Bi when alloyed with the magnetic non-superconducting elements Fe, Co, Ni, Cr, and Mn shows superconductivity in the range 3–6 K (Matthias et al., 1958, 1959, 1963, 1967). This is quite uncommon, since normally magnetic impurities destroy superconductivity (Abrikosov and Gor'kov, 1960).

Irradiation-induced disorder has been found to induce superconductivity at 3.2 K in a thin Pd film, which is otherwise highly paramagnetic and non-superconducting (Stritzker, 1979). The T_c of Be increases from about 0.03 K to 9 K in the amorphous state (Bergmann, 1976). Bulk Pt, which is otherwise non-superconducting, shows superconductivity at 0.02 K when prepared as very fine compressed particles (Schindler et al., 2002). The metal Mo behaves similarly, and its T_c is enhanced from 0.9 K to 9 K when formed as a thin amorphous film (Meyer, 1975).

Although one finds no apparent regularity in the occurrence of superconductivity and the placement of the element in the periodic table, Matthias (1957) could detect certain useful traits that led him to what are known as *Matthias' rules*, depicted in Figure 7.6. These qualitatively relate T_c to the valence electrons per atom, e/a. Accordingly, for non-transition elements, T_c is a smooth increasing function of the valence electrons per atom, while for transition metals, in stable crystalline form, the behaviour shows peaks in T_c for e/a values of 3, 5, and 7. For most metals, the rules work fairly well.

7.1.2 Alloys and compounds of transition metals

7.1.2.1 *Crystalline alloys and compounds*

Following Onnes' discovery, several thousands of crystalline metallic alloys and compounds with varying crystal structures were found to be superconducting; a list was compiled by Roberts (1978). Up to the mid-1970s, several important superconductors were discovered by Matthias by following his empirical rules. For alloys and compounds formed with transition metals, the peaks corresponding to the optimal T_c occur at e/a values of 4.75 and 6.5, instead of 3, 5, and 7 for pure metals. Here, the e/a represents the mean values of all the constituents forming the

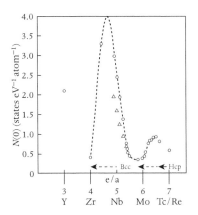

Figure 7.7 Measured values of the density of states for transition metals and their alloys show double peaks as a function of their e/a in accordance with Matthias' empirical rules of Figure 7.6 (after Morin and Maita, 1963).

stoichiometric alloy or compound. These empirical rules seem to be consistent with the BCS theory. For most superconductors, Morin and Maita (1963) found (Figure 7.7) the electron–phonon interaction parameter V to be essentially constant, and a large part of the variation in T_c is due to a variation either in the density of states at E_F, $N(0)$, or in the Debye frequency ω_D. Interestingly, the measured values of $N(0)$ for a large number of transition metal alloys are found to vary with e/a, showing the characteristic peaks in accordance with Matthias' rules.

Table 7.1 lists some of the prominent superconductors formed with transition metals. Until the discovery of HTS cuprates in 1986, the material having the highest T_c (23.2 K) was Nb_3Ge, in a thin-film form (Testardi et al., 1974), possessing the *A-15* or, alternatively, the so-called *β-tungsten* crystal structure. In the A-15 crystal structure of A_3B (e.g. Nb_3Sn), the B (= Sn) atoms form a body-centred cubic (bcc) sublattice, while each of the cube faces carry two A (= Nb) atoms, forming

Table 7.1 Superconductivity in metallic systems (some representative materials)

Type	T_c (K)	$\mu_0 H_{c2}$ (T) (at 4.2 K)
A-15 structure		
Nb_3Sn	18.5	24
Nb_3Al	19.0	30
Nb_3Ga	20.3	34
Nb_3Ge	23.2	38
V_3Ga	16.8	22
V_3Si	17.1	25
$Nb_3(Al,Ge)$	21.0	44
A-2 structure		
Nb–45 to 47wt%Ti	9.0	12
Nb–25 to 30wt%Zr	10.0	8
C-15 structure		
V_2Hf	10.0	20
$V_2(Hf,Zr)$	9.8	24
$V_2(Hf,Ta)$	10.2	26
Chevrel phases		
$PbMo_6S_8$	12.7	54
$Pb_{0.5}Eu_{0.3}Gd_{0.2}Mo_6S_8$	14.4	71
Intercalated layered compounds		
TaS_2	3.5	20

From Narlikar, A. V. and Ekbote, S. N. *Superconductivity and Superconducting Materials*. South Asian Publishers, New Delhi (1983), p. 20.

A-15 structure

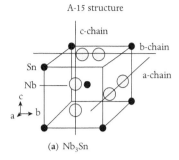

(a) Nb_3Sn

C-15 structure

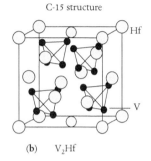

(b) V_2Hf

B-1 structure

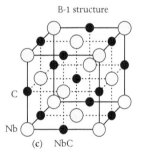

(c) NbC

Figure 7.8 (a) The A-15 crystal structure of Nb_3Sn. (b) The cubic C-15 structure of V_2Hf. (c) The B-1 (NaCl-type) structure of NbC.

three densely packed linear chains of transition metals, running along the three orthogonal axes (Figure 7.8(a)). These chains cause narrowing of the d-band at E_F and a sharp rise in $N(0)$ at E_F that favours its relatively large value of T_c. Because of the linear chain structure, the A-15s are considered as *quasi-one-dimensional superconductors*.

Other important metallic systems exhibiting superconductivity belong to the A-2, C-15, and Chevrel phase structures (Table 7.1). Transition metal alloys such as Mo–Re, Nb–Zr, and Nb–Ti have T_c of around 10 K. These are disordered alloys possessing a simple bcc (A-2) structure with the lattice points randomly occupied by the constituent metals. The most notable superconductor of this class is the ductile Nb–45 to 50%Ti alloy that is used for winding superconducting magnets to produce magnetic fields up to about 10 T at 4.2 K. Today, for almost all superconducting magnet applications—from small laboratory solenoids to large MRI systems and for gigantic accelerator and fusion reactor magnets—Nb–Ti has proved to be the unrivalled choice. In fact, many of the wide range of superconductor applications mentioned in Chapter 1 use Nb–Ti conductors.

The intermetallic compounds of stoichiometry A_2B, such as V_2Hf, $V_2(Hf, Zr)$, and $V_2(Hf, Ta)$, listed in Table 7.1 are the so-called *Laves phases* with cubic C-15 structure. They have a moderate T_c of around 10 K, but can remain superconducting in magnetic fields exceeding 20 T. The crystal structure of A_2B compounds (e.g. V_2Hf) comprises B (= Hf) atoms occupying the positions of the diamond structure, with the A (= V) atoms forming the corner-sharing tetrahedral network (Figure 7.8(b)).

Another important class of superconducting compounds of transition metals comprises the ternary molybdenum chalcogenides, also called the *Chevrel phases* (Chevrel et al., 1971), with T_c of around 10–16 K, described by the stoichiometry MMo_6X_8, where M = Pb, Sn, Eu, Gd, etc. and X = S, Se, or Te. They are noted for their particularly high critical field of around 40–70 T which surpasses the 20–40 T critical field of the A-15 superconductors. Their crystal structure, consists of X atoms arranged in rhombohedral positions enclosing a small cluster of Mo atoms forming octahedra. Superconductivity is believed to arise from the d-band of Mo clusters at the Fermi level. Since the clusters are localised in all three directions, the Chevrel-phase compounds have been considered as quasi-zero-dimensional. These chalcogenides, when formed with a lanthanide as M-element, show an interesting interplay of superconductivity with magnetism and also display the unusual phenomenon of field-induced superconductivity (FIS).

Intercalated layered compounds of the type TaS_2 and $NbSe_2$ are yet another important class of transition-metal-based metallic superconductors noted for their high critical field. These possess strongly bound molecular sheets with chalcogenides above and below the transition

Table 7.2 Superconducting metallic glasses of transition metals

Metallic glass	T_c(K)	$\mu_0 \, dH_{c2}/dT$ (T K^{-1})
$Zr_{70}Pd_{30}$	2.4	2.65
$Zr_{70}Be_{30}$	2.8	2.38
$Zr_{75}Rh_{25}$	4.55	2.63
$(Mo_{0.6}Ru_{0.4})_{80}P_{20}$	6.18	2.55
$(Mo_{0.8}Ru_{0.2})_{80}P_{20}$	7.31	2.45
$(Mo_{0.6}Ru_{0.4})_{90}B_{10}$	7.10	2.54
$(Mo_{0.8}Ru_{0.2})_{80}P_{10}B_{10}$	8.71	2.42

Data assembled from Narlikar and Ekbote (1983).

metal atoms, that is, S–Ta–S, Se–Nb–Se, etc. These compounds can be intercalated with organic molecules without losing superconductivity, which can modify the separation between the layers to display a large anisotropy and critical field. In TaS$_2$ intercalated with pyridine, T_c is small, about 3.5 K, but the critical field parallel to the sheets is 20 T (Morris and Coleman, 1973). Interestingly, their normal state exhibits charge-density waves (Wertheim et al., 1976), which has made them attractive compounds for basic research.

7.1.2.2 Amorphous and glassy alloys

Some instances of superconductivity occurring in pure elements in the amorphous state were described earlier. In most cases, such as pure Mo or some of its alloys, where T_c shows enhancement in the amorphous state, there is increase in $N(0)$ that accounts for the observed rise in T_c. Superconducting metallic glasses and phosphate glasses of binary alloys of Nb, V, Ta, Mo, Zr, Rh, Pd, Ru, etc. prepared by rapid quenching from the molten state show superconductivity up to about 9 K. Some of the representative data (Johnson and Poon, 1975; Johnson et al., 1975, 1978b; Togano and Tachikawa, 1975; Graebner et al., 1977; Hasegawa and Tanner, 1977; Domb and Johnson, 1978), compiled by Narlikar and Ekbote (1983), are depicted in Table 7.2.

Interestingly, superconducting transition metal alloys that in the crystalline state show two distinct peaks in T_c for e/a values of 4.75 and 6.5 in the amorphous state exhibit only a single broad maximum for $e/a = 6.5$ (Figure 7.9), with the first peak at 4.75 no longer being distinct (Collver and Hammond, 1973, 1977). On becoming amorphous, the alloys of V, Nb, and Ta from group V show a general decrease in T_c, while those of Mo (group VI) and Re (group VII) show a distinct increase, which is attributed to a change in N(0).

The metallic glasses such as $(Mo_{0.6}Ru_{0.4})_{80}P_{20}$ and $(Mo_{0.8}Ru_{0.2})_{80}$ P$_{10}$B$_{10}$ (Table 7.2), which are extreme type II superconductors with κ-values of about 100, possess attractive superconducting and

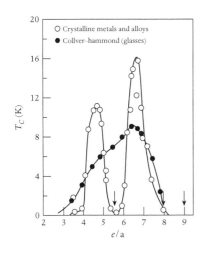

Figure 7.9 T_c dependence on e/a for crystalline and amorphous metallic alloys. The former, depicted by the open circles, is the same as Matthias' rules of Figure 7.6. The behaviour for the amorphous state, shown by the filled circles, is after Collver and Hammond (1973, 1977).

mechanical properties with regard to their possible use in superconducting electromagnets (Johnson et al., 1978a,b). Their upper critical field of 16 T lies between those of Nb-Ti and Nb_3Sn, but with respect to J_c they fall short of expectations. The J_c of $(Mo_{0.8}Ru_{0.2})_{80}P_{10}B_{10}$, for instance, in the self-field is reasonable, at 10^9 A m^{-2}, but drops steeply to $<10^7$ A m^{-2} in an applied field of 3–4 T. The problem with these materials is the lack of pinning centres of 4–5 nm in size needed for effective pinning. The disorder present in the metallic glasses is on a scale too fine for vortex lines to 'see', and therefore, no pinning occurs. Otherwise, these materials have an advantage over A-15s in exhibiting a greater tolerance against irradiation-induced disorder, making them more suitable for nuclear fusion applications.

Table 7.3 Superconducting carbides and nitrides (B-1 structure)

Compound	T_c (K)
$NbC_{0.977}$	11.1
$TaC_{0.987}$	9.7
TiN	5.6
VN	8.2
ZrN	8.9–10.7
NbN	16.0
HfN	6.2

From Matthias, B. T. et al. *Rev. Mod. Phys.* 35 (1963) 1.

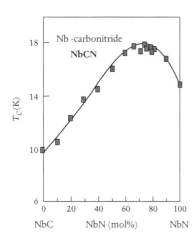

Figure 7.10 Niobium carbonitride NbCN, with $T_c \approx 18$ K, is a pseudobinary formed by alloying NbC and NbN (after Shulishova, 1966).

7.1.3 Superconducting semiconductors

By studying the band structure, Cohen (1964) was able to show the possibility of superconductivity occurring in semiconducting compounds possessing carrier concentration much smaller than that in typical metals. In self-doped germanium telluride $GeTe_{1+x}$, the critical temperature is raised to 0.3 K for a carrier concentration of 1.5×10^{13} m^{-3}. Similarly, the work of Schooley et al. (1965) showed the superconductivity of $SrTiO_3$ to depend sensitively on the carrier concentration, and when the material was optimally doped by reduction, the T_c realised was about 0.5 K.

7.1.4 Ceramic carbides and nitrides of transition metals

Carbides and nitrides of transition metals, also known as interstitial compounds, exhibit superconductivity in the range of 5 K to about 19 K, the highest value of 18.5 K being for niobium carbonitride NbCN. The binary compound NbN, discovered in 1941 (Aschermann et al., 1941), possesses an optimum T_c of 16 K. Their crystal structure is of the rock-salt type, denoted by B-1, depicted in Figure 7.8(c). T_c values of some of representative materials of this class are listed in Table 7.3. The T_c of these compounds depends sensitively upon carbon and nitrogen content and is optimum for the stoichiometric composition. For instance, $NbC_{0.977}$ has a T_c of 11.1 K while $NbC_{0.7}$ is found to be non-superconducting above 1.05 K (Matthias et al., 1963). Figure 7.10 shows the pseudobinary NbCN formed by alloying NbC and NbN (Shulishova, 1966). Niobium carbide and nitride have critical field values that exceed 10 T, while NbCN has been synthesised in flexible fibre form with a T_c of 18.5 K and a critical field of 25 T (Dietrich et al., 1981), making it commercially attractive.

7.1.5   Superconducting hydrides and deuterides

Apart from superconducting nitrides and carbides, hydrides and deuterides (i.e., materials formed with hydrogen H and deuterium D) with the NaCl structure have proved interesting superconductors since the early 1970s. These are metastable superconductors formed by implanting hydrogen (or deuterium) ions at low temperatures in Pd and its alloys with other noble metals. To make the material superconducting, the H:Pd ratio is required to be 0.7 or more (Skoskiewicz, 1972). Stritzker and Buckel (1972) could raise this ratio to above 1.0 and found optimum T_c values of 9 K and 11 K, respectively, for Pd–H and Pd–D. Pure Pd is otherwise non-superconducting owing to its strong magnetic fluctuations, and the observed T_c values are ascribed to reduced spin fluctuations and to enhanced electron phonon coupling from optical H and D phonon modes. Surprisingly, however, in some cases, relatively lower T_c values for the hydride in comparison with the deuteride suggests the occurrence of a negative (or inverse) isotope effect, which is contrary to the conventional phonon-mediated mechanism of superconductivity, and this is what initially made these materials interesting. The highest T_c recorded for such materials (Figure 7.11) is 16.6 K for $Pd_{0.55}Cu_{0.45}$–H (Buckel and Stritzker, 1973).

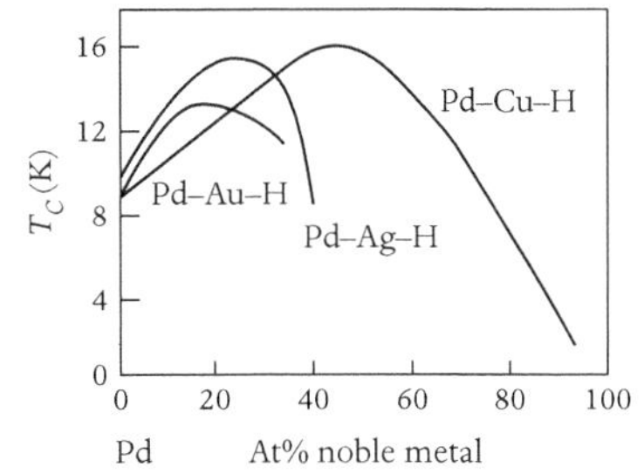

Figure 7.11 T_c variation of hydrogen-implanted Pd–noble metal alloys (after Stritzker and Buckel, 1973).

7.1.6   Rare-earth-based ternary borides, stannides, and quaternary borocarbides

Along with the ternary molybdenum chalcogenides (or Chevrel phases) mentioned above, in the early 1970s, ternary borides and stannides were also discovered showing an interesting interplay of superconductivity with long-range magnetic order, which made them particularly exciting in terms of their physics. The ternary borides include RRh_4B_4 (where the rare earth R = Nd, Sm, Er, or Tm), while the stannides are RRh_xSn_y (where R = La or Er and $x \approx 1$–1.5 and $y \approx 3.5$–4.5). The magnetic order in these materials can be ferromagnetic (F), antiferromagnetic (AF), spin glass (SG), oscillatory, or weakly ferromagnetic. In general, these materials can have $T_c > T_m$, $T_c < T_m$, or $T_c \approx T_m$. When the long-range order is antiferromagnetic, weakly ferromagnetic, or of oscillatory type, superconductivity can coexist with the magnetic order. When $T_c > T_m$, and the magnetic order formed at lower temperature is ferromagnetic (Narlikar and Ekbote, 1983), for example in $ErRh_4B_4$ ($T_c = 8.7$ K and $T_m = 0.93$ K) and $HoMo_6S_8$ ($T_c = 2.0$ K and $T_m = 0.61$ K), the material first becomes superconducting and then, on further cooling to T_m, it reverts to the normal state, with the superconducting state giving way to the non-superconducting ferromagnetic order. Such behaviour is referred to as *reentrant*, and the material is called a *ferromagnetic superconductor*.

Table 7.4 T_c and T_N values for Ni-based quaternary borocarbides

Compound	T_c (K)	T_N (K)
YNi$_2$B$_2$C	15.5	—
DyNi$_2$B$_2$C	6	11
HoNi$_2$B$_2$C	8.5	8
ErNi$_2$B$_2$C	11	6.5
TmNi$_2$B$_2$C	10.6	1.5
YLu$_2$B$_2$C	16.5	—

From Gupta, L. C. *Phil. Mag.* B 77 (1998) 717.

Table 7.5 Some prominent heavy fermion superconductors

Ce-based	U-based	Miscellaneous
CeCu$_2$Si$_2$	UBe$_{13}$	PrOs$_4$Sb$_{12}$
CeCu$_2$Ge$_2$	UPt$_3$	PuCoGa$_5$
CePd$_2$Si$_2$	URu$_2$Si$_2$	PuRhGa$_5$
CePd$_2$Ge$_2$	UPd$_2$Al$_3$	YbAl$_3$
CeRu$_2$Si$_2$	UNi$_2$Al$_3$	Pd$_2$SnYb
CeRh$_2$Si$_2$	UGe$_2$	NpPd$_5$Al$_2$
CeIn$_3$	URhGe$_2$	
CePt$_3$Si		
CeNi$_2$Ge$_2$		
CeCoIn$_5$		
CeRhIn$_5$		
CeIrIn$_5$		
CeNiGe$_3$		
CeNi$_3$Ge$_5$		

Quaternary borocarbides, discovered in 1993 (Mazumdar et al., 1993; Nagarajan et al., 1994), display an interesting interplay of superconductivity and antiferromagnetism (below the Néel temperature T_N). The Ni-based borocarbides with the chemical formula RNi$_2$B$_2$C (R = rare earth: Dy, Ho, Er, Tm, or Lu) have T_c in the range 6–17 K, with YNi$_2$B$_2$C becoming superconducting at 15.5 K. T_c is enhanced to 23 K when Ni is replaced by Pd. The crystal structure (described in Chapter 12) is body-centred tetragonal, with the electrical conduction taking place in Ni$_2$B$_2$ sheets. The T_c and T_N values are very close, exhibiting all possible combinations $T_c > T_N$, $T_c < T_N$, and $T_c \approx T_N$, as shown in Table 7.4.

7.1.7   Heavy fermion superconductors

Heavy fermion (HF) superconductors are lanthanide- and actinide-based intermetallic compounds, mainly of Ce and U (Table 7.5), where the heavy electron state originates from strong Coulomb repulsion in the partially filled 4f- and 5f-shells and their hybridisation with the conduction band. Heavy fermion compounds are strongly correlated materials with an effective electron mass 10–1000 times larger than the free or bare electron mass.

Such heavy quasiparticles form pairs in the superconducting state. CeCu$_2$Si$_2$ was the first HF superconductor discovered, in 1979 (Steglich et al., 1979), and was soon followed by a number of others, namely UBe$_{13}$, UPt$_3$, URu$_2$Si$_2$, UPd$_2$Al$_3$, and UNi$_2$Al$_3$, all showing an interesting coexistence of superconductivity with antiferromagnetic order. The preferred structure types for HF superconductors are cubic, tetragonal, and hexagonal. The superconducting state in HF systems mostly occurs at a relatively low $T_c < 2$ K, and that, quite often, only with the application of external pressure, which counters the magnetic order resulting from localised f-electrons. There are a few exceptions to the generally low T_c of HF materials, for example NpPd$_5$Al$_2$, with $T_c = 4.9$ K (Aoki et al., 2007), and PuCoGa$_5$ (Sarrao, 2002), with an ambient-pressure T_c of 18.5 K, which is the highest noted for this class.

Among HF materials, UGe$_2$ and URhGe$_2$ are of particular interest because they exhibit coexistence of superconductivity and ferromagnetism and are the prime contenders for the *p-wave pairing* that represents a spin-parallel coupled superconducting state. Some of these compounds also manifest the phenomenon of FIS.

7.1.8   Organic superconductors

The possibility of superconductivity occurring at room temperature or even above in long macromolecules with side chains was first conceived by Little (1964), but for many years the idea remained theoretically disputed, let alone experimentally verified. Organic materials tend to be insulators, but nevertheless efforts continued to look for superconductivity in polymeric materials, both inorganic and organic, and the first breakthrough came with the discovery of superconductivity with T_c of 0.37 K (Greene et al., 1975) in the inorganic polymer polysulfur nitride $(SN)_x$. Within a few years thereafter, the first organic quasi-one-dimensional compound $(TMTSF)_2PF_6$, known as a Bechgaard salt, was found to be superconducting (Jerome et al., 1980), with a T_c of 1.35 K under a pressure of 0.65 GPa. Here TMTSF is tetramethyltetraselenafulvalene and PF_6 is hexafluorophosphate. Replacing PF_6 by AsF_6, SbF_6, TaF_6, ClO_4, etc., led to a similar behaviour, except for the ClO_4 salt, which showed superconductivity at 1.2 K under ambient pressure. Since then, more organic systems have been found and T_c has crossed the 10 K mark. Among these are BEDT-TTF, DMET, MDT-TTF, and $M(dmit)_2$. Of these, $(BEDT-TTF)_2X$ (where BEDT-TTF denotes bis(ethylenedithio)tetrathiafulvalene), which is generally abbreviated as $(ET)_2X$, has yielded the highest T_c of about 12.8 K (under a pressure of 0.3 kbar) for $X = Cu[N(CN)_2]Cl$.

7.1.9   Hydrated sodium cobaltate

This superconductor, discovered in 2003 (Takada et al., 2003), with chemical formula $Na_xCoO_2 \cdot yH_2O$ (with $x = 0.35$ and $y = 1.3$) and a critical temperature of 4.7 K, is of interest because its normal state carries strong ferromagnetic fluctuations and the ensuing superconductivity in CoO_2 planes is unconventional. The structure consists of alternate layers of CoO_2 and Na (Figure 7.12), where the charge carriers are provided by Na atoms. The presence of intercalated water molecules between the CoO_2 and Na layers enhances the separation between the layers and the c-parameter of the unit cell, which seems to be essential for inducing quasi-two-dimensionality and superconductivity. Incidentally, sodium cobaltate stands out as the first superconductor containing water that is crucial for its superconductivity.

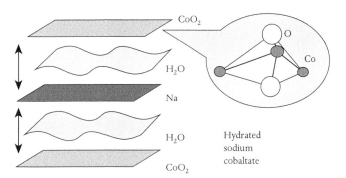

Figure 7.12 Schematic illustration of the structure of sodium cobaltate intercalated with water layers.

7.1.10 Superconducting non-oxide perovskites, metallonitride halides, pyrochlore oxides, and bismuth sulfide-based layered compounds

Non-oxide perovskites or *anti-perovskites* are ABO_3-type perovskite compounds where A = Mg, Cd, or Zn, B = C, and the anion O is replaced by a cation Ni, with superconductivity being due to Ni-d-band states. Interest in these superconductors (Table 7.6) stems from the fact that in $MgCNi_3$, for example, E_F is located at an electron count < 0.5 electron per formula unit above the peak maximum, which indicates that the compound is very close to the ferromagnetic instability. Consequently, some exotic effects are anticipated in $MgCNi_3$, although these have yet to be observed. All three materials of this family listed in Table 7.6 exhibit conventional singlet s-wave superconductivity.

Metallonitride halides, with the chemical formula MNX, where M = Hf, Zr, or Ti and X = Cl, Br, or I, are layered compounds that exhibit superconductivity when the halogen layers are intercalated with suitable charge-carrier dopants (Li, Na, K, etc.) and organic molecules, such as tetrahydrofuran (THF), to make the compound two-dimensional. As may be seen from Table 7.6, some of these superconductors, despite being conventional, have an appreciable T_c and, as we will see in Chapter 21, exhibit an unusual doping dependence of their superconductivity. The β-pyrochlore oxides, with chemical formula AOs_2O_6 (A = alkali metal: Cs, Rb, or K), are unusual systems where the alkali metal is located in an oversized pyrochlore cage, with superconductivity resulting from its rattling about inside the cage—this effect is maximum, and the T_c highest, for K, which shows the largest rattling. These are thus rare examples of *rattling-mode-induced superconductivity*, which exhibits both conventional and unconventional traits.

Table 7.6 Superconducting non-oxide perovskites, metallonitride halides, β-pyrochlore oxides, and bismuth sulfide-based superconductors

Non-oxide perovskites	T_c (K)	β-Pyrochlore oxides	T_c (K)
$MgCNi_3$	7	$CsOs_2O_6$	3.3
$CdCNi_3$	3	$RbOs_2O_6$	6.3
$ZnCNi_3$	3	KOs_2O_6	9.6
Metallonitride halides	T_c (K)	BiS_2-based compounds	T_c (K)
$Li_{0.48}(THF)_{0.3}HfNCl$	25.5	$Bi_4O_4S_3$	9 (onset)
$Na_{0.28}NCl$	22	$LaO_{0.5}F_{0.5}BiS_2$	10.6
$Li_{0.16}ZrNCl$	14	$NdO_{0.7}F_{0.3}BiS_2$	5.2

Data from Chapter 21 of this book.

The BiS_2-based superconductors (Table 7.6) are of recent origin, having been discovered in mid-2012. All three listed materials are layered superconductors with common BiS_2 layers in which superconductivity arises. In their chemical formulae and crystal structures some of these compounds share similarities with the HTS pnictides that make them promising materials. Their T_c has, however, remained low, although it may be too early to speculate about what can be achieved in this regard. Nevertheless, these materials, as we will find in Chapter 21, do show some intriguing features that are yet to be understood.

7.2 High-temperature superconductors (HTS)

7.2.1 HTS cuprates

The first of this most exciting family of oxide superconductors with a T_c of about 30 K, generally known as HTS cuprates, was $La_{1.85}Ba_{0.15}CuO_4$, discovered by Bednorz and Müller (1986). By partially substituting La^{3+} by Ba^{2+} in otherwise-insulating La_2CuO_4, they could change the ratio of Cu^{3+} to Cu^{2+} in a controlled way. For the above stoichiometry, which was close to the boundary of the metal–insulator transition, the material with holes as charge carriers exhibited (hole) superconductivity at 30 K . Subsequently, by replacing Ba by Sr, a still higher T_c of 36–38 K was realised (Takagi et al., 1987). This material is termed LSCO or La-214, while this family of HTS is referred to as R-214 (where R is a rare-earth ion such as La, Nd, Pr, Sm, Eu). If for Sr^{2+} (or Ba^{2+}), Ce^{4+} is used to partially replace R^{3+}, the nature of the charge carriers is changed from holes to electrons, and electron superconductivity results, albeit in a slightly lower temperature range of 22–28 K. The copper–oxygen plane of the structure is believed to be the most relevant for both normal-state conduction and superconductivity. Within a short time of Bednorz and Müller's discovery, new cuprate systems were synthesised with still higher T_c values, the most prominent including $RBa_2Cu_3O_7$ (R-123, also called RBCO, where R = any rare-earth ion except Ce, Pr, Pm, or Tb) with $T_c \approx 91$ K (Wu et al., 1987), $Bi_2Sr_2CaCu_2O_8$ (Bi-2212) with $T_c = 96$ K, $Bi_2Sr_2Ca_2Cu_3O_{10}$ (Bi-2223) with $T_c = 107$ K (Maeda et al., 1988), $Tl_2Ba_2Ca_2Cu_3O_{10}$(Tl-2223) with $T_c = 125$ K (Sheng and Hermann, 1988), and $HgBa_2Ca_2Cu_3O_8$ (Hg-1223) with $T_c = 135$ K (Schilling et al., 1993). Among these, with the application of pressure, the T_c of Hg-1223 was enhanced further from 135 K to 164 K (Gao et al., 1994), which stands out as the highest temperature for the occurrence of superconductivity yet reached. There are many more layered cuprate systems that have been discovered during the last 20 years, with T_c values in the range of 40–120 K and possessing perovskite or related structures. Some of the prominent HTS cuprate families are listed in Table 7.7.

Table 7.7 List of some prominent superconductors in the HTS category with their optimum critical temperatures

HTS cuprates	Critical temperature (K)
La-214	38
La-2126	60
Y-123	92
Nd-123	95
Y-247	95
Y-124	80
Bi-2212	95
Bi-2223	110
Bi-2234	110
Tl-1201	50
Tl-1223	133
Tl-2223	128
Hg-1212	128
Hg-1223	134
Hg-1223 with Tl	138
Hg-1223 with pressure	164
Fe-based pnictides	**Critical temperature (K)**
Nd-1111 (F)	52
Sm-1111 (F)	55
$Gd_{1-x}Th_xFeAsO$	56
Ba/K-122	38
Cs/Sr-122	37
FeSe with pressure ($\sim$7 GPa)	$\sim$30
Doped fullerides	**Critical temperature (K)**
Rb_3C_{60}	$\sim$30
$(NH_3)K_3C_{60}$	28
Cs_3C_{60} (with pressure)	30 (40)
Bismuthates	**Critical temperature (K)**
$(Ba/K)BiO_3$	34
$(Ba/Rb)BiO_3$	29
Magnesium diboride	**Critical temperature (K)**
MgB_2	39

From Chapters 15, 16, 20, and 21 of this book.

7.2.2   Strontium ruthenates and ruthenocuprates

Following the discovery of high T_c in copper oxide superconductors in 1986, much effort was directed at the study of other transition metal oxides to check if they were superconducting. Maeno et al. (1994) found that strontium ruthenate Sr_2RuO_4, a layered perovskite with identical crystal structure to La-214, with the CuO_2 layer being replaced by RuO_2, was superconducting, albeit with a low $T_c = 1.5$ K. As with La-214, the conducting charge carriers in Sr_2RuO_4 originate from the hybridisation of d-orbitals (of Ru/Cu) with p-orbitals of O. Instead of having a nearly filled Cu $3d$-shell with only one hole state, we now have the formal valence of ruthenium as Ru^{4+}, leaving four electrons in its $4d$-shell.

In this chapter we have already mentioned many systems displaying interplay of low-temperature superconductivity with magnetism. Ruthenocuprates are hybrids of HTS cuprates and strontium ruthenate that show an interesting interplay of high-temperature superconductivity and magnetic order. Basically, there are two stoichiometric forms of ruthenocuprates, with general formulae $RuSr_2RCu_2O_8$ (i.e., Ru-1212) (Bauernfeind et al., 1995) and $RuSr_2(R_{1+x}Ce_{1-x})Cu_2O_{10}$ (i.e., Ru-1222) (Felner et al., 1997), with the rare earth R being Sm, Eu, or Gd. These materials exhibit weak ferromagnetism, due to RuO_2 layers, at about 150 K and high-temperature superconductivity, due to CuO_2–R–CuO_2 stacks, below 70 K.

7.2.3   Ferro-pnictide and related superconductors

Pnictides are compounds containing P, As, Sb, or Bi, while oxypnictides or pnictide oxides additionally contain oxygen. Such compounds formed with Fe, that is, ferro-pnictides or ferro-oxypnictides, are among the recently discovered superconductors of considerable importance. Some of the high-T_c pnictides are listed in Table.7.7. Surprisingly, despite the presence of Fe, which, owing to its ferromagnetism, is commonly regarded as deleterious for superconductivity, these materials possess noticeably high T_c values. The study of these compounds started with the announcement by Kimihara et al. (2006) of a new Fe-based superconductor, LaFeOP, possessing a T_c of 6 K, which was raised to 26 K in $LaFe(O_{1-x}F_x)As$ by the same group (Takahashi et al., 2008). By replacing the larger La ion with a smaller rare-earth ion such as Ce, Pr, Nd, Sm, Eu, or Gd, the critical temperature was further increased, reaching its highest value of 56 K for $Gd_{1-x}Th_xFeAsO$ (Wang et al., 2008)—a critical temperature that is to date second only to the high-T_c cuprate superconductors, which makes these new systems particularly important. Besides oxypnictides and pnictides of Fe, some of the ferro-chalcogens, such as FeSe and $FeTe_xS_x$, have also been found to be superconducting, with the result that one may broadly identify the following families of *Fe-based superconductors* (FBS) as existing today:

1. LnFeAs(O, F) or LnFeAsO$_{1-x}$, with T_c up to 56 K, referred to as 1111 materials.
2. (Ba, K)Fe$_2$As$_2$ and related materials with pairs of iron arsenide layers, referred to as 122 compounds, with T_c values ranging up to 38 K. These materials also superconduct when Fe is replaced with Co.
3. LiFeAs and NaFeAs, with T_c up to around 20 K. These materials superconduct close to stoichiometric composition and are referred to as 111 compounds.
4. FeSe with small off-stoichiometry or tellurium doping, with $T_c \approx$ 25 K.

The upper critical fields of these materials are around 50–100 T.

7.2.4 Superconducting bismuthates

High-T_c bismuthates were discovered soon after the advent of HTS cuprates in the late 1980s. The perovskites BaBiO$_3$ and BaPbO$_3$ were respectively known to be insulating and metallic, with the latter becoming superconducting when cooled below 0.5 K. Partial substitution of Bi for Pb in the latter, to give BaPb$_{1-x}$Bi$_x$O$_3$, enhanced T_c significantly to 13 K for $x = 0.25$ (Sleight et al., 1975). When Ba in the above perovskite was partially replaced by K to form Ba$_{1-x}$K$_x$BiO$_3$, commonly called BKBO, T_c showed (Table 7.7) a dramatic increase to >30 K for $x = 0.4$ (Cava et al., 1988a). Its crystal structure is the typical cubic perovskite. T_c is sensitive to x, and for $x > 0.4$ the material forms multiphases. Despite its high T_c, it is still a conventional phonon-mediated BCS superconductor. Interestingly, in BKBO, when the potassium content is low, charge-density waves (CDWs) are observed.

7.2.5 Superconducting intercalated graphite and doped fullerides

Intercalated graphite and doped fullerides are popularly included among the organic superconductors. We mentioned both of them earlier while describing the superconductivity of pure elemental carbon. Superconductivity induced in intercalated graphite has been known for a long time (Hannay et al., 1965; Belash et al., 1987a). More recent work has shown a considerably higher T_c of 37 K obtained by doping graphite with sulfur (da Silva et al., 2001). The occurrence of induced superconductivity in the intercalated graphite has also been examined theoretically (Csányi et al., 2005).

Fullerides were discovered in early 1991 (Kroto et al., 1985; Kratschmer et al., 1990). Thin films of fullerene C$_{60}$ molecules form a close-packed face-centred cubic (fcc) structure, but, in the absence of free charge carriers, the films are insulating. The work of Haddon et al. (1991) showed that doping them with alkali metals (Li, Na, K, Rb, or Cs) brought a significant improvement in their conductivity. Interestingly, some of these doped films with the composition M_3C_{60}, where $M = K$, Rb, or Cs, exhibit superconductivity in the temperature range covering LTS and HTS; that is, K_3C_{60} has $T_c = 18$ K (Hebard et al., 1991), Rb_3C_{60} has $T_c = 29$ K (Rosseinsky et al., 1991), Cs_3C_{60} has $T_c = 30$ K (Kelty et al., 1991) (and, under pressure, about 40 K), and $(Cs, Rb)_3C_{60}$ has $T_c = 33$ K (Tanigaki et al., 1991).

7.2.6 Magnesium diboride

The discovery of superconductivity at 39 K (Table 7.7) in MgB_2 (Nagamatsu et al. 2001) was a most unexpected event, recalling the excitement of the HTS cuprate era. The material had been known since the early 1950s, and it was almost unbelievable that during all these years its superconductivity, and at such a high temperature, had been completely missed. Surprisingly, despite this high T_c, the superconductivity of MgB_2 is of the conventional phonon-mediated s-wave type. The crystal structure comprises honeycomb or graphite-like layers of B with intervening hexagonal layers of Mg, stacked one above the other (Figure 7.13) along the c-axis. Interestingly, MgB_2 is the first multiband superconductor to be discovered. It has two distinct bands, with energy gaps Δ_σ and Δ_π associated respectively with the σ- and π-bands, with $\Delta_\sigma / \Delta_\pi = 2.7$. Its critical field varies from 15 T to >40 T, depending on its processing. MgB_2 is currently considered as a potential material to replace Nb–Ti and Nb_3Sn, which have been universally used for high-field superconducting electromagnets over the last four decades or more.

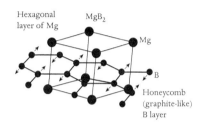

Figure 7.13 Structure of MgB_2.

7.2.7 Superconductivity induced through the proximity effect

Superconductivity, with low or high T_c, can be induced as a metastable or transient state through what is known as the *proximity effect*. This effect, first discovered by Holm and Meissner (1932) and subsequently extensively studied by de Gennes (1966), occurs when a superconductor (S) having a paired electron state is placed in good electrical contact with a normal non-superconducting material (N) in an unpaired electron state. At the boundary between the two materials, the paired state in S cannot change abruptly and infinitely rapidly to the unpaired state of N. Instead, the paired state from the superconductor side diffuses and is carried across the boundary into the normal material. This happens over a mesoscopic length scale, called the *Thouless range*, beyond which the pair correlation is destroyed. In very clean metals such as Cu or for electrically conducting amorphous thin films at low temperatures, this length can be a few hundreds of micrometres, which, in insulators, is reduced to nanometre size. As a result of the diffusion of superconducting pairs from S into N, the latter manifests signs of weak superconductivity. There is simultaneous diffusion of normal electrons from N to S, which causes a lowering of the superconducting gap in S in close proximity to the boundary. The suppression of T_c of S is maximal when N is a ferromagnet, as its internal magnetic field tends to weaken the electron pairing in S. This is commonly referred to as the *inverse proximity effect* (Bergeret et al., 2004; Tollis et al., 2005).

In recent years, there has been a considerable revival of interest in the proximity effect as a means to realise novel optoelectronic applications. The effect makes it possible to induce transient superconductivity practically in thin films of any material, resulting in unique combinations of diverse properties that do not otherwise normally exist. Proximity superconductivity has been reported in a variety of novel materials, including graphene (Heersche et al., 2007; Hayashi et al., 2010), DNA molecules (Yu et al., 2001, 2004, 2009), and various nanowires and nanobelts of Zn, Al, Au, Co, and Pb as described by Singh et al. (2012). Interestingly, high-temperature superconductivity with $T_c = 80$ K has been realised (Zareapour et al., 2012) in the topological insulators Bi_2Se_3 and Bi_2Te_3 by keeping them in proximity to HTS Bi-2212. These recent results hold considerable promise for future device applications.

7.3　Summary

A rich variety of materials today exhibit superconductivity of both LTS and HTS types, and the most prominent ones have briefly been summarised in this chapter. Wherever appropriate, we have tried to highlight their unusual or significant features from the view point of research and applications. Various families of LTS and HTS superconductors mentioned in this chapter form the mainstay of this book, with separate chapters providing more details. As the reader may have noted, although the more abundantly available LTS materials, possessing $T_c < 30$ K, in general follow the conventional BCS phonon mechanism, there are a number of low-T_c systems, such as heavy fermion superconductors, organic superconductors, and quaternary borocarbides, that are potential candidates for unconventional superconductivity. Equally surprising is the fact that there are superconductors such as BKBO, doped fullerides, and MgB_2 that possess $T_c > 30$ K and thus belong to the HTS class but whose superconductivity is still mainly of conventional type, albeit with some rather uncommon features. The HTS continue to provide a major challenge on the applications front, where, for the last 40 years, Nb–Ti and Nb_3Sn have been the front runners and unrivalled champions. New superconducting systems with novel properties are continually being discovered at intervals of every few years, and there appears every possibility that at any point of time an exceptionally different and exciting superconducting system with unexpected new traits or possessing a much higher T_c than known to date may yet emerge. Equally exciting is the revival of interest in the proximity effect for novel device applications of transient superconductivity.

CHAPTER 8

Niobium–zirconium and niobium–titanium alloys

Superconducting Nb–Zr and Nb–Ti alloys arrived on the scene during the 1960s. They were the harbingers of the excitement and revolution in magnet technology that we have seen reach its fruition today. Nb–Zr stands out as the first ductile high-field system to have been produced commercially in the form of long flexible wires in 1962. The potential of Nb–Ti had not yet been realised, and it was only after the mid-1960s that it was to supersede Nb–Zr. By then, some of the A-15 superconductors, such as Nb_3Sn, V_3Si, and V_3Ga, had revealed their higher T_c and H_{c2}, and Nb_3Sn in wire form had shown an appreciable J_c, but their inherent brittle nature had made mechanical handling difficult.

For a superconductor to become a viable choice for making high-field solenoids, it has to meet several constraints. A large H_{c2}, a substantial J_c in high magnetic fields, and a reasonably high value of T_c, of course, remain the prime requisites, and these three parameters are depicted for different superconductors in Figure 8.1. But clearly they alone do not suffice, and, in addition:

1. It should be both technically and economically feasible to produce several kilometres of continuous lengths of the material either in a flexible tape or in wire form.
2. The fabrication process must be compatible with creating and optimising flux-pinning centres.
3. Since superconducting properties have to remain nearly invariant over long lengths, chemical composition and microstructure should not vary significantly over the conductor length.
4. Finally, to stabilise and protect the magnet against the adverse effects of flux jumps, the conductor is required to be fabricated as a composite containing copper, as discussed in Chapter 6.

Except for the stability criterion 4, the above criteria were reasonably met by Nb–Zr when it was first introduced in the early 1960s. The stability criterion was, however, not known at that time and the Nb–Zr conductors produced for coil winding had not been stabilised. Their magnet performance was therefore seriously marred by flux

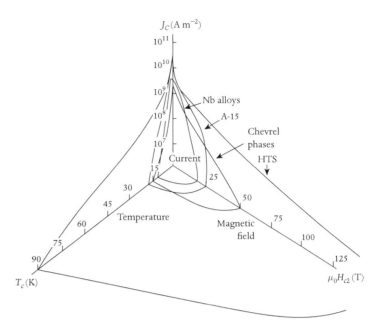

Figure 8.1 The three most important parameters, namely critical temperature, upper critical field, and critical current density, which are mutually dependent, are schematically displayed here in a 3D perspective for different superconducting systems.

jumps. Magnets would behave most erratically and would frequently quench unexpectedly at low currents, abruptly dissipating their large stored energy. Transient voltages and temperatures generated during quenching were often large enough to cause irreparable damage to the coil windings. The problems arising were sufficiently serious to jeopardise the future of superconducting magnet technology. The challenge thus posed was effectively met with the advent of fully stabilised Nb–Ti conductors. Today, we find stabilised Nb–Ti conductors universally employed in all superconductor magnet applications worldwide, and they form the base of a billion dollar superconductor wire industry. Before discussing the Nb–Ti system, for their obvious scientific interest we present below a short account of Nb–Zr superconductors, even though these are now largely of historical curiosity.

8.1 The niobium–zirconium system

Nb–Zr was first conceived by Kunzler (1961a, b) as a potential superconducting system for making high-field magnets. The crystal structure is disordered bcc with the lattice sites randomly occupied by both species of atoms. The significant advantage of Nb–Zr alloys is that they can be fabricated in a sufficiently flexible wire form to be wound into the smallest solenoid without difficulty. Further, their commendable tensile strength, exceeding 16 000 kg cm^{-2}, adds to their suitability for winding and handling.

Figure 8.2 depicts the Nb–Zr phase diagram (Rogers and Atkins, 1955). Nb and Zr form a complete series of bcc solid solutions

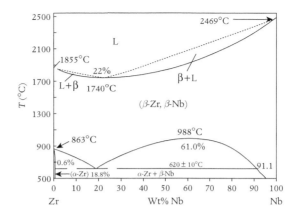

Figure 8.2 Phase diagram of the Nb–Zr system (after Rogers and Atkins, 1955).

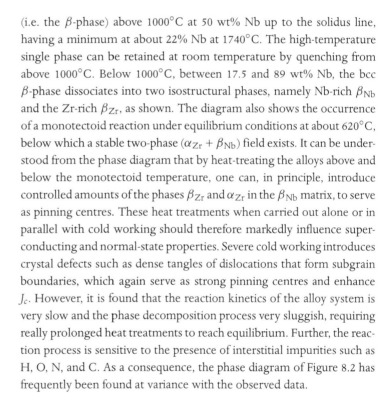

Figure 8.3 Compositional dependence of (a) normal-state resistivity and (b) critical temperature and upper critical field for Nb–Zr and Nb–Ti alloys (after Berlincourt and Hake, 1963).

(i.e. the β-phase) above 1000°C at 50 wt% Nb up to the solidus line, having a minimum at about 22% Nb at 1740°C. The high-temperature single phase can be retained at room temperature by quenching from above 1000°C. Below 1000°C, between 17.5 and 89 wt% Nb, the bcc β-phase dissociates into two isostructural phases, namely Nb-rich β_{Nb} and the Zr-rich β_{Zr}, as shown. The diagram also shows the occurrence of a monotectoid reaction under equilibrium conditions at about 620°C, below which a stable two-phase ($\alpha_{Zr} + \beta_{Nb}$) field exists. It can be understood from the phase diagram that by heat-treating the alloys above and below the monotectoid temperature, one can, in principle, introduce controlled amounts of the phases β_{Zr} and α_{Zr} in the β_{Nb} matrix, to serve as pinning centres. These heat treatments when carried out alone or in parallel with cold working should therefore markedly influence superconducting and normal-state properties. Severe cold working introduces crystal defects such as dense tangles of dislocations that form subgrain boundaries, which again serve as strong pinning centres and enhance J_c. However, it is found that the reaction kinetics of the alloy system is very slow and the phase decomposition process very sluggish, requiring really prolonged heat treatments to reach equilibrium. Further, the reaction process is sensitive to the presence of interstitial impurities such as H, O, N, and C. As a consequence, the phase diagram of Figure 8.2 has frequently been found at variance with the observed data.

8.1.1 General features

Figure 8.3(a) shows the compositional dependence of the normal-state resistivity ρ_n (just above T_c), while (b) depicts T_c and H_{c2} data for Nb–Zr and Nb–Ti alloys as functions of Zr and Ti content (Berlincourt and Hake, 1963). It is interesting that, up to 60% of alloying, the resistivity behaviour of the two is not very different. In the case of Nb–Zr alloys, both T_c and H_{c2} display peaks corresponding to around 20 and 50 wt% of Zr, respectively. The critical current density, which is controlled by the microstructure, is found to be optimum for 25–35 wt% Zr, which represents the composition range of conductors used for coil windings. The starting material is often in the form of a rod, vacuum-annealed for several hours at 1200°C and then quenched to retain the single β-solid-solution phase. It is sufficiently ductile and strong to be swaged and cold-drawn down to the lowest final wire diameter required for the coil winding. The cold drawing alone markedly enhances the J_c of the starting material. For instance, 92% reduction increases J_c by a factor of 60–100. Even further J_c enhancement to a value exceeding 10^9 A m^{-2} at 5 T and 4.2 K results from intermediate heat treatment at 500–700°C for a few hours, followed by further cold drawing (Hake et al., 1962). To achieve cryostatic stability, wire has been electroplated with copper, although co-processing the material with copper has also been tried with limited success. Typically, around 250 μm Nb–Zr core wire carried a Cu sheathing of about 65 μm (Milne and Ward, 1972). To prevent interlayer diffusion between the alloy and the Cu sheath during the heat treatment, an extra 5 μm thick Nb layer was added between them by inserting a suitably thick Nb cylinder at the time of co-processing.

Despite strong flux pinning and the high J_c of Nb–Zr alloys, the nature of the pinning centres resulting from heat treatment has surprisingly remained a matter of debate (Waldron, 1969; Milne, 1972; Milne and Ward, 1972). Transmission electron microscope studies (Milne, 1972) of cold-drawn wires of Nb–25%Zr, showing significantly enhanced J_c, revealed the characteristic dense tangles of crystal dislocations known to cause strong flux pinning in Nb alloys (Narlikar and Dew-Hughes, 1964, 1966). After heat treatment for several hours at 500–700°C, which causes a further rise in J_c, the following structural features can develop: (i) polygonisation of dislocation structure; (ii) precipitation of β_{Nb}; (iii) precipitation of α_{Zr}. The precipitates in (ii) and (iii) were found to be quantitatively too small to provide strong pinning. Moreover, the precipitates of β_{Nb}, being weakly superconducting, as discussed in Chapter 6, should have yielded a characteristic *peak effect*, which has not been observed. On the other hand, these heat treatments do give rise to polygonisation (Milne, 1972) and rearrangement of dislocations, which are known to enhance pinning in Nb alloys (Narlikar and Dew-Hughes, 1966). The presence of small quantities of β_{Nb} and α_{Zr} should, however, further help in making the dislocation structure denser during post-annealing cold drawing, which promotes J_c.

Figure 8.4 depicts typical J_c/H data for Nb–Zr alloys (Wong, 1962). For the sake of comparison, the curves for pure Nb and 10 wt%Zr, which possess much lower critical fields, are also shown. Nb–33 wt%Zr has a higher upper critical field and critical current density than 25%Zr alloy, but its conductor processing is mechanically harder. Until the late 1960s, when these alloys were finally replaced by Nb–Ti, they were extensively used in commercial superconducting magnets for producing magnetic

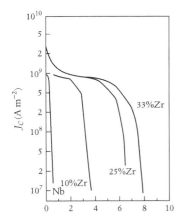

Figure 8.4 $J_c(H)$ performance of Nb–Zr samples in comparison with commercial Nb. The 10% Zr alloy has a much lower critical field for applications (after Wong, 1962). Both 25% and 33% alloys were used for magnets in the 1960s, although the latter was more difficult to work with.

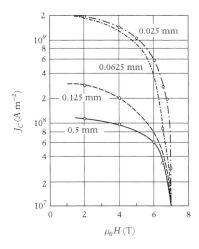

Figure 8.5 The low diameters of very fine wires of Nb–25%Zr, produced by heavy cold working, result in a raised J_c, which finally becomes saturated (after Olsen et al., 1962).

fields in the range of 4–6.5 T at 4.2 K and 5–8 T at 2.0 K. Because the conductors were not adequately stabilised, to save the magnets from quenching, they had to be energised slowly and kept away from vibrations and mechanical shocks during operation. Although fabrication of fine multifilamentary composite wires was problematic, ultrafine wires in small lengths, produced by large mechanical reductions, had exhibited commendable J_c values with decreasing diameter (Olsen et al., 1962), as shown in Figure 8.5. The increase in J_c is ascribed to a very high dislocation density resulting from heavy cold working, but beyond a certain level of reduction, both become saturated.

Nb–Zr alloys suffered from inherent drawbacks. They posed problems with regard to the use of high-conductivity sheaths of Cu, Al, Cu–Ni, etc. for their stabilisation. The quality of the bond between the normal metals and Nb–Zr is an important factor for the stable performance of the wire in a coil. With Nb–Zr, the co-processing approach was only partially successful. A much better bond could be obtained with electroplating, but this was expensive. Another problem with Nb–Zr was that the presence of zirconium hydride in the as-cast structure made the alloy harder and difficult to work with. The Nb–Ti alloys were found to have higher critical fields and current densities and at the same time possessed better ductility than Nb–Zr alloys. By simple co-extrusion with Cu or Cu–Ni, it was possible to get cladding with the good electrical and thermal bond needed for stability and protection. It became a relatively easy matter to form Nb–Ti alloys as multifilamentary composites in a Cu or Cu–Ni matrix for achieving adiabatic stability (Chapter 6). Because of these advantages, Nb–Zr alloys were soon replaced by Nb–Ti alloys.

8.2 The niobium–titanium system

8.2.1 Phase configurations

Like Nb–Zr, the alloys of Nb–Ti are disordered bcc (A-2 structure) with the lattice sites randomly occupied by the two elements. The equilibrium phase diagram of the system is shown in the main part of Figure 8.6 (Hansen, 1958; Brown et al., 1964), while the Ti-rich side of the diagram is depicted in the inset. Nb and Ti form single-phase bcc solid solutions, that is, of the β-phase, above 885°C (for 20 at% or 32 wt% Nb alloy, this temperature is about 525°C), below which the hexagonal close-packed (hcp) α_{Ti} phase can result. The equilibrium phases at room temperature are β and α, but the reaction takes a very long time to complete. The normal precipitates of α_{Ti} in the superconducting β-phase matrix are the mainstay of strong flux pinning in Nb–Ti systems (Pfeiffer and Hillmann, 1968; Larbalestier and West, 1984). Besides the two equilibrium phases, the system has three metastable martensitic phases, α', α'',

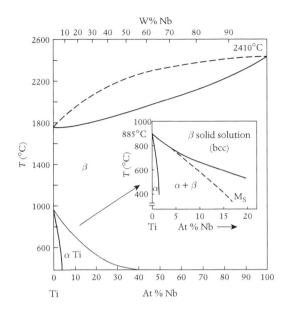

W% Nb

Figure 8.6 Phase diagram of the Nb–Ti system. The inset depicts an enlarged view of the Ti-rich side of the diagram. (After Brown et al. (1964).)

and ω, but since these are metastable, they do not appear in Figure 8.6. Of these phases, α' is confined only up to 13 wt% Nb and is not of practical interest for superconductivity, while the range of α'' extends from 13 to about 47 wt% Nb. On quenching, the α-phase transforms martensitically (indicated by the dashed line M_S in the figure) to a highly strained metastable α''-phase having orthorhombic structure, which, under electron microscopy, shows up as heavily micro-twinned lamellar plates. The phase is, however, unstable and with ageing at about 300°C gives way to the formation of the ω-phase (hcp) (Hatt and Rivlin, 1968), which, as we will see, is weakly superconducting and has interesting implications for superconducting behaviour (Baker and Sutton, 1969). At low temperatures, β-to-α'' transformation can take place under stress and give rise to microstructural instability. The stress-induced martensitic transformation in Nb–47 wt%Ti alloy is considered important (Obst et al., 1980) since it can lead to the phenomenon of *training* in which a superconducting solenoid is found to sustain higher currents after repeated normalisations. A Nb–Ti conductor, when wound into a coil, experiences magneto-mechanical forces that may result in microstructural instability mentioned previously. The ω-phase and the fine plates of ω formed intragranularly as *Widmanstätten* α_{Ti} tend to make the material mechanically very hard for wire drawing and are to be avoided. This can be achieved by heavy deformation and large prestrain of the conductor before the precipitation reaction (Buckett and Larbalestier, 1987).

8.2.2 General features and materials processing

Figure 8.3(b) shows the compositional dependence of T_c and H_{c2} for the alloys of the Nb–Ti system. As may be noted, the optimum T_c of about

9.8 K corresponds to the presence of 24 wt% Ti, while the peak $\mu_0 H_{c2}$ of 12 T at 4.2 K is realised for about 45 wt% Ti, which rises to 14.5 T at 1.8 K.

For practical applications, the most popular composition range for making Nb–Ti conductors is one containing 45–50 wt% Ti, which has its T_c slightly lower than the optimum. The advantage with this composition is that it yields the arrays of α-precipitates needed for flux pinning. With the above composition, magnetic fields up to 9 T can be readily produced at 4.2 K. This limit is raised to 13 T at 2 K by ternary additions, partly replacing Nb by Ta, Zr, or Hf (Table 8.1).

Alloys containing 45–50 wt% Ti are sufficiently ductile and, with heavy deformation resulting from wire drawing, they develop a fine-grained structure with high-angle grain boundaries along with a dense cellular dislocation structure (i.e., low-angle or subgrain boundaries). Consequently, besides arrays of α_{Ti}, both of these are potential pinning entities giving rise to additional pinning. The heat treatment for α_{Ti} precipitation is generally carried out between 370 and 420°C for several hours. A single heat treatment in general is inadequate to realise a sufficient volume fraction of the precipitates for optimum flux pinning, and therefore the cold drawing and annealing operations are carried out in repeated succession. The total prestrain in the material existing before annealing is found to be important for further precipitation of α_{Ti} (Lee et al., 1989). In practice, doing this 3–6 times helps in building up an approximately 20% volume fraction of α_{Ti} particles in Nb–47 wt%Ti, which can promote a J_c in excess of 3×10^9 A m^{-2} at 5 T and 4.2 K. Interestingly, despite the presence of the 20% volume fraction of α_{Ti}, the material continues to remain sufficiently ductile for wire drawing.

Table 8.1 Upper critical field of selected multielement Nb–Ti alloys

Alloy (wt%)	$\mu_0 H_{c2}$ (T) at 4.2 K	$\mu_0 H_{c2}$ (T) at 2.2K
Nb–30Ti	11.4	
Nb–40Ti	11.7	
Nb–47Ti	11.5	14.2
Nb–49Ti	12.2	
Nb–22Ti–22Zr	13.6	
Nb–45Ti–16Zr	12.4	
Nb–43Ti–25Zr	11.6	15.5
Nb–41Ti–13Hf	11.0	14.5
Nb–47Ti–20Ta	13.8	15.8
Nb–38Ti–26Ta–6Hf	11.3	15.3
Nb–39Ti–24Ta–6Zr	13.1	

From Glowacki, B. A. *Frontiers in Superconducting Materials* (ed. A. V. Narlikar). Springer, Berlin (2005).

The precipitation reaction causes a local depletion of Ti from the adjoining β-phase matrix and creation of concentration gradients surrounding α_{Ti} particles. The presence of concentration gradients can enhance the pinning due to α_{Ti} (Bormio-Nunes et al., 2007). In order to enhance the volume fraction of the pinning entities beyond 20%, Heussner et al. (1997) incorporated artificial pinning centres (APCs) in the form of Nb particles and showed that J_c could be raised significantly further. This approach of adding APCs, as we will find in later chapters, has proved particularly popular to increase the J_c of HTS cuprates and MgB_2.

Because of their better ductility, alloys richer in Nb, with compositions $\leq$45 wt% Ti, have also been used for Nb–Ti conductors. Below 40 wt% Ti, the β-phase is free from α_{Ti} and the prominent pinning centres are the densely formed cellular dislocation structures. By heating the cold-drawn Nb–Ti for a few hours or even less (depending on the temperature) at 350–550°C, precipitation of dissolved interstitial oxygen or nitrogen, in the form of fine particles or domains of niobium or titanium oxide or nitride, may occur, providing additional pinning centres for vortex lines (Critchlow et al., 1971; Witcomb and Narlikar, 1972). The significant effect of heat treatment on J_c of cold-drawn Nb–45 wt%Ti wires (Critchlow et al., 1971) is shown in Figure 8.7. Wires of lower diameter produced by heavy cold working have a higher dislocation density with a densely formed cell structure that contributes to J_c (Hampshire and Taylor, 1972). The heat treatment may simultaneously cause dislocation rearrangement and a refinement (Figure 8.8) of the cellular dislocation structures (Witcomb and Narlikar, 1972) and thereby further enhance flux pinning.

A particularly interesting situation results from the formation of ω-precipitates in Ti-rich 32 wt% Nb alloy, discussed by Baker and Sutton (1969). As mentioned earlier, this phase results after a short-time ageing of the martensitic α'' phase formed by quenching the β-phase. Also,

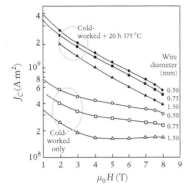

Figure 8.7. Effect of heat treatment in raising the critical-current performance of Nb–45 wt%Ti samples with different wire diameters produced by cold working (after Critchlow et al., 1971).

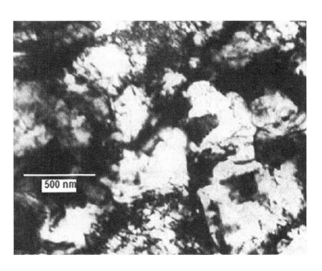

Figure 8.8 Densely formed dislocation cell structure in heavily cold-worked Nb–40 at%Ti, heat-treated at 500°C for 30 minutes, as revealed by TEM, showing a refined dislocation cell structure (after Witcomb and Narlikar, 1972).

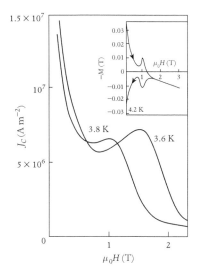

Figure 8.9 Peak effects in the critical current (main figure) and in the magnetisation behaviour (inset) of 32 wt% Nb alloy, aged for 2 minutes at 330°C, containing weakly superconducting ω-phase (after Baker and Sutton, 1969).

Brammer and Rhodes (1967) had found that, in Ti–34 wt% Nb, a diffuse ω-phase forms if the quenching of the β-phase is slow. The transformation of the α'' phase takes place with its partial reversal to the β-phase along with the formation of ω and β_e-phases, that is, $\alpha'' \rightarrow \beta + \omega + \beta_e$. The ω-phase is a Nb-deficient phase and, to compensate for that, it is encased by a shell of superconducting Nb, which is designated as the β_e-phase. The fine particles of this composite $\omega + \beta_e$ phase that are present in the β-phase matrix serve as weakly superconducting pinning entities and exhibit enhanced pinning only after they turn normal at higher temperature and magnetic field, as long as the matrix remains superconducting. As a result, the material exhibits unusual peak effects in its magnetisation and critical current behaviour, where, over a limited range of magnetic field and temperature, the critical current density and magnetisation exhibit increases with rising temperature and magnetic field. This uncommon behaviour is illustrated in Figure 8.9.

8.2.3 Pinning interactions

As described in Chapter 6, the pinning mechanisms comprise core pinning and magnetic pinning, the former arising from a variation in the thermodynamic critical field H_c and the latter from a variation in the Ginzburg–Landau parameter κ. As we have seen, Nb–Ti is a particularly interesting system to explore pinning mechanisms, since it offers a rich variety of microstructures that include different kinds of precipitates and phases such as α, β, α', α'', and ω. These alloys also contain various types of intercrystalline boundaries, such as high- and low-angle grain boundaries, dense dislocation cell structures, and interstitially ordered domains, all of which constitute important pinning centres. Interestingly, some of these potential pinning centres may be simultaneously present in the same sample. The elementary pinning interactions of these defects in Nb–Ti alloys have been extensively discussed by Dew-Hughes (1971, 1974). Flux pinning in 48% Ti alloys showing exceptionally high J_c was examined by Meingast and Larbalestier (1989). Their studies showed that in the situation of optimum pinning, a vortex line interacted with clusters of precipitates, instead of individual precipitates, and there were two distinct pinning mechanisms operating, which had different field and temperature dependences. In the low-field region near T_c, core pinning was dominant, but in the high-field region at low temperatures, the pinning that occurred resulted from a variation in κ, that is, $\Delta\kappa$ pinning (see Chapter 6). For lower-prestrained samples with lower J_c, instead of two separate pinning mechanisms, there were two types of pinning centres, namely α-precipitates and grain boundaries, which were responsible for different field and temperature dependences of the observed pinning. Figure 8.10 displays the high-$J_c(H)$

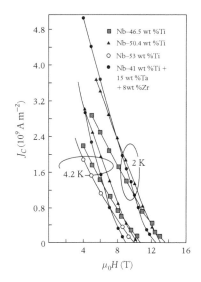

Figure 8.10 High-$J_c(H)$ performance exhibited by commercial composite wires of Nb–Ti with slightly varying composition containing α-Ti particles (after Larbalestier, 1981).

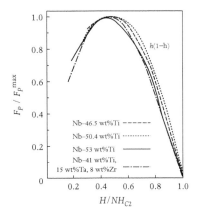

Figure 8.11 Reduced pinning force as a function of reduced magnetic field for the commercial composite wires of Nb–Ti of Figure 8.10 (after Larbalestier, 1981). All the curves closely follow the $h(1-h)$ scaling of the global pinning force exerted by normal particles.

behaviour of some commercial Nb–Ti wires of slightly varying compositions, in which the pinning is primarily due to normal α-Ti precipitates (Larbalestier, 1981).

In Nb–Ti superconductors with $\Delta\kappa$ pinning, the corresponding pinning function is conveniently described by $h(1 - h)$, where $h = H/H_{c2}$ is the reduced field (Dew-Hughes, 1974). As described earlier, the heavily deformed structure of Nb–Ti filaments comprises grains, subgrains (i.e., dislocation cell structure) and non-superconducting α-Ti precipitates, all elongated in the direction of drawing, and the Lorentz force acts on the flux vortices so as to drive them across the subgrains and normal to particle boundaries. The theory and experimental data are in good mutual accord (Dew-Hughes, 1987). The above scaling seems to hold in a situation where the density of pins is less than the density of flux lines. In the above expression for the pinning function, the h term arises because the increase in the density of flux lines is accompanied by an increase in the total length of the line pinned. The second term $(1 - h)$ manifests a decrease in the superconducting order parameter with increasing field. Figure 8.11 is an illustration of $h(1 - h)$ scaling exhibited by the commercial composite samples of Nb–Ti of Figure 8.10 (Larbalestier, 1981) containing α-Ti precipitates, and the observed scaling is in accord with the expected behaviour (Campbell and Evetts, 1972) of flux pinning due to normal particles.

8.2.4 Conductor fabrication

The finer details and tricks followed in the industrial fabrication of Nb–Ti conductors are guarded by commercial secrecy, but the essential steps may be briefly outlined. The alloys are fabricated by continuous melting in an electron-beam furnace or an arc furnace with consumable electrodes, at a residual pressure of 10^{-5} Torr. Since the electrode used with the electron-beam furnace (Figure 8.12(a)) does not carry any electrical current, it may be formed simply by stacking together Nb and Ti

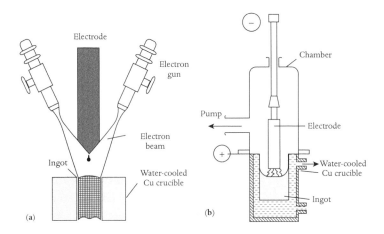

Figure 8.12 Commonly used techniques for melting Nb–Ti alloys: (a) continuous melting using an electron-beam furnace; (b) arc melting.

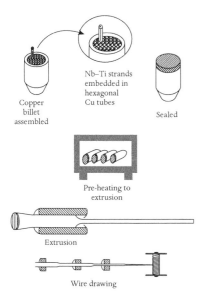

Copper
billet
assembled

Nb–Ti strands
embedded in
hexagonal
Cu tubes

Sealed

Pre-heating to
extrusion

Extrusion

Wire drawing

Figure 8.13 Prominent steps in the fabrication of multifilamentarey Nb–Ti wire.

rods in the required proportion without any welding process. In the case of an arc furnace (Figure 8.12(b)), since the electrode has to carry a large melting current, welding is necessary. Composition variations caused by evaporation of Ti in vacuum and the presence of undissolved Nb particles are common flaws of the melting process, but can be overcome by repeated melting of the ingot with extra Ti. An ingot typically weighs 5–50 kg. The ingots are machined into rods and, after cleaning, they are tightly inserted straight into hexagonally shaped tubes of oxygen-free high-conductivity (OFHC) copper with their inner surfaces precleaned. These tubes are then tightly packed inside a copper can for extrusion (Figure 8.13). In order to increase the copper-to-superconductor ratio, extra copper rods may be inserted into the can, which is subsequently sealed under vacuum using an electron-beam welder. The extrusion is carried out (Strauss et al., 1977) in the temperature range 500–650°C. Temperatures in excess of 650°C are avoided to prevent the formation of the intermetallic $(NbTi)_2Cu$ between the Nb–Ti and Cu. The presence of this brittle phase causes breaking of the filaments during extrusion. If the temperature drops below 500°C, the precipitation of α-Ti makes the material hard and at the same time makes the bonding between Cu and Nb–Ti poor. After the initial extrusion, the conductors are cold-drawn to the final size. As discussed earlier, the optimal heat treatment involves annealing at about 400°C for a few hours, which is carried out in repeated succession with the cold-drawing operation. Twisting (Chapter 6) of the filaments to reduce coupling losses is performed just before the final draw. In practice, it is common to have a Cu:Nb–Ti ratio of 1:1 or 2:1 and typically a wire of diameter of 0.5–2 mm carrying 50–5000 Nb–Ti filaments of diameter ranging from 5 to 50 μmr. If required, such wires can be bunched together in a specific conductor configuration. The design and architecture of the conductor is dictated by the application. In situations that demand lighter conductors or a greater transparency to nuclear radiation, instead of Cu, high-purity Al is considered for conductor stabilisation. This has, however, proved problematic, since co-processing the alloy with aluminium fails to yield a good metallurgical bond between them, leaving the stabilisation unsatisfactory.

8.3 Summary

This chapter has focused on the structure, properties, and conductor development of Nb–Ti alloys, which today stand out as the front runners in the superconductor industry. Substantial credit for the present-day success of the Nb–Ti system goes to the early experience gained with Nb–Zr, which was the first high-field superconductor to be commercially developed into long flexible wires for winding high-field superconducting magnets. The initial studies made on Nb–Zr coils proved invaluable in that they led to various novel approaches for achieving conductor stability. Later, these were successfully exploited in making the stabilised Nb–Ti conductors that are commercially available today. The importance of the materials processing and the influence of microstructure in realising strong flux pinning and high current densities have been discussed.

CHAPTER 9

A-15 superconductors

Until the arrival of the HTS cuprates in 1986 (Bednorz and Müller, 1986), the superconductor with the highest known T_c of 23.2 K was Nb_3Ge (Testardi et al., 1974), belonging to the A-15 structural class. During the period from 1950 to 1970, several A-15 superconductors were discovered, with T_c in the range of 15–22 K. The materials of this family, with the chemical formula A_3B, are also referred to as beta-tungsten (β-W) compounds. This term was coined to describe the structure of what was erroneously thought of as a new form of pure elemental tungsten, discovered in 1931 (Hartmann et al., 1931), which was, however, more correctly identified later (Morocom et al., 1974) as an oxygen-deficient tungsten oxide W_3O_δ, with the subscript δ indicating the oxygen deficiency. The A-15 superconductors, especially those formed with Nb or V as A-element and Al or Ga (of group IIIA) or Ge or Sn (of group IVA) as the B-element, are noted among LTS for their particularly high T_c (15–23 K) and a large upper critical field of 20–40 T (Table 7.1). Furthermore, as they possess a large current-carrying capacity, these A-15s are technologically important for producing much larger magnetic fields beyond the reach of the Nb–Ti alloys discussed in Chapter 8. They are generally referred to as the second generation of LTS conductors, with Nb–Ti alloys being the first generation. However, A-15s have not replaced Nb–Ti conductors in superconducting technology, and, for magnetic fields up to 10 T, the latter continue to be economically the most viable and, in fact, the only practical choice.

Nb$_3$Sn wires are used as inner windings of Nb–Ti magnets to supplement their field strength. Indeed, Nb_3Sn conductors are being routinely used for producing magnetic fields from 10 to 21 T in small laboratory magnets, and have been employed in such large installations as tokamaks in fusion research and high-energy particle accelerators. They find use in the high-field gigahertz class of nuclear magnetic resonance (NMR) machines. Furthermore, the thermodynamic critical field of some A-15s is significantly large, at around 0.5–1.0 T, which makes them potential candidates for AC applications such as AC power transmission cables, pulsed synchrotrons, and the resonance cavities used in high-energy linear accelerators. It should be noted that all the A-15s without exception are very hard and brittle substances, and novel approaches are required to fabricate them as wires or tapes.

For extended discussions and reviews on the basic aspects of A-15s, the reader is referred to Weger and Goldberg (1973), Testardi (1973, 1975), and Müller (1980).

9.1   Crystal structure, stoichiometry, and ordering

A-15 materials are intermetallic compounds, and many of them, when they are fully stoichiometric, possess valence electron-to-atom (e/a) ratios of about 4.75 and 6.5, as per Matthias' rules (Chapter 7). These e/a values, in general, also favour a higher coefficient of electronic specific heat γ and the corresponding peaks in the density of states (DOS) $N(0)$, which are instrumental to their enhancement of T_c. In many systems, the ideal stoichiometry is rarely reached and the observed T_c values are suboptimal. Figure 9.1(a) gives the crystal structure of the binary A-15 phase, where the A-atoms are located in pairs on the cube faces, along the three edges, of the bcc lattice formed by the B-atoms. The structure is routinely described as of Cr_3Si type with the cubic space group $Pm3n$ and the chemical formula A_3B. The A-atom is always a transition metal, while B may or not be a transition metal, and their ionic radii should be at least within 15% of each other. The separation of the neighbouring A-atoms on the cube faces represents the shortest distance $a/2$, a being the lattice parameter. The crystal structure is therefore characterised by the presence of three sets of densely packed orthogonal chains of transition metal atoms (Figure 9.1(b)). These chains are a unique feature of the ordered A-15 lattice and make the material *quasi-one-dimensional* with unusual characteristics.

Table 9.1 provides a list of superconducting binary A-15 phases. The compounds where both A and B are transition metals are shown in italics. Some of the listed phases are formed only in the metastable state produced by rapid quenching, under pressure, or by thin-film processing.

Unlike the disordered Nb–Ti and Nb–Zr alloys, which are ductile, the A-15s, being ordered intermetallic compounds, are intrinsically hard and brittle, requiring special fabrication routes to produce them in wire and tape forms for applications.

By 'ordering' it is meant that the right species of atoms occupy the right lattice sites. Consider an arbitrary phase composition $A_{1-X}B_X$ that in its ordered state has A- and B-atoms respectively occupying α- and β-sites. The fractions of α- and β-sites are respectively represented by M_α and M_β, such that $M_\alpha = 1 - M_\beta$ and for the stoichiometric composition, $1 - X = M_\alpha$ and $X = M_\beta$. If R_α and R_β are the fractions of α- and β-sites rightly occupied respectively by A- and B-atoms, and

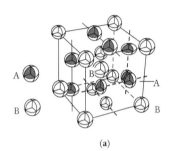

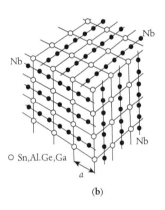

Figure 9.1 (a) A-15 crystal structure of the binary A_3B compounds. The B-atoms form a bcc structure, while the A-atoms are located in pairs on the cube faces, forming chains parallel to three cube edges, as depicted in (b).

Table 9.1 Critical temperatures and lattice parameters of binary A-15 compounds

A-15 phase*	T_c (K)	Lattice parameter a (nm)
Cr_3Ir	0.17	0.4678
Cr_3Os	4.03	0.4699
Cr_3Rh	0.07	0.4673
Cr_3Ru	3.4	0.4683
Mo_3Al	0.58	0.4950
Mo_3Ga	0.76	0.4943
Mo_3Ge	1.8	0.4933
Mo_3Ir	8.5	0.4968
Mo_3Os	13.1	0.4969
Mo_3Pt	4.6	0.4989
Mo_3Re[c]	15	0.4980
Mo_3Si	1.7	0.4888
Mo_3Tc	13.4	0.4934
Nb_3Al	19.1	0.5187
Nb_3Au	11.5	0.5200
Nb_3Bi[b]	3.0	0.5320
Nb_3Ga	20.7	0.5171
Nb_3Ge[c]	23.2	0.5139
Nb_3In[b]	9.2	0.5303
Nb_3Ir	3.2	0.5135
Nb_3Os	1.0	0.5131
Nb_3Pt	11.0	0.5155
Nb_3Rh	2.6	0.5137
Nb_3Sb	2.0	0.5262
Nb_3Si [a]	4.4	—
Nb_3Sn	18.5	0.5289
Ta_3Au	0.55	0.5224
Ta_3Ge[c]	8.0	—
Ta_3Sn	8.3	0.5276
Ta_3Sb	0.7	0.5259
Ti_3Ir	4.6	0.5009
Ti_3Pt	0.5	0.5033
Ti_3Sb	6.5	0.5222

(Continued)

Table 9.1 *(Continued)*

A-15 phase*	T_c (K)	Lattice parameter a (nm)
$V_3Al^{(c)}$	14	0.4830
V_3As	0.2	0.4874
V_3Au	3.2	0.4876
V_3Ga	15.9	0.4819
V_3Ge	6.0	0.4760
V_3Ir	1.7	0.4785
V_3Pd	0.08	0.4828
V_3Pt	3.7	0.4814
V_3Re	8.4	—
V_3Rh	1.0	0.4779
V_3Sb	0.8	0.4941
V_3Si	17.0	0.4722
V_3Sn	3.8	0.4984
$W_3Re^{(c)}$	11.4	0.5018
W_3Si	1.2	0.4910
Zr_3Au	0.9	0.5482
$Zr_3Bi^{(b)}$	3.4	—
Zr_3Pb	0.76	0.5656
$Zr_3Sn^{(a)}$	0.92	0.5650

*Compounds where both elements are transition metals are shown in italics. The super-scripts appearing after some of the materials indicate that they were synthesised by (a) rapid quenching, (b) high-pressure synthesis, or (c) thin-film processing.

Data collected from Flukiger, R. *Concise Encyclopedia of Magnetic and Superconducting Materials* (ed. J. E. Evetts). Pergamon, Oxford (1992), p. 1 and Narlikar, A. V. and Ekbote, S. N. *Superconductivity and Superconducting Materials*, South Asian Publishers, Delhi (1983).

W_α and W_β the fractions wrongly occupied by B and A atoms, then $R_\alpha + W_\alpha = R_\beta + W_\beta = 1$. The degree of structural order is expressed by the Bragg–Williams *long-range order* (LRO) parameter S, defined by

$$S_A = R_\alpha - W_\beta = \frac{R_\alpha - (1 - X)}{M_\beta},$$

$$S_B = R_\beta - W_\alpha = \frac{R_\beta - X}{1 - M_\beta},$$

where S_A and S_B are the order parameters for A (α-site) and B (β-site) atoms, respectively. With the A-15 unit cell, there are six α-sites and two β-sites, which give $M_\alpha = 0.75$ and $M_\beta = 0.25$. For the stoichiometric composition, the fully ordered state corresponds to $S_A = S_B = S = 1$, and for a completely disordered material, $S = 0$. In X-ray and neutron diffraction studies, the long-range order is manifested by the appearance of superstructure lines. The structure factor for these lines, determined by their intensity, is proportional to the order parameter. The measured value of S is a pointer to the success achieved with the experimental conditions in realising the ordered A-15 phase.

9.2 Long-range order and T_c

There is extensive evidence that the critical temperature of A-15 superconductors increases with long-range order (Blaugher et al., 1969). This is true in general for all the superconducting intermetallic compounds, such as Laves phases, Chevrel phases, quaternary borocarbides, and magnesium diboride. A-15 compounds have, however, received the most attention for their ordering behaviour, mainly because of their large T_c. Any disorder or adulteration that interferes with the integrity of A-chains has the greatest impact in degrading T_c. This can happen when an A-atom is substituted by a vacancy, a B-atom, or an impurity atom. One of the most prominent causes of disorder is off-stoichiometry, which results in excess (or depleted) concentration of A- or B-atoms and gives rise to anti-site defects, which lower both S and T_c. The stoichiometry and the degree of LRO achieved depend upon several factors, including the characteristics of the material system, particularly its equilibrium phase diagram, the A-15 phase fields, the presence of competing phases, and the fabrication process followed (Müller, 1980; Flükiger, 1981).

Greater disorder with greater T_c depression may occur on irradiating the pristine compound with high-energy particles. Extensive work by Sweedler et al. (1974, 1979) on several A-15s, including Nb_3Ge, Nb_3Sn, Nb_3Al, and V_3Si, showed that T_c was rapidly degraded by a 1 MeV neutron irradiation dose $>10^{18}$ n m^{-2} (Figure 9.2), but the original value was recovered after annealing. Interestingly, the data points for different compounds follow a single curve in Figure 9.2, which calls for some caution in using A-15 superconductors in fusion reactors. In sharp contrast to these results, heavy-ion irradiation enhanced the T_c of Mo_3Ge and Mo_3Si from the pristine value of 1.5 K to >6 K (Lehmann et al., 1981). This is attributed to superconductivity of amorphous Mo (see Chapter 7) produced by irradiation.

When the B-atom of the A-15 structure is also a transition metal (the materials depicted in italics in Table 9.1), the T_c is generally lower and becomes less sensitive to LRO. Cr- and Mo-based compounds formed with Os and Ir as B-atoms are examples of this type, known as *atypical* materials. However, in the case of the V–Au system, an increase in the

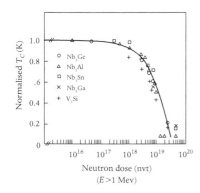

Figure 9.2 T_c normalised with respect to the corresponding value prior to neutron irradiation ($E > 1$ meV), as a function of irradiation dose for different A-15 superconductors (after Sweedler et al., 1974). The points lie on a common curve showing a rapid drop in T_c beyond a dose of 10^{18} nvt.

order parameter S from 0.8 to 0.99 dramatically enhances the T_c of V_3Au by over 400% from 0.7 K to 3.2 K (Hein et al., 1971).

Müller et al. (1972) addressed the basic question whether the T_c decrease resulting from lessening of LRO was a phononic process or an electronic one. Their heat capacity data showed that the phonons may contribute only 5% to the change of T_c, with the dominant effect being due to a change in the electronic DOS $N(0)$. Weger (1964) had previously pointed out that the densely packed chains of the transitional atoms made the d-band particularly narrow, with a large increase in the DOS $N(0)$ at E_F, which promoted a higher T_c. Later, self-consistent augmented plane-wave (APW) calculations of Klein et al. (1978) indeed revealed a sharp structure in the total DOS at E_F for V_3Ga, V_3Si, and Nb_3Sn, as shown in Figure 9.3 for Nb_3Sn. Any disruption in linear transition metal chains produced a broadening of the d-band at E_F and a decrease in $N(0)$ and T_c. Similar adverse effects followed if the A-atoms occupied B-sites or if the B-atoms were also transition metals: in these cases, there is overlap between the d-band of the atoms at the B-sites with the d-band of the chains, resulting in broadening of the latter and again a decrease in $N(0)$ and T_c.

The effect of any significant disorder on the normal-state resistivity is to enhance it. However, its positive impact (Section 5.3.2) on enhancing H_{c2} is generally outweighed by a more drastic decrease in T_c due to disorder, and therefore, the overall effect of disorder is to lower H_{c2}. The most logical approach to increase the upper critical field would therefore be to optimise LRO and enhance T_c.

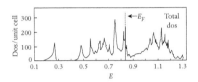

Figure 9.3 Calculated total DOS for Nb_3Sn, showing a sharp rise in the DOS at E_F (after Klein et al., 1978).

9.3 Structural instability at low temperature

An interesting feature common to Nb_3Sn (Mailfert et al., 1967) and V_3Si (Batterman and Barrett, 1964) is that these two A-15s at low temperatures below $T_L > T_c$ undergo a phase change from cubic to tetragonal structure via a martensitic reaction occurring respectively for T_L values of about 45 K and 21 K. The transition is never observed below T_c. Surprisingly, none of the other A-15s exhibits similar behaviour, although there are unconfirmed reports (Savitskii et al., 1973) of structural instability also in V_3Ga ($T_L = 20.6$ K), V_3Ge ($T_L = 10$ K), and V_3Sn ($T_L = 5.4$ K). In all these compounds of relatively high T_c, a soft phonon mode has been observed that is thought to be the driving force for the tetragonal distortion below T_L as well as for stabilising the high T_c. But, to date, the only established cases are V_3Si and Nb_3Sn. In the case of Nb_3Sn, the transition is of first order, while for V_3Si it remains disputed (Vieland et al., 1971). According to Flükiger et al. (1981), the A-15s showing structural instability are unique in meeting three important criteria: (i) a high electronic DOS $N(0)$ at E_F; (ii) a perfect LRO ($S = 1$); and (iii) a

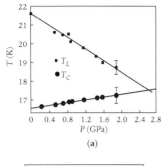

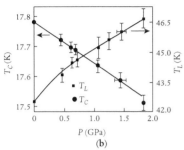

Figure 9.4 Pressure dependence of superconducting critical temperature T_c and martensitic transition temperature T_L in (a) V_3Si (after Chu and Testardi, 1974) and (b) Nb_3Sn (after Chu, 1974).

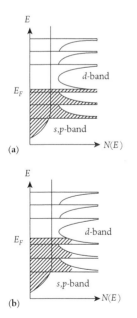

Figure 9.5 (a) There is a sharp DOS at E_F when the A-15 structure is ordered. (b) The effect of disorder is to cause rounding-off the peak and to decrease the DOS at E_F. (After Labbé and Friedel (1966a, b).)

stable stoichiometric composition. A deviation from any of the above, however small, makes the transition undetectable. For a perfect stoichiometric sample, the transition to the tetragonal phase is inevitable; the presence of defects and disorder allows the cubic phase to be retained below T_L, albeit with a slightly suppressed T_c. In the event of structural change, there is a marked softening of the lattice above T_L, accompanied by a considerable decrease in the elastic constants with decreasing temperature. The effect of the transformation is only marginal on T_c, which is reduced by a fraction of a degree. It shows up by a linear increase in T_c, with M_β becoming saturated between 24.4 and 25 at% Sn (Junod et al., 1978).

Surprisingly, V_3Si and Nb_3Sn exhibit mutually opposite trends in their transformation behaviour. For example, the tetragonal distortion $c/a - 1 = +0.2\%$ (Batterman and Barrett, 1964) for the former, while it is -0.6% for the latter (Mailfert et al., 1967, 1969). This is manifested also in the pressure dependences of their T_L and T_c. For V_3Si, $dT_L/dP < 0$ and $dT_c/dP > 0$ (Chu and Testardi, 1974), as shown in Figure 9.4(a), while for Nb_3Sn the behaviours are just the opposite (Chu, 1974), as shown in Figure 9.4(b).

Labbé and Friedel (1966a, b) proposed a *linear chain model* to explain the relatively high T_c of the A-15s and their structural instability. The model emphasised the importance of the d-electrons of the transition metal ions forming the characteristic orthogonal chains. All the interactions between chains, or between A- and B-atoms respectively at α- and β-sites, are neglected. Figure 9.5(a) depicts the DOS curve of the linear chain model. Since the three chains along three different directions lead to an identical DOS, the system carries a threefold degeneracy. Further, there is a singularity in the DOS at E_F, that is, a high $N(0)$, which promotes an increase in T_c, but simultaneously favours the lowering of the energy of the system through a tetragonal structural change in which the degeneracy of the d-band structure is partly lifted. The structural distortion with suppression of degeneracy of the d-bands constitutes a Jahn–Teller transformation. Clearly, any interaction taking place between the linear chains or between the atoms at the α- and β-sites would lower the DOS and thereby decrease T_c and suppress the structural instability (Figure 9.5(b)).

9.4 Potential binary systems

Purely from superconductivity considerations. one may identify six prominent binary A-15 systems for high-field applications: Nb–Sn, V–Ga, V–Si, Nb–Al, Nb–Ga, and Nb–Ge. The A-15 compounds of these systems are noted for their particularly high T_c and H_{c2} (Figure 9.6). Interestingly, V_3Si and V_3Ga are of weak-coupling BCS type, while

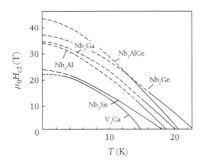

Figure 9.6 Variation of upper critical field with temperature for various high-field A-15 superconductors.

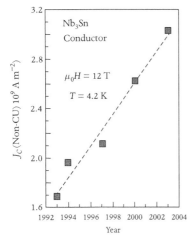

Figure 9.7 Progress made towards J_c enhancement of Nb_3Sn conductors since 1992 (after Parrell et al., 2004).

niobium-based Nb_3Sn, Nb_3Ge, etc. are strong electron–phonon-coupled superconductors. Multifilamentary conductors of Nb_3Sn and V_3Ga have been commercially produced for the last four decades, with their properties being continually improved with novel approaches, while Nb_3Al, noted for its higher critical temperature, upper critical field, and strain tolerance, today stands as a competitor for the third generation of LTS. As mentioned earlier, A-15s possess a larger specific heat coefficient γ, with V_3Ga having the largest value of 3.04 mJ cm^{-3} K^{-2}. This makes the thermodynamic critical field of V_3Ga larger, which (Chapter 6) promotes J_c. Indeed, at one point in time, V_3Ga conductors were noted for their comparatively rather large J_c and were chosen for winding high-field inserts for hybrid Nb_3Sn/Nb–Ti magnets for fields above 18 T. However, of late, instead of V_3Ga, Ti-alloyed Nb_3Sn is more commonly used in inserts for fields up to 21 T. Figure 9.7 shows the progress made over the years with Nb_3Sn conductors in enhancing J_c. For still higher fields ($\sim$30 T, 4.2 K), high-T_c cuprates are used for winding inserts for Nb_3Sn magnets.

Interestingly, there are two more binary A-15s, namely Nb_3Ga ($T_c = 20.7$ K) and Nb_3Ge ($T_c = 23.2$ K), that are noted for their even higher critical temperatures. Also, the ternary superconductor $Nb_3(Al_{0.8}Ge_{0.2})$ possesses an equally attractive T_c of 21 K and, within the A-15 family, this ternary compound has the highest critical field, >40 T. However, making these materials in long lengths, in multifilamentary form, has proved problematic, although some success has been achieved with the ternary compound.

9.5 Pseudo-binaries

A ternary addition B' to the binary A-15 system A–B can help to raise the Gibbs energy of the competing phase (called the C-phase) and thereby make the ensuing pseudo-binary A-15 A_3BB' phase stable. Dew-Hughes and Luhman (1978) gave guidelines for this, according to which both the binary systems A–B and A–B' should tend to form A-15 compounds with lattice parameters not mutually differing by more than 3%. Further, B–B' should form a simple eutectic. Dew-Hughes and Luhman (1978) gave a list of potential pseudo-binary systems, including the well-known example of $Nb_3(Al_{0.8}Ge_{0.2})$. Instead of the binary Nb–Ge or Nb–Al systems, where the stoichiometric A-15 phase is not commonly realised, if the alloy is formed with composition $Nb_3Al_{0.8}Ge_{0.2}$ and is allowed to solidify slowly, the A-15 phase is readily formed with $T_c \approx$ 21 K. The ternary addition thus effectively helps to stabilise the stoichiometric and ordered A-15 phase, and it further enhances the normal-state resistivity to achieve a significantly larger upper critical field. The other similar examples of pseudo-binaries fulfilling the above criteria are Nb_3AlGa ($T_c = 19.4$ K), Nb_3AlSi ($T_c = 19.2$ K), Nb_3AlSn ($T_c = 18.6$ K),

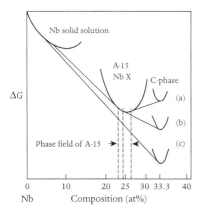

Figure 9.8 The relative stability of the competing phase (C-phase) in relation to A-15, as determined by the Gibbs function ΔG, which controls the A-15 phase field and its stoichiometry.

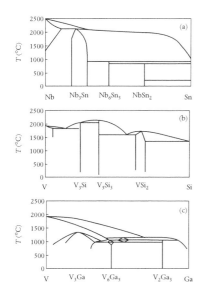

Figure 9.9 Composition–temperature binary-phase diagrams of (a) Nb–Sn, (b) V–Si, and (c) V–Ga (after Dew-Hughes, 1975).

Nb_3GaSn ($T_c = 18.35$ K), and Nb_3InSn ($T_c = 18.3$ K). However, none of these has to date been developed into a practical conductor.

9.6 A-15 phase formation

The phase diagrams of different A-15s have been extensively discussed by many authors (e.g. Dew-Hughes, 1975; Müller, 1980; Flükiger, 1981). In the majority of cases, the formation temperatures of the A-15 phases tend to be well above 1500°C, and, as Dew-Hughes (1975) pointed out, their equilibrium stoichiometry is dictated by the relative stability, in the form of the Gibbs function ΔG, of the competing phases. Figure 9.8 shows, for a hypothetical binary system Nb–X, the ΔG functions for the solid solution Nb-X, the A-15 Nb_3X phase, and a competing phase, the C-phase, which could be either Nb_5X_3, NbX_2, or Nb_6X_5, depending on the system. Equilibrium stoichiometry is identified by drawing tangents to the ΔG curves as shown in the figure for three different stability situations of the C-phase. When it is less stable, having a higher ΔG than the A-15, the range of homogeneity of the latter will include the stoichiometric Nb_3X composition. Any rise in the stability of the competing phase, which corresponds to lowering of its ΔG, will push the A-15/C-phase boundary towards the Nb-rich side of the composition, and the resulting A-15 phase will become off-stoichiometric $Nb_{3+\delta}X_{1-\delta}$, having a lower T_c. If the competing phase is more stable than the A-15 phase, then the latter may not form at all. In a diffusion reaction between Nb and X, the competing phases serve as diffusion barriers preventing the formation of the A-15, Nb_3X, phase.

Figures 9.9 and 9.10 depict the equilibrium phase diagrams of the six prominent systems referred to at the beginning of Section 9.4. Of these, the metallurgy Nb–Sn (Figure 9.9) with respect to A-15 phase formation is uniquely simple, which explains why Nb_3Sn stands out as the most exploited material of the A-15 family. Above 930°C, it is the only stable phase existing over a range of homogeneity that includes the correct stoichiometry A_3B. At lower temperatures, although two other phases Nb_6Sn_5 ($T_c = 2.07$ K) and $NbSn_2$ ($T_c = 2.68$ K) are present, Nb_3Sn is still the most stable phase of the system. The situation is more or less equally favourable for V_3Si and V_3Ga (Figure 9.9), which exist over a range of composition that includes the correct A_3B stoichiometry. It is true that in the V–Ga system, there are Ga-rich intermediate phases present at lower temperatures that are of comparable stability to V_3Ga (Dew-Hughes and Luhman, 1978), and they effectively retard the growth of the A-15 phase, but this problem is overcome with the addition of Cu (Tachikawa et al., 1967), which absorbs Ga from the other two phases and promotes the growth of the A-15 phase.

The remaining three systems, Nb-Al, Nb-Ge, and Nb-Ga (Figure 9.10), however, have the drawback that their temperature of A-15

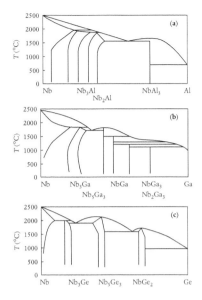

Figure 9.10 Composition–temperature binary-phase diagrams of (a) Nb–Al, (b) Nb–Ga, and (c) Nb–Ge (after Dew-Hughes, 1975).

formation is much higher where thermal disorder is excessive, and they suffer from the adverse effect of the competing phases discussed above. Consequently, for realising their optimum T_c, non-equilibrium fabrication routes such as sputtering, chemical vapour deposition (CVD), evaporation, and rapid heating, quenching, and transformation (RHQT) processes are mandatory. If a molten alloy of Nb–Ge, prepared with nominally correct stoichiometry, is allowed to cool slowly, a T_c of only 6 K results (Matthias et al., 1963). But, if the melt is splat-quenched, a disordered A-15 structure forms, with a very broad transition from 17 to 6 K (Matthias et al., 1965). The onset T_c of 17 K corresponds to a Ge-deficient composition of 23 at% Ge. For the binary Nb_3Ge, T_c values in excess of 22 K have been reached, primarily through thin-film routes such as sputtering (Testardi et al., 1974) and CVD (Braginski and Roland, 1974; Braginski et al., 1978; Cernusko et al., 1981). With the Nb–Ga system, the right stoichiometry of Nb_3Ga can be achieved by argon-jet quenching, but to get the optimum $T_c = 20.7$ K of the ordered state, an additional long-time annealing at $650°C$ is required (Müller, 1980). Nb_3Ga has also been successfully made using CVD (Webb et al., 1971; Vieland and Wickland, 1974). For realising stoichiometrically ordered Nb_3Al conductors, carrying a large J_c in strong magnetic fields, one needs to follow a complex processing schedule, involving a RHQT reaction (Takeuchi, 2000; Glowacki, 2005), to be described in Chapter 10.

9.7 Upper critical field and paramagnetic limitation

The large $H_{c2}(0)$ of A-15s is basically (see Section 5.3.2) a consequence of their intrinsic high T_c and high γ, since, the theoretical value of $H_{c2}(0)$ is proportional to the product of γ, T_c, and the normal-state resistivity ρ_N. As discussed in Section 5.3.2, if the normal state is weakly paramagnetic, the upper critical field is paramagnetically limited to a value $\mu_0 H_p(0) = 1.84T_c$ (T). The experimental value of the upper critical field is therefore given by whichever of $H_{c2}(0)$ and $H_p(0)$ is the lesser, provided that the two values are widely different (Berlincourt and Hake, 1963). When they are similar in magnitude, Maki (1964) showed that the maximum value of the paramagnetically limited upper critical field is given by $[\mu_0 H_{c2}(0)]_{max} = 1.31T_c$ (T). Table 9.2 gives these upper critical field values of potential A-15 systems along with experimental estimates obtained from extrapolation of values measured at 4.2 K. As may be seen, for the two vanadium-based compounds $\mu_0 H_{c2}(0) > \mu_0 H_p(0) > [\mu_0 H_{c2}(0)]_{exp}$, and both of them are paramagnetically limited. On the other hand, for Nb-based A-15s, the experimental values are very close to the values calculated without paramagnetic influence. This is a distinct advantage of Nb-based A-15s over V-based ones.

Table 9.2 Upper critical field of potential A-15 superconductors with their paramagnetically limited values

A-15	T_c (K)	$\mu_0 H_{c2}(0)$ (T)	$\mu_0 H_p(0)$ (T)	$[\mu_0 H_{c2}(0)]_{max}$ (T)	$[\mu_0 H_{c2}(0)]_{exp}$ (T)
V3Ga	14.8	34.9	27.2	19.5	25.0
V3Si	16.9	34.0	31.1	22.0	24.0
Nb3Sn	18.0	29.6	33.1	23.4	28.0
Nb3Al	18.7	32.7	34.7	24.3	33.0
Nb3Ge	22.5	37.1	41.3	29.3	38.0
Nb3Ga	20.2	34.1	37.2	26.2	34.0
Nb3Al0.7Ge0.3	20.7	44.5	38.0	26.9	43.5

Based on Dew-Hughes, D. *Cryogenics* 15 (1975) 435.

9.8 Critical current density and the nature of pinning centres in A-15s

As discussed previously in Chapter 4, the maximum current density, that is, the depairing current density J_d, that a superconductor can sustain at 0 K is given by $\mu_0 H_c(0)/\lambda(0)$, where $\mu_0 H_c(0)$ is the thermodynamic critical field and $\lambda(0)$ is the penetration depth at 0 K. Because A-15s possess a high H_c and a not too large λ, their depairing current density can be large. Taking the thermodynamic critical field as 1 T and $\lambda(0) = 10^{-7}$ m gives $J_d = 5 \times 10^{12}$ A m^{-2}, which is about 100 times larger than the highest value recorded in any A-15 at zero field. The higher upper limit of J_d thus provides sufficient scope for improving the experimental values of the critical current density through optimisation of the microstructure.

There is a weight of evidence that in pure A-15s and, in fact, in most intermetallic superconductors such as magnesium diboride and quaternary borocarbides, the naturally occurring defects in greatest abundance are grain boundaries, which provide effective pinning to enhance J_c. Unlike Nb–Ti and Nb–Zr alloys, these are ordered materials that are highly brittle, in which creation of dislocations costs energy, and whose microstructure is devoid of dense dislocation structures to contribute to flux pinning. The grain boundaries are disordered regions approximately 5–10 lattice constants (i.e., 2.5–5.0 nm) wide and are comparable to and, in fact, a little smaller than the coherence length ξ (e.g. $\xi \approx 6$ nm for Nb$_3$Sn), which makes them potential pinning entities and creates no weak links suppressing J_c. As the grain diameter d decreases, the volume fraction of the boundaries rises and pinning is enhanced linearly (Figure 9.11) as $1/d$ (Scanlan et al., 1975; West and Rawlings, 1977; Livingston, 1978; Fischer, 2002). In comparison with Nb–Ti alloys, the grains in A-15s are at least an order of magnitude smaller; the d values

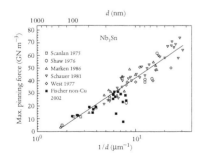

Figure 9.11 The maximum pinning force density is inversely proportional to the average grain size d and increases linearly with $1/d$ (after Fischer, 2002).

in the former, even after heavy cold working, are usually 1–2 μm, while in A-15s they are 0.1–0.2 μm or even smaller. These values are closer to the sizes of the subgrains or of the dislocation cell structure observed respectively in polygonised and heavily cold-worked Nb–Ti alloys, which manifest very strong flux pinning (Narlikar and Dew-Hughes, 1966).

Besides grain size refinement, other potential structural features enhancing the J_c of A15s are fine precipitates of a second phase formed during processing (Carlson et al., 1977; Braginski et al., 1978), particles, including artificial pinning centres (APCs) such as carbon nanotubes or nanodiamonds introduced as additives (Nembach, 1970; Benz, 1968; Enstrom et al., 1970, 1972; Enstrom and Appert, 1974), and defects and disorder induced by radiation damage (Cullen, 1968; Bode and Wohlleben, 1967). Here the additives have a twofold effect: first, their presence promotes grain refinement and thereby enhances flux pinning and J_c, and, second, fine particles and precipitates themselves serve as pinning centres. Although a large degree of disorder is deleterious for A-15 superconductivity, the presence of a small amount of local dis-order, such as introduced by ternary additions (e.g. a few at% of Ti or Ta) to Nb_3Sn and V_3Ga, markedly enhances J_c (Kamata et al., 1983; Sekine et al., 1983; Howe and Fancavilla, 1980; Tachikawa et al., 1979). Irradiation with neutrons (5×10^{21} m^{-3}), protons, or deuterons, in general, gives rise to local normal islands of 10 nm size that provide pinning as normal particles to enhance J_c. But, with a higher radiation dose, the pinning is reduced owing to lowering of T_c by disorder.

The critical Lorentz force in Nb_3Sn is found to obey a scaling law described by the pinning function $h^{1/2}(1 - h)^2$ (Kramer, 1973), where $h = H/H_{c2}$. As may be noted, this function differs from the function $h(1 - h)$ for Nb–Ti alloy, discussed in Chapter 8. Here the $(1 - h)^2$ term reflects some flux-shearing process of the flux line lattice (FLL), where the relevant c_{66} modulus, for instance, varies as $(1 - h)^2$ when h is high. The difference in the pinning functions of the two types of materials is related to the difference in their pinning entities. In the case of an Nb_3Sn conductor, prepared by the bronze process, for example, there are columnar grains present with their axes normal to the filaments. The Lorentz force is directed parallel to the boundaries to drive the flux vortices along them rather than across them, which results in flux shear, originating from the transport of flux lattice dislocations (FLDs) (Dew-Hughes, 1987). The model is found to be consistent with the above pinning function and further predicts the observed inverse dependence of J_c on grain diameter.

Comparing the two scaling laws $h(1 - h)$ and $h^{1/2}(1 - h)^2$, respectively for Nb–Ti and Nb_3Sn, as h approaches 1, the slope of the former function approaches 1, while the slope of the latter approaches 0. Accordingly, close to the upper critical field, Nb–Ti conductors can still sustain an appreciable current density, but in Nb_3Sn the current density is vanishingly low. This calls for a suitable modification in the pinning entities of Nb_3Sn to enhance J_c near the upper critical field (Dew-Hughes, 1999).

9.9 Strain sensitivity

A-15 compounds are sensitive to strain, and their properties markedly deteriorate in its presence. Unlike Nb–Ti alloys, the brittle Nb_3Sn, for instance, can withstand only around 0.2% of elongation strain as against around 10% for Nb–Ti. Since superconducting wires for practical use are always of composite type in which thin compound layers are embedded in a host of material such as copper, stainless steel, or tantalum, all differing in Young's modulus, thermal expansion coefficient, and Poisson's ratio, the A-15 layers are subject to considerable strain during coil-winding and cooling-down operations. The strain has an analogous effect to disorder in causing deterioration of properties. Strain also results from the presence of high magnetic fields owing to Lorentz forces arising in magnet operations. The critical current and the upper critical field are the most relevant parameters for a superconducting magnet, which can unexpectedly be quenched if these parameters are exceeded during its operation. Figures 9.12(a) and (b) respectively illustrate the strain sensitivities of J_c (Flükiger et al., 1984) and H_{c2} (Ekin, 1984) for various A-15 superconductors. As may be seen, Nb_3Sn and V_3Si, under strain, exhibit a greater decrease in their superconducting properties, while Nb_3Al stands out for its comparatively low strain sensitivity. Figure 9.13(a) shows how J_c and I_c in different magnetic fields (at 10 K) respond to axial strains and Figure 9.13(b) shows the axial strain response of H_{c2} measured at different temperatures (Keys et al., 2002).

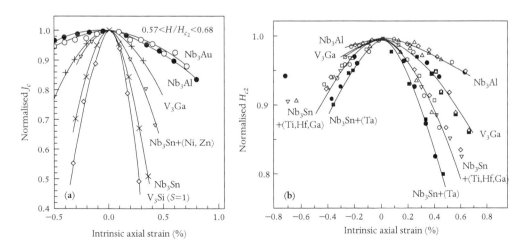

Figure 9.12 Strain sensitivities of (a) J_c (after Flükiger et al., 1984) and (b) H_{c2} (Ekin, 1984) of various A-15 superconductors.

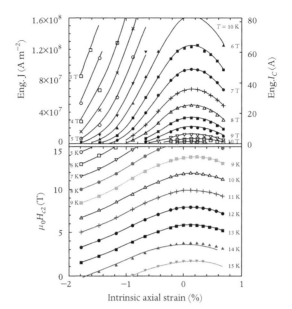

Figure 9.13 Compared with other A-15 superconductors, Nb$_3$Al is noted for having its properties least sensitive to axial strain. (a) depicts how the engineering J_c and I_c measured at different magnetic fields are influenced by axial strain. (b) shows the effect of strain on H_{c2} at different temperatures. (After Keys et al. (2002).)

These results are promising, giving support to Nb$_3$Al as a potential successor to Nb$_3$Sn.

9.10 Summary

In this chapter, we have described the prominent characteristics of A-15 superconductors, which constitute the second generation of LTS conductors and are used extensively to supplement and boost the magnetic fields of Nb–Ti magnets from 10 T to more than 20 T. Their noticeably high T_c and critical field among LTS materials originate from their unusual crystal structure and, as with most intermetallic superconductors, both stoichiometry and long-range order are crucial for optimal superconductivity performance. Potential A-15 systems suited for applications have been discussed and the challenges posed in their formation briefly outlined.

Conductor development of A-15 superconductors

Historically, the first conductor of the A-15 family to be produced (Kunzler, 1961) was an Nb_3Sn wire prepared by packing a stoichiometric mixture of Nb and Sn powders inside an Nb tube and drawing it to the final diameter. Heating the wire at $930°C$ led to Nb_3Sn formation, but turned the wire too hard and brittle for magnet winding. Consequently, the winding process had to be completed with the wire in the unreacted or green state prior to heat treatment. The process is known as *powder-in-tube* (PIT) and the approach to winding as *wind-and-react*. This was obviously inconvenient, particularly for larger magnets, where practical considerations necessitated a *react-and-wind* approach for commercially produced conductors.

The fabrication processes that have proved commercially viable for making the react-and-wind type of conductors with considerable lengths of Nb_3Sn and V_3Ga are mainly of three broad types:

(i) liquid–solute diffusion;
(ii) chemical vapour deposition (CVD);
(iii) solid state diffusion.

Both (i) and (ii), which were introduced during the 1960s, soon found themselves obsolete for Nb_3Sn and V_3Ga, being superseded by (iii). However, because these methods hold promise with some of the other A-15 systems where the solid state diffusion approach does not work, we briefly describe them first.

10.1 Liquid–solute diffusion

In the liquid–solute diffusion process (Benz, 1966), first used at General Electric, USA (the process is also known as the GE process), flexible Nb_3Sn was made by diffusing molten Sn into a thin Nb tape at $930–970°C$, when Nb and Sn reacted to form a thin layer of the A-15 phase on the Nb tape surface. The reaction was carried out in an inert atmosphere of Ar or He inside a sealed quartz tube carrying a graphite boat containing molten tin. The moving Nb ribbon entered

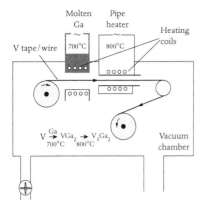

Figure 10.1 The experimental set-up used by Tachikawa et al. (1967) for producing V_3Ga tape by the liquid–solute diffusion route.

one end of the tube and was guided underneath the surface of the molten tin bath and was collected on a rotating reel at the other end of the tube. The diffusion reaction was stopped when the Nb substrate was surrounded by a Nb_3Sn layer, which in turn was covered by unreacted Sn. The critical current of such a tape was markedly improved by doping it with fine particles of zirconia. Typically, 10 minutes of reaction produced a compound layer, 10 μm thick, that sustained a current density in excess of 10^9 A m^{-2} at 10 T and 4.2 K (Benz, 1968). Tachikawa et al. (1967) successfully followed a similar approach for making V_3Ga by passing a moving V tape through a molten Ga bath at 700°C (Figure 10.1). The first phase formed was VGa_2, which was converted to the V_3Ga_2 phase by passing it through a pipe heater at 800°C followed by a further resistive heating produced by passage of electric current. To transform the V_3Ga_2 phase into V_3Ga, the ribbon was next coated with a 10 μm layer of Cu and was annealed at 700°C for 10 hours. The Cu layer absorbed the excess Ga, leaving behind the V_3Ga phase.

10.2   CVD process

The CVD process for Nb_3Sn was developed at RCA, USA (Hanak et al., 1964), and involved simultaneous reduction of gaseous mixture of Nb and Sn chlorides on a moving stainless steel or HASTELLOY ribbon (Figure 10.2(a)). The gaseous chlorides were obtained by passing chlorine into separate chlorinators containing Nb bars and molten Sn, respectively, at 900 and 800°C. The gaseous mixture of metal chlorides and hydrogen chloride, in the presence of hydrogen, entered the deposition chamber at 700°C, where they interacted to form an Nb_3Sn layer on the moving substrate. The flow rates of Nb and Sn chlorides were in the ratio of 3:1 to achieve the correct stoichiometry of the deposit, following the chemical reaction

$$3\,NbCl_4 + SnCl_2 = Nb_3Sn + 14\,HCl \qquad (10.1)$$

The gaseous reaction product HCl was exhausted from the chamber.

CVD has not worked for making V_3Ga tapes, but, more importantly, the process has been successfully used for making Nb_3Ge (Braginski and Roland, 1974; Braginski et al., 1978; Asano et al., 1985) and Nb_3Ga (Webb et al., 1971; Vieland and Wickland, 1974) from their respective chlorides, although long tapes have not been produced suitable for applications. The experimental arrangement of Asano et al. (1985) for Nb_3Ge tape is schematically depicted in Figure 10.2(b).

The brittle nature of A-15s posed a major hurdle with the co-processing approach that had successfully worked with Nb–Ti. The A-15 ribbons produced using the above methods were laminated with

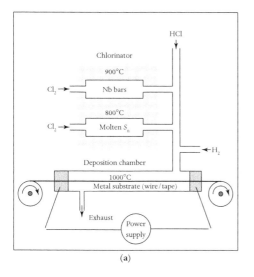

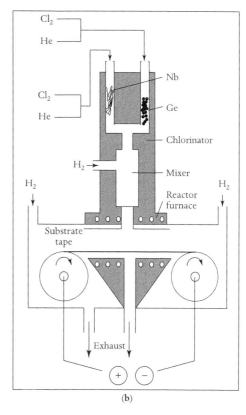

Figure 10.2 Schematic illustrations of chemical vapour deposition techniques: (a) the RCA process for Nb_3Sn wire/tape (after Hanak et al., 1964): (b) for Nb_3Ge (after Asano et al., 1985).

stainless steel for strengthening and were stabilised by electroplating them with copper, but the stability achieved this way was hardly adequate. Consequently, the advent of the bronze process, based on solid state diffusion, in 1970 (Kaufman and Pickett, 1970; Howlett, 1970; Tachikawa, 1970) proved particularly exciting, since it could lead to fully stabilised conductors of Nb_3Sn and V_3Ga in multifilamentary form. Several ingenuous modifications of the bronze process soon emerged that brought the commercial production of considerable lengths of these two A-15s on a firm footing.

10.3 The bronze process and formation of A-15 phase by solid state diffusion

The phase transformation of bcc Nb (respectively V) into Nb_3Sn (V_3Ga) of A-15 structure caused by solid state diffusion of Sn (Ga) into Nb (V) has led to several potential routes for fabricating stabilised multifilamentary Nb_3Sn (V_3Ga) conductors in a copper-alloy matrix for practical applications. This approach, commonly known as the *bronze process*, works also with V_3Si (Livingston, 1977) and V_3Ge (Tachikawa

143

et al., 1977), as well as with Nb_3Ge (Hopkins et al., 1978). This interesting development began about four decades ago (Kaufman and Pickett, 1970; Howlett, 1970; Tachikawa, 1970) and continues to be a front runner in A-15 conductor fabrication. It has led to numerous offshoots and modifications having specific advantages over some of its intrinsic drawbacks. The most prominent of these developments are the *internal tin process*, the *modified jelly-roll* (or *modified Swiss-roll*) *method*, the *in situ Cu–Nb alloy method*, and the *powder metallurgy approach*. These processes are in fact essentially different ways of producing *diffusion couples* comprising a large number of Nb (or V) filaments in a Cu or Cu-alloy (Sn- or Ga-bronze) matrix.

For making multifilamentary Nb_3Sn, the procedure is as illustrated in Figure 10.3 A billet of Sn-bronze, containing 8 at% Sn in Cu (which represents its solubility limit at ambient temperature) with the desired distribution of Nb strands is extruded and drawn to the final size. For V_3Ga, a matrix of Ga-bronze containing the optimum 19 at% Ga is prepared. The billet is prepared following the procedure described in Chapter 8 for Nb–Ti, with the difference that the tubes used are of bronze and not of copper, and instead of Nb–Ti, pure Nb (or V) rods are inserted in them. The tubes are packed inside another bronze tube or a can of larger diameter. Alternatively, the billet may be prepared by drilling holes in a bronze cylinder, into which Nb (or V) rods are

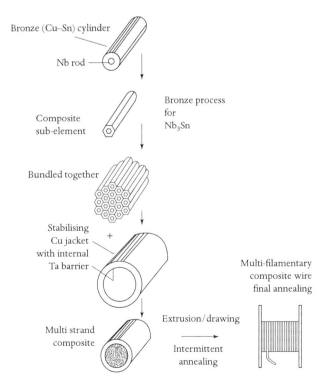

Figure 10.3 Principal steps in the bronze process for Nb_3Sn (schematic).

tightly inserted; the billet is then vacuum-sealed. The bronze should not contain phosphorus as impurity, since this serves as a poison for Sn diffusion onto Nb. The composite billet is extruded and drawn with intermittent annealing for 1–2 hours at 450°C, after successive 50–75% reductions, to remove the work hardening resulting from heavy deformation. These composite wires are suitably cut and restacked in another bronze billet, which is subjected to a further reduction to enhance the filament number and reduce their size to 3–5 μm. For the stabilisation, the composite should have an outermost jacket of oxygen-free high-conductivity (OFHC) copper that is protected against diffusion of Sn- (or Ga-) bronze by a few micrometres thick concentric Ta barrier separating the bronze from the copper. This is achieved by inserting the bronze composite in a concentric Ta tube, placed inside a tube of OFHC copper, just before the final extrusion. Typically, a composite wire of 1 mm diameter may contain 10 000–50 000 Nb filaments embedded in a bronze matrix having an outer copper jacket and a Ta barrier between the two. For converting the filaments from Nb (respectively V) to Nb_3Sn (V_3Ga), the composite wire is reacted at 700–800°C (600–650°C) for several hours and then at 650°C (550°C) for several days, during which Sn (Ga) from the bronze diffuses onto the Nb (V) filaments to form Nb_3Sn (V_3Ga) layers at their interfaces (as shown in Figure 10.4 for Nb_3Sn formation). The diffusion reaction is stopped after sufficiently thick layers are formed and there is still some unreacted Nb left in each filament, which provides necessary strength and ductility to the composite.

For a conductor to carry a large critical current I_c, the compound layer should be sufficiently thick, while for strong flux pinning and a high critical current density J_c, the grain size of the A-15 layer formed should be small. Consequently, the processing strategy to be adopted is one that will promote faster layer growth accompanied by slower grain growth. If the growth of the layer is sufficiently fast and the rate of

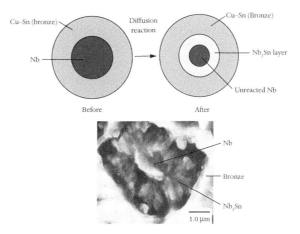

Figure 10.4 Formation of the A-15 compound layer at the bronze–Nb interface after the diffusion reaction (illustration on the right), leaving behind unreacted Nb. The SEM image shows the compound layer formed at the bronze–Nb interface of an individual Nb strand of a multifilamentary bronze composite (Narlikar, 1987).

growth of the grains sufficiently low, the layer will be completed before the grains have become too large. The higher the Sn or Ga concentration in the bronze, the larger is the driving force for diffusion and the faster is the layer growth. By choosing a lower reaction temperature, the rate of grain growth can be reduced. Because of the solubility limit of about 8 at% for Sn in Cu, one cannot have a larger concentration of Sn to achieve a faster layer growth and a larger transport current. As we will shortly see, the above constraint is overcome in the internal tin process, which makes 20 at %Sn available for diffusion, leading to greatly accelerated layer growth. In the case of V_3Ga, the solubility limit of Ga in Cu is already higher (around 19 at%) and moreover its decomposition temperature is 1300°C as against 2100°C for Nb_3Sn; both these factors favour a higher growth rate and a lower reaction temperature for V_3Ga, promote a higher J_c.

Besides these considerations, an approach that favours high J_c is to introduce a few atom percent of additives such as Ti, Ta, Zr, Mn, or Hf in the Nb (or V) core or in the bronze matrix. In either case (Kamata et al., 1983; Sekine et al., 1983; Howe and Fancavilla, 1980; Tachikawa et al., 1979), the additives become alloyed with the A-15 compound layer to exhibit a markedly enhanced J_c in high fields (Nembach, 1970). With 3 at% Ti addition, for example, the compound layer of $(Nb-3at\%Ti)_3Sn$ has a much higher J_c of 3×10^8 A m^{-2} at 15 T and 4.2 K. The additives tend to refine the grain structure and additionally enhance H_{c2} by raising the normal-state resistivity ρ_n, without drastically lowering T_c or the specific heat coefficient γ of the compound layer. Ti addition has evolved as an established approach to enhance J_c, and such Ti-doped conductors are used for high-field inserts in hybrid $Nb_3Sn/Nb-Ti$ magnets producing fields in excess of 20 T. Figure 10.5 and its inset respectively depict the significant enhancement occurring in the overall J_c and the compound-layer J_c of a bronze-processed Nb_3Sn conductor in which the Nb core was alloyed with a small concentration (2–4%) of Ti (Asano et al., 1986). Similarly, V_3Ga formed with 0.5 at% Mg added to a Cu–19 at%Ga matrix and 6 at% Ga to the V core has been found to sustain an exceptionally high J_c of 1×10^9 A m^{-2} at 20 T and 4.2 K (Tachikawa et al., 1979). Ga has a solubility of 8 at% in V at 600°C, which prompted Tachikawa et al. to pre-add Ga to the core to achieve faster layer growth. Addition of Mg to the matrix further lowers its melting point and promotes faster Ga diffusion.

However, the bronze process has two drawbacks. First, with Nb_3Sn, one is restricted to a rather low solubility limit of 8 at% Sn in Cu, which leads to slower layer growth and faster grain growth. Second, with the bronze matrix, intermittent annealing at 450°C is necessary at frequent intervals to relieve work hardening and to sustain the drawability of the wire for further reduction. Apart from adding to the production cost and time, these heat treatments, even at a relatively low temperature of

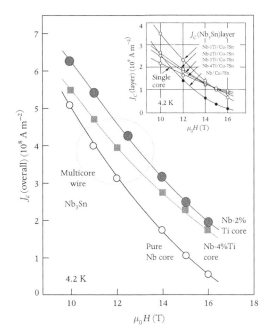

Figure 10.5 Overall J_c of 160-core bronze-processed Nb_3Sn wire in which the Nb cores were alloyed with 2 and 4 at%Ti. The inset shows the J_c behaviour of single-core wire; the J_c values plotted are of the compound layer. Both the overall and layer J_c are markedly raised by Ti alloying of the Nb core. (After Asano et al., 1986).

450°C, are found to cause Sn to react with the Nb filaments to adversely influence their ductility during mechanical processing.

10.4 Thermodynamics and kinetics of compound-layer formation in the bronze process

There are four types of kinetics involved with A-15 formation by solid state reaction (Narlikar, 1987; Narlikar and Dew-Hughes, 1987):

(i)   nucleation;
(ii) grain growth;
(iii) layer growth;
(iv) ordering kinetics.

The nucleation kinetics is of importance for spatial location of the transformed phase. The critical current density J_c is controlled by (ii), and the overall critical current I_c by (iii), while T_c, which is controlled by the long-range order (LRO), is determined by (iv). Interestingly, all four process come into operation with the onset of the diffusion reaction.

10.4.1   Thermodynamics of A-15 formation

The thermodynamics of the bronze process was discussed in detail by Dew-Hughes and Luhman (1978) and Narlikar and Dew-Hughes (1985,

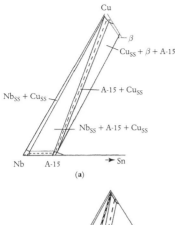

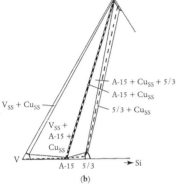

Figure 10.6 (a) Phase diagram of Cu–Nb–Sn, showing that the tie-line route passes through the phase field of Nb₃Sn. (b) In the case of Cu–V–S, where there are two tie lines present, only the one from lower Si content passes through V₃Si. As a result, the latter forms only when the Si content is smaller than stoichiometric (Dew-Hughes and Luhman, 1978).

1987). In Chapter 9, we pointed out the importance of the relative stability of competing phases in A-15 formation. The stability of a phase is measured by its *stability index* (SI). For an arbitrary composition $A_x B_{1-x}$ with a melting temperature T_m,

$$SI = \frac{T_m}{T_A - (T_A - T_B)(1 - x)},$$

where T_A and T_B ($T_A > T_B$) are the melting temperatures of the constituent metals A and B. For phase stability, $SI \geq 1$ and should be higher than for any other phase of the system. While this is fulfilled for Nb₃Sn and V₃Ga, with other potential systems, the V₅Si₃, Nb₅Ge₃, Nb₃Ga₂, and NbAl₃ phases have higher SI values and are thus more stable than the corresponding A-15 phases, which are therefore not favoured.

The bronze process works for a given A₃B provided that in the ternary ACuB phase diagram (Dew-Hughes and Luhman, 1978) there exists a direct diffusion path between pure A and the Cu–B bronze that passes through the A-15 phase having the highest SI value. This holds for Nb₃Sn, V₃Ga, and V₃Si, but not for Nb₃Ga, Nb₃Ge, and Nb₃Al. In the case of Nb₃Sn (Figure 10.6(a)) and V₃Ga, the tie line paths from Cu–Sn and Cu–Ga to Nb–Sn and V–Ga pass directly through the phase fields of A-15s, which are readily formed owing to their higher SI values. In the NbCuGe, NbCuGa, and NbCuAl systems, the tie paths are instead intercepted by other more stable phases that prevent their A-15 formation. In the case of Nb₃Ge, however, the solid state diffusion process works if the Ge-bronze matrix is replaced by an Ag–Ge alloy matrix (Hopkins et al., 1978). Interestingly, in the VCuSi system, there are two tie lines: one coming from a low-Si bronze going directly through the V₃Si phase, and the other from a higher-Si bronze going through a more stable V₅Si₃ phase (Figure 10.6(b)). Consequently, for V₃Si, the bronze route works only when the Si content of the bronze is low (Livingston, 1977). Attempts to form pseudobinary A-15s using hosts of quaternary diffusion couples, have, however, proved unsuccessful (Dew-Hughes and Luhman, 1978).

10.4.2 Kinetics of A-15 formation in bronze process

10.4.2.1 *Nucleation kinetics*

In the case of NbCuSn, nucleation of the first A-15 layer takes place at the bronze–Nb interface, while subsequent layers form at the Nb₃Sn–Nb interface. The driving force for nucleation arises from (a) the free-energy difference between Nb₃Sn and Nb and (b) the release of stored energy of heavily deformed Nb filaments by the formation of strain-free Nb₃Sn grains in a process akin to recrystallisation. The negative contribution of volume energy ΔG_v, favouring Nb₃Sn formation, is countered by two positive contributions: (i) the surface energy ΔG_s associated with the

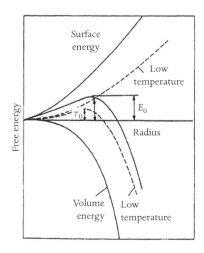

Figure 10.7 Energies involved in A-15 phase nucleation. E_0 is the nucleation barrier.

creation of the interface between Nb_3Sn and Nb and (ii) the volume contribution of the strain energy ΔG_m caused by the 37% volume expansion that occurs when Nb of shorter lattice parameter is replaced by Nb_3Sn of longer parameter. For a particle of small radius, the positive surface energy term dominates, whereas when the radius exceeds a critical value r_0, the overall free energy becomes negative and the transformation to Nb_3Sn is favoured (Figure 10.7). The initial grain size (when the neighbouring nuclei grow and impinge on each other) of the A-15 layer formed at a lower reaction temperature would be smaller than that of a layer formed at a higher temperature. The junction points of grain boundaries and dislocation cell walls of Nb filaments are expected to serve as nucleation sites for A-15 formation.

10.4.2.2 Grain growth kinetics

Grain growth in metallic systems generally follows a parabolic growth law

$$d_t^2 - d_0^2 = k't, \tag{10.2}$$

where d_0 is the initial grain diameter, d_t the grain diameter after time t, and k' the grain growth constant.

Effect of melting and reaction temperatures
The larger the melting temperature of the A-15 compound, the smaller is the grain growth at a fixed reaction temperature. Also, the larger the melting temperature, the higher is the reaction temperature for a comparable grain growth and grain sizes (Livingston, 1977).

Effect of cold work
Finer filaments formed with greater cold work develop smaller A-15 grains by diffusion reaction.

Effect of additives
Additives in the bronze matrix or filaments can enhance layer growth, which allows less time for grains to grow. Also, added impurities may pin down the grain boundaries and retard grain growth.

Grain size distribution
The grains closer to the bronze interface are formed before the interior grains, and thus there is a radial decrease in their size towards the filament centre.

Grain morphology
Enhanced delivery of Sn atoms due to either a higher reaction temperature or greater Sn content in the matrix favours the nucleation of new equiaxed grains, while their slower transport at a lower temperature or smaller Sn content causes them to become attached to the

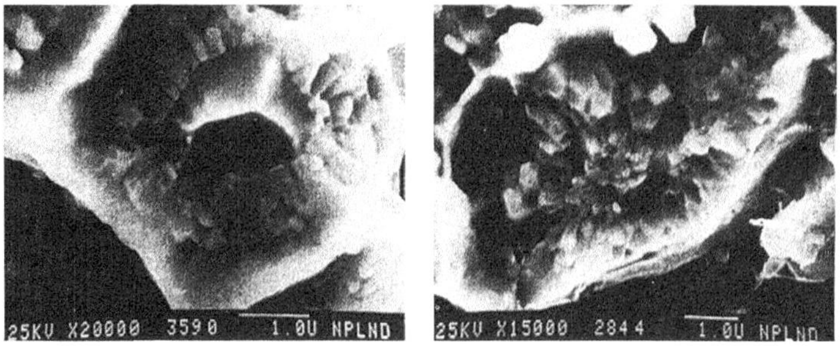

Figure 10.8 SEM images showing (a) columnar grains near the bronze interface and (b) equiaxed grains formed when Nb is fully transformed to Nb_3Sn.

(a)　　　　　　　　　　　　　　　(b)

existing grains along the Sn gradient, making the grains (Figure 10.8(a)) columnar (Okuda et al., 1983). The grains at the Nb_3Sn–Nb interface are columnar, while those formed in the middle or outer regions of the layer are equiaxed (Figure 10.8(b)): the latter are also comparatively larger in diameter.

10.4.2.3 Layer growth kinetics

Unlike the two kinetics described in the preceding subsections, there are abundant experimental data pertaining to layer growth kinetics, which follow the layer growth equation

$$R = kt^n, \tag{10.3}$$

where R is the layer thickness formed at a fixed reaction temperature after a time t, n is the numerical time exponent, and k is the reaction rate constant. The growth of the compound layer can be sub-parabolic ($n < 0.5$), parabolic ($n = 0.5$), or super-parabolic ($n > 0.5$), and, interestingly, diffusion couples formed with the same components can yield all three growth laws under different situations (Reddi et al. 1978, 1983; Narlikar and Dew-Hughes, 1985, 1987). In fact, n is found to vary from 0.15 to 1.0. An enhanced growth rate can occur owing to an increase in k and/or n. Reddi et al. (1978, 1983) and Agarwal et al. (1984) developed analytical models of layer growth kinetics of bronze-processed Nb_3Sn and V_3Ga. The layer growth (of Nb_3Sn) depends upon three sequential steps:

1. transport of Sn from the bronze to the bronze–Nb_3Sn interface;
2. transport of Sn across the Nb_3Sn layer to the Nb_3Sn–Nb interface;
3. the reaction of Sn with Nb to form Nb_3Sn.

When the three rates are different, the rate of layer growth will be controlled by the slowest of the three. Step 3 is the fastest, being essentially instantaneous, and consequently it will be the slower of the other two steps (i.e. either step 1 or step 2) that will emerge as the rate-controlling step for compound layer formation.

Growth controlled by diffusion of Sn through the bronze matrix

The flux of Sn atoms at the bronze–Nb$_3$Sn interface determines the overall layer growth. As the compound layer grows, it gives rise to depletion of Sn atoms and a corresponding depletion distance in the matrix, measured from the bronze–Nb$_3$Sn interface. The analytical models envisage three possibilities:

(i) *Small depletion distance*. This gives rise to a parabolic growth law

$$R = k_1 t^{1/2}. \tag{10.4}$$

This applies when the bronze matrix has a higher Sn content and the reaction time t is small.

(ii) *Large depletion distance*. In this situation, the growth predicted is super-parabolic:

$$R = k_2 t^{2/3}. \tag{10.5}$$

Clearly, when the bronze has a smaller Sn content and the reaction time is large, n is predicted to increase from 0.5 to 0.67. Results of Suenaga et al. (1974), Dew-Hughes et al. (1976), Luhman and Suenaga (1977), and Dew-Hughes and Suenaga (1978) on monofilamentary Nb$_3$Sn all substantiate this contention.

(iii) *Depletion distance exceeding the thickness of the bronze matrix*. The predicted growth law is linear:

$$R = k_3 t. \tag{10.6}$$

This corresponds to prolonged diffusion in which the Sn concentration at the outer rim of the bronze matrix begins to deplete and the corresponding n value approaches unity. Reddi et al. (1983) found $n = 0.9$ when the bronze had the lowest Sn content of 2 wt%. When the growth is super-parabolic and the growth rate is increased by adding impurities to the matrix or the filaments, the obvious explanation is an increase in the reaction rate constant, which is related to the diffusivity of Sn in bronze. The presence of additives increases the diffusivity of Sn, which is manifested by the formation of *Kirkendall voids* at the, bronze–Nb$_3$Sn interface (Berthel et al., 1978; Suenaga, 1981).

Growth controlled by diffusion of Sn through the compound layer

The analytical models examine two possibilities: (1) the bulk diffusion through the layer and (2) grain boundary diffusion through the layer. The latter involves two situations: (i) zero grain growth and (ii) grain growth.

1. *Bulk diffusion through the layer*. Bulk diffusion dominates over grain boundary diffusion when the reaction temperature $\geq T_m / 2$, where T_m is the melting temperature. Clearly, for Nb_3Sn and V_3Si, the reaction temperature is too small for bulk diffusion, but in the case of V_3Ga, the possibility of bulk diffusion exists, for which the models predict a parabolic growth law

$$R = k_4 t^{1/2}, \tag{10.7}$$

where the reaction rate constant k_4 is related to the bulk diffusivity of Sn through the layer and the layer will grow faster at higher reaction temperature. Most layer growth studies on V_3Ga reveal parabolic growth. However, as we will soon see, parabolic growth can also occur also in grain boundary diffusion.

2. *Growth controlled by grain boundary diffusion*. The analytical models examine three possibilities.

(i) *Zero grain growth*. When the grain diameter d in the layer remains invariant throughout the diffusion reaction, parabolic growth is again predicted:

$$R = k_5 t^{1/2}, \tag{10.8}$$

where k_5 is related to the grain boundary diffusivity of Sn atoms through the compound layer. The models further show that a higher Sn content in the matrix provides a larger driving force for diffusion through the grain boundaries, which enhances the growth rate.

(ii) *Grain growth imposed on grain boundary diffusion*. Consider an arbitrary grain growth described by

$$d_t = d_i + Gt^m, \tag{10.9}$$

where d_i is the initial nucleation grain diameter, d_t is the grain diameter after reacting at a fixed temperature for a time t, G is a temperature-dependent coefficient, and m is a numerical time exponent.

Case A

When the initial grain size is small, the analytical models predict grain-growth-controlled layer growth, given by

$$R = k_6 t^n, \tag{10.10}$$

where $n = (1 - m)/2$. Whenever the layer growth is sub-parabolic, the above model is the only choice. Parabolic grain growth (i.e. $m = 0.5$) yields $n = 0.25$. For slower grain growth, $0.25 < n < 0.5$. Faster grain growth occurring at elevated temperatures would lead to smaller n values, which is consistent with observations for both Nb_3Sn (Larbalestier et al., 1975) and V_3Ga (Critchlow et al., 1974; Tachikawa

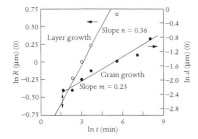

Figure 10.9 Interdependence of layer growth kinetics and grain growth kinetics of Nb₃Sn formation in the bronze route (after Agarwal and Narlikar, 1985b).

et al., 1972). The interrelation between grain growth kinetics and layer growth kinetics has been fully confirmed by transmission and scanning electron microscopic studies by Agarwal and Narlikar (1985a) on bronze-processed Nb₃Sn (Figure 10.9). More detailed considerations (Narlikar, 1987; Narlikar and Dew-Hughes, 1985, 1987) show that the rate constant k_6 responds in a complicated way to the exponent m, the grain boundary diffusivity, and the coefficient G in (10.9), whose overall effect determines the ultimate layer growth.

Case B

When the initial grain size d_i is large, further grain growth has less effect in reducing the grain boundary channels for diffusion, and, as with the situation of zero grain growth (equation (10.8)), the layer growth is parabolic.

10.4.2.4 Ordering kinetics

As discussed in Section 9.1, by 'ordering' is meant the right species of atoms occupying the right crystallographic sites, and in the case of A-15 superconductors it directly controls the critical temperature T_c. The kinetics of ordering of the A-15 layer in the bronze process has, however, received only limited attention. Dew-Hughes (1980) performed model calculations to explain the observed rise in T_c of disordered bulk binary A-15s after annealing. The ordering process considered was the vacancy-assisted hopping of the two species of atoms from wrong lattice sites to the correct ones and the rate of ordering was governed by the slower of the two species. Accordingly, when Sn atoms are slower than Nb atoms, the rate of ordering is governed by first-order kinetics, while for slower Nb atoms, the kinetics is of second order. Experimental results of Winkel and Bakker (1985) had shown the bulk diffusivities of Sn and Ga to be smaller than those of Nb and V, which favoured the first order kinetics for ordering.

Figure 10.10 depicts the results of Agarwal and Narlikar (1985b), showing the T_c of multifilamentary bronze-processed Nb₃Sn formed after a short duration of diffusion reaction at different temperatures. The use of long annealing times is of little help when attempting to understand the ordering process, since the outermost grains formed initially have already attained a near-optimum T_c. The fact that that the unannealed sample shows a T_c higher than that of Nb indicates that Sn is reacting with Nb during intermittent annealings (at 450°C) carried out to remove work hardening. Agarwal et al. (1986) found the ordering behaviour to be complex, governed by a duplex process involving a sequential second- and first-order kinetics. Formation of Nb₃Sn at the start of the reaction requires Sn atoms to move faster than Nb, that is, second-order kinetics, but instead of their moving directly to Sn vacancies, they tend to form anti-site disorder. As the Sn concentration in the

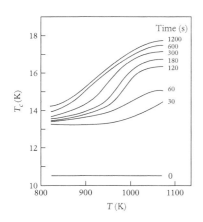

Figure 10.10 Time dependence of T_c of a Nb₃Sn layer formed after a short duration of diffusion reaction at different temperatures (after Agarwal and Narlikar, 1985b).

outermost layer rises, the mobility of Sn becomes slower in comparison to Nb, which now moves faster from the wrong sites to the correct ones, following first-order kinetics. Simultaneously, grain boundary diffusion of Sn atoms is believed to continue as before towards unreacted Nb to form the next layer of Nb_3Sn, grains which become ordered by the above duplex process. The estimated values of the activation energies of ordering for Nb and Sn are respectively 1.46 and 1.13 eV, which are consistent with the results of Dew-Hughes (1980).

10.5 Modifications of the bronze process

10.5.1 Internal tin process

In this process, which is mainly due to Hashimoto et al. (1974), all the components involved (e.g. Cu, Nb, Ta, and Sn) are ductile and the mechanical processing can therefore be carried out until the final draw without any intermittent annealing. It is shown schematically in Figure 10.11. The starting billet is formed by inserting Nb rods in a Cu cylinder that has a cylindrical hole at its centre into which a rod of Sn (or of Sn-rich Cu alloy) is inserted and that serves as a large reservoir of tin. Thus, the Sn content available for diffusion is much more than the solubility limit of 8 at% in the conventional bronze process. The method is also known as the rod-in-tube (RIT) process. The billet is extruded and drawn into hexagonal monofilamentary sub-elements to be used for reassembly. After cutting into short lengths, the sub-elements are reassembled into another Cu tube, whose inside is protected against Sn diffusion by a cylindrical Ta barrier. Before subjecting it to the final diffusion reaction, as part of the homogenisation process, the composite is subjected to a long heat treatment of several days, first at 200°C and then

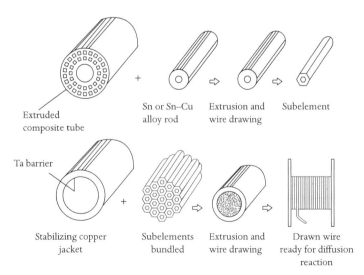

Figure 10.11 Steps involved in the internal tin diffusion process.

at 340°C. The final diffusion reaction is carried out for 50–100 hours at 700–750°C to form multifilamentary Nb_3Sn. In this way, excluding the outer Cu stabilizer, a large J_c of 1×10^9 A m^{-2} at 10 T and 4.2 K has been realised (Zeitlin et al., 1985). As with the bronze process, additives introduced in the Nb core (such as 0.2–0.37% Ti) or in the Sn rod (1.7% In) led to an enhanced J_c of 3×10^8 A m^{-2} at 15 T and 4.2 K (Yoshizaki et al., 1985), while Suenaga et al. (1985) could get a high J_c of 1.4×10^8 A m^{-2} at 18 T and 4.2 K by adding 1.5% Ti to the Sn.

10.5.2 Modified jelly roll (or Swiss roll) process

In the modified jelly roll (MJR) process for Nb_3Sn, which is due to McDonald et al. (1983) and the Teledyne-Wah Chang Company, Albany, USA, metal sheets of Nb and Cu are rolled into a spiral form, resembling a Swiss (jelly) roll, around an Sn rod and this assembly is then inserted into a copper tube and drawn into hexagonal composites (Figure 10.12). These are then cut and reassembled in a copper tube with a Ta barrier and drawn to the final size, without any need for intermediate annealing. In this way, Nb_3Sn conductors have been produced with J_c of (5–7) $\times 10^8$ A m^{-2} at 10 T and 4.2 K (Smathers et al., 1985) and the process has emerged as a potential route for making the Nb_3Al conductors to be described later.

10.5.3 ECN process

This process, which is essentially similar to the previously mentioned Kunzler PIT method, emerged from work carried out at the Netherlands Energy Research Foundation at Petten (Veringa et al., 1983, 1984). An Nb tube filled with $NbSn_2$ with about 3% Cu is embedded in a Cu tube and is extruded into a hexagonal shape. The presence of Cu lowers the reaction temperature below 700°C. The short lengths of the extruded hexagonal tube are assembled in a Cu tube and cold-drawn

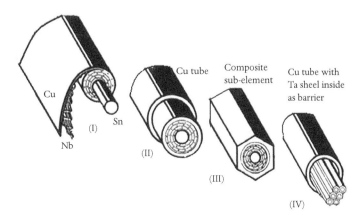

Figure 10.12 Modified jelly-roll process.

to the final size without intermediate annealing. The composite wire is diffusion-reacted at 650–700°C. The heat treatment results in the formation of Nb_6Sn_5 as a precursor phase, and the A-15 forms from the supply of Sn from the core material as well as from the precursor phase. This results in two grain sizes for Nb_3Sn: the smaller grains through the diffusion reaction with Nb and the larger grains in the phase formed from the conversion of Nb_6Sn_5. The coarse grains provide effective pinning in lower fields and the finer grains in higher fields. The process successfully yielded a commendably high J_c of 1×10^9 A m^{-2} at 10 T and 4.2 K (Hornsveld, 1988).

10.5.4   In situ Cu–Nb alloy method

In this technique, due to Tsuei (1973), the filaments are realised in situ by alloying Nb in a Cu–Sn alloy followed by cold working. Ductile alloys of Cu–Nb–Sn formed with 8 at% Sn and 10–20 at% Nb when cold-drawn, the undissolved Nb particles becoming elongated to form fine filaments along the conductor length. The discontinuous filaments too are able to exhibit superconducting connectivity through the proximity effect. A diffusion reaction carried out at 600–750°C for several hours results in the formation of an Nb_3Sn layer around each filament. Increases in the concentrations of Nb and Sn seriously lower the drawability of the composite. However, this problem is tackled by using a Cu–Nb alloy to make the wires, which, after being coated externally with Sn, are heat-treated for several days at 200–300°C before performing the final diffusion reaction at 600–750°C. In the approach of Roberge and Fishey (1977), the ternary alloy is rapidly quenched from the liquid state to produce Nb dendrites, which form fine filaments with subsequent cold drawing. The filaments are next converted to Nb_3Sn as described above. The in situ method works also with V_3Ga (Nagata et al., 1977) and with both Nb_3Sn and V_3Ga, exhibiting J_c values comparable to those of the commercial bronze-processed conductors. However, to date, long lengths of in situ A-15 conductors have not been produced commercially.

10.5.5   Powder metallurgy: cold and hot processing and infiltration methods

In this approach, following the *cold processing* approach (Flükiger et al., 1979), fine powders of hydride–dehydride Nb mixed with Cu in the stoichiometry of Cu–30 at%Nb are cold-extruded in a Cu–1.8%Be tube, which results in fine Nb filaments. The composite wire is externally coated with Sn and diffusion-reacted at 650°C for 24 hours to transform Nb to Nb_3Sn, with the overall J_c of the wire being 1.8×10^8 A m^{-2} at 14 T and 4.2 K. By increasing the Nb content to 36% and following a multiple strand bundling procedure for realising submicrometer-size filaments, J_c was nearly doubled (Otubo et al., 1983).

Instead of cold extrusion, when *hot extrusion* was tried, it led to oxygen hardening of Nb and poor filamentary structure (Schultz et al., 1975). The problem was tackled by adding a third component such as Al, Zr, Hf, Mg, or Ca in a small concentration of 0.5–2% to the powder mixture of Cu and Nb (Bormann et al., 1979). These elements possess a larger binding enthalpy than Nb for O and during hot extrusion their oxides are formed, with Nb remaining unaffected. This way, the hot extrusion led to very high J_c values of 1×10^8 A m^{-2} and 4×10^8 A m^{-2}, respectively, for Nb_3Sn and V_3Ga at 16 T and 4.2 K. Flükiger et al. (1979), on the other hand, performed all the mechanical operations at ambient temperature. After an area reduction ratios $R > 300$, they could obtain self-supporting Cu–Nb wires or ribbons, which they subsequently diffusion-reacted to form Nb_3Sn, possessing a very high J_c, comparable to the above results of Bormann et al. (1979) obtained by hot extrusion.

The third approach of powder metallurgy is the *infiltration route* due to Hemachalam and Pickus (1975). Hydride–dehydride Nb powder was compressed into a rod under low pressure and sintered at $2250°C$ in vacuum to produce an interconnected porous network in 15–20% volume fraction. The rod was then infiltrated with Sn by dipping it in a molten Sn bath at $350–400°C$ and jacketed in a Monel tube with a Ta barrier. The rod was extruded into a fine wire and heat-treated at $950°C$ to produce Nb_3Sn. Such a wire could sustain a J_c of 1×10^8 A m^{-2} at 20 T and 4.2 K (Pickus et al., 1980). Interestingly, the infiltration approach has worked successfully for materials such as Nb_3AlGe (Pickus et al., 1976) and Nb_3AlSi (Quinn, 1977) that could not be fabricated using the solid state diffusion processes described earlier.

10.6   Fabrication of Nb_3Al conductor

Among A-15s, Nb_3Al has long been noted for its very high current density in high magnetic fields and for possessing a greater strain tolerance than both Nb_3Sn and V_3Ga (Ceresara et al., 1975; Akihama et al., 1980). However, as discussed earlier, its binary phase diagram shows that the stoichiometric phase forms only at an elevated temperature ($1940°C$) and cannot be realised routinely by the conventional bronze process. Processing at very high temperature results in thermal disorder and a large grain growth detrimental for high J_c. During the last decade, however, there has been significant progress in realising stabilised Nb_3Al conductors (Takeuchi, 2000; Glowacki, 2005).

Broadly, multifilamentary Nb_3Al conductors can be processed by four separate approaches: (1) low-temperature processes ($<1000°C$), such as jelly roll, powder metallurgy, clad-chip extrusion, and rod-in-tube, some of which have already been discussed for Nb_3Sn; (2) high-temperature processing ($>1800°C$) using electron-beam or laser irradiation; (3) the rapid heating, quenching, and transformation (RHQT) method, where the quenching is carried out from 1900°C; (4) DRHQ, that is, the double rapid heating, quenching process.

10.6.1 Low-temperature processing

The processing comprises two steps as described previously with Nb_3Sn. The first step involves creation of a filamentary structure with the elemental constituents Nb and Al in a fine state of subdivision, which, in the next step, is converted into Nb_3Al at a diffusion temperature below 1000°C. In the jelly roll approach, the composite billet is prepared by winding a Nb sheet and an Al (or Al-alloy) sheet together onto an OFHC copper rod (or a Nb rod), which is inserted into a Cu tube for extrusion and cold drawing in hexagonal form. The pieces of hexagonal composite wire are restacked into a second billet, which is extruded and cold-drawn, without intermittent annealing, to form the multifilamentary wire. During mechanical processing, Nb attains a much greater hardness than Al, which adversely affects the drawing process. To circumvent this, Al sheet used in the process is alloyed with 1–2% of Mg, Zn, Ag, or Cu so that its hardness is increased to match that of Nb. In low-temperature processing, the diffusion reaction is performed at 700–900°C for many hours to form Nb_3Al.

Instead of pursuing the jelly-roll approach, one may use rod-in-tube (RIT), powder-in-tube (PIT), or clad-chip extrusion (CCE). In RIT, an Al-alloy rod (containing about 1–2% Mg, Zn, Ag, or Cu) is first inserted into a Nb tube, which is inserted into an OFHC Cu tube and the assembly is extruded and cold-drawn to form monofilamentary wire. The pieces of this wire are assembled in a second billet to form a multifilamentary wire. In PIT, a mixture containing hydride-dehydride Nb powder and Al powder in the stoichiometric ratio is filled into a Cu–Be alloy tube (or a Nb tube), which is again inserted into a Cu tube for extrusion, and multifilamentary wire is made in the next step as described above. In CCE, a Nb sheet having both surfaces clad with Al is cut into small rectangular pieces, which are used to fill a Cu tube for extrusion. The multifilamentary composite wire is made as described above. In all these approaches, the final thickness/diameter of Al in contact with Nb is maintained below about 100 nm to complete the reaction in a shorter time, giving less time for grains to grow. A short diffusion distance between Nb and Al is very important for achieving a better quality of Nb_3Al. However, low-temperature processing, in general, fails to deliver the right stoichiometry of Nb_3Al and consequently the conductor performance is sub-optimal (Takeuchi, 2000) in respect of T_c (around 15.5 K) and J_c (6×10^8 A m^{-2} at 12 T/4.2 K and 4×10^8 A m^{-2} at 18 T/4.2 K). On the other hand, Nb_3Al fabricated by low-temperature processing shows better strain tolerance than Nb_3Sn

(Zeritis et al., 1991; ten Haken et al., 1996). Consequently, despite their relatively low J_c, jelly-roll-processed Nb_3Al conductors have been preferred for winding ITER (International Thermonuclear Experimental Reactor) magnets (Takeuchi, 2000; Glowacki, 2005).

10.6.2 High-temperature processing

To realize stoichiometric Nb_3Al, it is necessary to have the diffusion reaction of Nb/Al composite take place at a higher temperature (1900°C) produced by application of a high-power electron/laser beam. For this, the composites are prepared as described above using the PIT or RIT approach, but without including Cu to avoid the formation of undesired ternary Nb–Al–Cu compounds during the high-temperature reaction. Instead of Cu tube, a Nb tube is preferred. The composite wire, moving from one reel to the other, is subjected to a short-time high-temperature reaction under electron/laser-beam irradiation to form the stoichiometric compound. The wire is subsequently heat-treated at 700°C to achieve long-range order, which results in a T_c of 18.5 K and a J_c of 1×10^8 A m^{-2} at 26 T and 4.2 K (Takeuchi, 2000). The J_c values in low fields are low owing to enhanced grain growth occurring during the high-temperature reaction.

10.6.3 RHQT method

The rapid heating and quenching technique (from 1900°C), followed by the transformation reaction (below 1000°C) provides a better alternative for realising multifilamentary long lengths of highly stoichiometric and ordered Nb_3Al with fine-grained structure (Iijima et al., 1997). Rapid heating (with a heating time of about 0.1 s) and quenching of Nb–25 at%Al wire from 1900°C to 50°C results in a metastable supersaturated bcc solid solution $Nb(Al)_{ss}$ in which the grains formed are small owing to quenching. The material formed is ductile and its strands can be easily clad and drawn with Cu in the wire form. To transform it into the A-15 phase, it is subjected to a further annealing at 700–900°C. The composite wire is made using the jelly roll process as described above, with the difference that Nb and Al foils are wound around a Nb rod instead of a Cu rod. The outer Cu jacket that was added for stabilisation is removed by etching to avoid ternary Nb–Al–Cu compounds formed by alloying with molten Cu at 1900°C. High-temperature heating can be performed by a laser or an electron beam. However, in the case of long wires, the rapid heating to 1900°C can be more conveniently achieved resistively in vacuum using the set-up depicted in Figure 10.13. As shown, the wire moving on the reels passes in the last stage through a molten Ga bath at 50°C, where it is quenched from the high temperature. The Ga bath also serves as

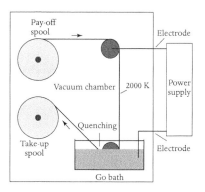

Figure 10.13 Schematic representation of the RHQT set-up.

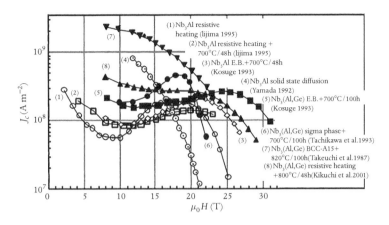

Figure 10.14 Transport $J_c(H)$ behaviour of Nb_3Al and $Nb_3(Al, Ge)$ prepared following the low-temperature, high-, temperature and RHQT approaches (after Glowacki, 2005). RHQT (curves 7 and 8) yields the best performance.

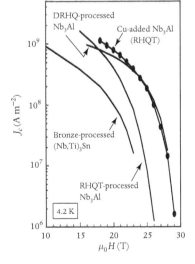

Figure 10.15 Superior performance of DRHQ-processed Nb_3Al over an RHQT sample (after Iijima et al., 2002). For comparison, the $J_c(H)$ behaviour of bronze-processed $(Nb, Ti)_3Sn$ is shown. The RHQT sample of Nb_3Al performs better than the DRHQ sample when Cu is added.

an electrode for the resistive heating. The transformation reaction for forming Nb_3Al is carried out by heating the wire for 10 hours at $800°C$. The T_c and $H_{c2}(4.2 K)$ of RHQT wire (about 17.8 K and 26 T) are, however, found to be slightly lower (by 0.7 K and 4 T) than for the high-temperature-processed samples (formed without rapid quenching) described above, but its J_c of up to 22 T is significantly higher. Figure 10.14 shows the transport J_c/H curves for Nb_3Al and $Nb_3(AlGe)$ samples prepared by the three types of processes described earlier, where the general superiority of RHQT approach can be seen. Transmission electron microscopy of RHQT samples has revealed the presence of a large stacking fault density and very small subgrains of diameter 80–150 nm. The latter are believed (Kikuchi, 2001a) to be responsible for the higher J_c, while the former possibly degrades T_c and H_{c2}. The RHQT process has been applied also for the fabrication of Nb_3Ga wire (Inoue et al., 2003).

10.6.4 DRHQ approach

The double RHQ (DRHQ) approach (Kikuchi et al., 2001b) was developed to meet the challenge posed by the stacking fault issue mentioned above. In this, the standard RHQT approach is followed, but instead of low-temperature transformation heat treatment, a repeat RHQ process at a higher temperature ($2000°C$) produces a disordered A-15 phase, and this is followed by long-range ordering heat treatment at $800°C$ for 12 hours. The material thus produced is free from stacking faults and with a higher $T_c \approx 18.4$ K. DRHQ has been found also to yield a higher J_c than RHQT (Figure 10.15). However, the addition of Cu to the RHQT-processed Nb_3Al improves T_c up to 18.3 K, H_{c2} up to 29.7 T, and J_c in higher fields, because the near-stoichiometric Nb_3Al is formed during the RHQT itself (Iijima et al., 2002) and thus no DRHQ is required.

10.7   Summary

The bronze process for A-15 superconductors, based on solid state diffusion, which was developed in the early 1970s, was a milestone in the conductor development of these highly complex brittle materials. The basic thermodynamics and various kinds of kinetics responsible for the compound layer formation in the bronze process have been discussed. Various interesting modifications of the bronze process have been presented that circumvent some of its intrinsic limitations. Progress made in the fabrication of Nb_3Al conductors has been assessed. Clearly, Nb_3Al does not merely have to serve as a better substitute for Nb_3Sn, but must also meet a more challenging onslaught from both LTS and HTS quarters. First, the Chevrel-phase superconductors with critical fields exceeding 50 T, as we will see in the next chapter, have long been potential contenders to A-15s as the possible third generation of LTS conductors. Second, we have now numerous HTS systems, including various HTS cuprates, iron-based pnictides, and magnesium diboride, all posing a major threat to the future of LTS.

CHAPTER 11

Chevrel-phase superconductors

The Chevrel phases are ternary molybdenum chalcogenides, with the general formula $M_yMo_6X_8$, where M is a metal ion and X a chalcogen (S, Se, or Te). This interesting family of materials and their unusual crystal structure were discovered by Chevrel et al. (1971), while their superconductivity, with T_c values up to 15 K, was first reported by Matthias et al. (1972). Subsequently, Fischer et al. (1973, 1974), Odermatt et al. (1974), and Foner et al. (1974) found that many members of this family possessed exceptionally high upper critical fields, surpassing the 20–40 T range of A-15 superconductors, which captured immediate attention. Superconductivity is observed for compounds containing M = Li, Na, Sc, Pb, Sn, Cu, Ag, Zn, Cd, and almost all lanthanides, except Ce, Pm, and Eu. Among the ternary compounds, $PbMo_6S_8$ (referred to as PMS) is noted for its highest T_c of 15 K and its upper critical field at 4.2 K exceeds 50 T. Close to it is SMS (where the first S stands for Sn), with T_c of around 14 K and a critical field between 40 and 50 T. These parameters are further raised to about 17 K and 55–70 T for $Pb_{0.7}Eu_{0.3}Gd_{0.2}Mo_6S_8$ and $PbGa_{0.4}W_{0.4}Mo_6S_8$ (Alekseevskii et al., 1978). Interestingly, nanocrystalline samples of PMS are found to exhibit a critical field of 100 T (Niu and Hampshire, 2003) and in this respect this class of LTS materials have emerged as competitors to HTS cuprates. The striking advantage of the Chevrel phases is that their anisotropy factor is <2, in comparison with 5–100 or more for HTS and 2.5 for MgB_2. Disregarding the anisotropy aspect, the coherence length ξ of PMS is around 2.5 nm which lies between those of Nb_3Sn ($\xi \approx 3$ nm) and the HTS cuprates (e.g. $YBa_2Cu_3O_{7-\delta}$ with $\xi \approx 1.5$ nm). Table 11.1 gives some of the superconducting parameters of typical high-field superconductors, including PMS, while Figure 11.1 compares the upper critical field behaviour of different high-field superconductors at different temperatures. Clearly, PMS seems the logical choice to succeed A-15s for the third generation of LTS conductors for ultrahigh-field applications. Figure 11.2 shows $H_{c2}(T)$ data measured (Alekseevskii et al., 1978) for different members of the Chevrel-phase family.

Table 11.1 Superconducting parameters compared for representative high-field superconductors (there is extensive variation in the literature and the values given here are averages to allow mutual comparison)

High-field superconductors	T_c (K)	Upper critical field $\mu_0 H_{c2}$ at 0 K (T)	Penetration depth λ at 0 K (nm)	Coherence length ξ at 0 K (nm)	Ginzburg–Landau parameter κ at 0 K
Nb–40% Ti alloy	9.6	14	130	4.5	29
Nb_3Sn	18.3	24	100	3.1	32
$PbMo_6S_8$	12–15	50–56	~240	~2.5	~125–260
$YBa_2Cu_3O_{7-\delta}$	91	~200–500	~150	~1.5	~100–250
MgB_2	39	~15–40	~185	~5–6	~35
$SmFeAsO_{0.8}F_{0.2}$	55	~50–150	~200	~4	~50

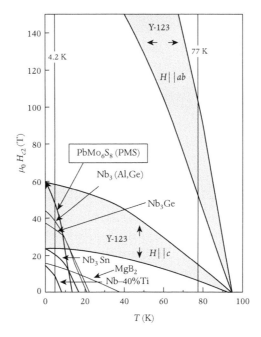

Figure 11.1 Comparison of $H_{c2}(T)$ data for the superconductor PMS with those of other high-field superconductors (after Seeber et al., 1989; Narlikar and Ekbote, 1983).

Besides the aforesaid technological advantage, the Chevrel-phase superconductors synthesised with the M element from the rare-earth group exhibit rather bizarre properties. For example, the magnetic rare-earth ions in Chevrel phases are not deleterious to superconductivity as is commonly found with the majority of other systems. In fact, in contrast, there are instances where the rare-earth ions enhance superconductivity and can lead to the extraordinary phenomenon of *field-induced superconductivity* (FIS). More excitingly, the rare-earth-doped Chevrel phases have opened up a stimulating new area of basic physics, concerning the interplay of magnetic and superconducting orders

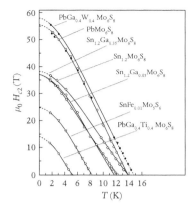

Figure 11.2 Upper critical field as a function of temperature for different compounds of the Chevrel-phase family (after Alekseevskii et al., 1978). It should be noted that the data from different research groups differ.

and their coexistence. A comprehensive review of Chevrel phases, covering their unusual normal-state and superconducting properties, may be found elsewhere (Fischer, 1978).

11.1 Crystal structure and stoichiometry

All the Chevrel-phase compounds, with the chemical formula $M_yMo_6X_8$, comprise Mo_6X_8 units as basic building blocks of the rhombohedral-hexagonal structure with space group $R(-3)$ depicted in Figure 11.3. Each unit is in the form of a distorted cube with X-atoms located at its corners while the six Mo atoms are placed near the centres of each of its six faces to form a dense *cluster* in the shape of a distorted Mo octahedron. The intracluster distance, for example, varies from 0.267 nm for $Cu_{3.6}Mo_6S_8$ to 0.276 nm for Mo_6S_8 and the intercluster distance from 0.308 nm for Mo_6S_8 to 0.366 nm for $PbMo_6Se_8$. The Mo 3d-orbitals are sufficiently extended to favour metallic bonding. Both intra- and intercluster distances are important in controlling T_c. Such pseudo-cubes are stacked in an almost-cubic unit cell, with each of its one-eighth unit cells being occupied by a Mo_6X_8 unit, as shown in Figure 11.3(b). The sides of the unit cell are about 0.65 nm and those of the Mo_6S_8 cluster about 0.38 nm. The M-elements enter the structure to become located within the three-dimensional channels separating the neighbouring pseudo-cubes where there is sufficient space available, either in proximity to the origin of the unit cell or between the neighbouring planes of the pseudo-cube along the rhombohedral axis. Broadly, in $M_yMo_6X_8$, there exist two situations: (i) M large and (ii) M small. When M is large, such as for Na, Pb, Sn, and various rare earths, its concentration $y \approx 1$ and such materials are stoichiometric. On the other hand, for small M such as Li, Mg, Cu, and Zn, $y \approx 1$–4 and such compounds are non-stoichiometric. Near the origin, there are six closely spaced statistical sites, effectively serving as a single site to accommodate

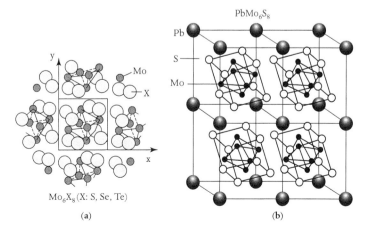

Figure 11.3 The binary compound Mo_6X_8 (X = S, Se, or Te) shown in (a) forms the basic building block of the ternary molybdenum chalcogenide (the Chevrel-phase structure) shown in (b). The Mo octahedra that can be seen here are the most important part of the structure responsible for superconductivity.

Mo_6X_8 (X: S, Se, Te)

(a)

$PbMo_6S_8$

(b)

a large atom M. In a non-stoichiometric compound, to accommodate up to 4 small M atoms, say of Cu, statistically there are 12 sites possible. Six inner sites are placed around the origin, while the outer six are located along the rhombohedral axis, approximately on the planes of the pseudo-cube surrounding the origin. The rhombohedral structure of the stoichiometric compounds is stable down to low temperatures and their rhombohedral angle is slightly smaller than 90°, generally between 89° and 90°, which makes the structure pseudo-cubic. The non-stoichiometric compounds have their rhombohedral angle larger than 90°, between 92° and 95°, and these materials generally develop triclinic disorder on cooling, or when their concentration y increases. As the size of M decreases, the delocalisation at the statistical sites is enhanced, the pseudo-cube becomes more compressed, and the rhombohedral angle increases above 90° (Yvon, 1979). Their density of states (DOS) at E_F is low and they generally do not exhibit superconductivity.

Tables 11.2 and 11.3 show superconducting ternary molybdenum sulfides and selenides, respectively, with their lattice parameters and T_c values, where non-stoichiometric compounds are printed in italics. Table 11.4 gives these quantities for binary, pseudo-binary, and halogen-doped phases. Te-based ternary compounds are difficult to form and, as may be noted from the tables, none exhibits superconductivity, with three exceptions, namely $Mo_6Te_6I_2$ ($T_c = 2.6$ K), $Mo_4Re_2Te_8$ ($T_c = 3.5$ K), and Mo_5RuTe_8 ($T_c \approx 2$ K). In these, superconductivity occurs in non-superconducting Mo_6Te_8 only after partial substitution of Te by I or of Mo by Re or Ru (Chevrel, 1981).

While Mo_6Se_8 and Mo_6Te_8 can be formed easily by reacting the two components together, an indirect approach has to be followed to synthesise Mo_6S_8: first, $M_yMo_6S_8$ (M = Cu, Mg, or Zn) is synthesised, and then M is washed out of the ternary phase with mineral acids (Chevrel, 1981). Mo_6S_8 is unstable above 470°C, while the other two compounds can exist up to about 1100°C.

11.2 Occurrence of superconductivity in Chevrel phases

11.2.1 Phononic structure

In general, for any conventional superconductor, one would like to know whether the superconductivity has its origin in its phononic structure or electronic structure. Interestingly, the Chevrel phases possess unusual lattice properties with low-frequency modes, as revealed by the low-temperature heat capacity and inelastic neutron scattering measurements (Bader et al., 1976a, b). These are attributed to the presence of tightly bound Mo_6X_8 molecules that are weakly coupled to each other and to the M-atom (e.g. Pb or Sn). In particular, there is

Table 11.2 Lattice parameters and critical temperatures of ternary molybdenum sulfides

| Compound* | Rhombohedral lattice parameters | | T_c (K) | $\mu_0 H_{c2}$ at 0 K (T) |
	a_R (nm)	α_R (°)		
$Li_2Mo_6S_8$	0.662	94.53	5.5	—
$Na_2Mo_6S_8$	0.653	89.83	8.6	—
$Mg_{1.14}Mo_{6.6}S_8$	0.651	93.58	3.5	—
$ScMo_6S_8$	—	—	3.6	—
$Y_{1.2}Mo_6S_8$	0.644	89.53	2.3	0.5
$LaMo_6S_8$	0.651	88.90	7.1	~6
$PrMo_6S_8$	0.649	89.00	4.0	—
$NdMo_6S_8$	0.648	89.06	3.5	—
$Sm_{1.2}Mo_6S_8$	0.647	89.20	2.9	—
$Gd_{1.2}Mo_6S_8$	0.647	89.30	1.4	0.12
$Tb_{1.2}Mo_6S_8$	0.646	89.42	1.65	0.20
$Dy_{1.2}Mo_6S_8$	0.645	89.50	2.1	0.13
$Ho_{1.2}Mo_6S_8$	0.645	89.53	2.2	—
$Er_{1.2}Mo_6S_8$	0.644	89.66	1.85	0.12
$Tm_{1.2}Mo_6S_8$	0.644	89.80	2.1	—
$Yb_{1.2}Mo_6S_8$	0.649	89.40	9.1	—
$Lu_{1.2}Mo_6S_8$	0.643	89.90	2.0	—
$Cu_{1.8}Mo_6S_8$	0.650	94.93	10.8	14–16
$Cu_{3.6}Mo_6S_8$	0.657	95.56	6.4	5–8
$Ag_{1.6}Mo_6S_8$	0.648	91.95	9.1	—
$Zn_{1.1}Mo_6S_8$	0.649	94.68	3.6	—
$CdMo_6S_8$	0.652	92.82	1.5	—
$Sn_{1.2}Mo_6S_8$	0.652	89.63	14.2	30–40
$PbMo_6S_8$	0.654	89.47	15.0	45–60

Compounds in italics are non-stoichiometric.
From Narlikar, A.V. and Ekbote, S.N. *Superconductivity and Superconducting Materials*. South Asian Publishers, New Delhi (1983).

a sharp peak between 4 and 5 meV associated with Pb or Sn, which is absent in the binary Mo_6Se_8 phase, and this part of the spectrum shows low-temperature softening. It was therefore suggested that this softening was responsible for the higher T_c observed for $PbMo_6S_8$ and $SnMo_6S_8$. However, this explanation had to be rejected following the subsequent observation that similar low-frequency softened modes were also present in various low-T_c Chevrel phases (Kimball et al., 1976).

11.2.2 Electronic structure

Superconductivity of Chevrel-phase compounds is considered to be primarily due to the electronic structure of the mutually well-separated

Table 11.3 Lattice parameters and critical temperatures of ternary molybdenum selenides

Compound*	Rhombohedral lattice parameters		T_c (K)	$\mu_0 H_{c2}$ at 0 K (T)
	a_R (nm)	α_R (°)		
$Y_{1.2}Mo_6Se_8$	0.671	89.23	6.2	—
$LaMo_6Se_8$	0.680	88.96	11.4	45–55
$Pr\,Mo_6Se_8$	0.677	88.80	9.2	21
$NdMo_6Se_8$	0.676	89.07	8.2	—
$Sm_{1.2}Mo_6Se_8$	0.673	88.93	6.8	—
$Gd_{1.2}Mo_6Se_8$	0.673	89.13	5.6	—
$Tb_{1.2}Mo_6Se_8$	0.671	89.13	5.7	—
$Dy_{1.2}Mo_6Se_8$	0.670	89.27	5.8	—
$Ho_{1.2}Mo_6Se_8$	0.670	89.30	6.1	—
$Er_{1.2}Mo_6Se_8$	0.670	89.33	6.2	—
$Tm_{1.2}Mo_6Se_8$	0.669	89.40	6.3	—
$Yb_{1.2}Mo_6Se_8$	0.678	89.37	5.8	—
$Lu_{1.2}Mo_6Se_8$	0.669	89.80	6.2	—
$Cu_{1.2}Mo_6Se_8$	0.679	94.91	5.9	—
$Ag_{1.2}Mo_6Se_8$	0.672	91.37	5.9	—
$Sn_{1.5}Mo_6Se_8$	0.678	89.60	4.8	—
$Pb_{1.2}Mo_6Se_8$	0.681	89.23	3.6	$\sim$4

Compounds in italics are non-stoichiometric.
From Narlikar, A.V. and Ekbote, S.N. *Superconductivity and Superconducting Materials*. South Asian Publishers, New Delhi (1983) and Decroux, M. and Seeber, B. *Concise Encyclopedia of Magnetic and Superconducting Materials* (ed. J. Evetts). Pergamon Press, Oxford (1992), p. 61.

Table 11.4 Lattice parameters and critical temperature of binary, pseudo-binary, and halogen-substituted superconducting Chevrel phases

Compound	Rhombohedral lattice parameters		T_c (K)	$\mu_0 H_{c2}$ at 0 K (T)
	a_r (nm)	α_R (°)		
Mo_6S_8	0.643	91.34	1.8	—
Mo_6Se_8	0.666	91.58	6.3	8–18
Mo_6Te_8	0.710	92.60	n.s.	—
$Mo_6S_6Br_2$	0.650	94.43	13.8	—
$Mo_6S_6I_2$	0.656	94.50	14.0	—
Mo_6Se_7Br	0.667	92.60	7.1	—
Mo_6Se_7I	0.672	93.62	7.6	—
$Mo_6Te_6I_2$	0.709	93.43	2.6	—
$Mo_4Re_2Te_8$	0.703	93.04	3.5	—

From Narlikar, A.V. and Ekbote, S.N. *Superconductivity and Superconducting Materials*. South Asian Publishers, New Delhi (1983).

dense Mo octahedrons, which are therefore analogous to the three dense orthogonal linear chains of transition metal atoms of the quasi-one-dimensional A-15 structure (Chapter 10). This makes the Mo d-band at E_F narrow, which enhances both $N(0)$ at E_F and T_c. Since the range of coherence ξ is very small ($\sim$2.5 nm) and nearly isotropic, the Chevrel-phase superconductors are considered *quasi-zero-dimensional*. Application of pressure reduces the cell volume and decreases the inter-cluster spacing. For many Chevrel-phase superconductors, T_c in the high-pressure range is indeed found to decrease linearly with pressure (Figure 11.4) (Shelton, 1975), and in the case of the PMS superconductor, $dT_c/dP = -1.6$ K GPa^{-1}.

The role of the M-atoms in superconductivity is essentially twofold. First, as they are placed within the channels separating the neighbouring Mo octahedrons, their presence through their ionic size can influence both intercluster and intracluster spacing and thereby affect T_c in the manner described above. For instance, the non-superconducting Mo_6S_8 has a relatively larger intracluster spacing of 0.278 nm and a smaller intercluster spacing of 0.308 nm in comparison with PMS, where the Pb ions are large, having respective spacings of 0.270 nm and 0.327 nm. The T_c is thus very sensitive to both inter- and intracluster spacing.

Second, charge transfer from M, X, and halogen atoms (if they have been doped at the X-site) to Mo can give rise to a change in the position of E_F and thereby affect T_c and other electronic properties. Band calculations (Mattheiss and Fong, 1977; Andersen et al., 1978; Jarlborg and Freeman, 1980) suggest that for a stable $M_yMo_6X_8$ formation, the Mo cluster should carry 20–24 d-electrons. In the binary Mo_6S_8 unit, with S^{2-}, which is unstable and non-superconducting, owing to charge transfer (holes) from S, the cluster has 20 d-electrons, while in PMS, which possesses the highest T_c of 15 K, this is raised to 22. Thus, the effect of Pb^{2+} in the ternary compound is not only to increase the intercluster spacing, but also to modify the charge on the clusters, which sensitively influences the position of E_F at the d-band. A similar effect occurs

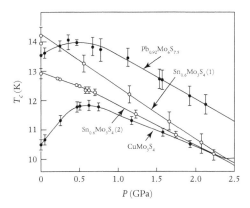

Figure 11.4 Pressure dependence of T_c of some Chevrel-phase superconductors (after Shelton et al., 1975).

through charge transfer from halogens, such as Br^- and I^- introduced to partly replace X^{2-}. For example, both $Mo_6S_6Br_2$ and $Mo_6S_6I_2$ have 22 d-electrons on their Mo clusters and possess relatively high T_c values of 13.8 K and 14 K, respectively. Likewise, we have previously mentioned examples of $Mo_4Re_2Te_8$ ($T_c = 3.5$ K) and Mo_5RuTe_8 ($T_c \approx 2$ K), where the Mo clusters are mixed, but which still exhibit superconductivity when their d-electrons add up to 22. On the other hand, $Mo_2Re_4S_8$ and $Mo_2Re_4Se_8$ with 24 d-electrons on their mixed clusters are semiconducting. A valence electron concentration (VEC) of 3.66 electrons per Mo atom of the cluster seems necessary for superconductivity.

An interesting situation is when M is magnetic, belonging to either the rare-earth group or the $3d$ transition metal group. While a minute concentration ($\sim$0.5 at%) of Fe or Mn from the $3d$ group is found to quench superconductivity completely, the magnetic rare-earth atoms (except Ce and Eu) can exist as part of the lattice with the material showing superconductivity. Being large, the rare-earth atom occupies the site at the cube centre, from which the Mo atoms are separated by a distance >0.423 nm. This distance is too large for the spins of the rare-earth atom to affect the conduction electrons of Mo. On the other hand, the magnetic $3d$ atoms, such as Fe or Mn, are of smaller radius and their occupancy at the 12 statistical sites, as discussed above, is at a much shorter distance of about 0.28 nm from the Mo atoms, from where they can readily destroy superconductivity.

11.3   Synthesis of bulk samples

11.3.1   Solid state reaction process

For synthesising bulk samples, the powder metallurgy route is followed, involving a two-step reaction procedure (Seeber, 1998; Zheng et al., 1995). For PMS, high-purity powders of the elements Pb, Mo, and S, mixed in a nominal composition of $PbMo_6S_8$, are sealed under vacuum in a clean silica tube, which is heated for 4 hours at 450°C. The furnace temperature is then increased to 650°C at a rate of 33°C/h and heated for 8 hours, after which it is rapidly cooled to room temperature. The reacted powder is thoroughly ground and compressed into pellet form. It is again vacuum-sealed in a silica tube and reacted for 44 hours at 1000°C to form the PMS compound. PMS is stable over a large temperature range of 450–1650°C (Decroux et al., 1993). One may therefore use higher reaction temperatures, although this will lead to an increase in the final grain size from a few hundred nanometres to about a hundred micrometres. The compounds produced are often contaminated by oxygen released by the silica tube, which degrades T_c (Hinks et al., 1983). This can be avoided by using a glove box (Zheng et al., 1995).

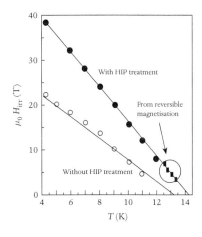

Figure 11.5 There is a marked improvement in H_{irr} due to hot isostatic pressing (HIP) treatment (after Ramsbottom and Hampshire, 1997).

11.3.2 Hot isostatic pressing (HIP) treatment

The samples synthesised by the procedure just described are porous, with a poor intergrain connectivity and large grain size, and therefore suffer from the weak-link and granularity effects, all leading to a low J_c. The problem is circumvented (Zheng et al., 1995) by hot isostatic pressing (HIP). For this, the pellet formed as above is wrapped in Mo foil, serving as a barrier to avoid the compound reacting with the container, and is vacuum-sealed in a stainless-steel tube. HIP treatment is carried out at 0.1–0.2 GPa at a temperature of 700–1100°C for 8–10 hours. To avoid oxygen contamination, the process is carried out using a glove box. The sample thus produced possesses a near-optimal T_c and a sharp resistive transition of width $\Delta T_c \approx 0.2$ K. The vast improvement in the irreversibility field H_{irr} of PMS resulting from HIP treatment is shown in Figure 11.5 (Ramsbottom and Hampshire, 1997).

11.4 Upper critical field

The dense Mo clusters that are responsible for the superconductivity of Chevrel phases are also important for their very high critical fields (Fisher, 1978). As already discussed, the presence of these clusters makes the material low-dimensional, the range of coherence ξ being reduced in all three directions. H_{c2} is related to ξ by $H_{c2} = \phi_0/2\pi\xi^2$, where ϕ_0 is a quantum of magnetic flux. In terms of the microscopic theory, $\xi \propto v_F/T_c$, where v_F is the Fermi velocity, which is small for these materials with their high DOS. More importantly the Chevrel phases are extremely dirty materials, possessing a short electron mean free path and a high value of normal-state resistivity ρ_n. As discussed in Chapter 5 (equation (5.19)), a large normal-state resistivity corresponds to a higher upper critical field given by $\mu_0 H_{c2}(0) = 3.09\gamma\rho_n T_c$ (T), where γ is in J m^{-3} K^{-2} and ρ_n in Ω m. Careful measurements on single crystals of $Cu_x Mo_6 S_8$ and $Mo_6 Se_8$ (Flükiger et al., 1978, 1979) revealed a ρ_n in the range of $(1–2) \times 10^{-5}$ Ω m at room temperature and a very short electron mean free path of about 4 nm, which are responsible for the high upper critical field. For nanocrystalline samples, the resistivity is several times larger, although their T_c is lowered to around 12 K and so is their γ but their critical fields shoot up to the region of 100 T (Niu and Hampshire, 2003). Very small grain size contributes significantly to flux pinning and to enhancing the irreversibility field H_{irr} and J_c.

11.5 Critical current density: inherent problems and progress in raising J_c

Achieving an acceptably large critical current density J_c with Chevrel-phase superconductors in high magnetic fields has remained a challenge for the last 30 years. This may be ascribed to various intrinsic factors, some of which are shared with HTS cuprates. The Chevrel-phase superconductors, such as PMS or SMS, and also many of the HTS, are noted for their ultrahigh upper critical field H_{c2} and correspondingly very small coherence length ξ (Table 11.1). Owing to the very small ξ, coupled with the fact that these materials are fabricated as sintered compacts using the powder metallurgy route, their intergrain connectivity is generally poor, which gives rise to serious granularity and weak-link problems degrading J_c (Portis et al., 1988; Decroux et al., 1990; Zheng et al., 1995; Seeber, 1998; Metskhvarishvili, 2009). In impure samples, the presence of impurities at the grain boundaries further aggravates these problems. The large grain size of the sample also needs to be dealt with. Most of these challenges are, however, effectively met by applying HIP processing during fabrication. Although their anisotropy (anisotropy factor $\gamma = 1.20$) is small, it is still non-trivial since it effectively contributes to lowering the H_{c2} of polycrystalline samples by 20% and simultaneously reduces the high-field J_c, which even otherwise is only moderate for these compounds, as will be explained.

PMS is an extreme type II superconductor and typically possesses a very large Ginzburg–Landau parameter κ of 125–260 Table 11.1), depending on the way the sample has been processed, and for nano-grained samples, produced by prolonged milling for 200 hours, followed by HIP processing and annealing, κ increases to >500 (Niu and Hampshire, 2003). This represents the largest κ value known for any superconductor. As discussed in Chapter 6, for a superconductor with large κ, both core pinning and magnetic pinning are equally likely, and the problem with Chevrel-phase superconductors is that both types of pinning interactions are inherently weak. For core pinning, the pinning force varies as $\frac{1}{2}\xi\mu_0 H_c^2$, while with magnetic pinning it is given by $\phi_0 H_{c1}/\lambda$ (Narlikar and Ekbote, 1983). Because of the short ξ and small thermodynamic critical field H_c of Chevrel-phase superconductors, the core-pinning contribution is weak, and, similarly, the low value of the lower critical field H_{c1} and a large penetration depth λ make also the magnetic pinning poor. Alternatively, if we consider $\Delta\kappa$ pinning (Chapter 6), which holds for a variety of pinning entities such as normal particles, grain boundaries, and dislocation cell structures, J_c varies as $H_{c2}(T)\Delta\kappa/\kappa^3$. Although for normal particles embedded in a large-κ matrix $\Delta\kappa$ is optimum, the overall pinning effect resulting from $\Delta\kappa$ is significantly diminished owing to the κ^3 factor in the denominator. Compared with Nb_3Sn, whose κ is 4–5 times smaller than that of PMS, the pinning effects in the latter are expected to be around 50–100 times smaller. Consequently, despite a higher $H_{c2}(T)$, the net J_c of PMS is only moderate in comparison with that of Nb_3Sn. The only way to deal with this situation would be to optimise the metallurgical factors

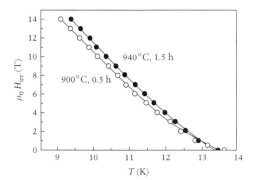

Figure 11.6 Improvement in the irreversibility field of PSMS resulting from an increase in temperature and duration of HIP treatment (after Cheggour et al., 1998).

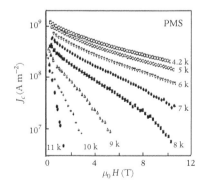

Figure 11.7 High $J_c(H)$ values of a PMS sample, deduced from magnetisation behaviour, exhibiting a current density $J_c(10\ \mathrm{T}, 4.2\ \mathrm{K}) = 8 \times 10^8$ A m^{-2} (after Zheng et al., 1997).

involved in sample processing to yield a more homogeneous sample of improved intergrain connectivity embedded with strong flux-pinning centres. Finally, as pointed out earlier, the pair-breaking (depairing) current density of a superconductor, which represents the upper limit of J_c, is proportional to H_c/λ. Small H_c and large λ make these materials possess intrinsically low J_c, and therefore the strategies just described have inherent limitations for Chevrel-phase superconductors.

The J_c values reported for short samples are moderately large and promising. For instance, a zero-field $J_c \approx 1 \times 10^9$ A m^{-2} at 4.2 K has been commonly found for hot isostatically pressed (HIP) PMS (Zheng et al., 1995; Seeber, 1998). A threefold increase has been reported for a sample with the nanosized grains produced by a long milling procedure prior to HIP (Niu and Hampshire, 2003). Although the above J_c may appear high, for the reasons discussed above, it is still an order magnitude lower than the zero-field J_c for Nb–Ti or Nb$_3$Sn at 4.2 K. The HIP-processed samples, however, behave better under high fields, which may be ascribed to increase in H_{c2} and decrease in granularity and weak links (Seeber, 1998). For PMS, typically the J_c values found by different groups at 4.2 K are 5×10^8 A m^{-2} at 10 T (Seeber et al., 1989), (1 − 2) × 10^8 A m^{-2} at 20 T (Seeber et al., 1989; Bouquet et al., 1994; Zheng, 1995), and 1×10^8 A m^{-2} at 24 T (Kubo et al., 1993). Interestingly, the performance of PMS is markedly improved by adding or partly substituting Pb by Sn (Rickel et al., 1986; Cheggour et al., 1998; Selvam et al., 1995; Capone et al., 1990). For Pb$_{0.6}$Sn$_{0.4}$Mo$_6$S$_8$, when samples were HIP-processed at a higher temperature for a longer time (940°C for 1.5 hours compared with 900°C for 0.5 hour), the irreversibility field (Figure 11.6) was noticeably enhanced (Cheggour et al., 1998). Gd substitution in (Pb, Gd)Mo$_6$S$_8$ improved T_c and H_{c2} but degraded J_c (Leigh et al., 1999; Zheng et al., 1997). The high-J_c performance of a pristine PMS sample is depicted in Figure 11.7 (Zheng et al., 1997). As may be seen, at 4.2 K, J_c(zero field) $> 1 \times 10^9$ A m^{-2}, which is lowered to 2×10^8 A m^{-2} at 10 T. The corresponding degraded figures for the Gd-doped compound Pb$_{0.8}$Gd$_{0.2}$Mo$_6$S$_8$ were found to be 8×10^8 A m^{-2} in

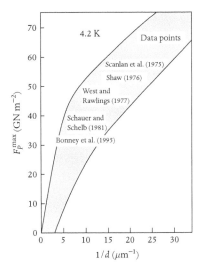

zero field and 2×10^7 A m^{-2} at 10 T. The observed J_c decrease in the Gd-doped sample is possibly due to grain boundary contamination by Gd (Zheng et al., 1997). Because of the small ξ (<3 nm), the conventional lattice defects in the compound are too coarse and only the fine-grained structure proves effective in flux pinning at low magnetic fields. As with Nb$_3$Sn (Chapter 10), the pinning force in polycrystalline Chevrel-phase compounds is inversely proportional to the grain diameter d. When the maximum pinning force determined for both Nb$_3$Sn and SMS (Bonney et al.1995) is plotted as a function of $1/d$, the data points for the two compounds lie mutually close within the grey region in Figure 11.8. This substantiates the importance of small grains in the J_c enhancement of these compounds (Niu and Hampshire, 2004) (Figure 11.9).

There are indications that at fields exceeding 10 T, the pinning is provided by intragrain structures such as planar defects and interstitial oxide particles whose exact nature remains unclear (Bonney et al., 1995). Neutron irradiation also gives rise to defects that improve J_c, but the pinning-force density F_p as a function of reduced field $h = H/H_{c2}$ does not follow any scaling law for irradiated samples (Rosel and Fischer, 1984). In most other cases, however, the observed field dependence (Zheng et al., 1995; Bonney et al., 1995) of the pinning force may be fitted to the function $h^{1/2}(1-h)^2$, with $h = 0.2$. This is similar to the behaviour of A-15 superconductors (see Chapter 9) and is indicative of grain boundary pinning. These studies further exhibit the occurrence of a second peak in the pinning curve, corresponding to the intragranular pins becoming effective in higher fields at lower temperatures, as shown in Figure 11.10 for SnMo$_6$S$_8$ (Bonney et al., 1995).

Figure 11.8 Plot of maximum pinning force (4.2 K) versus inverse of grain diameter for polycrystalline samples of SnMo$_6$S$_8$ (SMS) and Nb$_3$Sn (after Bonney et al., 1995). The data points (references mentioned) for both compounds fall within the grey region. Thus, their critical current density would increase on lowering the grain size.

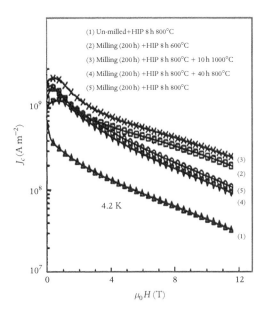

Figure 11.9 High-$J_c(H)$ behaviour of PMS samples (nanograined) subjected to prolonged milling followed by processing under varied thermal conditions (after Niu and Hampshire, 2004).

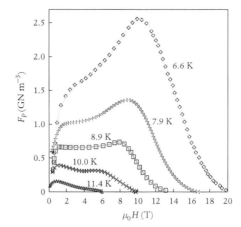

Figure 11.10 Pinning force F_P as a function of H at various temperatures ranging from 6.6 to 11.4 K, showing the appearance of two distinct peaks: one at lower temperature and higher magnetic field (intragranular pinning centres) and the other at higher temperature and lower magnetic field (grain boundary pinning) (after Bonney et al., 1995).

It is relevant to mention that PMS may be strained by up to 0.65% without irreversibility in J_c, compared with a figure of 0.5% for Nb_3Sn (Seeber, 1998). One may therefore expect to be able to achieve an increase in the acceptable bending strain to make a change over from the *wind-and-react* technique to the more convenient *react-and-wind*.

11.6 Conductor development of Chevrel-phase compounds

The starting materials for the powder metallurgy process can be either prereacted PMS powder or an unreacted stoichiometric mixture of the constituents. The powder mixture is isostatically cold-pressed and machined in cylindrical form for extrusion. Typically, as with A-15s, the PMS conductor is required to be fabricated in a complex composite form where it is finally left surrounded by a barrier layer of Mo embedded in a stainless-steel matrix. Accordingly, a billet is required to be properly assembled with different components before it is extruded and drawn into wire. The purpose of the Mo layer is to prevent the contamination and degradation of the PMS compound during thermal treatments. Additionally, for thermal stability, a Cu layer is also sometimes introduced between the Mo barrier and the outer stainless-steel shield. On the other hand, the Mo layer, apart from its role as a barrier, effectively also serves the purpose of Cu in providing thermal stability. The role of the stainless-steel matrix is to provide stress compensation during the cooling to 4.2 K. Compared with the PMS phase, the Mo layer possesses a much smaller coefficient of thermal expansion, with the result that during cooling the PMS phase is subjected to tensile strain, developing cracks. The outer stainless-steel matrix combats the above problem and gives mechanical stability and strength to the PMS conductor. Owing to the stainless-steel matrix, the yield strength of the PMS conductor is around 750 MPa at 4.2 K, which is about three times larger than that of a bronze-processed Nb_3Sn conductor and ideal to withstand the large magnetomechanical forces produced in ultrahigh-field solenoid windings. Alternative barrier materials such as Nb and Ta do not require hot processing, but above 800°C they react with PMS phase to drastically reduce its T_c.

The final extrusion billet therefore comprises an outer can of stainless steel, with an internal diffusion barrier of Mo, Nb, or Ta, into which is inserted the isostatically pressed and machined powder serving as the core. The billet is vacuum-sealed and is reduced in diameter by extrusion and final wire drawing. If the wire has been drawn with its core made from preformed PMS powder (prereacted PMS), it is subjected to recovery annealing at 700°C to improve or restore the intergrain connectivity that is mandatory for high J_c. On the other hand, when the powder used contains the unreacted components, the heat treatment is performed at about 1000°C to form the PMS phase. Chevrel-phase wires of 0.4 mm diameter with length in the kilometre range have been successfully fabricated, while the feasibility of multifilamentary wires has also been demonstrated (Seeber, 1998). Unfortunately, at present, the optimised short-sample J_c values cannot be maintained over the kilometre length of the conductor wire. The main reasons for this are chemical inhomogeneities, stoichiometry, the grain boundary structure, granularity, and microcracks, all of which change over the length of the conductor wire. The average of T_c and H_{c2} distributions are the bulk or *effective* values T_c^* and H_{c2}^*, which are generally found to be significantly degraded from their respective optimised values (Seeber et al., 1989; Seeber, 1998). This clearly calls for conductor processing to yield kilometre lengths of wire that would be stoichiometric and homogeneous, possessing sound intergrain connectivity and embedding strong flux-pinning centres. This is a major challenge that Chevrel-phase compounds must face for them to become the third generation of LTS conductors for ultrahigh-field applications.

11.7 Nature of superconductivity of Chevrel-phase compounds

Finally, it is relevant to ask whether the superconductivity of Chevrel-phase compounds is conventional or unconventional. Ever since their discovery, they have generally been considered as BCS singlet superconductors with s-wave pairing and, at least for $M_yMo_6X_8$ where M is a non-lanthanide, there seemed little reason to expect anything otherwise. However, materials such as PMS, SMS, and PSMS are all noted for

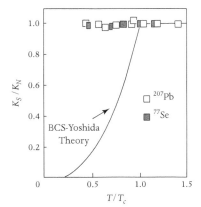

Figure 11.11 The ^{207}Pb and ^{77}Se NMR Knight shift for Pb$_{1.12}$Mo$_6$Se$_{7.5}$ does not decrease as the temperature is lowered below T_c, which is an indication of unconventional superconductivity behaviour (after Sano et al., 1980).

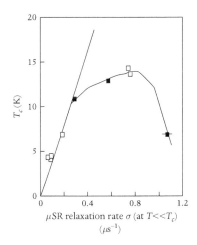

Figure 11.12 T_c as a function of μSR relaxation rate in PMS (after Uemura, 1991). At very low temperatures, T_c increases linearly with σ, which has also been found for HTS cuprates.

their ultrahigh critical fields of 40–100 T, far exceeding their Pauli limit, which is a common trait of unconventional superconductivity. Also, NMR studies of the ^{207}Pb and ^{77}Se Knight shift (Sano et al., 1980) in Pb$_{1.125}$Mo$_6$Se$_{7.5}$ have shown that it remains unchanged below T_c (Figure 11.11), in contradiction with the BCS theory. In addition, for a conventional type II superconductor in the mixed state, where the magnetic field inside is inhomogeneous, the NMR spin-echo decay has a Lorentzian form and the pertinent decay time is generally found to be much longer in the superconducting state than in the normal state. The measured data for Chevrel phases are just the opposite. On the other hand, μSR studies (Uemura, 1991) found the temperature dependence of the penetration depth $\lambda(T)$ in PMS and other Chevrel phases to be consistent with conventional *singlet* pairing. However, Uemura (1991) observed the μSR relaxation rate $\sigma = n_s/m^*$ (the ratio of pair density to effective mass) at low temperatures for different Chevrel-phase superconductors to vary linearly (Figure 11.12) with T_c. Similar behaviour was previously reported for unconventional HTS cuprates (Seaman et al., 1990). This shows a similarity between the Chevrel phase and HTS. The isotope effect measurements on Chevrel-phase superconductors are too scant to help in resolving the issue of their superconducting order parameter. However, in the case of the binary compound Mo$_6$Se$_8$, Culetto and Pobell (1978) found the isotope effect exponent for both Mo and Se to be about 0.27, suggesting that the phonon modes associated with both Mo and Se contribute equally to the conventional superconductivity of the compound.

An indication of unconventional superconductivity in PMS with the presence of nodes in the superconducting gap has come from scanning tunnelling spectroscopy (STS) studies (Dubois et al., 2007) of carefully prepared surfaces of single-crystalline samples. The shape of their zero-bias conductance spectra at low temperatures, below T_c, shows the presence of low-energy excitations, and when the observed shape is

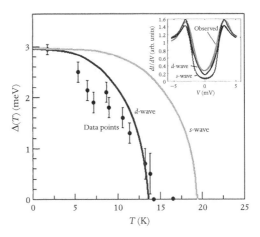

Figure 11.13 The main figure shows the temperature dependence of the superconducting gap, with the observed data compared with the expected behaviour for s- and d-wave superconductivity. The date points are closer to d-wave than to s-wave behaviour. The inset depicts the observed zero-bias tunnelling spectra at 1.8 K (the innermost spectrum). The outermost spectrum is the expected spectrum for the s-wave case, while the middle one is for the d-wave case, which lies closer to the observed spectra. (After Dubois et al. (2007).)

compared with the expected fits for s- and d-wave pairing (Figure 11.13), it is closer to the latter. The temperature dependence of the superconducting gap for PMS, deduced from the observed spectra at different temperatures, is shown in the main part of Figure 11.13. The data points lie closer to the d-wave than to the s-wave. Interestingly, the gap coefficient $2\Delta/k_B T_c = 4.9$ measured for PMS is very close to that of HTS. On the other hand, subsequent STS studies by Petrović et al. (2011) on SMS and PMS at subkelvin temperatures have revealed two distinct superconducting gaps of 3 meV and 1 meV for the former and 3.1 meV and 1.4 meV for the latter. The spectra of both materials are found to be consistent with an anisotropic two-band BCS s-wave gap function. The existence of two gaps is corroborated by specific heat measurements (Petrović et al., 2011), giving credence to multiband superconductivity in Chevrel-phase compounds.

11.8 Summary

Chevrel-phase superconductors are potential materials for ultrahigh-field applications. They are noted for their extremely high critical fields. These materials have some similarities with HTS cuprates and also face some of the same hurdles for practical applications. Superconductivity in Chevrel-phase compounds was previously thought to be of conventional singlet s-wave type, although experimental data exhibit features of both conventional and unconventional superconductivity. Interestingly, recent studies have found them to be d-wave superconductors, and even multiband superconductivity has been manifested in PMS and SMS compounds in STS studies. A full understanding of these unusual materials is still awaited.

The inherent theoretical and practical problems facing conductor development using these superconductors have been analysed in terms of their short coherence length, low thermodynamic critical field, and unusually large Ginzburg–Landau parameter. Chevrel-phase superconductors possess an intrinsic low J_c, which poses a major problem for conductor development in practical applications.

CHAPTER 12

Rare-earth-based ternary superconductors and quaternary borocarbides

Ever since the discovery of superconductivity, for various reasons superconductivity and magnetism have been considered inimical to each other. The occurrence of the Meissner effect and the existence of a critical field are the two fundamental characteristics that respectively show that a superconductor expels an external magnetic field from its interior and that if the magnetic field is sufficiently large, it quenches the superconducting phenomenon abruptly by penetrating the sample. As discussed in Chapter 2, the magnetic field acts in two ways to destroy superconductivity. First, through the paramagnetic effect, it attempts to align the electron spins of the conventional singlet Cooper pairs along the field direction and in this way destroys their antiparallel spin correlation. This situation, as discussed in Chapter 5, is responsible for the paramagnetically limited critical field H_p (in T) $= 1.84T_c$ (in K). Second, through the orbital effect, the imposed magnetic field gives rise to a Lorentz force and, as the paired electrons possess opposite momenta, the force acts in the opposite directions on the two electrons, breaking the pair apart.

If the material contains ferromagnetic impurities, the singlet superconductivity may be suppressed by Zeeman (or paramagnetic) effect of their effective internal (local) magnetic field arising from strong exchange interaction of the electronic and localised spins. The effective internal exchange field H_{ex} that acts on the conduction electron spins can be very large, typically equivalent to a temperature of 1000 K and therefore much larger than T_c. Early experiments on dilute magnetic alloys of elemental superconductors Zn, Sn, In, and so on had revealed that a concentration of a few percent of magnetic impurities such as Cr, Mn, and Fe was enough to completely destroy their superconductivity (Matthias and Corenzwit, 1955; Matthias, 1957). This was explained theoretically by Abrikosov and Gor'kov (1960), who were able to further predict the appearance of a *gap-less region* just before the superconductivity was quenched. But the alloys studied were too dilute to produce any ferromagnetic ordering to investigate its possible competition with superconductivity. Moreover, when magnetic impurities are randomly distributed, the magnetic state at low temperatures is of short-range spin-glass type (Fischer and Peter, 1973). This rules out the possibility of realising a homogeneous system in which it would be possible to check whether superconductivity and magnetic order were really occurring in the same region of the sample. It would be interesting to study a homogeneous superconducting system formed with magnetic ions as its regular sublattice and ask what will happen at lower temperatures: will the magnetic ions give rise to a long-range magnetic order, and of which type—ferromagnetic, antiferromagnetic, oscillatory/weakly ferromagnetic, or any other—and, if so, how would superconductivity respond to the advent of such a state? Many of the binary alloys of Ti, Zr, and Bi with d-band magnetic elements such as Cr, Mn, Fe, Co, and Ni surprisingly had shown superconductivity being favoured by the presence of magnetic elements (Alekseevskii et al., 1952, Matthias et al., 1963, 1967; Jayaram et al., 1987). This was attributed to shifting of the average number of valence electrons per atom to 5, which, as per Matthias' rules (Chapter 7), led to higher T_c through increase in the density of states (DOS) at E_F. The alloyed elements in the crystal lattice, however, were no longer found to be in the magnetic state (Jayaram et al., 1987) and there existed no long-range magnetic order at low temperatures to compete with superconductivity.

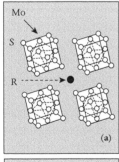

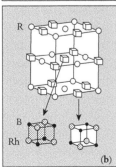

Figure 12.1 Crystal structures of (a) Chevrel-phase RMo_6S_8 and (b) RRh_4B_4.

12.1 LTS systems with magnetic order

A major breakthrough came in the mid-1970s with the discovery of rare-earth-based ternary systems (Matthias et al., 1977; Fischer et al., 1975), including chalcogenides, borides, and stannides, where for some of the compositions there was interesting interplay occurring between superconductivity and long-range magnetic order (Ishikawa and Fischer, 1977; Fertig et al., 1977; Moncton et al., 1978; Ott et al., 1978; Bulaevskii et al., 1981). In this chapter, we start with these intermetallic ternary systems. Here, the chalcogenides showing magnetic order comprise Chevrel phases (discussed in Chapter 11 for their unique high-field properties) of the type RMo_6X_8 (where X = S or Se and R = rare earth Gd, Tb, Dy, Er, etc.). The ternary borides and stannides are respectively represented by RT_4B_4 (where T = transition metal Rh or Ir) and RT_xSn_y (with $x \approx 1 - 1.5$ and $y \approx 3.5 - 5$), although of these, the Rh-containing materials have proved particularly interesting from the view point of the interplay. Some of the silicides of the type $R_2T_2Si_5$ also showed similar behaviour. The crystal structures of RMo_6X_8 and RT_4B_4 are illustrated in Figure 12.1(a) and (b). In these ternary systems, the magnetic order stems from the $5f$-orbitals of R-ions while superconductivity largely comes from the $4d$-electrons derived from the small clusters of Rh_4B_4 and Mo_6X_8 ions, as previously discussed in Chapter 11 for the Chevrel-phase materials.

In this chapter, we have also included the quaternary boro-carbides with the chemical formula RNi_2B_2C, discovered in 1993 (Mazumdar et al., 1993), which demonstrate an exciting interplay of low-temperature superconductivity with antiferromagnetic order. This class of materials has various points of similarity with the afore-mentioned ternary systems, although in the quaternary systems both magnetic order and superconductivity occur at comparatively higher temperatures than in the ternary systems, which has made them par-ticularly interesting and convenient for exploring their exciting basic physics.

12.2 The interplay

The interplay of superconductivity and magnetism is commonly described in terms of their characteristic temperatures T_c and T_m, where the latter corresponds to the temperature of onset of the long-range magnetic order. Broadly, the ensuing magnetic order, in principle, can be of ferromagnetic (FM), antiferromagnetic (AFM), quasiferromagnetic (QFM), oscillatory, or spin-glass (SG) type. There are three possibilities: $T_c > T_m, T_c < T_m$, and $T_c \approx T_m$. In the case of FM, T_m is the Curie tem-perature, while with AFM, it is the Néel temperature. If the long-range order is FM, there is a macroscopic internal field, with a parallel spin alignment, that destroys superconductivity and turns the material nor-mal. Such materials are called *ferromagnetic* or *reentrant superconductors*. The reentrant phenomenon is very unusual in that the material loses its superconductivity on cooling to lower temperatures (Riblet and Winzer, 1971; Maple et al., 1972). Of the two competing phenomena, ferromag-netism is much stronger than superconductivity, and it was Ginzburg (1956) who first asserted that the two orders cannot coexist, except when the internal ferromagnetic field is smaller than the critical field.

If the long-range order is AFM, superconductivity is rarely destroyed and the two phenomena can readily coexist. These are known as *antifer-romagnetic superconductors*. The situation is similar when the long-range order is quasiferromagnetic, oscillatory, or SG type having low exchange fields.

12.3 Various ternary materials and their interplay behaviour

Table 12.1–12.3 list the rare-earth-based ternary chalcogenides, borides, and stannides with their T_c and T_m values as collected by Subba Rao

Table 12.1 Ternary chalcogenides with their T_c and T_m values

Compound	T_c (K)	T_m (K)	Magnetic order
$NdMo_6S_8$	3.6	0.85	AFM
$EuMo_6S_8$	—	0.6/0.2	SG/AFM
$Gd\,Mo_6S_8$	1.1	0.85	AFM
$TbMo_6S_8$	1.45	0.90	AFM
$DyMo_6S_8$	2.05	0.40	AFM
$HoMo_6S_8$	1.82	≈ 0.7	FM
$ErMo_6S_8$	1.9	0.15	AFM
$YbMo_6S_8$	9.2	2.6	AFM
$GdMo_6S_8$	5.6	0.75	AFM
$TbMo_6Se_8$	5.7	1.03	AFM
$ErMo_6Se_8$	6.0	1.07	AFM
$HoMo_6Se_8$	5.5	0.8	Oscillatory

AFN, antiferromagnetic; FM, ferromagnetic; SG, spin-glass type.
Data collected by Subba Rao and Shenoy (1981).

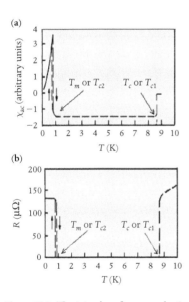

Figure 12.2 The interplay of superconductivity and magnetic order in $ErRh_4B_4$ is revealed by measurements of both magnetic susceptibility and electrical resistivity as functions of temperature (after Fertig et al., 1977). Near 1 K, there is a hysteretic region, below which there is the reentrant normal state.

and Shenoy (1981). The established cases of reentrant behaviour are $HoMo_6S_8$, $ErRh_4B_4$, and $ErRh_{1.1}Sn_{3.6}$. The resistivity and magnetic susceptibility characteristics of $ErRh_4B_4$ (Fertig et al., 1977) are shown in Figure 12.2(a) and (b), respectively. The material first becomes superconducting at $T_c = 8.7$ K, also known as the *upper critical temperature* T_{c1}, and on further cooling to $T_m = 0.93$ K, superconductivity is lost with the advent of ferromagnetic order. This temperature is also known as the *reentrant temperature* or the lower critical temperature T_{c2}. There is a narrow temperature range just above T_m or T_{c2}, below 1 K, over which the resistivity and susceptibility behaviours are hysteretic, which is indicative of the first-order transition occurring to the ferromagnetic normal state below T_m. Similar behaviour is observed for $HoMo_6S_8$ (Ishikawa and Fischer, 1977), where the hysteretic region is still narrower. Neutron diffraction studies show (Lynn, 1981) that in this region the two states coexist, although the magnetic state is oscillatory. In the case of $HoMo_6Se_8$ (Table 12.1), where the exchange interaction is comparatively weaker, $T_c = 5.5$ K and $T_m = 0.8$ K, and the coexistence of superconductivity and oscillatory ferromagnetic order persists down to the lowest temperatures studied. A *spontaneous vortex state* was also suggested (Kuper et al., 1980), but has not been confirmed to date. Some of these materials (e.g. $TmRh_4B_4$) exhibit reentrant behaviour at T_m under an imposed magnetic field of about 0.1–1 T (Alekseevskii et al., 1973; Hamaker et al., 1979; Ott et al., 1980; Maple, 1981), depicted in Figure 12.3 (Maple, 1981).

The oscillatory structure may be described in terms of domains with alternating magnetic moments. This situation where the neighbouring

Table 12.2 Ternary borides with their T_c and T_m values

Compound	T_c (K)	T_m (K)	Magnetic order
$NdRh_4B_4$	5.36	1.3	AFM
$SmRh_4B_4$	2.72	0.87	AFM
$GdRh_4B_4$	—	5.62	ΦM
$TbRh_4B_4$	—	7.08	ΦM
$DyRh_4B_4$	—	12.03	ΦM
$HoRh_4B_4$	—	6.56	ΦM
$ErRh_4B_4$	8.7	0.93	ΦM
$TmRh_4B_4$	9.86	0.4	AFM
$LuRh_4B_4$	11.76	—	—
YRh_4B_4	11.34	—	—

Data collected by Subba Rao and Shenoy (1981).

Table 12.3 Ternary stannides with their T_c and T_m values

Compound	T_c (K)	T_m (K)	Magnetic order
$LaRh_xSn_y$	3.2	—	—
$EuRh_xSn_y$	—	11.0	ΦM
$GdRh_xSn_y$	—	11.2	ΦM
$TbRh_{1.1}Sn_{3.6}$	—	3.8	ΦM
$DyRh_{1.1}Sn_{3.6}$	—	2.1	ΦM
$HoRh_{1.2}Sn_{3.9}$	—	1.7	ΦM
$ErRh_{1.1}Sn_{3.6}$	1.36	0.46	ΦM
$TmRh_{1.3}Sn_{4.0}$	2.3	—	—
$YbRh_{1.4}S_{4.6}$	8.6	—	—
$LuRh_{1.2}Sn_{4.6}$	4.0	—	—

Data collected by Subba Rao and Shenoy (1981).

domains are pointing in opposite directions, when viewed on the larger scale of range of coherence, is essentially analogous to AFM order. On lowering the temperature below T_m, the system reduces its energy by getting rid of the walls separating the neighbouring domains, with the result that all the magnetic moments point in the same direction, characteristic of collinear ferromagnetic order, which is antagonistic to superconductivity, and it stays as the ground state of the system.

The confirmed cases of AFM superconductors are RMo_6S_8 with R = Nd, Gd, Tb, Dy, Er, or Yb; RMo_6Se_8 with R = Gd, Tb, or Er; and RRh_4B_4 with R = Nd, Sm, or Tm. The notable effect on superconductivity is a local depression of the upper critical field, in the form

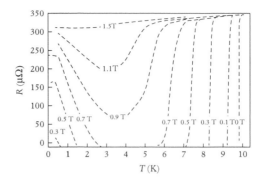

Figure 12.3 The field-induced reentrant behaviour observed in TmRh$_4$B$_4$ (after Maple, 1981).

of a kink, near the AFM ordering temperature, which alters the temperature dependence of the critical field, as shown for representative samples of the above types in Figure 12.4 (Thomlinson et al., 1981; Maple et al., 1982). Figure 12.4(c) also shows similar behaviour observed for a compound belonging to the quaternary borocarbide family to be described in Section 12.4 (Schmidt et al., 1996a). In the case of stannides (Table 12.3), most of the compositions show either superconductivity or magnetic order, but not both. All the above materials exhibit AFM order at $T_m < T_c$ without destroying superconductivity in zero magnetic field. Table 12.1 and Table 12.2 give their T_c and T_m values.

There are some isolated examples where the magnetic order precedes superconductivity (i.e. $T_m > T_c$) and superconductivity occurs in a magnetically ordered lattice. It was theoretically realised (Morozov, 1980) that in AFM superconductors non-magnetic impurities cause pair breaking as effectively as magnetic impurities in conventional non-magnetic superconductors. This explains why many cases with $T_m > T_c$ are not reported. In most such cases, T_c may already have been suppressed by non-magnetic impurities inadvertently introduced

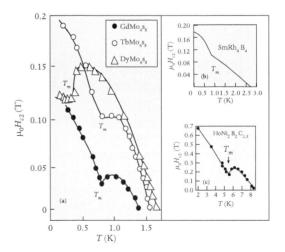

Figure 12.4 Anomalous $H_{c2}(T)$ behaviour, displaying a kink at T_m, of (a) different Chevrel-phase compounds (after Thomlinson et al., 1981), (b) SmRh$_4$B$_4$ (after Maple et al., 1982), and (c) HoNi$_2$B$_2$C (after Schmidt et al., 1996a).

during sample processing. Clearly, the behaviour is sample-dependent, which calls for carefully prepared samples free from non-magnetic impurities. For example, in the pseudoternary system $Ho(Rh_{1-x}Ir_x)\,B_4$, Ku et al. (1980) showed that both T_m and T_c could be monitored by changing x. For example, for $x = 0.7$, $T_c = 1.6$ K and $T_m = 2.7$ K, and below 1.6 K superconductivity and antiferromagnetism coexist. Other such materials are the ternary silicides, such as $Er_2Fe_3Si_5$ ($T_c \approx$ 0.9 K and $T_m \approx 2.5$ K) (Noguchi and Okuda, 1994) and $Tb_2Mo_3Si_4$ ($T_c \approx$ $1 - 2$ K and $T_m \approx 19$ K) (Aliev et al., 1994). In Section 12.4, we will see that in the quaternary borocarbide $DyNi_2B_2C$, the AFM order occurs at 11 K and, below 6 K, coexists with superconductivity. Also, in another system Y_9Co_7, quasiferromagnetic order is formed just above 5 K, and, similar to oscillatory order, coexists with superconductivity below 1.7 K (Kolodziejczyk et al., 1980).

Among ternary superconductors, the likely cases with $T_m = T_c$ are YRh_4B_4 and $Y_{1-x}Er_xRh_4B_4$ (Tse et al., 1979). They are of special interest because the AFM order formed is believed to be among their itinerant electrons and it occurs at the same temperature as T_c. Using pulsed NMR, Tse et al. studied the nuclear spin echo of [11]B in conjunction with the magnetic susceptibility of these two ternary compounds and detected a significant loss in the spin echo intensity at $T_c = T_m$, a considerable increase in the static magnetic susceptibility on lowering the temperature to T_c, and identical field dependences of T_c snd T_m. The decrease in nuclear spin-echo intensity is generally taken to indicate formation of itinerant antiferromagnetism (Borsa and Lecander, 1976). These results could not be explained in terms of p-wave pairing and seemed to fit better with the possible coexistence of itinerant electron antiferromagnetism with superconductivity.

12.3.1 Magnetic-field-induced superconductivity (FIS)

The presence of magnetic order in superconductors has led to a fascinating novel phenomenon in which superconductivity is induced by the application of a strong magnetic field. This unusual effect, which is called *field-induced superconductivity (FIS)*, is in contrast to the behaviour of conventional superconductors, where an applied magnetic field destroys superconductivity. The pertinent mechanism responsible for this unusual effect, experimentally observed in some ternary molybdenum chalcogenides of the Chevrel-phase family (Meul et al., 1984a, b; Rossel et al., 1985), is the so-called *exchange field compensation effect*, suggested more than 20 years earlier by Jaccarino and Peter (1962). As may be seen in the main part of Figure 12.5, the normalised resistance R/R_N of $Eu_{0.75}Sn_{0.25}Mo_6S_{7.2}Se_{0.8}$ (Meul et al., 1984a, b) measured as a function of increasing magnetic field H at T $= 0.37$ K exhibits a sequence of transitions from superconducting (S) to normal (N) at lower magnetic

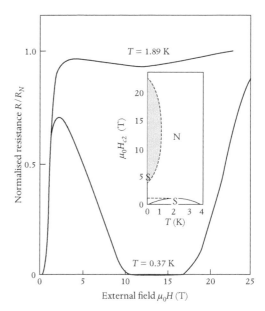

Figure 12.5 Field-induced superconductivity in $Eu_{0.75}Sn_{0.25}Mo_6S_{7.2}Se_{0.8}$ showing a sequence of S–N–S–N transitions occurring in increasing magnetic field at 0.37 K (after Meul et al., 1984a).

fields, and then again to S, and finally back to N at much higher magnetic fields. The second induction of S occurring at a higher magnetic field constitutes the unusual behaviour. The phase diagram in the $H - T$ plane obtained from a series of similar measurements of R/R_N versus H, carried out at different temperatures (Meul et al., 1984a, b; Rossel et al., 1985), is depicted in the inset of Figure 12.5. As may be seen, instead of the conventionally observed single S-domain in low fields, $Eu_{0.75}Sn_{0.25}Mo_6S_{7.2}Se_{0.8}$ exhibits two S-domains: one at lower fields and another induced at higher fields.

Broadly, for FIS to occur, the material is required to be a type II superconductor in which the upper critical field $H_{c2}(T)$ is determined by the Pauli limiting field $H_p = 1.84T_c$ and further it must carry magnetic moments that are antiferromagnetically coupled to the conduction electron spins. The pertinent magnetic moments in $Eu_{0.75}Sn_{0.25}Mo_6S_{7.2}Se_{0.8}$ are due to Eu^{2+} and their coupling with the conduction electrons gives rise to an exchange field H_j that, like the applied magnetic field, acts on the conduction electrons. Since the sign of the coupling is negative, the direction of the applied field H is opposite to that of H_j and consequently the effect of the latter is compensated by the former. The net magnetic field in the material is therefore $H_T = H - |H_j|$. Because H_j is rather large, about 30–50 T, the compensation effect takes place at very high magnetic fields. The aforementioned sequence of S–N–S–N transitions in increasing field follows from the relative values of $|H_T|$ and $|H_p|$ (Maple, 1985), and superconductivity is destroyed when $|H_T| > |H_p|$. In zero applied field, the compound is initially superconducting. As H is raised, the Eu magnetic moments become aligned along the field, direction and at a relatively low field the exchange field H_j attains a high saturation value $H_j^{\max}$. The lower-field transition to the normal state occurs when H_T is negative and crosses H_p on the negative side. With further increase in H, the net magnetic field starts decreasing, and when it falls below H_p, superconductivity reappears. The higher-field transition to the normal state takes place when the net magnetic field has turned positive and exceeds H_p on the positive side. Clearly, for FIS to occur, the value of H_{c2} (orbital critical field) has to be sufficiently large $\left(> H_p\right)$, since the orbital effects are not compensated by the Jaccarino–Peter model. Originally, the model had conceived FIS to occur in a weakly ferromagnetic material, but this remains to be experimentally corroborated. The model has also been theoretically applied to FIS in paramagnetic systems (Schwartz and Gruenberg, 1969; Fischer, 1972).

12.4   Quaternary borocarbides

The quaternary borocarbides, with composition RNi_2B_2C, discovered in the first half of the 1990s (Mazumdar et al., 1993; Nagarajan et al., 1994, 1998; Cava et al., 1994a), possess a rich variety of unusual features. The materials in this broad category can, however, be described by a general formula $(RC)_m(T_2B_2)_n$, where R is a non-magnetic rare-earth ion (Y, Lu, Dy, Ho, Er, or Tm) and T is a transition metal (Ni, Pd, or Pt). As may be noted, the R ions are heavier lanthanides having a relatively small ionic size. The compounds RNi_2B_2C, corresponding to $m = n = 1$ and $T = Ni$, perhaps represent the most interesting group, with the largest number of members, belonging to this family. Despite having a high Ni content and a significantly large concentration of magnetic rare-earth ions, Dy, Ho, Er, and Tm, the materials show unusually high and mutually close values of T_c and T_m, of the order of 10 K. Since the long-range order formed is antiferromagnetic, T_m refers to the Néel temperature. For ternary superconductors, T_m is typically 1 K and T_c is an order of magnitude larger. The much larger T_m of quaternary borocarbides indicates that the interaction taking place is of RKKY (Ruderman–Kittel–Kasuya–Yosida) type, mediated through conduction electrons. A more intimate interplay between superconductivity and magnetic order is therefore expected in these materials.

The quaternary borocarbides display an interesting interplay of superconductivity with antiferromagnetic order, manifesting all combinations of T_c and T_m: $T_c > T_m$, $T_c < T_m$, and $T_c \approx T_m$. On substituting non-magnetic Y or Lu for R (Cava et al., 1994b), the compounds showed only superconductivity ($T_c = 16.5$ K and 15.5 K, respectively), with no magnetic order. The same situation was found with Sc, where the material ($T_c \approx 15$ K) formed only after quenching from high temperature. It was however, in a metastable state where superconductivity disappeared on annealing (Ku et al., 1994). Interestingly, with Yb, a moderate heavy fermion state was detected in YbNi$_2$B$_2$C (Dhar et al., 1996), with no superconductivity or magnetic order.

When Ni was replaced by paramagnetic Pd, the T_c of Y$-$Pd$-$B$-$C, in a multiphase system, shot up to 24 K (Cava et al., 1994b; Hossain et al., 1994), which until the discovery of a T_c of 39 K in MgB$_2$ in 2001 was the highest T_c known for any intermetallic superconductor.

The Pt-based quaternary borocarbides RPt$_2$B$_2$C, on the other hand, are formed only with lighter R ions, of larger radius, namely La, Ce, Pr, Nd, and Y (Cava et al., 1994c), and, more recently, the coexistence phenomenon has been found for NdPt$_2$B$_2$C (Dhar et al., 2002; Paulose et al., 2003). The phase purity of the samples is improved by partial replacement of Pt by Au (Nagarajan et al., 2005), and for both NdPt$_{1.5}$Au$_{0.6}$B$_2$C and NdPt$_{2.1}$B$_2$$_4C_{1.2}$ (Dhar et al., 2002; Paulose et al., 2003) T_c(oneset), while $T_m \approx 1.7$ K, below which the two phenomena coexist.

A few compounds (RC)$_m$(T$_2$B$_2$)$_n$ are known with $m = 2$, $n = 1$ (i.e. RTBC) (Siegrist et al., 1994; Massalami et al., 1995; Hossain et al., 1998; Chang et al., 1996), of which LuNiBC exhibits superconductivity at 2.9 K (Gao et al., 1994b). The compounds Lu$_2$NiBC$_2$ and Y$_2$NiBC$_2$, with $m = 4$ and $n = 1$, are, however, found to be non-superconducting (Zandbergen et al., 1994; Rukang et al., 1995), but Y$_3$Ni$_4$B$_4$C$_3$, with $m = 3$ and $n = 2$, shows multiphase superconductivity (Kito et al., 1997) with T_c values of 11 K and 3 K.

12.5 Crystal structure and related aspects

Figure 12.6 depicts the crystal structure of RNi$_2$B$_2$C, which has a body-centred tetragonal structure with space group $I4/mmm$ (Siegrist et al., 1994). The structure comprises layers of R–C separated by corrugated Ni$_2$B$_2$ sheets with two formula units of RNi$_2$B$_2$C per unit cell running parallel to the basal plane of the structure. The NiB$_4$ tetrahedron is considered important and is distortion-free for LuNi$_2$B$_2$C, which possesses the highest stable T_c (16.5 K) among these compounds. The presence of a lighter and larger lanthanide ion like La results in a distorted tetrahedron and no superconductivity is observed. Interestingly, the Ni–Ni distance (0.245 nm) is shorter than in elemental Ni (0.250 nm), which gives the material a stronger metallic character. The Ni$_2$B$_2$ sheets, sandwiched between the insulating RC sheets, are electrically conducting and give rise to the d- and sp-electrons for superconductivity, while the magnetic order stems from the f-orbitals of the magnetic R ions. The structure has some resemblance to that of the HTS cuprates, where, for example in the Y-123 system, the conducting Cu–O layers sandwich the insulating Y sheets. Like the HTS cuprates, the structure depicted in Figure 12.6 is anisotropic, having a lattice parameter ratio $c/a \approx 3$,

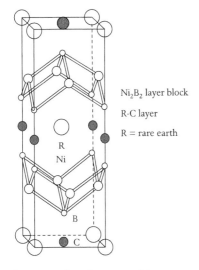

Ni$_2$B$_2$ layer block

R-C layer

R = rare earth

Figure 12.6 Crystal structure of the quaternary borocarbide RNi$_2$B$_2$C (after Siegrist et al., 1994).

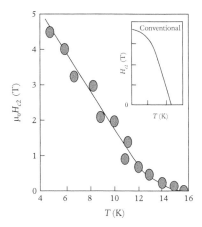

Figure 12.7 Unconventional $H_{c2}(T)$ behaviour of YNi_2B_2C; the inset depicts the typical conventional behaviour (after Nagarajan and Gupta, 1998).

although the actual anisotropy of various superconducting parameters (e.g. $\lambda, \xi,$ and H_{c2}) is comparatively much smaller. This is because the material, despite its layered structure, is still three-dimensional with regard to its electronic band structure (Ravindram et al., 1996), as against HTS cuprates, which are quasi-two-dimensional. Unexpectedly, however, single-crystal studies in the superconducting state reveal a small anisotropy in the square basal plane of the tetragonal lattice whose origin is related to the intrinsic anisotropy of the superconducting gap function and of the Fermi surface (Nagarajan et al., 2005). The gap function is believed to be of $s + g$ type with point nodes in the $\langle 100 \rangle$ directions. Such a gap function does not normally follow from the conventional phononic mechanism.

A conventional superconductor, in general, shows the isotope effect with $T_c \propto m^{-\alpha}$, with the exponent $\alpha \approx 0.5$. For ^{10}B and ^{11}B, the estimated values of the exponent α for YNi_2B_2C and $LuNi_2B_2C$ are respectively found to be 0.21 and 0.11 (Nagarajan et al., 2005), which are indicative of only a feeble isotope effect. The energy-gap values measured give a gap coefficient $\Delta/k_BT_c \approx 2 - 3.5$ instead of the characteristic value of 1.76 from the weakly coupled BCS theory. This makes these superconductors strongly coupled (Nagarajan et al., 2005). The T_m values of these materials are too large to be explained in terms of dipole–dipole coupling among R-moments. Instead, the pertinent exchange interaction is believed to be of the RKKY type, mediated through the conduction electrons, which are also responsible for superconductivity.

Quaternary borocarbides are type II superconductors with a Ginzburg–Landau parameter κ, typically of about 15, with $\xi \approx 10$ nm and $\lambda \approx 150$ nm, with $H_{c2}(0) \approx 10$ T. The temperature dependence of H_{c2} shows (Nagarajan and Gupta 1998) a positive curvature (Figure 12.7), characteristic of the grain-boundary weak-link effect. The inset in Figure 12.7 depicts the parabolic-like behaviour of a conventional superconductor. The grain-boundary width of polycrystalline arc-melted samples is found to exceed ξ, which causes the boundary to serve as a weak link. This has been clearly demonstrated (Khare et al., 1996) by the natural grain boundaries in bulk samples of YNi_2B_2C manifesting rf-SQUID behaviour (Figure 12.8) owing to their Josephson weak-link effect. Note that the oscillations are absent at the resonant frequency $f_0 = 19.5$ MHz and appear only at slightly lower or higher frequencies.

12.6 Coexistence and interplay of T_c and T_m

12.6.1 Macroscopic studies

Table 12.4 depicts T_c and T_m values of RNi_2B_2C for different R. For R = Dy, Ho, Er, or Tm, there is a coexistence of superconductivity and

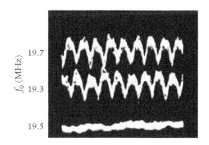

Figure 12.8 SQUID-like behaviour displayed by grain boundary weak links in YNi_2B_2C (after Khare et al., 1996).

Table 12.4 T_c and T_m values for superconducting quaternary Ni-containing borocarbides

Material	T_c (K)	T_m (K)	Magnetic structure[*]
YNi_2B_2C	15.5	—	—
$LuNi_2B_2C$	16.5	—	—
$DyNi_2B_2C$	6.0	11.0	Commensurate AFM
$HoNi_2B_2C$	8.5	8.0	Commensurate AFM/c-MM/a-MM
$ErNi_2B_2C$	11.0	6.5	a-MM (<2.3 K: WFM)
$TmNi_2B_2C$	10.6	1.5	ab-MM
$YbNi_2B_2C$	—	—	Heavy fermion behaviour

[*]a-MM, a-axis modulated incommensurate magnetic structure; c-MM, c-axis modulated incommensurate spiral magnetic structure; ab-MM, incommensurate antiferromagnetic order modulated along the (110) plane; WFM, weak ferromagnetic state.
From Nagarajan et al. (2005) and Schmidt and Braun (1998).

antiferromagnetic order, although with Er, below 2.3 K there is a weak ferromagnetic order. The magnetic structures as revealed by neutron diffraction studies (Goldman et al., 1994; Sinha et al., 1995; Zarestky et al., 1995; Chang et al., 1996b) are also given in the table and are shown in Figure 12.9. For R = Dy, the material shows $T_c < T_m$, while for R = Er or Tm, $T_c > T_m$, and for R = Ho, $T_c \approx T_m$. On the other hand, for R = Pr, Nd, Sm, Gd, and Tb (which are not discussed here and are not shown in Table 12.4), only magnetic order has been reported respectively below $T_m \approx 4$ K, 5 K, 10 K, 19 K, and 15 K (Nagarajan et al., 2005).

Figure 12.10(a) illustrates the typical coexistence behaviour in $ErNi_2B_2C$ (Gupta et al., 1994) as revealed through heat capacity measurements. While the occurrence of superconductivity at 11 K is indicated by a small kink, the more pronounced anomaly at 6.5 K corresponds to the onset of AFM ordering. Since the resistivity of the sample still remains zero below this temperature (Figure 12.10(b)), there is coexistence of the two phenomena. Below 2.3 K, a weak ferromagnetic

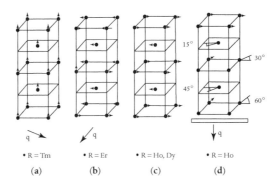

Figure 12.9 Magnetic structures of RNi_2B_2C, for R = Tm, Er, Ho, and Dy, as determined by neutron diffraction (after Schmidt and Braun, 1998).

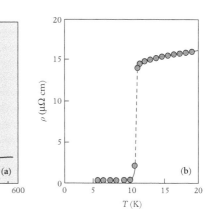

Figure 12.10 (a) Heat capacity measurements on $ErNi_2B_2C$ showing anomalies corresponding to both T_c and T_m (after Gupta, 1996), respectively, at 11 K and 6.5 K. (b) Resistivity measurements on the same sample exhibit the zero-resistance state at all temperature below 10 K. These observations give credence to the coexistence of the two orders below 6.5 K.

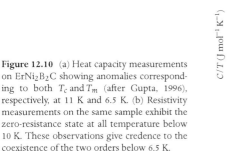

Figure 12.11 Double reentrant transition in $HoNi_2B_2C$ (Gupta, 1998).

(WFM) state has been suggested (Canfield et al., 1996), which coexists with superconductivity. The interplay of the two phenomena is manifested (Eisaki et al., 1994; Gupta, 1998) in an interesting way in $HoNi_2B_2C$, for which $T_c \approx T_m \approx 8$ K (Figure 12.11). The magnetic susceptibility behaviour of a sample in the temperature range 10–4 K shows (Gupta, 1998) that it first becomes superconducting at about 8.5 K, partially re-enters the normal state at about 5 K, and again becomes superconducting just below this temperature. The double reentrant superconducting transition observed here has been explained in different ways. Neutron diffraction studies show two types of magnetic structure. First, below 8.5 K, there is a commensurate AFM structure that consists of ferromagnetic moments in the ab planes, with the adjacent AFM-coupled sheets along the c-axis, which can coexist with superconductivity. At about 5 K, the structure becomes an incommensurate spiral structure that suppresses superconductivity and causes the reentrant behaviour. On further lowering the temperature, the structure returns to the commensurate AFM type that allows superconductivity to coexist with long-range AFM order (Everman et al., 1996; Lynn et al., 1997; Müller et al., 1997). Additionally, Nagarajan et al. (2005) have pointed out that the properties of $HoNi_2B_2C$ are sensitive to the stoichiometry and impurities, which can seriously interfere with the superconducting characteristics and cause the reentrant behaviour (Schmidt et al., 1995).

Among this class of materials, $DyNi_2B_2C$ represents the unique example of $T_c < T_m$, that is, an AFM-ordered lattice undergoing a superconducting transition on cooling. Again, as already mentioned, the non-magnetic impurities serve as potential pair-breakers to suppress T_c. For example, the T_c of this material is depressed when non-magnetic Lu or Y partly replaces magnetic Dy (Cho et al., 1996).

The quaternary borocarbides have been extensively studied for pressure effects on their T_c and T_m (Schmidt and Braun, 1994; Alleno et al., 1995; Carter et al., 1995; Looney et al., 1995; Bud'ko et al., 1996;

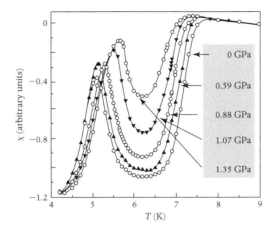

Figure 12.12 Pressure dependence of the low-temperature magnetic susceptibility behaviour of $HoNi_2B_2C$ (after Schmidt and Braun, 1998).

Uwatoko et al., 1996). In general, the pressure derivatives of T_c are small, about $(\pm)0.2 - 0.6$ KGPa^{-1}. The derivatives are found to be linear and negative for R = Y, Tm, Er, and Ho (Schmidt and Braun, 1994; Looney et al., 1995; Uwatoko et al., 1996), linear and positive for R = Lu (Schmidt and Braun, 1994), and positive quadratic for Lu (Alleno et al., 1995), although the opposite signs have also been noted. A particularly interesting case is that of the pressure studies (Schmidt et al., 1995; Schmidt and Braun, 1998; Carter et al., 1995) on reentrant $HoNi_2B_2C$ described above (Figure 12.12). Applying a greater pressure enhances and shifts the reentrant maximum to higher temperatures and the upper T_c is lowered at a rate of $dT/dP \approx -0.34$ KGPa^{-1}. On the other hand, Carter et al. (1995) found T_c to increase under pressure, which could be related to the difference in sample composition.

These materials have been subjected to extensive substitution studies in which R and Ni are partially replaced by different lanthanides and transition metals to see their effects on T_c and T_m. Because of their relatively high T_c values, $LuNi_2B_2C$ ($T_c = 16.5$ K) and YNi_2B_2C ($T_c = 15.5$ K) have received a greater attention in such studies. The results showing changes in T_c on partial replacement of Lu by Yb (Luo, 1996), Lu by Gd and Dy (Cho et al., 1996), and Y by Ho (Müller, 1996) are shown in Figure 12.13. Similarly, the effects of partial replacement of Ni by Co in different R-based borocarbides are shown in Figure 12.14 (Felner, 1998a; Schmidt et al., 1995, 1996a; Schmidt and Braun, 1998). Figure 12.14(a) also shows the effect of partly substituting the Ni site of YNi_2B_2C by other transition metal cations (Ru, Fe, Cu, and Pd).

12.6.2   Interplay at the microscopic level

The relatively high T_c and T_m values of quaternary borocarbide superconductors have provided the first evidence (Yaron et al., 1996) of

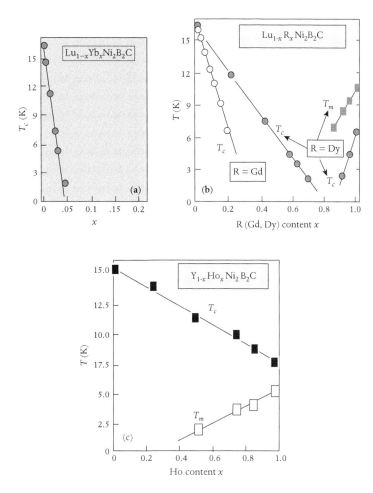

Figure 12.13 Substitutional studies at the rare-earth site of $LuNi_2B_2C$, where (a) Lu is partially replaced by Yb (after Luo, 1996), (b) Lu is partially substituted by Gd and Dy (after Cho et al., 1996), and (c) Y in YNi_2B_2C is partially substituted by Ho (after Müller, 1996).

coexistence of superconductivity and magnetism on a microscopic scale through studies of the flux line lattice (FLL). The main experimental techniques employed have been small-angle neutron scattering (SANS) and scanning tunnelling microscopy (STM) on single-crystal samples. In non-magnetic YNi_2B_2C and $LuNi_2B_2C$, above H_{c1}, the FLL at low fields (<0.05 T) is hexagonal, transforming into a square one in higher fields (>0.1 T) (Yethiraj et al., 1997; Eskildsen et al., 1997). Similar behaviour is observed in magnetic superconductors, where the presence of magnetic order produces additional effects. In the magnetic superconductor $ErNi_2B_2C$, at $T < T_m$, owing to the local magnetic order, the square FLL is found to rotate away from the direction of the external field (Yaron et al., 1996; Eskildsen et al., 1997) along the a-axis, remaining confined to the ac plane. Below 2.3 K, when the magnetic order becomes weakly ferromagnetic, the internal field adds to the external field and causes a deviation of the square FLL from the applied field direction. The observed strong coupling between FLL order and magnetic order in $TmNi_2B_2C$ may lead to new FLL dynamics in this class of materials.

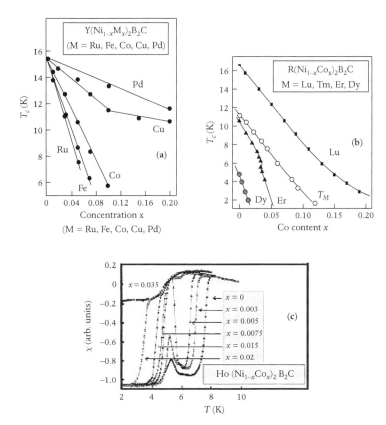

Figure 12.14 The effects of (a) partial substitution of Ni by other transition metals in YNi_2B_2C (after Felner, 1998), (b) substitution of Ni by Co in RNi_2B_2C with R = Lu, Tm, Er, and Dy, and (c) substitution of Ni by Co in $HoNi_2B_2C$ (after Schmidt and Braun, 1998).

12.6.3 De Gennes scaling factor and T_m and T_c

Figure 12.15(a) presents a plot of T_m values of both superconducting and non-superconducting borocarbide samples formed with R ranging from La to Lu (Gupta, 1998). As may be seen, the values follow the de Gennes scaling factor $DG = (g_J - 1)^2 J(J + 1)$, where g_J is the Landé g-factor and J is the total angular momentum of the R^{3+} ion. Surprisingly, however, the above scaling is expected to hold only when the magnetic moments are uncorrelated, as in the Abrikosov–Gor'kov (1960) theory, whereas in four superconductors formed with Dy, Ho, Er, and Tm, pronounced R-moment correlations exist. However, if one extends this linear scaling to $TbNi_2B_2C$, which exhibits AFM order at $T_m \approx 15$ K but in which no superconductivity has been reported down to the millikelvin range (Gupta, 1996), one expects the material to turn superconducting at 2 K. Similarly, $YbNi_2B_2C$ shows neither superconductivity nor magnetic order down to the lowest temperatures studied. However, by assuming a fixed valency of 3+ for the Yb ions, the de Gennes scaling yields $T_c \approx 12$ K. The anomalous absence of T_c observed is ascribed to the possible formation of a heavy fermion state in this material. The T_c values decrease as the radius of the R-ion increases. In Figure 12.15(b), both

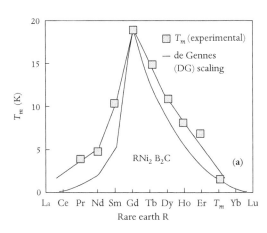
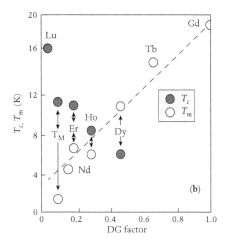

Figure 12.15 (a) T_c and T_m values of RNi$_2$B$_2$C for R ranging from La to Lu (after Gupta, 1998). (b) T_c and T_m as functions of the normalised de Gennes scaling factor DG (after Felner, 1998).

T_c and T_m are plotted, with the DG scaling factor normalised to DG = 1 for the Gd-based sample (Felner, 1998a). As may be seen, a larger T_c corresponds to a smaller T_m, which indicates that T_c variation is being influenced by local moments, giving rise to the observed interplay.

12.6.4 Heavy fermion behaviour of YbNi$_2$B$_2$C

The reported value of the coefficient of electronic specific heat γ for YbNi$_2$B$_2$C is about 200 mJ mol^{-1}K^{-2} (Dhar et al., 1996), as against a few mJ mol^{-1} K^{-2} for normal metallic systems. Since the pertinent heat capacity behaviour as a function of temperature shows an upturn, γ is expected to be temperature-dependent, and its extrapolated value below 2 K is even larger, about 600 mJ mol^{-1}K^{-2}, which is corroborated experimentally by single-crystal data (Yatskar et al., 1996). This is indicative of the (moderate) heavy fermion character, with 4f-electrons of Yb hybridised with the conduction electrons to form a dense Kondo system. This is a spin fluctuation system (Dhar et al., 1996), in which ^{170}Yb Mossbauer measurements further confirmed that the fluctuations continue down to 70 mK, when no magnetic order due to Yb ions is formed. The estimated Kondo temperature T_K of 11 K is comparable to the T_c of this class of materials. The absence of superconductivity in YbNi$_2$B$_2$C is therefore attributed to its moderate heavy fermion character in which superconductivity is destroyed owing to the competition between the Kondo interactions and superconductivity. This is substantiated by the fact that 5 at% Yb substitution for Lu in LuNi$_2$B$_2$C, as previously illustrated in Figure 12.13(a), completely quenches its superconductivity. A similar huge depression of T_c is caused by Yb substituting for Y in YNi$_2$B$_2$C. Inelastic neutron

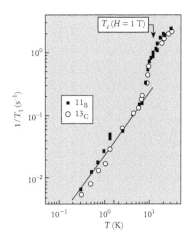

Figure 12.16 Nuclear spin–lattice relaxation rate $1/T_1$ of ^{11}B and ^{13}C as a function of temperature in YNi$_2$B$_2$C, manifesting the absence of a Hebel–Slichter peak below T_c and the presence of linear behaviour at $T < T_c$ (after Iwamoto et al., 2000).

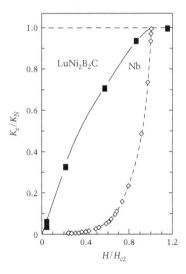

Figure 12.17 Magnetic-field dependence of the electronic thermal conductivity of single-crystalline LuNi$_2$B$_2$C, normalised to its value at H_{c2}, measured at $T = 70$ mk (after Boaknin et al., 2002). For comparison, the characteristic exponential behaviour for the conventional superconductor Nb is also shown.

scattering measurements (Boothroyd et al., 2003) give further credence to the Kondo lattice and heavy fermion nature of YbNi$_2$B$_2$C. Clearly, further studies of Yb-doped RNi$_2$B$_2$C samples promise new insights into the possible interplay of both superconductivity and magnetic order with the heavy fermion behaviour.

12.6.5 Unusual superconductivity traits of borocarbides

Besides some uncommon features of quaternary borocarbides already discussed, these materials display a number of other traits of unconventional superconductivity. The NMR studies by Iwamoto et al. (2000) show (Figure 12.16) that the nuclear spin–lattice relaxation rate $1/T_1$ of ^{11}B and ^{13}C in superconducting YNi$_2$B$_2$C decreases rapidly with temperature without manifesting the Hebel–Slichter peak below T_c and at still lower temperatures the decrease is found to be linear. These characteristics are considered typical of an unconventional superconductor like HTS cuprate that shows d-wave pairing. However, such a pairing symmetry is not supported by other experiments. The magnetic-field dependence of the electronic thermal conductivity of single-crystalline LuNi$_2$B$_2$C measured at very low temperatures ($T \approx 70$ mK), normalised to its value at H_{c2}, is shown in Figure 12.17 (Boaknin et al., 2001). Also displayed for comparison is the typical exponential behaviour of a conventional LTS, Nb. The curve for LuNi$_2$B$_2$C is markedly different, as also reported by others for YNi$_2$B$_2$C and LuNi$_2$B$_2$C (Nohara et al., 2000; Izawa et al., 2001a). Detailed angular-dependence studies of thermal conductivity of YNi$_2$B$_2$C by Izawa et al. (2001a) revealed that the thermal conductivity was sensitive to the field angle θ with respect to the direction of the thermal flow, which could be accounted in terms of the $s + g$ wave pairing symmetry with a point node. In contrast, for d-wave pairing with line nodes, the thermal conductivity behaviour is expected to be insensitive to θ. *Directional point contact spectroscopy* (DPCS) has been proved by Raychaudhuri et al. (2004) to be an effective tool for probing the pairing symmetry of borocarbide superconductors. Their studies on YNi$_2$B$_2$C single crystals showed the energy gap, when measured with the tunnel current along the c-direction, to be 4–5 times greater than along the a-direction, which rendered support to the presence of $s + g$ pairing. However, further studies by the same group (Mukhopadhyay et al., 2005) seemed to question this above contention, with the unusual behaviour seeming to be due to *multiband superconductivity*.

12.7 Summary

This chapter has focused on novel superconductors in the form of ternary borides, stannides, chalcogenides, and quaternary borocarbides, which have had a key role in initiating and promoting exciting research on the coexistence of superconductivity and magnetism. In these important categories of low-temperature superconductors, the interplay of superconductivity with different kinds of ensuing magnetic order have been discussed in terms of their characteristic temperatures T_c and T_m. This chapter has presented a wide variety of complex interplays that include the coexistence of superconductivity with antiferromagnetism (and other weak magnetic orders) and the exclusion of superconductivity from the collinear ferromagnetism, leading to a re-entrance transition and the formation of an oscillatory ferromagnetic order in the superconducting state. Such interplays also show up at the microscopic level in the form of flux line lattice studies of magnetic and non-magnetic borocarbide superconductors. Interestingly, Yb-based borocarbide shows a moderate heavy fermion character. The symmetry of the superconducting order parameter remains a matter of debate, with the possible options appearing to be $s + g$ wave pairing and multiband superconductivity.

CHAPTER 13

Heavy fermion superconductors

In heavy fermion (HF) compounds, where the prerequisite is a large concentration of magnetic ions forming periodic lattice sites, the possibility of superconductivity does indeed sound remote, as the magnetism should prohibit the formation of Cooper pairs. It therefore emerged as a great puzzle in 1979 when $CeCu_2Si_2$ was discovered (Steglich et al., 1979) as the first superconductor belonging to the HF class, showing unconventional superconductivity at ambient pressure. Since then, superconductivity has been discovered in more than 30 HF materials and many more of their substituted phases, all showing novel and intriguing characteristics. In many such materials, superconductivity and magnetism emerge as close partners where, surprisingly, one may not exist without the other. The emerging superconductivity is unconventional in terms both of a non-phononic coupling mechanism and of the symmetry of the wavefunction. This chapter focuses on these novel systems and their intriguing properties. There are a number of excellent review articles (Stewart, 1984; Thalmeier et al., 2005; Maple et al., 2008, 2010; Riseborough et al., 2008; Pfleiderer, 2009) and a few books (Goll, 2006; Hewson, 1993; Misra, 2009) covering the topic of HF superconductivity.

13.1 Discovery of HF superconductors

Historically, the first observation of superconductivity in an HF system was made in UBe_{13} by Bucher et al. (1975), four years before $CeCu_2Si_2$, but it was then erroneously attributed to precipitated U-filaments, with no direct link to its HF character. The field of HF superconductors thus really began in 1979 and their discovery has proved to be a landmark in establishing unconventional superconductivity as a general phenomenon. The large low-temperature specific heat, which is a prominent characteristic of HF behaviour, is depicted in Figure 13.1 for $CeCu_2Si_2$ (Lang et al., 1991) and UBe_{13} (Ott et al., 1984); the insets show the crystal structures of the two compounds.

The advent of HF superconductors is schematically presented in Figure 13.2. Table 13.1 gives the crystal structure and other relevant parameters of the listed materials. Superconductivity generally occurs below 3 K; the higher values, such as 18.5 K for $PuCoGa_5$, 8.7 K for

PuRhGa$_5$, and 4.9 K for NpPd$_5$Al$_2$, being rare exceptions. Further, in many cases, external pressure, magnetic field, or dopants are found to be mandatory to get the system tuned for superconductivity. The magnetic order (Table 13.1) formed in HF superconductors can be spin-density wave (SDW), antiferromagnetic (AFM), or ferromagnetic (FM), and T_m may exceed T_c by an order of magnitude or more.

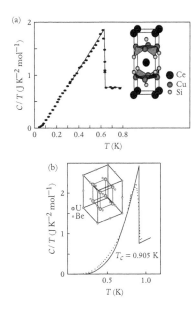

Figure 13.1 Large electronic specific heats of (a) superconducting CeCu$_2$Si$_2$ (after Lang et al., 1991) and (b) superconducting UBe$_{13}$ (after Ott et al., 1984).

13.2 Quantum phase transition and quantum critical point

The unusual HF character of these materials causes the electronic states to have their energy scales smaller by several orders of magnitude than those of ordinary metals. This, at 0 K, favours the occurrence of a so-called *quantum phase transition* (QPT) between competing nearly degenerate states. The transition is continuous, that is, of second order, and takes place at a quantum critical point (QCP). In fact, HF superconductors have emerged as ideal prototypes for studying QPTs and QCPs, where the ground state of the system is determined through an effective tuning of the controlling parameters. In *conventional phase transitions*, which take place at a finite temperature, the change of state results from thermal fluctuations. On the other hand, QPTs at absolute zero are due to quantum fluctuations (i.e. the Heisenberg uncertainty principle), tuned by non-thermal controlling parameters, for example the application of external pressure, magnetic field, chemical doping (which corresponds to internal pressure), or a change in the electron density realised through a change in valence.

Two types of quantum criticality have been identified in HF superconductors: *conventional* and *unconventional* (Gegenwart et al., 2008; Si and Steglich, 2010). In the former, the heavy quasiparticles maintain their integrity at the QCP on either side of it. The pertinent magnetic

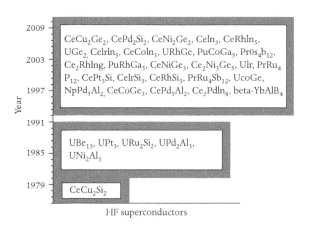

Figure 13.2 Discovery of HF superconductors in the last three decades.

Table 13.1 Prominent HF superconductors with characteristic parameters

Compound	Structure type	γ (mJ mol^{-1} K^{-2})	m^*/m_e	T_c (K)	P_c (kbar)	T_m (K)	Magnetic order*
$CeCu_2Si_2$	$ThCr_2Si_2$ $I4/mmm$	700–1000	351	0.49–0.7		0.8	SDW/AFM
$CePd_2Si_2$	$ThCr_2Si_2$ $I4/mmm$	62	97	0.43	28	10	AFM
$CeRh_2Si_2$	$ThCr_2Si_2$ $I4/mmm$	27–80	118	0.40	6	35	AFM
$CeCu_2Ge_2$	$ThCr_2Si_2$ $I4/mmm$		37	0.64–2	70–165	4.15	AFM
$CeNi_2Ge_2$	$ThCr_2Si_2$ $I4/mmm$	350	75	0.23–0.4	23		
$CeIn_3$	$AuCu_3$ $Pm3m$	140	191	0.18	25	10.1	AFM
$CeCoIn_5$	$HoCoGa_5$ $P4/mmm$	290	121	2.3			AFM
$CeIrIn_5$	$HoCoGa_5$ $P4/mmm$	720	100	0.40			AFM
$CeRhIn_5$	$HoCoGa_5$ $P4/mmm$	200	104	2.17	17	3.8	AFM
Ce_2RhIn_8	Ho_2CoGa_8 $P4/mmm$	400	94	1.1–2.0	23	2.8	AFM
Ce_2PdIn_8	Ho_2CoGa_8 $P4/mmm$			~0.70		10.0	AFM
$CePt_3Si$	$CePt_3B$ $P4/mm$	390	140	0.75	8	2.2	AFM
$CeRhSi_3$	$BaNiSn_3$ $I4/mm$	120	140	0.7–1.0	20	1.6	AFM
$CeIrSi_3$	$BaNiSn_3$ $I4/mm$	120	105	1.6	25	5.0	AFM
$CeNiGe_3$	$SmNiGe_3$ $Cmmm$	40	120	0.48	65	5.5	AFM
$CeCoGe_3$	$SmNiGe_3$ $Cmmm$	32		0.69	55	21.12	AFM
$Ce_2Ni_3Ge_5$	$U_2Co_3Si_5$ $Ibam$	90	145	0.26	36	4.8/4.2	AFM
$CePd_5Al_2$	$ZrNi_2Al_5$ $I4/mmm$	56		0.57	108	3.9/2.9	AFM
$PrOs_4Sb_{12}$	$LaFe_4P_{12}$ $Im3$	350–500	44	1.85			AFQ
$PrRu_4Sb_{12}$	$LaFe_4P_{12}$ $Im3$	59		1.3			
$PrRu_4P_{12}$	$LaFe_4P_{12}$ $Im3$			1.8			
UBe_{13}	$NaZn_{13}$ $Fm3c$	900	351	0.90			
UPt_3	$SnNi_3$ $P63/mmc$	420	265	0.55		5.0–6.0	AFM
URu_2Si_2	$ThCr_2Si_2$ $I4/mmm$	65	184	1.5		17.5	AFM, HO
UNi_2Al_3	$PrNi_2Al_3$ $P6/mmm$	120	147	1.0		4.5	SDW/AFM
UPd_2Al_3	$PrNi_2Al_3$ $P6/mmm$	150–200	187	1.9		14.0	AFM
UGe_2	$ZrGa_2$ $Cmmm$	32	235	0.8	10	52	FM
UIr	$PdBi$ $P2_1$	49	137	0.14	26	46	FM
URhGe	$TiNiSi$ $Pnma$	160	281	0.3		10	FM
UCoGe	$TiNiSi$ $Pnma$	57		0.8		3	FM
U_6Fe	Mn_6Fe $I4/mcm$	280	109	3.7			
$PuCoGa_5$	$HoCoGa_5$ $P4/mmm$	80	47	18.5			
$PuRhGa_5$	$HoCoGa_5$ $P4/mmm$	70	40	8.7			
$NpPd_5Al_2$	$I4/mmm$	200		4.9			
β-$YbAlB_4$	$ThMoB_4$ $Cmmm$	300		0.08			

*AFM, antiferromagnetic; AFQ, antiferro-quadrupolar; FM, ferromagnetic; HO, hidden order; SDW, spin-density wave.
Data compiled from Maple, M. B., Bauer, E. D., Zapf, V. S., and Wosnitza, J. *Superconductivity*, Vol. I: *Conventional and Unconventional Superconductors* (eds K. H. Bennemann and J. B. Ketterson) Springer-Verlag, Heidelberg (2008), p. 639 and Pfleiderer, C. (2009) *Rev. Mod. Phys.* 81 (2009) 1551.

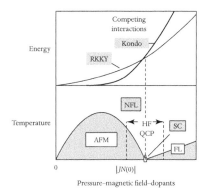

Figure 13.3 Doniach diagram (after Doniach, 1977).

ordering temperature for SDW (the Néel temperature, T_N in $CeCu_2Si_2$), AFM (i.e. T_N in $CePd_2Si_2$), or FM (i.e. the Curie temperature T_C in UGe_2) is suppressed at $T = 0$ K by tuning the controlling parameters. In the vicinity of the critical pressure P_c, critical magnetic field, critical stoichiometry, or electron concentration at which QCP occurs, the HF compound exhibits marked deviations from the conventional *Fermi-liquid* (FL) behaviour, which is referred to as *non-Fermi-liquid* (NFL). In some compounds, the superconducting transition occurs in close proximity to the QCP. The situation is represented by the phase diagram of Figure 13.3 based on Doniach's (1977) considerations involving the temperature and the controlling parameters that tune the composite interaction parameter $|JN(0)|$. The Kondo interaction (Kondo, 1964) varies as $\exp[-1/|JN(0)|]$, while the *Ruderman–Kittel–Kasuya–Yosida* (RKKY) interaction varies as $\left[|JN(0)|\right]^2$, where $N(0)$ is the density of states (DOS) of the conduction band at E_F. As a result of the competition between the two interactions, the magnetic ordering temperature increases initially with increasing $|JN(0)|$ and, after passing through a peak, becomes zero at the critical value $\left[|JN(0)|\right]_c$, which corresponds to the QCP. Both NFL and superconductivity may result at the QCP.

In the unconventional criticality (Si et al., 2001; Coleman et al., 2001), the heavy quasiparticles disintegrate at the QCP, termed the *Kondo-destroying (KD) QCP*. This causes the f-electrons to become localised and decoupled from the conduction-band states at the QCP, where the Fermi surface is reconstructed. The latter is revealed by a sudden change in the Hall coefficient occurring at the QCP. Extensively studied cases of unconventional criticality are $YbRh_2Si_2$ and $CeCu_{5.9}Au_{0.1}$ (Friedemann et al., 2009), but neither turn superconducting. The recently discovered HF superconductor β-$YbAlB_4$ ($T_c \approx 80$ mK) (Nakatsuji et al., 2008) has been mentioned as a possible candidate for unconventional criticality (Friedemann et al., 2009).

13.3 General features of anomalous normal state and unusual superconductivity

Both HF superconductivity and its NFL normal state from which the former evolves are remarkably different from their conventional counterparts. Since the heavy charge carriers in these materials are slow, possessing a low Fermi velocity, they are not able to readily escape their own polarisation cloud, which makes the conventional phonon-mediated pairing mechanism unfeasible.

The standard FL theory predicts the metallic resistivity $\rho(T)$ to saturate as AT^2. In HF materials, however, the resistivity deviates from such quadratic behaviour, with the exponent varying from 1 to 1.5, which is characteristic of NFL behaviour. The magnetic susceptibility in the HF state is typically 10–100 times larger than in ordinary metals. The NFL normal state is generally regarded as a signature of ensuing unconventional superconductivity with anisotropic order parameter, possessing line or point nodes on the Fermi surface where the superconducting gap locally vanishes.

The behaviours of various superconducting parameters in HF compounds are in general indicative of unconventional superconductivity with anisotropic gap parameter, characterised by a power-law ($\propto T^n$) temperature dependence (e.g. T^3 or T^2) of various properties below T_c, such as specific heat, thermal conductivity, ultrasonic attenuation, and inverse spin–lattice relaxation rate in nuclear resonance (NMR/NQR) studies. In conventional isotropic s-wave superconductivity (the spin-singlet state), these properties exhibit an exponential temperature dependence of the form $\exp(-\Delta/k_B T)$. In conventional superconductors, nuclear resonance spectroscopic studies show that the temperature dependence of the spin–lattice relaxation rate, as well as its characteristic exponential behaviour, exhibits a so-called *Hebel–Slichter peak* (Hebel and Slichter, 1959), which is not observed for unconventional superconductivity. In the power-law behaviour of unconventional superconductors, the exponent $n = 3$ corresponds to the vanishing of the gap at point nodes while $n = 2$ is indicative of line nodes.

For a conventional superconductor, the microscopic theory gives $\Delta C/\gamma T_c = 1.43$ for the weak-coupling limit and 1.6 for the strong-coupling limit. For HF superconductors, this ratio is generally ~ 1, although much larger values have also been observed. For example, for $CeCoIn_5$ (Petrovic et al., 2001a; Zapf et al., 2001), the reported value of is in the range 4.5–5.0.

A large effective mass m^* has an immediate effect on the characteristic lengths λ and ξ of HF superconductors, the former increasing and the latter decreasing as m^* increases (see Chapter 5, equations (5.1) and (5.8)). As a consequence, the HF superconductors tend to be extreme type II with Ginzburg-Landau parameter $\kappa > 50$. Further, with increasing m^*, the lower critical field $H_{c1} \propto \lambda^{-2} \propto 1/m^*$ decreases while the upper critical field $H_{c2} \propto \xi^{-2} \propto m^{*2}$ increases, which makes the former small (10^{-3} T) and the latter very large ($\sim 1-10$ T), considering the intrinsically low T_c values of these materials. Similarly, the increased effective mass also gives rise to a large initial slope ($-dH_{c2}/dT$) at T_c. Table 13.2 gives H_{c2} values with their initial slopes for different HF superconductors. As may be seen, the values are anisotropic with respect to the crystallographic unit cell.

The HF superconductors are expected to manifest a strong Pauli-limited upper critical field (see Chapter 5, Section 5.3.2). If the measured value of H_{c2}, however, significantly exceeds the paramagnetic limit, this is taken as an indication of spin-triplet superconductivity. Another

Table 13.2 Upper critical field (extrapolated to 0 K) and its initial slope dH_{c2}/dT at T_c for HF superconductors[*]

Compound	$\mu_0 H_{c2}^{ab}(0)$ (T)	$\mu_0 H_{c2}^{c}(0)$ (T)	dH_{c2}^{ab}/dT at T_c (T K^{-1})	dH_{c2}^{c}/dT at T_c (T K^{-1})
$CeCu_2Si_2$	0.45, 1.7[a]	—	-5.8[*]	—
$CePd_2Si_2$	0.7	1.3	-12.7	-16.0
$CeCu_2Ge_2$	2.0	—	-11.0	—
$CeNiGe_3$	1.55	—	-10.8	—
$CeIn_3$	0.45	0.45	-3.2	-2.5
$CeCoIn_5$	11.6–11.9	4.95	-24	-8.2
$CeIrIn_5$	1.0	0.49	-4.8	-2.54
$CeRhIn_5$	—	10.2	—	-15.0
$CeRh_2Si_2$	—	0.28	—	-1.0
$CePt_3Si$	3.6, 2.6	4.0	$-8.5, -7.2$	-8.5
$CeIrSi_3$	9.5	>30.0	-13	-20.0
$PuCoGa_5$	—	—	-10	-8.0
$PuRhGa_5$	27.0	15.0	-3.5	-2.0
$NpPd_5Al_2$	3.7	14.3	-6.4	-3.1
Ce_2RhIn_8	5.4	—	-9.2	—
$Ce_2Ni_3Ge_5$	0.7	—	—	—
$CePd_5Al_2$	0.25	—	-1.04	—
$PrOs_4Sb_{12}$	2.3	—	-1.9	—
$PrRu_4Sb_{12}$	0.2	—	—	—
$PrRu_4P_{12}$	2.0	—	—	—
UPt_3	2.1	2.8	-7.2	-4.4
URu_2Si_2	3.0	14.0	-5.3	-14.5
UPd_2Al_3	3.9	3.3	-5.45	-4.6
UNi_2Al_3	0.9	0.35	-1.14	-0.42
UBe_{13}	14.0	—	-45	—
UIr	0.0265	—	—	—
β-$YbAlB_4$[b]	0.025	—	—	—
Ce_2PdIn_8[a]	4.8	—	-14.3	—

[*]Empty entries (—) represent either isotropic cases or the unavailability of reliable data.
Data from Pfleiderer, C. *Rev. Mod. Phys.* 81 (2009), 1551 and from [a] Kaczorowski, D., Pikul, A. P., Gnida, D., and Tran, V. H. *Phys. Rev. Lett.* 103 (2009), 027003 and [b] Matsumoto, Y., Kuga, K., Karaki, Y., Tomita, T., and Nakatsuji, S. *Phys. Stat. Sol.* B247 (2010), 720.

important general feature of HF superconductivity is its extreme sensitivity to impurities and the material's stoichiometry. While in conventional superconductors T_c remains sensitive only to magnetic impurities, in HF superconductors, both types of impurities are equally deleterious to T_c. This serves as a pointer to identify the nature of unconventional superconductivity. Since the T_c values of HF superconductors are generally very low, special care is called for to achieve the right stoichiometry and to avoid all kinds of impurities, the presence of which could lead to superconductivity being readily obscured.

13.4 Short description of various HF superconductors

Following Pfleiderer (2009), the HF superconductors of Table 13.1 may be categorised as (i) those in which superconductivity occurs in proximity to antiferromagnetism, (ii) superconducting antiferromagnets, (iii) superconducting ferromagnets, (iv) those in which superconductivity occurs in proximity to ferromagnetism, and (v) miscellaneous types in which superconductivity occurs in proximity to localisation, mixed valence, or polar order.

13.4.1 Superconductivity in proximity to antiferromagnetism

13.4.1.1 CeM_2X_2 (M = Cu, Pd, Rh, Ni and X = Si, Ge)

The HF materials $CeCu_2Si_2$, $CeCu_2Ge_2$, $CePd_2Si_2$, $CeRh_2Si_2$, and $CeNi_2Ge_2$, with the general chemical formula CeM_2X_2 and of structural type $ThCr_2Si_2$, with tetragonal $I4/mmm$ space-group symmetry, are materials in which superconductivity occurs in proximity to antiferromagnetism. Of these, $CeCu_2Si_2$, being the first ambient-pressure HF superconductor discovered (Steglich et al., 1979), has received the most attention.

The crystal structure (inset in Figure 13.1(a)) of $CeCu_2Si_2$ is the same as for its sister compounds mentioned above. This material is noted for its many ground-state phases with vastly different physical characteristics. These could be superconducting (S-phase), magnetic (A-phase), or both (A/S), and there is yet one more magnetic phase (B-phase) found in a high magnetic field. The magnetic phases are identified as antiferromagnetic SDW type (Tayama et al., 2003). A slight excess of Cu during sample processing results in a single S-phase, while its deficiency promotes the formation of the A-phase. This behaviour is intimately related to a small change in the unit-cell volume caused by chemical pressure. This is corroborated by studies carried out by partial replacement of the smaller Si ion by a larger Ge ion (a negative pressure), which favours the formation of the A-phase, while a positive hydrostatic pressure promotes formation of the S-phase (Krimmel and Loidl, 1997).

The T_c of the S-phase of $CeCu_2Si_2$ is 0.68 K, which is just below the $T_N \approx 0.8$ K of the SDW A-phase. The two states essentially compete with each other and there is no microscopic coexistence. The mutual proximity of the two states, however, favours NFL behaviour, with thermal conductivity, thermal expansion, specific heat, and penetration depth at low temperature all manifesting a T^2 dependence, characteristic of line nodes at the Fermi surface (Goll, 2006, and references therein). Similarly, NMR and NQR measurements indicate an absence

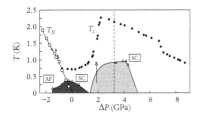

Figure 13.4 Temperature–pressure phase diagram of $CeCu_2(Si_{1-x}Ge_x)_2$ (after Thalmeier et al., 2005).

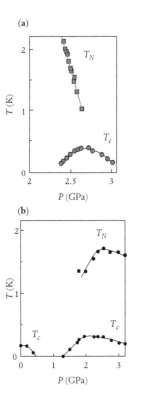

Figure 13.5 Temperature–pressure phase diagrams of (a) $CePd_2Si_2$ and (b) $CeNi_2Ge_2$ (after Grosche et al., 2000).

of the Hebel–Slichter peak, suggesting unconventional superconductivity, and display a T^3 dependence of the spin–lattice relaxation rate, again corroborating the presence of line nodes. The upper critical field $\mu_0 H_{c2} \approx 0.45$ T is Pauli paramagnetically limited (Goll, 2006), which suggests an even-parity order parameter possibly with d-wave symmetry.

The application of pressure P is found to enhance T_c of the S-phase in $CeCu_2Si_2$ from about 0.7 K to about 2.25 K at $P \approx 2.5$ GPa, with the effect becoming pronounced for P greater than about 2 GPa (Thomas et al., 1993). As the pressure is further increased, T_c starts to decrease, and at about 7 GPa exhibits a steeper drop followed by a moderate decrease. Both Yuan et al. (2003) and Thalmeier et al. (2005) have discussed the pressure effects on superconductivity and magnetism of $CeCu_2Si_2$ doped with a small concentration of Ge. Two distinct superconducting phases appear in the form of two separate domes (Figure 13.4). One at lower pressure (<2 GPa) appears in proximity to SDW order, while the other, at higher pressure (<5 GPa), appears at the valence transition of Ce from f^{3+} to f^{4+}. In both instances, superconductivity is unconventional, being mediated by SDW fluctuations in the former and by charge-density wave (CDW) fluctuations in the latter (Steglich, 2005).

A magnetic field applied along the a-axis in the basal plane of $CeCu_2Si_2$ suppresses superconductivity when $\mu_0 H > \mu_0 H_{c2} \approx 0.45$ T and it restores the magnetic A-phase (Bruls et al., 1994; Steglich et al., 2001). Above $H \approx 6.4$T, the A-phase is suppressed and the B-phase is formed, the characteristics of which remain to be fully explored.

At ambient pressure, $CeCu_2Ge_2$ exhibits AFM order below $T_N = 4.15$ K, with an ordered moment of $0.74\mu_B$. It exhibits superconductivity, with $T_c = 0.64$ K, only when the sample is subjected to a pressure $P \approx 7$ GPa at which the AFM state is suppressed, corresponding to an AFM QCP (Jaccard et al., 1992). When $P > 13$ GPa, T_c rapidly starts increasing, reaching an optimum value of about 2 K at $P \approx 16$ GPa, when X-rays confirm a valence transition of Ce (Onodera et al., 2002). At $P \approx 20$ GPa, superconductivity is no longer observed (Vargoz and Jaccard, 1998). The upper critical field of $CeCu_2Ge_2$ of about 2 T is Pauli paramagnetically limited, indicating an even-parity order parameter (Kobayashi et al., 1998).

Like $CeCu_2Ge_2$, $CePd_2Si_2$ is antiferromagnetic at ambient pressure, with $T_N \approx 10$ K and an ordered moment $0.62\mu_B$ (van Dijk et al., 2000). T_N shows a linear decrease with pressure (Grier et al., 1984; Grosche et al., 1996, 2000), which extrapolates to zero at a critical pressure $P_c \approx 2.8$ GPa (Figure 13.5(a)). This corresponds to an AFM QCP, with the sample showing superconductivity at an optimum $T_c \approx 0.4$ K (Thompson et al., 1986). The resistivity behaviour near P_c is of NFL type, with a power-law temperature dependence $\Delta\rho \propto T^{1.2}$, suggestive of strong spin fluctuations reflecting two-dimensionality. These fluctuations are considered to be responsible for the unconventional superconductivity of the material (Grosche et al., 1996; Mathur et al., 1998). The material, despite its low T_c, exhibits a relatively large anisotropic upper critical field of 0.7–1.3 T and a large initial slope in the basal plane of -12 T K^{-1}, which are characteristics of HF materials (Sheikin et al., 2001).

The sister compounds CeNi$_2$Ge$_2$ (Figure 13.5(b)) and CeRh$_2$Si$_2$ behave similarly to CePd$_2$Si$_2$. CeNi$_2$Ge$_2$ exhibits a T_c of 0.3 K at $P_c \approx$ 2 GPa (Grosche et al., 2000). Surprisingly, at high pressure, there are indications of another superconducting state (Grosche et al., 2000), whose origin, however, remains to be confirmed. In CeRh$_2$Si$_2$, the ambient-pressure AFM order forms below $T_{N1} \approx 36$ K, which when suppressed at $P_c \approx 0.9$ GPa yields superconductivity with optimum $T_c \approx 0.42$ K (Movshovich et al., 1996). This material shows a second ordering temperature $T_{N2} \approx 24$ K below which the ordering wavevector depicts a significant change. The ordered moment is very large, about 1.34–1.42μ_B, which is sensitive to the Ce site.

13.4.1.2 $Ce_nM_mIn_{3n+2m}$ ($M = Co, Ir, Rh$)

This series of HF superconductors of relatively recent origin (Hegger et al., 2000; Petrovic et al., 2001a) contains three extensively studied members, namely CeCoIn$_5$, CeIrIn$_5$, and CeRhIn$_5$ of the general formula Ce$_n$M$_m$In$_{3n+2m}$, corresponding to $n = m = 1$, which are also called Ce115. Their tetragonal crystal structure with space-group symmetry of $P4/mmm$ is depicted in Figure 13.6(b). The cubic CeIn$_3$ (space group $Pm3m$) is generally considered as the parent compound (with structure as in Figure 13.6(a)) of these three materials. While CeIn$_3$ exhibits superconductivity under pressure, with $T_c \approx 0.19$ K, both CeCoIn$_5$ and CeIrIn$_5$ are ambient-pressure superconductors with respective T_c values of 2.3 K and 0.4 K. The third member, CeRhIn$_5$, also exhibits a relatively high T_c of 2.1 K, albeit under an imposed pressure. There is a general correlation between T_c and unit-cell volume for these three materials. The Co-containing compound has the lowest unit-cell volume and possesses the largest T_c, and this fact is relevant to understanding pressure effects on the three materials. In comparison with CeIn$_3$, the relatively larger T_c of these ternary compounds is, however, attributed to their quasi-two-dimensional crystal structure, comprising alternate layers of CeIn$_3$ and MIn$_2$ stacked along the c-axis.

The characteristics of CeIn$_3$ (Mathur et al., 1998) are very similar to those of CePd$_2$Si$_2$ and other compounds discussed in Section 13.4.1.1. At ambient pressure, the material is antiferromagnetic ($T_N \approx 10$ K) with an ordered moment of 0.65μ_B/Ce atom, and with the application of pressure T_N monotonically decreases and at a critical pressure $P_c \approx 2.6$ GPa extrapolates to zero at the QCP (Figure 13.7). In a narrow region near P_c in the P–T phase diagram, the material becomes

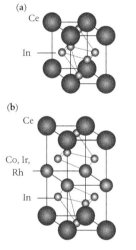

(a) Ce
In

(b) Ce
Co, Ir, Rh
In

Figure 13.6 Crystal structures of (a) cubic CeIn$_3$ and (b) tetragonal CeMIn$_5$, with M = Co, Ir, or Rh.

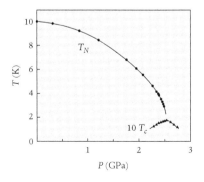

T_N

10 T_c

Figure 13.7 Temperature–pressure phase diagram of CeIn$_3$ (after Mathur et al., 1998).

superconducting at $T_c \approx 0.19$ K owing to a mediation through magnetic spin fluctuations. The resistivity just above T_c varies as $T^{1.6}$, which is indicative of spin fluctuations having a three-dimensional character. The unconventional superconductivity is corroborated by the absence of a Hebel–Slichter peak, and a theoretical analysis of data is consistent with d-wave symmetry (Fukazawa and Yamada, 2003). However, further experimental studies are still needed for confirmation of this.

While the cubic CeIn$_3$ is a bulk three-dimensional material, the ternary superconductors CeRhIn$_5$, CeIrIn$_5$, and CeCoIn$_5$, as already mentioned, are tetragonal and quasi-two-dimensional materials. In contrast to CeIrIn$_5$ and CeCoIn$_5$, which are ambient-pressure superconductors, CeRhIn$_5$ becomes superconducting only under pressure. It is an antiferromagnet at ambient pressure, with $T_N = 3.8$ K and a magnetic moment of $0.37\mu_B$/Ce atom (Hegger et al., 2000). Under an imposed pressure, T_N decreases, and at a pressure of 2 GPa, the sample shows bulk superconductivity with optimum $T_c \approx 2.1$ K (Chen et al., 2006; Knebel et al., 2006). At an intermediate pressure of 1.77 GPa, $T_N \approx T_c \approx 2.0$ K and the two states coexist (Knebel et al., 2006). The possibility of a second superconducting transition had previously been suggested at a higher pressure of 8.5 GPa (Muramatsu et al., 2001), which has, however, not been confirmed (Knebel et al., 2008). NQR measurements in the superconducting state (Mito et al., 2001) do not show the presence of a Hebel–Slichter peak, and the pertinent spin–lattice relaxation rate varies as T^3 (Figure 13.8). This provides credence to the presence of unconventional superconductivity with line nodes at the Fermi surface. This is consistent also with the low-temperature specific heat measurements showing a T^2 dependence below T_c (Fisher et al., 2002).

Among Ce-based HF superconductors, CeCoIn$_5$ is noted to possess the highest ambient pressure T_c of 2.3 K. It closely matches with the T_c of 2.1 K observed for CeRhIn$_5$ under a pressure of about 2 GPa, at which their unit-cell volumes also match. Because of this, CeCoIn$_5$ is considered as the itinerant version of the localised CeRhIn$_5$. Alternatively, under ambient pressure, CeRhIn$_5$ is equivalent to CeCoIn$_5$ possessing an expanded unit cell or under a state resulting from a negative pressure. Under a hydrostatic pressure, CeCoIn$_5$ exhibits a T_c maximum at 1.5 GPa, and superconductivity is fully suppressed at 3.7 GPa (Sparn et al., 2002; Sidorov et al., 2002; Knebel et al., 2004). Superconductivity in CeCoIn$_5$ evolves out of an unusual normal state (when $H > H_{c2}$) with characteristics of NFL behaviour, such as a linear $\rho(T)$ and a logarithmically diverging specific heat, both suggesting proximity to an AFM QCP. However, at much higher fields and low temperatures, the FL behaviour is restored. The H–T phase diagram of CeCoIn$_5$ (Figure 13.9) has been studied by Petrovic et al. (2001c) and Bianchi et al. (2003a). Interestingly, the QCP in this material occurs with the destruction of

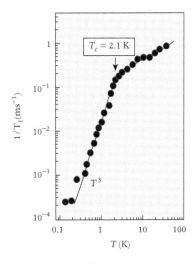

Figure 13.8 From ^{115}In NQR (Mito et al., 2001), the spin–lattice relaxation rate (for CeRhIn$_5$) shows no coherence peak below T_c (2.1 K under a pressure $P = 2.1$ GPa) and a T^3 dependence.

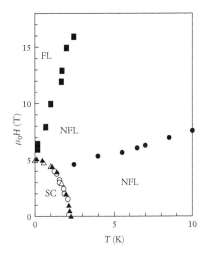

Figure 13.9 Magnetic field–temperature phase diagram of CeCoIn$_5$ (after Petrovic et al., 2001c).

superconductivity at H_{c2}, where the AFM order is hidden by superconductivity. Considering the case of CeRhIn$_5$, discussed earlier, it seems natural to expect a second QCP in CeCoIn$_5$ at negative pressures, as suggested by Zheng et al. (2001).

In CeCoIn$_5$, the unconventional nature of the superconducting gap with line nodes is indicated by several of its properties below T_c, such as the T^2 and T^3 temperature dependences of specific heat and thermal conductivity, the T^3 dependence of the spin–lattice relaxation rate and the absence of a Hebel–Slichter peak in nuclear resonance experiments (Kohori et al., 2001). Interestingly, the transition at H_{c2} in high magnetic fields is of first order (Izawa et al., 2001b; Bianchi et al., 2002). At ambient pressure, the material shows a relatively large upper critical field $\mu_0 H_{c2}^{ab} = 11.6$ T and $\mu_0 H_{c2}^c = 4.95$ T, along with equally large $dH_{c2}^{ab}/dT = -24$ T K^{-1} and $dH_{c2}^c/dT = -11$ T K^{-1}, which are characteristic of HF superconductors (Petrovic et al., 2001a; Ikeda et al., 2001). Also, it is an extreme type II superconductor with $k^{ab} = 108$ and $k^c = 50$ (Kasahara et al., 2005).

CeCoIn$_5$ is a Pauli paramagnetically limited superconductor that is quasi-two-dimensional and can be prepared with high purity with $\xi_0/l > 10$. These are the prerequisites for the formation of the unique FFLO state, and, as we will see in Section 13.5.3, CeCoIn$_5$ has emerged for the first time as a strong case manifesting this state.

CeIrIn$_5$, similar to CeCoIn$_5$, is an ambient-pressure superconductor, albeit with a lower bulk T_c of 0.4 K (Petrovic et al., 2001a,b) and weak antiferromagnetic characteristics. Interestingly, the resistively measured T_c is about 3 times larger, namely about 1.2 K, which is attributed to either the presence of preformed pairs or filamentary superconductivity (Bianchi et al., 2001). For both compounds, the spin susceptibility is suppressed below T_c, indicating singlet pairing. As with CeCoIn$_5$, there is substantial evidence for the unconventional superconductivity of CeIrIn$_5$ through NQR, specific heat, and thermal conductivity measurements (Kohori et al., 2001; Zheng et al., 2001; Izawa et al., 2001a; Movshovich et al., 2001). These findings are supportive of the presence of line nodes in the gap function. But, unlike CeCoIn$_5$, there is no convincing evidence of Pauli limitation of the upper critical field of CeIrIn$_5$ (Movshovich et al., 2002).

13.4.2 Superconducting antiferromagnets

13.4.2.1 *Large magnetic moments: UPd$_2$Al$_3$, UNi$_2$Al$_3$, and CePt$_3$Si*

In this group of HF superconductors, superconductivity arises deep within the AFM state carrying a relatively large magnetic moment of more than $0.1\mu_B$/U- or Ce-atom. Of these, UPd$_2$Al$_3$ and UNi$_2$Al$_3$ are isostructural, possessing the hexagonal PrNi$_2$Al$_3$ structure with the space-group symmetry $P6/mmm$. Typically, in the case of UPd$_2$Al$_3$,

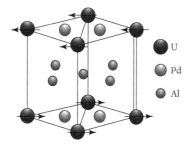

Figure 13.10 Crystal structure (hexagonal) of UPd_2Al_3. The AFM structure is indicated by arrows showing the magnetic moments at the U sites.

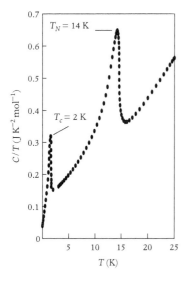

Figure 13.11 Heat capacity measurements showing the coexistence of T_c and T_N in UPd_2Al_3 (after Geibel et al., 1991).

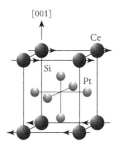

Figure 13.12 Crystal structure (tetragonal) of $CePt_3Si$. The AFM structure is indicated by arrows showing the magnetic moments at the Ce sites.

the structure (Figure 13.10) comprises alternate, oppositely oriented, FM layers of U atoms in the hexagonal basal plane that give rise to the AFM order below $T_N \approx 14$ K, with a large magnetic moment of $0.85\mu_B/U$ (Geibel et al., 1991; Krimmel et al., 1992; Paolasini et al., 1993). Superconductivity occurs below $T_c \approx 2$ K, where the two orders coexist. The two orders are mutually coupled, as the AFM Bragg peak shows a small decrease in intensity below T_c (Metoki et al., 1998). The coexistence is depicted in Figure 13.11 through low-temperature heat capacity measurements (Geibel et al., 1991). The nuclear relaxation rate $1/T_1$ (in NMR/NQR studies) shows the T^3 behaviour and the absence of a Hebel–Slichter peak, in accordance with the presence of unconventional superconductivity and line nodes (Matsuda et al., 1997). Both thermal conductivity and specific heat exhibit power law behaviour (Hiroi et al., 1997; Geibel et al., 1991), corroborating the presence of line nodes in the gap and d-wave pairing. UPd_2Al_3 exhibits strong Pauli paramagnetism, favouring the presence of the spin-singlet state (Watanabe et al., 2004b).

UNi_2Al_3 has similar characteristics, but with lower T_N and T_c values of about 4.5 K and about 1 K and a smaller magnetic moment of about $0.2\mu_B/U$-atom (Uemura and Luke, 1993; Schröder et al., 1994). However, interestingly, in contrast to UPd_2Al_3, it does not show Pauli paramagnetic limitation and its superconductivity, as Knight-shift measurements show, is believed to arise from triplet instead of singlet pairing (Ishida et al., 2002). Both UPd_2Al_3 and UNi_2Al_3 exhibit a phase change from hexagonal to orthorhombic structure, with disappearance of superconductivity, when they are subjected to a hydrostatic pressure of more than 25 GPa (Krimmel, 2000; Link et al., 1995). This is believed to be related to the pressure-induced valence change from U^{4+} to U^{5+} (Rueff et al., 2007), suggesting the possible importance of the U^{4+} state in superconductivity (Pfleiderer, 2009).

$CePt_3Si$, possessing a tetragonal crystal structure (Figure 13.12) with space group $P4mm$, shows AFM order with a magnetic moment of $0.16\mu_B/Ce$-atom below $T_N \approx 2.2$ K and turns superconducting below $T_c \approx 0.75$ K (Metoki et al., 2004; Amato et al., 2005). It exhibits several features indicative of unconventional superconductivity. The observed specific heat jump ΔC at T_c yields $\Delta C/\gamma T_c \approx 0.25$ (Bauer et al., 2005) instead of the BCS value of 1.43, which is indicative of the presence of nodes at the Fermi surface. This is corroborated by the observed T^3 dependence of the specific heat (Bauer et al., 2004). NMR relaxation measurements depict neither exponential nor T^3 dependence. On the other hand, a Hebel–Slichter peak is present, albeit of much reduced peak height than found for conventional superconductors. All these features indicate an unusual nature of superconductivity in $CePt_3Si$. The interest in this material is not just because of the coexistence of superconductivity with antiferromagnetism, but also because of its unusual crystal structure (Bauer et al., 2004), which has no inversion symmetry. $CePt_3Si$ was the first such compound to be discovered belonging to such a class, although a few more similar materials have subsequently been discovered. The upper critical field of $CePt_3Si$ exceeds the Pauli paramagnetic limit, which makes it a potential candidate for triplet pairing (Frigeri et al., 2004). This is contrary to what has been predicted for non-centrosymmetric superconductors (Anderson, 1984). Further, this material exhibits multiple superconducting phases, which is again something unusual and is expected only for an unconventional superconductor. These special features are briefly discussed in Sections 13.5.1 and 13.5.2.

13.4.2.2 Small magnetic moments: UPt_3 and URu_2Si_2

Among this category of HF superconductors, UPt_3, discovered in 1984 (Stewart et al., 1984) is perhaps the most extensively studied material and is now regarded as an established candidate for the presence of triplet pairing (Joynt and Taillefer, 2002). The material is hexagonal, having space group $P6_3/mmc$ and with unit cell as depicted in Figure 13.13. The material becomes antiferromagnetic below $T_N \approx 5$ K, with small magnetic moments of about $0.02\mu_B$/U-atom directed along the a-axis in the basal plane. The magnetic moments seem to be of rapidly fluctuating type and are detected only through neutron scattering (Keizer et al., 1999). Superconductivity is observed below $T_c \approx 0.54$ K. The normal state is antiferromagnetic, exhibiting heavy FL behaviour. The superconducting state is particularly unusual in that below 0.54 K it shows multiple superconducting phases A, B, and C, of which the C-phase is formed only in a magnetic field greater than 0.4 T. Superconducting A- and B-phases appear as split phases with their critical temperatures T_{c1} and T_{c2} differing only by about 50 mK. Under an imposed pressure of 0.3 GPa, the AFM order is destroyed and the two critical temperatures merge with each other into a single transition. This indicates that the original splitting was caused by the presence of AFM order. The formation of such multiple superconducting phases constitutes a special feature that, along with the phase diagram, is discussed in Section 13.5.2, and it corroborates the unconventional superconductivity of UPt_3. The material also shows the phenomenon of metamagnetism, described in Section 13.5.4. Various theoretical models developed (Joynt and Taillefer, 2002) support the existence of an odd-parity order parameter and possibly spin-triplet f-wave pairing (Sauls, 1994).

The AFM superconductor URu_2Si_2 possesses the same crystal structure as $CeCu_2Si_2$ (inset of Figure 13.1), and below 17.5 K, at ambient pressure, it exhibits AFM order with a very small magnetic moment of $0.03\mu_B$/U-atom (Broholm et al., 1991) and shows unconventional superconductivity below 1.4 K (Palstra et al., 1985; Maple et al., 1986; Schlabitz et al., 1986). The Knight-shift behaviour (Maple et al., 1986) suggests

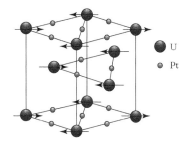

Figure 13.13 Crystal structure (hexagonal) of UPt_3. The AFM structure is indicated by arrows showing the magnetic moments at the U sites.

U
Pt

even-parity pairing. However, this material has attracted special attention because the AFM order formed below 17.5 K is just too small to account for the large entropy released at the transition and thus is incompatible with the magnitude of the specific heat jump observed. Clearly, this indicates the presence of some mysterious *hidden order*, whose nature and origin are topics of current studies. This special feature of URu_2Si_2 is briefly discussed in Section 13.5.5.

13.4.3 Superconducting ferromagnets: UGe_2 and $URhGe$

The HF ferromagnet UGe_2 possesses an orthorhombic crystal structure, with space group *Cmmm*. An interesting aspect of the crystal structure is that the U atoms form a zig-zag arrangement (Figure 13.14(a)) and the material prepared in single-crystal form exhibits unconventional superconductivity under pressure in a narrow range of 1 GPa $< P < 1.6$ GPa, with an optimum T_c of 0.7–0.9 K at $P \approx 1.3$ GPa (Saxena et al., 2000). As may be seen from the phase diagram in Figure 13.14(b) (Saxena et al., 2000; Huxley et al., 2001; Maple et al., 2008), the effect of pressure is to cause a gradual decrease in the ambient-pressure Curie temperature $T_C \approx 53$ K and no ferromagnetism is observed beyond the critical pressure $P_c \approx 1.6$ GPa, at which the material is paramagnetic. Interestingly, there is no superconductivity either above this pressure, which suggests that the two ordered states are intimately related and not mutually inimical, as commonly supposed. At a pressure of 1.3 GPa, when T_c is maximum, the FM order is still present, with a Curie temperature of about 30 K and an ordered magnetic moment of $1.4\mu_B$/U-atom. Thus, superconducting order is formed on the FM side, which makes UGe_2 a truly itinerant ferromagnetic superconductor formed under pressure. Apart from the presence of these two orders, the resistivity behaviour of the sample further shows (Huxley et al., 2001; Bauer et al., 2001) an anomaly at a temperature T_x, which gradually disappears with increasing pressure below about 1.2 GPa and whose origin is not clearly understood. It is believed to be related to the incipient SDW or CDW state. Alternatively, this phase has also been suggested as a second ferromagnetic phase FM2 with the temperature T_x as its Curie temperature T_{C2} (Maple et al., 2010). The fluctuations associated with SDW/CDW or FM states are believed to be responsible for superconducting *p*-wave pairing (Watanabe and Miyake, 2002). In the triplet state, the pairs formed are sufficiently robust against the pair-breaking effect of the large internal magnetic field of the FM state, which is about 100 T. But it is known that *p*-wave pairing is very sensitive to non-magnetic impurities, and it was therefore surprising that polycrystalline high-resistivity samples of UGe_2 did not exhibit a lower T_c than was found for single crystals (Blagoev et al., 1999; Suhl, 2001). This is indicative of the pairing being of *s*-wave

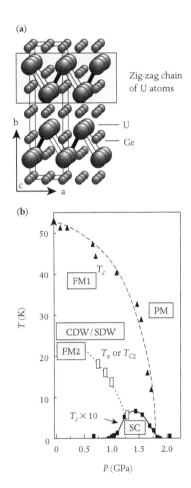

Figure 13.14 (a) Crystal structure of UGe_2. (b) Temperature–pressure phase diagram for UGe_2 (after Maple et al., 2008).

type. The issue of the pairing symmetry of UGe$_2$ remains to be fully established.

Another interesting example of a FM superconductor is URhGe, discovered by Aoki et al. (2001), which, like UGe$_2$, is orthorhombic but with a different space group, *Pnma*. As with UGe$_2$, the structure of URhGe comprises zig-zag U chains, and surprisingly their U–U atomic separation at ambient pressure closely matches that of UGe$_2$ corresponding to the pressure range of 1–1.6 GPa at which the latter shows superconductivity. URhGe single crystals do indeed exhibit ambient-pressure superconductivity and ferromagnetism. Below the Curie temperature $T_C \approx 9.6$ K, the material changes from paramagnetic to ferromagnetic, with an ordered magnetic moment of about $0.42\mu_B$/U-atom, and for $T < T_c \sim 0.25$ K it shows coexistence of superconductivity with itinerant ferromagnetism (Aoki et al., 2001; Prokes et al., 2002). The T_c value is very sensitive to the presence of impurities and rapidly decreases and disappears with increasing residual resistivity, which is an indication of unconventional superconductivity. Furthermore, in all directions, the upper critical field over-rides the Pauli limitation, which suggests the presence of *p*-wave pairing.

The ordered FM moment in URhGe is spontaneously aligned along the *c*-axis (Huxley et al., 2003). The field-induced magnetisation is smaller for fields applied along the *a*-axis than other axes of the crystal. If a magnetic field greater than 10 T is applied parallel to the *b*-axis, it causes the ordered moment to gradually rotate into the field direction. A first-order transition occurs when the spins jump into alignment with the *b*-axis. First, the initially formed superconducting state is quenched above a field of 2 T, and at sufficiently low temperatures a second pocket of superconductivity emerges (Lévy et al., 2005; Miyake et al., 2008) in the form of a dome, following an unusual phase diagram, depicted in Figure 13.15. The *field-induced superconductivity* (FIS) in the present scenario differs distinctly from that previously reported in some superconducting Chevrel phases, described in Chapter 12, as well as in a few organic superconductors (Chapter 14). In the latter two instances, the phenomenon is explained in terms of the Jaccarino–Peter (1962) model, where the applied field is essentially compensated by the internal field produced by magnetic ordering, such that the effective magnetic field inside the material is much lower. For these materials, this compensation effect is possible because the induced moments in them are parallel to the applied field. In URhGe, such a compensation effect seems unlikely as the FIS is observed over a range of angles from 30° to 50° to the field direction. The origin of FIS in URhGe is ascribed to the magnetic fluctuations associated with the change in the orientation of the magnetic moment and an enhancement of the effective mass at $H \approx H_R$.

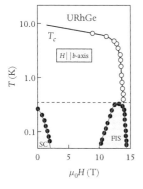

Figure 13.15 Field induced superconductivity (after Levy et al., 2005).

13.4.4 Superconductivity in proximity to ferromagnetism: UCoGe and UIr

UCoGe is isostructural to the orthorhombic URhGe, discussed in Section 13.4.3, with which it shares some similar characteristics. Polycrystalline samples become ferromagnetic below the Curie temperature $T_C \approx 3$ K, with an ordered moment of $0.03 \mu_B$/U-atom, and below $T_c \approx 0.8$ K superconductivity coexists with itinerant ferromagnetism (Huy et al., 2007). Single-crystal samples show a slightly lower T_C and manifest superconductivity below 0.6 K (Huy et al., 2008). As with URhGe, the magnetic moment of UCoGe is aligned along the c-axis, and for the field along the b-axis, single-crystal samples give a slight indication of an FIS-like situation developing in the material (Aoki et al., 2009). T_C decreases with magnetic field and is completely suppressed at 15 T. The effect of pressure is to lower T_C, which at a critical pressure $P_c \approx 1.5$ GPa vanishes, with the material becoming paramagnetic and the resistivity behaviour becoming NFL type. The T_c exhibits a small peak close to P_c, beyond which, in the paramagnetic state, T_c shows a small decrease (de Nijs et al., 2008). This is an indication of unconventional superconductivity. The upper critical field along both a- and b-axes exceeds the Pauli limit, which indicates the pairing to be of spin-triplet type.

UIr is a non-centrosymmetric material with a monoclinic crystal structure with space-group symmetry $P2_1$, and manifests superconductivity under pressure, at the border of ferromagnetism (Akazawa et al., 2004a, b). In the pressure range 0–3 GPa, the material exhibits three ferromagnetic phases FM1, FM2, and FM3, with FM1, at ambient pressure, having the highest Curie temperature $T_{C1} \approx 46$ K with an ordered moment of $0.5 \mu_B$/U-atom. The effect of pressure is to suppress FM1 at $P_{c1} \approx 1.7$ GPa. In the limit $T = 0$, FM2 and FM3 respectively exist in the pressure intervals of P_{c1} and P_{c2} (2.1 GPa) and of P_{c2} and P_{c3} (2.75 GPa). Beyond P_{c3}, UIr is paramagnetic. Superconductivity occurs only when the sample purity is high and at the edge of FM3 within a very small pressure interval of 2.6–2.75 GPa with an optimal $T_c \approx 0.15$ K (Akazawa et al., 2004b). As with UGe$_2$, there is no superconductivity when $P > P_{c3}$, that is, when the material is paramagnetic. All these features indicate unconventional superconductivity.

13.4.5 Miscellaneous unusual HF superconductors: UBe$_{13}$, PuCoGa$_5$, PrOs$_4$Sb$_{12}$, and β-YbAlB$_4$

The HF superconductivity in UBe$_{13}$, possessing the cubic NaZn$_{13}$ structure (Figure 13.1(b)) with space-group symmetry $Fm3c$, was discovered in 1984 (Ott et al., 1984). Since then, it has continued to attract attention due to its numerous unusual features, some of which remain elusive. High-quality single crystals of this compound have not yet been produced as with various other HFs, and therefore its superconducting anisotropic gap symmetry awaits full identification. Although the material is strongly paramagnetic, interestingly there is no distinct evidence of magnetic ordering in UBe_{13}. As regards its superconductivity, which emerges out of a highly unusual normal state, the material has two variants, one with $T_c \approx 0.9$ K and the other with $T_c \approx 0.75$ K, which are determined by processing (Langhammer et al., 1998). Among HF superconductors, UBe_{13} is noted for its largest effective mass of about 350 (Ott et al., 1984) and its exceptionally high normal-state resistivity (Maple et al., 1985) of about 250 mΩ cm at about 1 K, just above T_c. The initial temperature variation of H_{c2} near T_c is also very large, about -45 T K^{-1} (Maple et al., 1985), in conformity with the HF character. The pronounced specific heat anomaly ΔC gives $\Delta C/\gamma T_c \approx 2.5$ (Maple et al., 1985), which is much larger than the value of 1.43 found for conventional superconductivity. The unconventional superconductivity of UBe_{13} is corroborated by a host of properties, such as specific heat, nuclear relaxation, and penetration depth, all showing power-law temperature dependence. Of these, the specific heat (Ott et al., 1987) and penetration depth (Gross et al., 1986) indicate the presence of point nodes, while NMR relaxation reveals line nodes (MacLaughlin et al., 1987) and the absence of a Hebel–Slichter peak. Substantial further support for unconventional superconductivity comes from partial substitution of Th for U in UBe_{13} (Ott et al., 1985), leading to multiple superconducting phases with different order-parameter symmetries, which is discussed in Section 13.5.2. The effect of hydrostatic pressure (Aronson et al., 1989) is to suppress T_c, which extrapolates to 0 at 4 GPa. In a magnetic field, the normal state exhibits NFL behaviour, indicating possible proximity to a field-tuned AFM QCP and the mediation of pairing by AFM fluctuations. NMR and μSR data (Wälti et al., 2000) are consistent with odd-parity pairing, which also receives credence from the results of earlier Josephson effect studies (Han et al., 1986).

$PuCoGa_5$ (Sarrao et al., 2002) and its sister compound $PuRhGa_5$ (Wastin et al., 2003) possess the same crystal structure as $CeCoIn_5$ (Figure 13.6(b)). These two materials are particularly exciting owing to their remarkably high T_c values of 18.5 K and 8.7 K, respectively. Their effective mass, as determined from specific heat measurements, is, however, only moderately large, namely $m^*/m_e \approx 50$. Their normal-state magnetic susceptibility exhibits Curie–Weiss behaviour with an effective fluctuating moment of about $0.75\mu_B$/Pu-atom, which is close to $0.84\mu_B$, the value for localised Pu^{3+}. The $5f$-electrons in these compounds at E_F exhibit both localised and itinerant character (Joyce et al., 2003). The normal state of $PuCoGa_5$ manifests NFL behaviour at low temperatures. For example, in $PuCoGa_5$, $\rho(T)$ varies as $T^{4/3}$. Both compounds

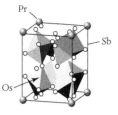

Figure 13.16 Crystal structure (cubic) of the skutterudite compound $PrOs_4Sb_{12}$, where the Os atoms are placed at the centres of the octahedra of Sb atoms.

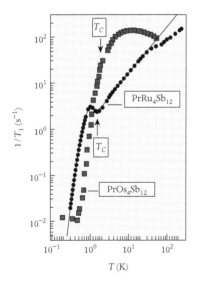

Figure 13.17 From ^{121,123}Sb-NQR studies, the temperature variation of the nuclear spin–lattice relaxation rate for $PrRu_4Sb_{12}$ follows that expected for typical weak-coupling s-wave superconductivity, that is, with a resonance peak below T_c and exponential behaviour (after Yogi et al., 2003). On the other hand, studies on $PrOs_4Sb_{12}$ (Kotegawa et al., 2003) have found no resonance peak, and the behaviour is unconventional.

are type II superconductors and, as to be expected from its high T_c, $PuCoGa_5$ possesses a large initial slope $dH_{c2}/dT \approx -5.9$ T K^{-1} and, from the theory of Werthamer et al. (1966), a large $\mu_0 H_{c2}(0) \approx 74$ T that far exceeds the Pauli paramagnetic value of 34 T. The unconventional superconductivity of these compounds is revealed by NMR/NQR studies showing a T^3 dependence of the spin–lattice relaxation rate and the absence of a Hebel–Slichter peak (Curro et al., 2005; Sakai et al., 2005). Knight-shift measurements indicate both these superconductors to be of spin-singlet d-wave type. The valence state of Pu is believed to vary between 4, 5, and 6 with 5.2 as its average valence in $PuCoGa_5$. The mixed valence renders support to valence fluctuations being the prominent cause of unconventional superconductivity in $PuCoGa_5$.

The HF superconductor $PrOs_4Sb_{12}$ possesses many notable features. It is the first praseodymium-based HF superconductor to be discovered (Bauer et al., 2002; Maple et al., 2001), belonging to the so-called *rare-earth skutterudites* (cubic structure) with general formula RT_4X_{12} (R = rare earth; T = transition metal Fe, Ru, or Os; and X = Sb, P, or As), which are known for manifesting a rich variety of correlated ground states. The material shows multiple superconducting phases (see Section 13.5.2) with different order-parameter symmetries, which is a pointer to its unconventional superconductivity. In comparison with most other HF superconductors, $PrOs_4Sb_{12}$ possesses a relatively large $T_c = 1.85$ K, and, above all, its HF behaviour and unconventional superconductivity are believed to result from quadrupolar fluctuations. The pertinent crystal structure (Figure 13.16) comprises a cage in which voids in T_4X_{12} octahedrons are filled by R-atoms. Since the atoms are loosely bound, they tend to rattle inside the cage, giving rise to anharmonic oscillations and tunnelling split states. The correlated properties are governed by the extent of cage filling and may lead to a host of different properties such as quadrupolar and magnetic order, mixed valence, the HF phenomenon, and unconventional superconductivity.

Superconductivity in $PrOs_4Sb_{12}$ involves HF quasiparticles with $m^*/m_e \approx 50$, as corroborated by measurements of specific heat and critical magnetic field. The specific heat shows (Bauer et al., 2002; Vollmer et al., 2003) twin (or split) anomalies at $T_{c1} \approx 1.85$ K and $T_{c2} \approx 1.75$ K, and their cumulated jump gives $\Delta C/\gamma T_c \approx 3$, which markedly exceeds the conventional BCS value of 1.43. The presence of a split transition is reflected also in thermal expansion of the material (Oeschler et al., 2003). Besides the split transition, lower critical field measurements indicate yet another superconducting phase at $T_{c3} \approx 0.6$ K (Cichorek, 2005). NMR/NQR measurements reveal the absence of a Hebel–Slichter peak (Figure 13.17), which indicates unconventional superconductivity, but the spin–lattice relaxation time shows an exponential rather than the T^3 dependence observed for most HF superconductors. This is indicative of an order parameter that is isotropic (Kotegawa et al., 2003). All these observations, along with the presence of multiple transitions and the large ΔC anomaly, are strong indications of unconventional multiphase HF superconductivity. Since the heavy fermions result from coupling with quadrupolar fluctuations, superconductivity too is believed to arise from a novel process involving exchange of such quantum quadrupolar fluctuations of the Pr ions. Interestingly, it is found that in proximity to superconductivity there is a polar order present above H_{c2} in the form of an antiferro-quadrupolar (AFQ) ordered phase (Aoki, Y. et al., 2002; Aoki, D. et al., 2007; Vollmer et al., 2003; Sugawara et al., 2005), indicating the presence of such fluctuations, which may be responsible for the observed unconventional superconductivity. The two sister compounds of $PrOs_4Sb_{12}$, namely $PrRu_4Sb_{12}$ and $PrRu_4P_{12}$, have been comparatively less studied. The former becomes superconducting at 1.3 K while the latter shows superconductivity at 1.8 K, but only under a pressure in excess of 11 GPa (Miyake et al., 2004). Surprisingly, however, $PrRu_4Sb_{12}$ has shown the characteristic peak below T_c in studies of the spin–lattice relaxation rate in ^{121,123}Sb-NQR (Figure 13.17) (Yogi et al., 2003), as well as other features of conventional s-wave superconductivity. The nature of superconductivity in these compounds therefore remains debated and is yet to be fully established.

The importance of the recently discovered HF compound β-$YbAlB_4$ lies in the fact that it is the first Yb-based HF superconductor to exhibit superconductivity (Nakatsuji et al., 2008), albeit at a very low $T_c \approx 80$ mK. Uniquely, the material exhibits a novel type of NFL state above T_c that manifests quantum criticality without any external tuning through pressure, magnetic field, or doping. There is, however, strong evidence of its arising from valence fluctuations of Yb (Okawa et al., 2010). The crystal structure of β-$YbAlB_4$ is orthorhombic, as depicted in Figure 13.18. As with most HF superconductors, the material shows extreme sensitivity of superconductivity to its state of purity, and its T_c is fully suppressed when the electron mean free path $l < \xi_0 \approx 50$ nm. This suggests that the superconductivity is an unconventional non-s-wave type. No magnetic order is seen at ambient pressures down to $T = 35$ mK. The NFL features show up in transport and thermodynamic properties (Nakatsuji et al., 2008). The breakdown of the FL behaviour is believed to take place in the mixed-valence state, in contrast to most other HF systems, which are magnetic Kondo lattice systems with integrated valence (Matsumoto et al., 2011). X-ray photoemission studies of β-$YbAlB_4$ (Okawa et al., 2010) confirm the Yb valence to be fractional and equal to about 2.75, which give credence to a pronounced mixed-valence quantum criticality for its unconventional superconductivity. $YbRh_2Si_2$ is another Yb-based compound that was expected to become superconducting. However, it has not shown superconductivity in the millikelvin temperature range, which is believed to be due

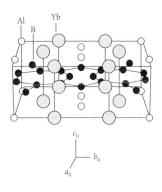

Figure 13.18 Crystal structure of β-$YbAlB_4$.

to its unconventional QCP in which its Kondo-destroying (KD) QCP coincides with an AF QCP (Steglich et al., 2010).

13.5 Special features of HF superconductors

A number of unusual features can be found among the HF super-conductors, namely non-centrosymmetric superconductors, multiple phases, the FFLO state, metamagnetism, and hidden order. We describe these in the following subsections.

13.5.1 Non-centrosymmetric superconductors

Some of the HF superconductors are strong contenders for unconventional triplet (p-wave) pairing. It was, however, pointed out by Anderson (1984) that for triplet pairing to occur, both time-reversal and inversion symmetry are necessary. As discussed in Chapter 2, the pairing electrons possess equal and opposite momenta, $\pm\mathbf{k}$. The absence of inversion symmetry causes the formation of a finite-momentum pair in which an electron with momentum $+\mathbf{k}$ pairs with an electron of momentum $-\mathbf{k} + \mathbf{q}$. It turns out that this prevents a simple p-wave pairing and instead the pairs formed are an admixture of spin-triplet and spin-singlet states. Consequently, p-wave pairing was considered to be excluded for materials without an inversion centre. While the crystal structures of most HF superconductors do possess inversion symmetry, there are a few exceptions that are non-centrosymmetric and yet exhibit the traits of spin-triplet pairing with p-wave symmetry. These include $CePt_3Si$, UIr, $CeRhSi_3$, $CeIrSi_3$, $CeCoGe_3$, $CeNiGe_3$, and $Ce_2Ni_3Ge_5$.

13.5.2 Multiple superconducting phases

Despite every care being taken in sample preparation, some of the HF superconductors, apparently fully stoichiometric, surprisingly are found to exhibit multiple phases, some of which are superconducting. The appearance of such multiple superconducting phases, which cannot be explained in terms of conventional s-wave paring, has led to a flurry of investigations into these exciting compounds. Their occurrence is generally taken as convincing evidence for a complex multicomponent superconducting order parameter. The formation of multiple super-conducting phases has been suggested in many HF systems, such as $CeCu_2Si_2$, $CeCu_2Ge_2$, $CeNi_2Ge_2$, $CeIrIn_5$, $CePt_3Si$, UPt_3, UBe_{13}, U(Be, Th)$_{13}$, UPd_2Al_3, and $PrOs_4Sb_{12}$, although in some of these cases the possibility has been debated. Here, one has to be careful to determine whether the multiple phases formed are indeed intrinsic to the system or are the consequences of locally altered composition or undetected impurities. For example, the first HF superconductor, $CeCu_2Si_2$ was for

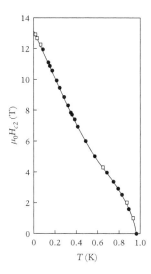

Figure 13.19 Temperature variation of the upper critical field of UBe$_{13}$, showing two distinct regions with linear and positive curvature (after Brison et al., 1989).

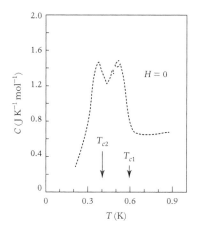

Figure 13.20 Two superconducting specific heat jumps in U$_{0.9669}$Th$_{0.0331}$Be$_{13}$ (after Ott et al., 1985).

a long time reported to have multiple ground states. Through a lot of careful sample preparation (Steglich et al., 2001), it has now been realised that this is a routine inhomogeneity problem due to its incongruent melting. On substituting 10% of Si by Ge in this compound, Yuan et al. (2003) could detect two superconducting phases as a function of pressure. The multiple superconducting phases have again been suggested in the sister compound CeNi$_2$Ge$_2$, which has an identical crystal structure. In this compound, the indications of superconductivity, seen at ambient pressure, disappear with pressure, but at higher pressures a second superconducting transition appears (Grosche, 1997). A similar situation occurs for CeIrIn$_5$.

Interestingly, UBe$_{13}$ forms in two types, *L-type* and *H-type*, the former having a lower T_c of about 0.75 K, in comparison with the value of about 0.90 K for the latter (Langhammer et al., 1998), and their origin is ascribed to indistinguishable stoichiometric variations due to metallurgical processing. Earlier observations of the temperature variation of H_{c2} of UBe$_{13}$ by Brison et al. (1989) (Figure 13.19) are in accord with this, showing two distinct regions where at very low temperatures the behaviour is near-linear, but above 0.5 K exhibits an unusual positive curvature. Interesting superconducting multiphase behaviour was detected in UBe$_{13}$ by Ott and his collaborators (Smith et al., 1984; Ott et al., 1985) by doping Th at the U site, that is, U$_{1-x}$Th$_x$Be$_{13}$ (for $0.01 < x < 0.06$), when they made a spectacular observation of two comparable specific heat jumps (Figure 13.20) at T_{c1}(optimum) $\approx$ 0.6 K and $T_{c2} \approx$ 0.4 K in close proximity to each other. The formation of two superconducting phases was further corroborated by pressure-dependence studies of the Th-doped samples (Lambert et al., 1986). Subsequent μSR measurements (Heffner et al., 1990) revealed the presence of very small magnetic moments (about $10^{-3}\mu_B$) just below T_{c2}. The phase diagram obtained from specific heat, lower critical field, magnetic susceptibility, and magnetisation measurements (Heffner et al., 1990) is depicted in Figure 13.21. The occurrence of the second transition is intriguing. It seems unlikely that a magnetic ordering such as SDW is responsible for such a large specific heat jump at T_{c2}, since the Fermi surface is already largely consumed by the first transition. The samples were structurally homogeneous (Smith et al., 1984) and the presence of two superconducting phases is considered as a pointer to the unconventional nature of the superconductivity.

The formation of multiple phases in UPt$_3$ is particularly intriguing and has been discussed in a comprehensive review by Joynt and Taillefer (2002). Here the two superconducting transitions observed are probably intrinsic features, as indicated by recent tunnelling experiments, revealing the presence of a complex superconducting order parameter resulting from a non-trivial coupling of superconductivity with the magnetic moments (Strand et al., 2010). Thus, in UPt$_3$, it is not inhomogeneity

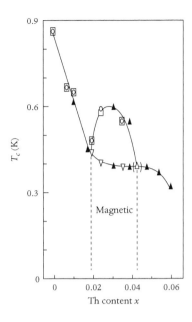

Figure 13.21 Phase diagram of $U_{1-x}Th_xBe_{13}$ (after Heffner et al., 1990). Here the open rectangles, open circles, and filled triangles respectively denote T_{c1} from magnetic susceptibility, magnetisation, and heat capacity measurements, while the downward-pointing and upward-pointing open triangles respectively depict T_{c2} from $H_{c1}(T)$ and specific heat studies.

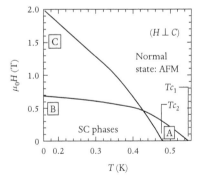

Figure 13.22 Schematic magnetic field–temperature phase diagram of superconducting phases in UPt_3 (after Joynt and Taillefer, 2002).

that results in this behaviour. This hexagonal HF superconductor that exhibits a triplet pairing is known to display at least three distinct superconducting phases, A, B, and C, where C forms above 0.4 T (Figure 13.22). Although early heat capacity observations were of a single broad T_c at about 0.5 K, later studies on improved samples resolved it into two peaks at temperatures T_{c1} and T_{c2}, separated by about 50 mK (Fisher et al., 1989). The situation is analogous to that in Th-substituted UBe_{13} discussed above. The specific heat jumps ΔC at the two split transitions, expressed in the form $\Delta C/\gamma T_{c1}$ and $\Delta C/\gamma T_{c2}$, are, respectively, 0.545 and 0.272 (Fisher et al., 1989), which together are smaller than the BCS value of 1.43, indicating the possible nodes in the gap.

In the case of the non-centrosymmetric superconductor $CePt_3Si$, initially a single specific heat anomaly was observed, corresponding to $\Delta C/\gamma T_c = 0.25$, that is, much smaller than the BCS value of 1.43 (Bauer et al., 2005), suggesting that only a very small part of the Fermi surface was involved with the superconducting transition. However, the situation was changed with the observation of two superconducting transitions for a sample synthesised at a higher temperature (Scheidt et al., 2005). In UPd_2Al_3, with $T_c \approx 2.0$ K, another phase is detected at 0.6 K through measurements of specific heat, thermal expansion, and elastic constants (Matsui et al., 1994; Sakon et al., 1994; Sato et al., 1994), which is generally believed to be a second superconducting phase, although this is still to be fully established.

Multiple superconducting phases are found in the HF superconductors $CeCoIn_5$ and $PrOs_4Sb_{12}$. In the former, as we will see in Section 13.5.3, there is strong evidence (Watanabe et al., 2004) of the formation of a novel superconducting FFLO phase at H_{c2} where the order parameter is spatially modulated, while in $PrOs_4Sb_{12}$ an unusual change in the symmetry of the gap function takes place, well inside the superconducting state, indicating the formation of two distinct superconducting phases possessing twofold and fourfold symmetries. More interestingly, $PrOs_4Sb_{12}$ has emerged as a unique system whose HF behaviour and superconductivity arise from quadrupolar fluctuations (Anders, 2002; Miyake et al., 2003; Matsuda, 2006).

13.5.3   The FFLO state

In mid 1960s, Fulde and Ferrell (1964) and Larkin and Ovchinnikov (1964) independently conceived the possibility of a novel form of unconventional superconductivity described by what is now known as Fulde–Ferrell–Larkin–Ovchinnikov (*FFLO*) *state*. As previously discussed in Chapter 5, in spin-singlet superconductivity, at H_{c2}, the phenomenon is suppressed by either the orbital effect (vortices) or the Zeeman effect (i.e. Pauli paramagnetism of the conduction electrons). Below a certain critical temperature $T_0 < T_c$, the phase transition from the superconducting state to the normal state is of first order when H_{c2} is determined by the Zeeman effect and of second order when H_{c2} is determined by the orbital effect. The FFLO state, which is predicted to evolve in a high magnetic field, has its origin in the paramagnetism of the conduction electrons, strongly coupled to the electron spins. The magnetic field polarises the spins, and the Fermi surface of up ($\uparrow$) and down ($\downarrow$) spins leads to a non-zero total momentum $\mathbf{q}$ associated with the paired state. As a result of the finite centre-of-mass momenta, the superconducting order parameter has an oscillatory component varying as $\exp(i\mathbf{q}.\mathbf{r})$, which gives rise to real-space modulation with a wavelength equal to the range of coherence ξ. In the FFLO state, which represents an inhomogeneous superconductivity, a superconductor essentially lowers its energy in a magnetic field by forming periodic regions of superconductors separated by domains of aligned spins. This novel superconducting state is expected to manifest unusual properties quite different from those of the conventional BCS pairing.

The occurrence of the FFLO state, however, requires that a number of conditions be satisfied:

1. The material must be in the form of extremely pure (to give a large electron mean free path l) and strongly anisotropic single crystals.
2. The Pauli paramagnetic effect must dominate the orbital effect, so that there is little destruction of superconductivity and so that the spin–orbit interaction is small.
3. The orbital H_{c2} must be large, so that ξ is small and the condition $l >> \xi$ is readily satisfied.
4. A quasi-two-dimensional structure of the material, with the magnetic field applied parallel to the conducting layers, is found to be more favourable for the formation of the FFLO state.
5. Anisotropies of the Fermi surface and superconducting gap function are important for the stability of the FFLO state.

Only relatively recently has it been realised that some of the above conditions are met in some HF and organic superconductors.

A more or less established observation of the FFLO state among HF superconductors is in CeCoIn$_5$. In this material, with $T_c = 2.3$ K, a host of measurements, including specific heat (Bianchi et al., 2003; Radovan et al., 2003), magnetisation (Radovan et al., 2003), thermal conductivity (Capan et al., 2004), penetration depth (Martin et al., 2005), NMR Knight shift (Kakuyanagi et al., 2005; Kumagai et al., 2006; Mitrović et al., 2006), ultrasound velocity (Watanabe et al., 2004c), and magnetostriction (Correa et al., 2007), have revealed that the transition at H_{c2}, below $T = T_0 < T_c$, to be of the first order. On further cooling to $T_{FFLO} < T_0$, the new FFLO phase is formed. The schematic phase diagram of FFLO formation is depicted in Figure 13.23. The triangular

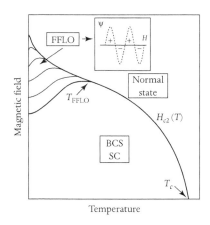

Figure 13.23 Schematic phase diagram of FFLO state formation.

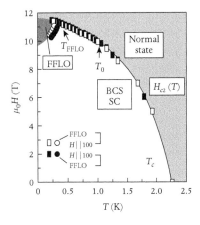

Figure 13.24 Phase diagram of FFLO state formation in CeCoIn₅ (after Bianchi et al., 2003).

region to the left of T_{FFLO}, which separates the homogeneous superconductivity region below from the normal state above, is the FFLO state characteristic of inhomogeneous superconductivity. The transition from the homogeneous superconducting state to the FFLO state, occurring at a lower magnetic field, is of second order, while that from FFLO to the normal state, occurring at a higher field, is of first order. The two transitions are observed in the specific heat measurements of CeCoIn₅ at 250 mK, with the applied magnetic field held parallel to the conducting planes (Radovan et al., 2003). Essentially similar observations were noted by Bianchi et al. (2003), whose suggested phase diagram of the FFLO state is depicted in Figure 13.24 and closely matches with Figure 13.23 and in general agreement with the results of Radovan et al. (2003). Other potential candidates for the FFLO state to be realised are UPd_2Al_3, URu_2Si_2, UBe_{13}, $PuCoGa_5$, $PuRhGa_5$, and $NpPd_5Al_3$.

13.5.4 Metamagnetism

The term 'metamagnetism' describes a dramatic increase in the magnetisation of the material with a small change in a relatively high external magnetic field. Since there is no broken symmetry involved, the transition from the low- to the high-magnetisation state is expected to be of first order. In HF superconductors, metamagnetic transitions have been observed in $CeRu_2Si_2$ (8 T), UPd_2Al_3 (18 T), UPt_3 (20.3 T), URu_2Si_2 (35–40 T), and $CeRhIn_5$ (45 T) at the respective *metamagnetic critical fields* H_{cmm} indicated in parentheses. In the metamagnetic transition, occurring at $H > H_{cmm}$, the large effective mass rapidly decreases with increasing field (Aoki et al., 1993; Julian et al., 1994), which manifests itself in a rapid change in a host of properties of HF materials, making them highly nonlinear. For example, the transition shows up in magnetic susceptibility (Frings et al., 1983), magnetoresistivity (Franse et al., 1985), magnetostriction (de Visser et al., 1987a), magnetocaloric effect (van der Meulen et al., 1990), specific heat (Müller et al., 1989), thermal expansion, sound velocity (Kouroudis et al., 1987), softening of the elastic constants (Bruls et al., 1990), and Hall effect measurements (van Sprang et al., 1988). On lowering the temperature, the transition generally becomes sharper. Figure 13.25(a) and (b) are typical examples of metamagnetic behaviour, showing jumps in the magnetic susceptibility of $CeRu_2Si_2$ at 8 T (Bruls et al., 1990) and in the specific heat of UPt_3 at about 20 T (Müller et al., 1989).

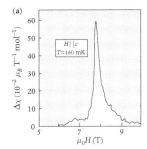

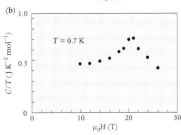

Figure 13.25 (a) Metamagnetic phase formation in CeRu₂Si₂, showing a jump in magnetic susceptibility at the critical field of 8 T (after Bruls et al., 1990). (b) Metamagnetic phase formation in UPt₃ at a field of about 20 T seen in specific heat measurements (after Müller et al., 1989).

The metamagnetic behaviour indicates that the system is possibly on the verge of an instability, or a spin split branch of quasiparticles crossing the Fermi surface. In URu_2Si_2, as many as three metamagnetic transitions (de Visser et al., 1987b; Bakker et al., 1993) are observed, at H_{cmm} values of 35.9 T, 36.1 T, and 39.7 T.

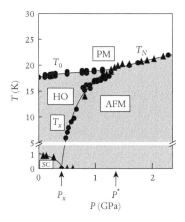

Figure 13.26 Temperature–pressure phase diagram of URu_2Si_2, showing the location of the paramagnetic, hidden order, superconductivity, and antiferromagnetic phase boundaries (after Hassinger et al., 2008). As may be seen, superconductivity exists only in the HO phase.

13.5.5 Hidden order

The compound URu_2Si_2 has continued to capture the attention of physicists because of the intriguing hidden order (HO) that it manifests at a temperature $T_0 \approx 17.5$ K (at ambient pressure). The material shows unconventional superconductivity below $T_c \approx 1.4$ K. The pertinent phase diagram (Hassinger et al., 2008) is depicted in Figure 13.26. The transition at T_0 is of particular interest: it takes place on cooling from a higher temperature as a second-order phase change and displays large thermodynamic effects with respect to specific heat, thermal expansion, etc. Anomalies show up also in resistivity and magnetic susceptibility measurements. Neutron diffraction has revealed AFM order below T_0, but with an ordered moment of only about $0.03\mu_B$/U-atom (Broholm et al., 1987). The most puzzling part of the transition is that the small magnitude of the AFM order can in no way account for the released entropy at the transition. Clearly, the observed anomalies are to be interpreted as a pointer to the formation of a hidden order that is invisible to neutrons or X-rays and is responsible for the entropy change.

Many different theoretical models for the HO state have been proposed where $5f$-electrons of U are considered essentially localised, or itinerant, or of dual nature, and different possibilities for the HO state have been suggested. These include unconventional density waves (Ikeda and Ohashi, 1998), isotropic CDW (Chandra et al., 2005), d-SDW (Ramirez et al., 1992a; Anderson, 1972), quadrupolar ordering (Santini and Amoretti, 1994), orbital antiferromagnetism (Mydosh et al., 2002), helicity order (Varma and Zhu, 2006). Mydosh et al. (2002) envisaged orbital AFM order resulting from circulating currents between U ions to account for the local fields and entropy loss observed at T_0. The local fields arise from current loops developing inside the material as the HO state is formed.

However, none of these models seems to adequately explain the observed data pertaining to the HO state. μSR measurements (Amitsuka et al., 2003) suggest that the very small magnetic moment of the AFM state observed at T_0 may not be intrinsic to the material but instead may simply represent a small volume fraction of a much larger magnetic moment. This is corroborated by pressure studies, which show that, under large hydrostatic pressure, the magnetic moment rises and finally becomes saturated at an atomic size moment of $0.4\mu_B$/U-atom (Amitsuka et al., 1999). As regards the superconducting transition at about 1.4 K, nuclear resonance studies (Kohori et al., 1996) have revealed a T^3 dependence of the spin relaxation rate and an absence of a Hebel–Slichter peak indicative of unconventional superconductivity with line nodes at the Fermi surface. The observed Knight shift (Brison et al., 1994) is indicative of even-parity pairing, and the T^2 behaviour of the specific heat (Maple et al., 1986) further corroborates the presence of line nodes. The superconducting gap coefficient $2\Delta/k_B T_c \approx 5-12$ (Naidyuk et al., 1996) is indicative of the strong-coupling character of the superconducting state. Interestingly, T_c decreases under pressure and finally vanishes below a pressure of 1.5 GPa. Consequently, superconductivity and the AFM state seem to be mutually competing with each other under pressure. Since the precise nature of the HO state remains unclear, its role in the occurrence of superconductivity in URu_2Si_2 to date remains mostly in dark.

13.6　Summary

The field of HF superconductivity has made triumphant progress over the last 30 years. This chapter has described major advances made with the discovery of a large variety of new HF superconducting systems, adding richness and new challenges to the physics and materials science of this growing field. Much of the understanding of HF superconductors has evolved through a new class of phase transition, a quantum phase transition (QPT), occurring at a quantum critical point (QCP), reached at the extrapolated absolute zero temperature by appropriately tuning the control parameters. Studies of QCPs have led to the realisation of complicated phase diagrams for various f-electron systems comprising magnetic and non-magnetic novel phases and exhibiting NFL normal states and, above all, unconventional superconductivity. Since most HF superconductors exhibit power-law temperature dependences for their low-temperature transport and thermodynamic properties, and many of them also show multiple superconducting phases, the possibility of conventional superconductivity for them is generally ruled out. In this chapter, various interesting categories of HF superconductors have been briefly discussed. Various special features described include non-centrosymmetric materials, occurrence of multiple superconducting phases, the FFLO state, hidden order, and metamagnetic transitions. The fact that in $PrOs_4Sb_{12}$ the crucial interaction is of quadrupolar origin suggests a new route to unconventional superconductivity. Similarly, the exciting observation of exceptionally high T_c values in recently discovered Pu-based HF superconductors strongly indicates that the field is wide open for unexpected developments resulting from the discovery of new systems. However, a major problem with unconventional superconductivity is that the phenomenon shows extreme sensitivity to defects, impurities, and stoichiometry. Materials processing therefore pose a clear challenge to the discovery of new unconventional superconductors.

CHAPTER 14

Organic superconductors

Superconducting materials based on carbon and its allotropes—graphite, diamond, fullerene, carbon nanotubes (CNTs), etc.—are all categorised in a single class called organic superconductors, which contains around 200 members whose T_c varies from less than 1 K to around 40 K. This broad category of superconductors includes (1) about 15 different families of charge-transfer salts, (2) doped C_{60} fullerides, and (3) graphite intercalation compounds (GICs). In addition, as mentioned in Chapter 7, boron-doped diamond also shows superconductivity around 4 K, while superconductivity has also been suggested, though not confirmed, in CNTs. There are several comprehensive extended papers and review articles on these materials that are worth mentioning (Jérome and Schultz, 1982; Dressel, 2000; Ishiguro and Tanatar, 2000; Wosnitza, 2000; Saito and Yoshida, 2000; Jérome and Pasquier, 2005; Han et al., 2005; Maple et al., 2008; Lang and Müller, 2008).

14.1 Evolution of organic superconducting salts

The possibility of organic superconductors was first conceived in the mid-1960s by Little (1964) and, excitingly, for a hypothetical one-dimensional polymeric organic molecule having polarisable side chains, his model predicted superconductivity above room temperature. His interesting idea was based on the isotope effect in conventional superconductors (Chapter 2), where the phenomenon arose through ionic polarisation of the crystal lattice caused by a moving electron, giving $T_c \alpha M^{-1/2}$, with M being the ionic mass. An electron moving through the central stem of the molecule caused an electronic polarisation of the side chains, and the pairing of the two electrons occurred via excitonic mediation. In this scenario, the virtual oscillations of the electronic charge of mass m_e in the side chains would mediate electron pairing with vastly increased T_c by a factor of $(M/m_e)^{1/2}$, leading to superconductivity at an unbelievable $T > 1000$ K! There were serious challenges to these ideas (Ferrel, 1964; Hohenberg, 1967; Dayan, 1981), since for a one-dimensional molecule the long-range order required for superconductivity is never formed. It may be worth mentioning that among polymeric materials, although of inorganic type, only $(SN)_x$ is known to show superconductivity, albeit with very low $T_c = 0.28$ K (Greene and Engler, 1980), and, moreover, its structure is completely different from what Little had proposed. Nevertheless, Little's model gave remarkable impetus to the search of organic superconductors, and, as a first step, led to the discovery (Heeger and Garito, 1975) of a quasi-one-dimensional, highly conducting organic crystal called TTF-TCNQ (tetrathiafulvalene-tetracyanoquinodimethane). This is a complex charge-transfer salt comprising the acceptor molecule TCNQ with TTF as its donor. But this material, although metallic, failed to show superconductivity and, below 53 K, owing to its quasi-one-dimensionality, exhibited a *Fröhlich–Peierls metal-to-insulator (M–I) transition* (Fröhlich, 1954; Peierls, 1955), generally called *Peierls instability* or *spin-Peierls (SP) transition*, caused by the mobility of density waves (CDW/SDW). Various other salts of this family, such as NMP-TCNQ, QN-(TCNQ)$_2$, and TTT-TCNQ, also behaved no differently in this respect (Rice, 1975).

Two separate strategies were followed for synthesising superconducting salts (Figure 14.1). In the first (Jérome et al., 1980), the smaller S atom in TTF was replaced by a larger Se atom to form a new planar donor molecule TMTSF (tetramethyltetraselenafulvalene). This substitution further lowered the on-site Coulomb repulsion energy U and enhanced hybridisation. TMTSF formed the basic building blocks, stacked into parallel columns, of the superconducting *Bechgaard salts*, with the chemical formula (TMTSF)$_2$X, synthesised with different inorganic counter-anions X such as PF$_6$, AsF$_6$, ReO$_4$, and ClO$_4$ (Table 14.1), whose T_c ranged from about 0.3 K to 3 K. With the exception of ClO$_4$, which yields ambient-pressure superconductivity, the others need to be hydrostatically pressurised to achieve superconductivity. The effect of pressure is to further increase the dimensionality

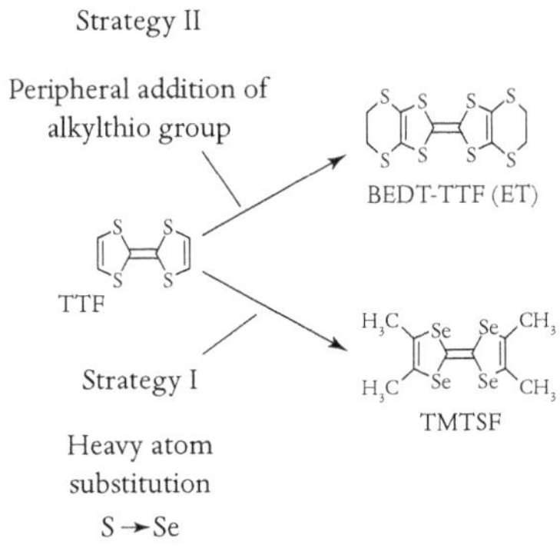

Strategy II

Peripheral addition of alkylthio group

BEDT-TTF (ET)

TTF

Strategy I

TMTSF

Heavy atom substitution

S → Se

Figure 14.1 Two approaches followed to synthesise low-dimensional organic superconductors.

Table 14.1 (TMTSF)$_2$X Compounds

Counter-anion X	Anion symmetry	T_{M-I} (K)	Electronic property*	T_c (K)	P_c (GPa)
PF$_6$	Spherical	12	SDW	1.2	0.65
AsF$_6$	Spherical	12	SDW	1.4	0.95
SbF$_6$	Spherical	17	SDW	0.38	1.05
TaF$_6$	Spherical	11	SDW	1.35	1.1
ReO$_4$	Tetrahedral (non-centrosymmetric)	177	Anion order–disorder at 177 K FISDW	1.2	0.95
FSO$_3$	Tetrahedral (non-centrosymmetric)	88	Anion order–disorder at 88 K	3.0	0.5
ClO$_4$	Tetrahedral (non-centrosymmetric)	–	Anion order–disorder at 24 K FISDW	1.3	0

*SDW, spin-density wave; FISDW, field-induced spin-density wave.
Data from Saito and Yoshida (2000).

Table 14.2 (TMTTF)$_2$X Compounds

Counter-anion X	T_{M-I} (K)	Characteristics*	T_c (K)	P_c (GPa)
PF$_6$	250	CO (70 K) SP (15 K)	1.8	5.4
Br	100	AFM ($T_N = 15$ K)	0.8	2.6
SbF$_6$	154	CO (154 K) $T_N = 15$ K	2.6	6.1
BF$_4$	210	Anion disorder (40 K)	1.38	3.35

*CO, charge ordering; SP, spin Peierls; AFM, antiferromagnetic order.
Data from Adachi et al. (2000), Jaccard et al. (2001), Balicas et al. (1994), and Auban-Senzier et al. (2003).

through higher intermolecular hybridisation, which suppresses the M–I transition. Although the TMTSF compounds thus produced possess an increased dimensionality in comparison with pristine samples, they are still quasi-one-dimensional and less susceptible to Peierls instability. The sulfur analogue of TMTSF is TMTTF (tetramethyltetrathiafulvalene), formed by replacing four Se atoms by S, which is again quasi-one-dimensional but of comparatively much lower dimensionality than TMTSF. In this, the Peierls instability becomes more important and calls for a larger pressure for achieving superconductivity. Table 14.2 lists (TMTTF)$_2$X superconductors. Both these types are included in the short form (TM)$_2$X.

In the second strategy, followed by Saito et al. (1982, 1983) and Kobayashi et al. (1987) (Figure 14.1), the dimensionality was increased by using alkylthio-substituted TTFs, which led them to a sulfur-based, two-dimensional donor molecule BEDT-TTF (bis-ethylenedithio-tetrathiafulvalene), in short form called ET. Here, the overlap between the π-orbitals of the adjacent ET molecules is enhanced by the presence of carbon and sulfur rings formed at the outer end of the pristine TTF molecule, because of which anisotropy is reduced and the Peierls instability is suppressed. Unlike TM molecules, ET are not planar, but the structure formed is layered, carrying planar stacks of ET molecules separated by sheets of mono-valent anion X. The family of (ET)$_2$X salts includes a rich variety of more than 50 superconductors (Saito and Yoshida, 2000), not merely in terms of having numerous counter-anions X(X = ReO$_4$, I$_3$, Cu(NCS)$_2$, Cu[N(CN)$_2$]Br, etc.), but also possessing varying arrangements of donor molecules, denoted by α, β, κ, etc., as shown in Figure 14.2. Of these, the members of the κ-group such as κ-(ET)$_2$Cu(NCS)$_2$, κ-(ET)$_2$Cu[N(CN)$_2$]Br, and κ-(ET)$_2$Cu[N(CN)$_2$]Cl are noted for their relatively high T_c values, exceeding 10 K (Oshima et al., 1988). The prominent compounds of the (ET)$_2$ family are listed in Table 14.3.

Besides the (TM)$_2$ and (ET)$_2$ families, there are at least a dozen more, where in most cases the donor molecules are derived as pro-totypes of the latter. For instance, selenium- and oxygen-substituted

α-phase β-phase κ-phase

Figure 14.2 Three phases of (ET)$_2$X salts. The thick lines denote planar ET molecules along the long axis while the rectangles and parallelepiped denote the cross-section of the unit cell in the two-dimensional conducting plane.

Table 14.3 Some prominent superconductors of $(ET)_2X$ family

$(ET)_2X$	T_c (K)	Pressure (GPa)	Anion symmetry
$(ET)_2ReO_4$	2.0	0.04	Tetrahedral
β_L-$(ET)_2I_3$	1.5	Ambient	Linear
β_H-$(ET)_2I_3$	8.1	Ambient	Linear
β-$(ET)_2IBr_2$	2.7	Ambient	Linear
κ-$(ET)_2 I_3$	3.5	Ambient	Linear
κ-$(ET)_2Cu(NCS)_2$	10.4	Ambient	Polymer
κ-$(ET)_2Cu[N(CN)_2]Br$	11.5	Ambient	Polymer
κ-$(ET)_2Cu[N(CN)_2]Cl$	12.8	0.03	Polymer
α-$(ET)_2KHg(SCN)_4$	0.1	0.25	Polymer
α-$(ET)_2NH_4Hg(SCN)_4$	1.1	Ambient	Polymer
β''-$(ET)_2SF_5CH_2CF_2SO_3$	5.2	Ambient	Polymer

Data from Maple et al. (2008) and Saito and Yoshida (2000).

variants of BEDT-TTF are respectively BEDT-TSF (bis-ethylenedithio-tetraselenafulvalene, or BETS in short) and BEDO-TTF (bis-ethylenedioxy-tetrathiafulvalene, or BO in short). Further, there are organic salts derived by incorporating asymmetric hybrids, such as DMET, MDT-TTF, and MDT-TSF. The structures of various donor molecules forming over 15 different systems of organic salts are summarised in Figure 14.3.

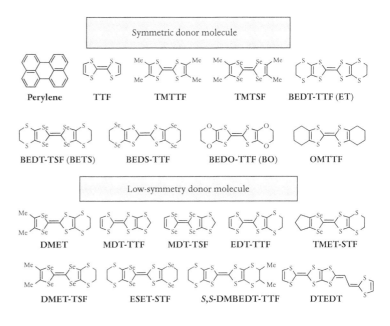

Figure 14.3 Various donor molecules in organic superconductors.

14.2 The $(TM)_2$ family of quasi-one-dimensional superconductors

This family of quasi-one-dimensional superconductors, called Bechgaard salts, comprises the Se-based $(TMTSF)_2X$ (Table 14.1) and its sulfur analogue $(TMTTF)_2X$ (Table 14.2); their molecular structures may be seen in Figure 14.3. At ambient pressure, $(TMTSF)_2PF_6$ becomes insulating at 12 K when there is an onset of itinerant antiferromagnetism (AFM) (Mortensen et al., 1981; Walsh et al., 1982; Torrance et al., 1982). This gives rise to a magnetic modulation in the form of spin-density waves (SDWs) of wave (nesting) vector $2\mathbf{k}_F$. This perturbation opens up a gap at E_F over the entire Fermi surface that turns the material insulating. This transition temperature steadily drops with applied pressure and finally vanishes with the appearance of superconductivity with a $T_c \approx 1.2$ K at a critical pressure $P_c \approx 0.6$ GPa. Further pressurisation at $P > P_c$ causes a rapid decrease of T_c (Greene and Engler, 1980) at a rate of $dT_c/dP \approx 1.0$ K GPa^{-1}. This is ascribed to stiffening of the lattice due to pressure, which enhances the average phonon frequency (Maple et al., 2008). The suppression of the magnetic state with pressure is ascribed to an increase in the dimensionality of the material due to inter-column overlap, and this effect is maximum when the pressure is unidirectional along the a-axis. Superconductivity occurs in the vicinity of the AFM state, where, close to P_c, the two phenomena are reported to coexist (Greene and Engler, 1980; Vuleti et al., 2002). As with various heavy fermion (HF) superconductors (Chapter 13), where superconductivity occurs in proximity to the AFM state, the materials of this class are considered potential candidates for unconventional superconductivity. Among $(TM)_2$ salts, $(TMTSF)_2ClO_4$ represents the only member of this family exhibiting superconductivity at ambient pressure (Bechgaard et al., 1981), and has therefore received more attention.

The Bechgaard salts are prepared in the form of fine needle-shaped black shiny crystals, following the electrochemical oxidation of TMTSF with an appropriate electrolyte to provide the required counter-anions. The crystal structure is triclinic (space group $P\bar{1}$) in which the planar brick-like molecules of TMTSF are stacked in columns along the a-axis. The counter-anions (e.g. PF_6) are located in between adjacent columns along the c-direction (Figure 14.4). Along the columns, the electrical conductivity is highest owing to the shortest Se–Se distances and greatest overlap of π-orbitals along this direction. The corresponding overlaps along the b- and c-directions are lower by factors of 10 and 300, respectively, making the material quasi-one-dimensional to manifest strong electron correlations.

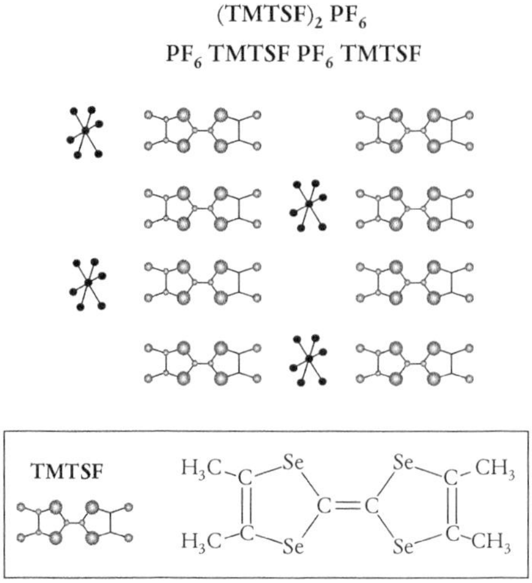

(TMTSF)$_2$ PF$_6$

PF$_6$ TMTSF PF$_6$ TMTSF

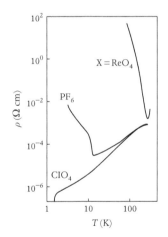

Figure 14.4 Crystal structure of (TMTSF)$_2$ PF$_6$.

Figure 14.5 Low-temperature resistivity behaviour of (TMTSF)$_2$X for different X (after Saito and Yoshida, 2000).

14.2.1   Role of counter-anions

The counter-anions X are necessary, since they promote charge transfer from the neutral organic molecule and give rise to molecular bands leading to conductivity. Figure 14.5 depicts the typical low-temperature resistivity behaviour of TMTSF salts with X = PF$_6$, ReO$_4$, and ClO$_4$ at ambient pressure (Saito and Yoshida, 2000). Whereas for ClO$_4$ the material remains metallic throughout, for ReO$_4$ and PF$_6$ the behaviour becomes insulating at temperatures below 177 K and 12 K, respectively. The cause of the M–I transition in the two cases is different. As may be seen from Table 14.1, the first four counter-anions, which include PF$_6$, are all centrosymmetric with spherical symmetry and for them the material exhibits an SDW insulating state that develops at lower temperatures of 10–20 K. However, the remaining three anions, namely, ReO$_4$, FSO$_3$, and ClO$_4$, possess non-centrosymmetric tetrahedral symmetry, which leads to an intrinsic disorder that suppresses the stability of the conducting phase. These anions can adopt two equivalent orientations that are nearly degenerate in energy. They are disordered at room temperature, with an equal occupation for both orientations, and on cooling the anion lattice becomes ordered and frozen, the ordering temperatures (column 4 of Table 14.1) for ReO$_4$, FSO$_3$, and ClO$_4$ being 177 K, 88 K, and 24 K, respectively. In the case of ReO$_4$ (below 177 K) and FSO$_3$ (below 88 K), for example, the superstructure associated with ordering involves doubling of all three lattice parameters ($2a \times 2b \times 2c$) Since this matches perfectly with the nesting vector of ($2a \times 2b$), such an ordering gives rise to an insulating state. However, by applying pressure, this superstructure is altered and ceases to match with the nesting vector, which restores the metallic state. For X = ClO$_4$, however, the superstructure formed below 24 K corresponds to ($a \times 2b \times 2c$), which does not match with the nesting vector and therefore does not interfere

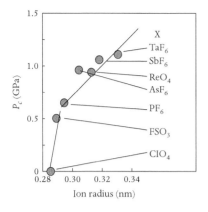

Figure 14.6 Dependence of the critical pressure P_c to suppress the metal–insulator transition on the anion radius of X in $(TMTSF)_2X$ compounds (after Saito and Yoshida, 2000).

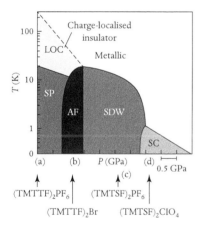

Figure 14.7 Temperature–pressure phase diagram of $(TMT)_2X$ (after Jérome, 1991).

with the pristine metallic state. But, if the sample of $(TMTSF)_2ClO_4$ is cooled very rapidly through 24 K, the room-temperature anion disorder may become frozen, resulting in the formation of an SDW state below 5 K, and the sample does not exhibit superconductivity (Takahashi et al., 1982; Kagoshima et al., 1983; Garoche et al., 1983). In $(TM)_2$ superconductors, the critical pressure required for superconductivity is found to increase with anion size or, alternatively, with the lattice parameter b, as depicted in Figure 14.6 (Saito and Yoshida, 2000). A larger anion size essentially lowers the inter-column overlap along the b-direction and thus needs a higher P_c for superconductivity.

14.2.2 Variety of ground states and generic phase diagram

Superconductivity is just one among many electronic states encountered in the $(TM)_2$ family under varied conditions. Different phases that are stabilised at various temperatures in the two isostructural systems depend on the donor molecule (i.e. TMTSF or TMTTF), the nature of the counter-anion present, and the pressure applied. Most of these salts are in an insulating regime at elevated temperatures (varying from 20 K to 250 K, depending on the material) for to various reasons, such as Mott–Hubbard transition, charge ordering (CO), or anion disorder (as discussed above for TMTSF). In general, at $T < 20$ K, as illustrated in the generic P–T phase diagram (Jérome, 1991) of Figure 14.7, a variety of ground states, such as SP, Néel AFM (or commensurate SDW), incommensurate SDW, and finally superconductivity, may form in succession, either with increasing pressure on a given compound or by changing the salts. The arrows in the figure indicate the positions of the representative salts at ambient pressure. The sequence of electronic states is SP, AFM, SDW phase with incommensurate modulation, and superconductivity.

The salt located at the arrow on the extreme left at (a) in Figure 14.7, which corresponds to a low applied pressure, is $(TMTTF)_2PF_6$. It is relatively close to being ideally one-dimensional. The materials at the arrows on its right are in order of increasing degree of quasi-one-dimensionality, with $(TMTSF)_2ClO_4$, located at the arrow on the extreme right at (d), possessing the greatest degree of quasi-one-dimensionality among all the $(TM)_2X$ compounds. Between $(TMTTF)_2PF_6$ and $(TMTSF)_2ClO_4$, we have depicted two representative compounds, $(TMTTF)_2Br$ and $(TMTSF)_2PF_6$, of intermediate dimensionalities, placed at the arrows at (b) and (c), respectively. The compounds that we have not shown, namely $(TMTTF)_2AsF_6$ and $(TMTTF)_2SbF_6$, are located between (a) and (b), while $(TMTSF)_2AsF_6$ lies between (b) and (c). This phase diagram shows that, below 20 K at ambient pressure, the salts at both (a) and (b) exhibit the SP transition, that at (c) the AFM/SDW transitions, and that at (d) the superconducting transition. The salts at (a) to

(c) become superconducting only under pressure after they have passed through the succession of ground states mentioned in the phase diagram. In the case of $(TMTTF)_2PF_6$, for example, experimental studies have demonstrated that this highly one-dimensional salt passes through the complete sequence of the above-mentioned ground states before becoming superconducting at a high pressure of 4.35 GPa (Wilhelm et al., 2001; Jaccard et al., 2001), which substantiates the generic nature of the above phase diagram for the $(TM)_2X$ system. Further, these measurements give strong indications of the presence of SDW correlations in the superconducting state formed after its suppression, which gives credence to an unconventional superconductivity mediated by magnetic fluctuations.

Apart from various physical states appearing in the phase diagram in Figure 14.7, this class of materials exhibit a novel phenomenon of *field-induced spin density waves (FISDW)*. The effect of a magnetic field is essentially to lower the dimensionality of the material, just as applied mechanical pressure enhances it (Gor'kov and Lebed, 1984). Consequently, if the magnetic field applied is sufficiently strong, it can counter the effect of imposed pressure that had originally suppressed the SDW state by increasing the dimensionality of the sample. In this way, the SDW state is restored by a magnetic field (Ishiguro et al., 1998), and its presence shows up in Shubnikov–de Haas (SdH)-like oscillations in the longitudinal resistivity and quantised steps observed in Hall effect in bulk single crystals.

14.2.3 Superconductivity of $(TM)_2X$ systems

In the $(TM)_2$ family, ClO_4-containing salts have been more extensively studied for superconductivity since they are free from the constraint of requiring an applied pressure. The system being highly anisotropic, many of the parameters need to be measured along the three crystallographic axes. Some of the characteristic parameters measured for $(TMTSF)_2ClO_4$ are given in Table 14.4.

The London penetration depth, measured along the *a-axis* (Schwenk et al., 1983), is quite large, about 40 μm, which, along with the very

Table 14.4 Superconducting parameters of $(TMTSF)_2ClO_4$

μ_0H_{c1} (mT) at 0.5 K	μ_0H_{c2} (T) at 0 K	ξ (nm)	γ (mJ mol^{-1} K^{-2})	Debye temperature θ (K)	ΔC (mJ mol^{-1} K^{-1})	$2\Delta_0$ (meV)
0.02 (*a*-axis)	2.8 (*a*-axis)	70.6–83.7 (*a*-axis)	10.5			
0.10 (*b*-axis)	2.1 (*b*-axis)	33.5–38.5 (*b*-axis)		213	21.4	0.44–3.68
1.0 (*c*-axis)	0.16 (*c*-axis)	2.03–2.27 (*c*-axis)				

Data from Saito and Yoshida (2000) and Lang and Müller (2008).

small coherence length (Table 14.4), makes this material extreme type II, with Ginzburg–Landau parameter $\kappa \approx 400$, which is consistent with a very large upper critical field and a very small lower critical field, as shown in Table 14.4. The thermodynamic critical field, as estimated (Garoche et al., 1982) from the condensation energy, is $\mu_0 H_c(0) = (44 \pm 2) \times 10^{-4}$ T which is in accord with the large κ value. The very low value of $\mu_0 H_c(0)$ makes flux pinning poor, and the measured critical current density $J_c(0.5 \text{ K})$ in zero field is indeed very low, about $(1 \pm 0.5) \times 10^3 \text{Am}^{-2}$. As discussed by Saito and Yoshida (2000), the specific heat anomaly ΔC at T_c for X = ClO$_4$ yields $\Delta C / \gamma T_c = 1.67$, a value fairly close to the BCS value of 1.43 for conventional superconductors. However, the measured values of the energy gap show a large variation, ranging from a near-BCS value of 0.44 meV to a value 10 times larger.

Various other measurements also seem to show a similar disparity, and therefore one cannot assert that the superconductivity of (TM)$_2$X is really conventional or unconventional, although the latter appears more plausible. Whereas conventional superconductivity is primarily suppressed by magnetic impurities, in unconventional superconductivity, as discussed in Chapter 13, the presence of non-magnetic impurities, defects, and disorder has a pronounced deleterious effect on T_c. For the non-centrosymmetric anion ClO$_4$, the quenched-in anion disorder (Takahashi et al., 1982; Kagoshima et al., 1983; Garoche et al., 1983) inhibits the superconductivity of (TMTSF)$_2$ClO$_4$. Similarly, disorder created by mixing either the donors (Coulon et al., 1982), that is, [(TMTSF)$_{1-x}$(TMTTF)$_x$]$_2$ClO$_4$, or the counter-anions (Tomić et al., 1983), such as (TMTSF)$_2$(ClO$_4$)$_{1-x}$(ReO$_4$)$_x$, markedly suppresses superconductivity for $x > 2$–3%. In addition, very small (<100 ppm) disorder produced by proton irradiation of (TMTSF)$_2$ClO$_4$ fully quenches its superconductivity (Choi et al., 1986), although it is not known if the disorder carries any magnetic component. These findings are, in general, consistent with unconventional superconductivity.

Isotope-effect studies performed by replacing hydrogen by heavier deuterium (Schwenk et al., 1984) revealed a considerably decreased T_c, which could not be accounted for in terms of the BCS theory. The Knight shift, which depends linearly on magnetic susceptibility, measured from ^{77}Se NMR in (TMTSF)$_2$PF$_6$ under pressure revealed (Lee et al., 2003) no decrease from its normal state value on cooling below T_c. This contradicted the conventional singlet-state pairing in favour of an unconventional triplet state. Similarly, the ^{1}H-NMR relaxation rate varied as T^3, and below T_c no Hebel–Slichter peak was observed (Takigawa et al., 1987), which suggested an unconventional pairing with line nodes at the Fermi surface. But, contrary to the above findings, the thermal conductivity measurements of Belin and Behria (1997) on (TMTSF)$_2$ClO$_4$ showed the electronic contribution to decrease rapidly

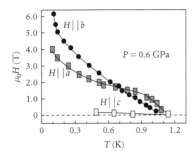

Figure 14.8 Temperature dependence of the upper critical field of $(TMTSF)_2PF_6$ measured with H along different directions a, b, and c (after Lee et al., 1997). Here the a-direction is along the TMTSF molecular chain, the b-direction is normal to a in the a–b plane, and the c-direction is normal to the a–b plane.

below T_c, which indicated the conventional node-less energy gap at the Fermi surface.

As may be seen from Table 14.4, the critical fields of $(TMTSF)_2ClO_4$ are markedly anisotropic. Extensive measurements of the upper critical field of $(TMTSF)_2PF_6$ under an imposed pressure of 0.6 GPa, with the field carefully aligned to the a- and b-axes, showed that μ_0H_{c2} increased as the temperature was lowered, but indicated no sign of any saturation (Figure 14.8) down to 0.1 K (Lee et al., 1997). The extrapolated $\mu_0H_{c2}(0)$ was three to four times larger than the Pauli limit $H_p(0)$. Similar behaviour was observed also with $(TMTSF)_2ClO_4$ and is popularly taken in support of the triplet pairing in $(TM)_2X$ salts.

14.3 The $(ET)_2$ family of quasi-two-dimensional superconductors

Some prominent members of this two-dimensional sulfur-based $(BEDT\text{-}TTF)_2X$ family are listed in Table 14.3. The monovalent counter-anions can be either organic or inorganic, and either discrete such as I_3, ReO_4, and AuI_2, possessing linear, planar, or octahedral symmetry, or polymerised such as $Cu(NCS)_2$, $KHg(SCN)_4$, $Cu[N(CN)_2]Br$, and $Cu[N(CN)_2]Cl$, where the metal ions are bridged by different ligands (Saito and Yoshida, 2000). The metallic character depends strongly on the ordering of the hydrogen bonds connecting the ethylene donors C_2H_4 at the outer ring of the ET molecule with the anion X. The ethylene CH_2 groups lie outside the molecular plane of ET and thereby make the ET molecule comparatively less planar than the TM molecule. Similar to anion ordering, discussed previously in $(TM)_2X$ compounds, which makes their properties sensitive to the cooling rate, there is ethylene ordering exhibited by $(ET)_2X$ salts and having analogous effects.

The structure of $(ET)_2X$ salts comprises alternate layers of conducting ET molecules separated by poorly conducting planes containing the chains of counter-anions X. The chemical structure of the ET molecule is depicted in Figure 14.9. The packing motifs of ET molecules within a layer may vary in different salts (Figure 14.2), which results in a variety of polymorphic phases of different structures, indicated by the Greek letters α, β, κ, θ, etc., all possessing different physical properties. As can be seen from Figure 14.2, in the β-phase the ET molecules are stacked in parallel columns forming a quasi-two-dimensional layer, while in the κ-phase the layer consists of dimers formed by pairs of ET molecules. The charge carriers can move more easily from column to column, or from dimer to dimer, along such layers than across them, which makes the electronic structure quasi-two-dimensional and highly anisotropic. Typically, the electrical resistivity perpendicular to the layers

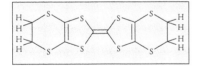

Figure 14.9 Chemical structure of the ET molecule.

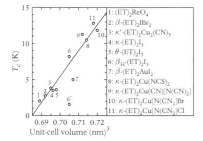

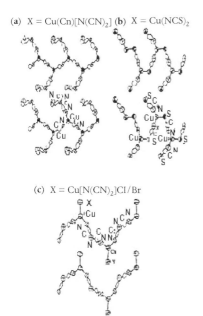

Figure 14.10 Plot of T_c versus unit-cell volume of κ-(ET)$_2$X superconductors (after Saito and Yoshida, 2000).

(a) X = Cu(Cn)[N(CN)$_2$] **(b)** X = Cu(NCS)$_2$

(c) X = Cu[N(CN)$_2$]Cl/Br

Figure 14.11 Anion structures of κ-(ET)$_2$X showing long zigzag chains of the various X (after Saito and Yoshida, 2000).

can be three to five orders of magnitude greater than within the layers. These materials have a strong resemblance to the HTS cuprates in having (a) low dimensionality, (b) low superfluid density, (c) low Fermi energy, (d) magnetic correlations, (d) unstable lattice, (e) numerous phase transitions above T_c, (f) normal conduction and superconduction mostly through holes, and (g) unconventional (possibly d-wave) superconductivity.

Historically, (ET)$_2$ReO$_4$ was the first member of this family that showed superconductivity (Parkin et al., 1983) at 2 K under a pressure of 0.04 GPa, while ambient-pressure superconductivity at 1.5 K was first reported (Yagubskii et al., 1984) for X = I$_3$. The latter was subsequently identified as the low-temperature phase β_L-(ET)$_2$I$_3$, since there seemed to be another, metastable, phase β_H-(ET)$_2$I$_3$ of higher T_c = 8.1 K that formed under a pressure of 0.04 GPa, although this was gradually suppressed at a rate of $dT_c/dP \approx -(8-14)$ K GPa^{-1} (Laukhin et al., 1985; Murata et al., 1985) when the pressure was increased. Further studies led to other superconducting phases based on the I$_3$ anion, including θ-(ET)$_2$I$_3$(T_c = 3.6 K), κ-(ET)$_2$I$_3$(T_c = 3.6 K), β-(ET)$_2$AuI$_2$(T_c = 4.9 K), and β-(ET)$_2$AuBr$_2$ (Wang et al., 1985; Kobayashi et al., 1986). The anion length, which is relatively long for I$_3$, helps to lower the intermolecular overlap along its length and promotes a higher T_c through an increased density of states at E_F. The search for longer and larger anions that enhanced the unit-cell volume led to T_c enhancement of ET salts (Figure 14.10): κ-(ET)$_2$Cu(NCS)$_2$ (T_c = 10.5 K), κ-(ET)$_2$Cu(CN)[N(CN)$_2$] (T_c = 11.2 K), κ-(ET)$_2$Cu[N(CN)$_2$]Br (T_c = 11.8 K), and κ-(ET)$_2$Cu[N(CN)$_2$]Cl (T_c = 12.8 K with pressure). The molecular structures of the anions forming these high-T_c salts are long and zig-zag in shape, as shown in Figure 14.11. Among all the families of organic charge-transfer salts known to date, the above κ-series of the (ET)$_2$X family stands out for their highest T_c values and has therefore been more extensively studied.

14.3.1 Phase diagram of κ-(ET)$_2$X and the pressure effect

Similar to the generic P–T phase diagram of (TM)$_2$X in Figure 14.7, Figure 14.12 presents a common phase diagram (Wzietek et al., 1996) for the κ-series of ET compounds, namely, κ-(ET)$_2$Cu[N(CN)$_2$]Cl, κ-(ET)$_2$Cu[N(CN)$_2$]Br, and κ-(ET)$_2$Cu(NCS)$_2$, where the pressure increases from left to right. These salts in that order are marked at the arrows shown from left to right in the figure. The location of each arrow indicates the ambient pressure for the corresponding compound in the P–T diagram. The phases shown are either for a given salt with increasing pressure or for a change in its counter-anion. The properties of κ-(ET)$_2$Cu[N(CN)$_2$]Cl reproduce the characteristics of the whole κ-series on varying the applied pressure to cover this generic phase

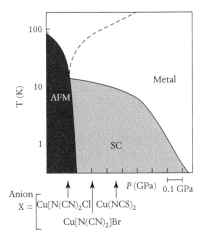

Figure 14.12 General P–T phase diagram of κ-$(ET)_2X$ compounds (after Wzietek et al., 1996). The arrows indicate the starting points for different compounds at ambient pressure.

diagram. The interesting observation is that superconductivity occurs in the proximity of insulating AFM order on increasing the pressure. In κ-$(ET)_2Cu[N(CN)_2]Cl$, the ambient-pressure phase is an insulating AFM, but on increasing the pressure to 0.03 GPa this gives way to superconductivity at the highest $T_c = 12.8$ K found for any organic salt. The insulating AFM ground state is presumably driven by a Mott–Hubbard transition due to a small bandwidth and a strong Coulomb interaction. The other two compounds, namely κ-$(ET)_2Cu[N(CN)_2]Br$ and κ-$(ET)_2Cu(NCS)_2$, are ambient-pressure superconductors and do not exhibit the insulating ground state. As indicated in the phase diagram, and found also for the $(TM)_2X$ salts, T_c decreases rapidly with increasing pressure (Sadewasser et al., 1997). The phase diagram in Figure 14.12, showing superconductivity occurring in the vicinity of an AFM state, is regarded as a strong pointer to the possible unconventional superconductivity of these salts.

The pressure dependence of various $(ET)_2X$ compounds is illustrated (after Lang and Müller, 2008) in Figure 14.13. The three κ-series compounds exhibit a T_c depression rate exceeding 30 K GPa^{-1}, which is greater than for most other superconductors, and exceeds by an order of magnitude the observed rate for $(TM)_2ClO_4$ discussed earlier. The pressure effect on T_c gives information about the volume dependence of the pairing interaction. In these organic systems, where the molecules are held together by a weak van der Waals force, the effect of applied pressure is primarily to lower the intermolecular spacing rather than the molecular radius, and consequently a highly compressed crystal lattice may result under pressure. This scenario therefore corroborates Figure 14.10, indicating a larger unit-cell volume, with a narrower electronic

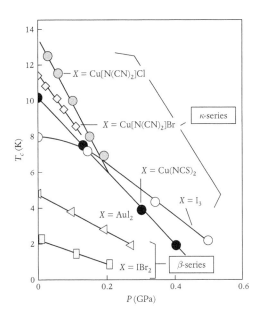

Figure 14.13 Pressure dependence of T_c for various $(ET)_2X$ superconductors (after Lang and Müller, 2008).

band with a high electronic density of states (DOS) at E_F, giving rise to a higher T_c before pressurisation.

14.3.2 Ethylene ordering

The ET compounds of the κ-series are found to show a glass-like transformation at a temperature $T_g \approx 80$ K. This was first noticed in heat capacity and thermal expansion measurements (Kund et al., 1993; Saito et al., 1999; Sato et al., 2001; Müller et al., 2002) of κ-(ET)$_2$Cu[N(CN)$_2$]Cl (at $T_g = 73$ K) and κ-(ET)$_2$Cu[N(CN)$_2$]Br ($T_g = 80$ K) and subsequently also in κ-(ET)$_2$Cu(NCS)$_2$ and in non-superconducting κ-(ET)$_2$Cu[N(CN)$_2$]I ($T_g = 84$ K). If the sample is rapidly cooled or quenched through about 80 K, the T_c and superconducting volume fraction are drastically lowered because of the ethylene group disorder (Kawamoto et al., 1997) that creates random potentials, which are deleterious for superconductivity. In the case of deuterated κ-(ET)$_2$Cu[N(CN)$_2$]Br, the heavy disorder is found to stabilise a magnetic state and suppresses superconductivity (Tanatar et al., 1999), as mentioned previously for (TM)$_2$X salts. On the other hand, if the sample is cooled through a lower temperature, say $T = 60$ K $< T_g$, superconductivity is still affected through the formation of a *pseudogap* in the normal state, which is believed to be related to ethylene ordering (Tanatar et al., 1999).

14.3.3 Superconductivity of (ET)$_2$X

14.3.3.1 *General features*

Superconducting parameters of some of the (ET)$_2$X salts are given in Table 14.5. It also shows the data for the BETS compound, which is simply the Se analogue of ET with X = GaClO$_4$, FeClO$_4$, etc. These compounds are extreme type II superconductors, possessing Ginzburg–Landau parameters κ exceeding 100. They are also highly anisotropic, with range of coherence in the conducting plane much larger than along the c-direction. Also, $H_{c2} >> H_{c1}$, which is often less than the Earth's magnetic field and therefore difficult to measure accurately. The thermodynamic critical field of the more extensively studied κ-series of salts is around 50–100 mT. Their mean free path l, as determined from de Haas–van Alphen (dHvA) studies, is about 20–30 nm (Maple et al., 2008), which makes $l/\xi \approx 5$–10, and these salts are therefore within the clean limit. This is one of the factors favouring the formation of the Fulde–Ferrell–Larkin–Ovchinnikov (FFLO) state.

14.3.3.2 *Fluctuation effects*

The short coherence length coupled with the low dimensionality and anisotropy results in pronounced fluctuation effects in these superconductors near T_c, which are seen in magnetisation, specific heat, and

Table 14.5 Superconducting parameters of some $(ET)_2$ and BETS superconductors

$(ET)_2X$	T_c (K)	$\mu_0 H_{c1}$ (mT)	$\mu_0 H_{c2}$(T) at 0 K	ξ (nm) at 0 K	λ (nm) at 0 K	κ (G–L) at 0 K
β_L-$(ET)_2 I_3$	1.5	0.005 (0.1 K along a-axis) 0.009 (0.1 K along b-axis) 0.036 (0.1 K along c-axis)	2.09 (a-axis) 2.48 (b-axis) 0.081 (c-axis)	63.3 (a-axis) 60.8 (b-axis) 2.90 (c-axis)	—	—
β_H-$(ET)_2 I_3$	8.1	—	25 ($//$) at 0.16 GPa) 2.7 ($\perp$) (0.16 GPa)	12.7 ($//$) 1.0 ($\perp$)	—	—
β-$(ET)_2 IBr_2$	2.7	0.39 (0 K, a-axis) 1.6 (0 K, b-axis)	3.36 (0 K, a-axis) 3.60 (0 K, b-axis) 1.5 (0 K, c-axis)	46.3 (a-axis) 44.4 (b-axis) 1.85 (c-axis)	—	—
β-$(ET)_2 AuI_2$	4.9	0.4 (0 K, a-axis) 2.05 (0 K, b-axis)	6.63 (0 K, a-axis) 0.51 (0 K, b-axis)	24.9 (a-axis) 1.92 (b-axis)	—	—
κ-$(ET)_2 Cu(NCS)_2$	8.7–10.4	0.07 (0.5 K, $//$) 20 (0.5 K, $\perp$)	24.5 (0.5 K, $//$) 5.5 (0.5 K, $\perp$)	2.9 ± 0.5 ($//$) 0.31 ± 0.5 ($\perp$)	~510 ($//$) ~40 000 ($\perp$)	~150 ($//$)
κ-$(ET)_2 Cu[N(CN)_2]Br$	11.0–11.8	3.0 (0 K, $\perp$)	30.6 (1.5 K, $//$) 7.4 (1.5 K, $\perp$)	2.3 ± 0.4 ($//$) 0.58 ± 0.1 ($\perp$)	~650 ($//$) ~38 000 ($\perp$)	~250 ($//$)
λ-$(BETS)GaClO_4$	5.0–6.0	—	12.0 (0 K, $//$) 3.0 (1.5 K, $\perp$)	1.43 ($//$)	~150 ($//$)	~100 ($//$)

Data from Lang and Müller (2008) and Saito and Yoshida (2000).

resistivity studies in the form of an enhanced width of the transition. The broadened transition can be quantitatively expressed in terms of the *Ginzburg number G*. Typically, for the κ-series of salts, G is of the order of 10^{-2}, which is similar to that for HTS cuprates, whereas for elemental superconductors, G is of the order of 10^{-8}. In the presence of a magnetic field normal to the ET layers, the fluctuation effects are more intense. An interesting behaviour due to these fluctuations is the appearance of a peak in the interlayer magneto-resistance close to T_c. Over a temperature range below T_c, the interlayer resistance in a decreasing magnetic field, normal to the layer, goes through a maximum (Oshima et al., 1988b, c; Su et al., 1999) just before the superconducting transition. Despite various explanations having been proposed (Pratt et al., 1993; Ito et al., 1994; Friemel et al., 1997; Zuo et al., 1996), the phenomenon still remains to be understood.

14.3.3.3 *Upper critical field and the FFLO state*

With quasi-two-dimensional superconductors, determination of H_{c2} from resistive measurements is problematic owing to pronounced fluctuation effects and vortex motion occurring in the layer if the applied magnetic field has a component perpendicular to the layer. In this situation, the destruction of superconductivity occurs through the orbital motion of electrons in the layer and resistive measurements yield an anomalously large curvature (Oshima et al., 1988b) of $H_{c2}(T)$ behaviour at low temperatures. These problems can be circumvented if the field is held strictly parallel to the layer. The $H_{c2}(T)$ data have been measured by different researchers (Nam et al., 1999; Lyubovskii et al., 1996; Shimojo et al., 1999; Tanatar et al., 1999b; Zou et al., 2000; Ohmichi et al., 1999; Ishiguro et al., 1998) for an applied field parallel to the layers. From these results, it is found that the κ-series of salts at low temperatures possess particularly large upper critical fields exceeding the paramagnetically limited value $H_p(0)$. $H_{c2}(T)$ curves for some of these compounds are illustrated (after Shimojo et al., 2003) in Figure 14.14. The lower branch for the sample with X = Cu(NCS)$_2$ represents the boundary of the superconducting mixed state with the FFLO state described below. In the case of κ-(ET)$_2$Cu[N(CN)$_2$]Br, the paramagnetic limit is exceeded even at 9 K. These materials thus become potential candidates for triplet p-wave pairing.

However, particularly unusual behaviour is exhibited by κ-(ET)$_2$ Cu(NCS)$_2$, which shows no saturation near 0 K, and $\mu_0 H_{c2}$ instead increases linearly on cooling near 0 K (Nam et al., 1999; Zou et al., 2000; Singleton et al., 2000). This is convincingly taken (Shimahara, 1997) as an indication of the formation of an FFLO state (Fulde and Ferrell, 1964; Larkin and Ovchinnikov, 1964), previously discussed in Section 13.5.3 for some of the HF superconductors. For highly anisotropic systems, such as the present organic salts, the calculations suggest (Shimahara, 1997) that FFLO formation can enhance $H_{c2}(0)$ to

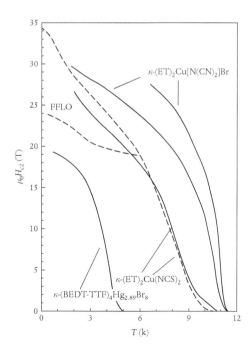

Figure 14.14 $H_{c2}(T)$ behaviour of some of the κ-(ET)$_2$X compounds. Note that for X = Cu(NCS)$_2$, the formation of an FFLO state is indicated at low temperatures in high fields.

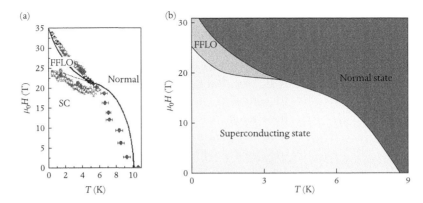

Figure 14.15 (a) Temperature dependence of the upper critical field of κ-$(BEDT)_2Cu(NCS)_2$ showing the formation of an FFLO state in high fields at low temperatures (triangular region) in between the superconducting and normal states (after Singleton et al., 2000). (b) Schematic phase diagram of FFLO state formation in $(BEDT)_2Cu(NCS)_2$ based on (a).

1.5–2.5 times $H_p(0)$. Various stringent criteria for stabilising the FFLO state were stated in Section 13.5, the majority of which are found to be effectively met in κ-$(ET)_2Cu(NCS)_2$. Various experimental studies (Singleton et al., 2000; Lortz et al., 2007) do indeed strongly support the formation of an FFLO state in this material. The results of Singleton et al. (2000) are shown in Figure 14.15(a) and the corresponding phase diagram of FFLO formation is depicted in Figure 4.15(b). λ-$(BETS)_2FeClO_4$ (Uji et al., 2006) and λ-$(BETS)_2GaClO_4$ (Tanatar et al., 2002) are other two-dimensional salts (based on the Se analogue of ET) where indications of an FFLO state have been reported.

14.3.3.4 *Field-induced superconductivity and coexistence with magnetism*

The organic superconductor λ-$(BETS)_2FeClO_4$ exhibits (Uji et al., 2001) the intriguing phenomenon of field-induced superconductivity (FIS), which we discussed previously for some of the HF superconductors in Chapter 13 and for Chevrel-phase superconductors in Chapter 12. The behaviour of the BETS compound is analogous to that of the latter, namely $Eu_xSn_{1-x}Mo_6S_8$, where FIS results from the field compensation effect of the Jaccarino–Peter (1962) model. In λ-$(BETS)_2FeClO_4$, the presence of Fe^{3+} ions makes this material an AFM insulator below 8.5 K (Kobayashi et al., 1993), whereas the isostructural salt λ-$(BETS)_2GaClO_4$ exhibits superconductivity at 6 K (Kobayashi et al., 1997). In the former case, by applying a magnetic field parallel to the conducting layers, the AFM insulating state can be quenched at a field above 10 T (Figure 14.16(a)), when the paramagnetic metallic state is restored (Uji et al., 2001). Further increase of the field to about 18 T (Uji et al., 2001, Balicas et al., 2001) at 0.8 K induces superconductivity, depicted resistively in Figure 14.16(b), which persists up to a field of 41 T. In this field range, the external magnetic field compensates the internal exchange field due to

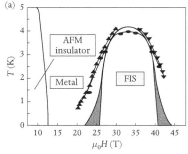

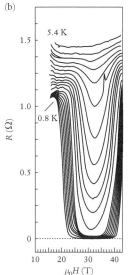

Figure 14.16 (a) *H–T* phase diagram of λ-(BETS)$_2$FeCl$_4$, showing that after suppression of the AFM insulating state in a low field, an applied field in the range of 18–41 T induces superconductivity (FIS). This is illustrated in (b) by resistive transition as the temperature is lowered below 5.4 K. (After Uji et al. (2001) and Balicas et al. (2001).)

the aligned Fe^{3+} moments, with the optimum $T_c \approx 4.2$ K occurring at an external field of about 33 T. Since the external field is held strictly parallel to the conducting layers, the orbital effect is strongly suppressed, and 42 T represents the Pauli paramagnetic limit. Clearly, FIS is destroyed by tilting the external field away from the conducting layers. Other interesting compounds showing superconductivity and magnetism include β''-(ET)$_4$[M(C$_2$O$_4$)$_3$]·H$_3$O·PhCN (with M = Fe or Cr, and the oxalate dianion C$_2$O$_4$) (Martin et al., 1997).

14.3.3.5 *Nature of the superconducting state of (ET)$_2$X salts*

Unlike with (TM)$_2$X salts, where the unconventional triplet state seems to be favoured, in (ET)$_2$X compounds, the observational data are less definitive in this respect. There seem to be as many experiments suggesting conventional pairing as there are otherwise. For instance, the temperature dependence of the penetration depth $\lambda(T)$ in κ-(ET)$_2$Cu(NCS)$_2$ and κ-(ET)$_2$Cu[N(CN)$_2$]Br has been measured using different techniques, such as AC susceptibility (Kanoda et al., 1990; Takahashi et al., 1991; Pinterić et al., 1999), magnetisation (Lang et al., 1992, 1993), microwave surface impedance (Achkir et al., 1993; Holczer et al., 1990; Klein et al., 1991; Dressel et al., 1993, 1994), high-frequency oscillations (Carrington et al., 1999), and muon-spin relaxation (Le et al., 1992; Harshman et al., 1990, 1994). While some of these have revealed an exponential decrease of $\lambda(T)$ in conformity with conventional coupling (Lang et al., 1992, 1993; Harshman et al., 1990, 1994; Holczer et al., 1990; Klein et al., 1991; Dressel et al., 1993, 1994), the others have found a power-law dependence, suggesting unconventional pairing (Kanoda et al., 1990, Takahashi et al., 1991, Le et al., 1992, Carrington et al., 1999, Achkir et al., 1993).

The NMR studies were performed on κ-(ET)$_2$Cu[N(CN)$_2$]Br in which ^{12}C atoms of the central carbon double bond of the donor molecule were substituted by ^{13}C (de Soto et al., 1995; Mayaffre et al., 1995; Kanoda et al., 1996). The Knight shift showed the spin susceptibility, below T_c, to gradually vanish with temperature, consistent with the spin-singlet state. The spin–lattice relaxation rate did not reveal a Hebel–Slichter peak and followed a T^3 dependence, which supports an unconventional pairing with anisotropic gap having line nodes at the Fermi surface. These observations favour *d*-wave pairing. Tunnelling studies (Arai et al., 2000) carried out on κ-(ET)$_2$Cu(NCS)$_2$ yielded a gap coefficient $2\Delta/k_B T_c \approx 8.7-12.9$, which is large in comparison with the BCS weak-coupling value of 3.52 and the strong-coupling value of 4.5. The observed dI/dV–V characteristics of the tunnelling measurements were consistent with the *d*-wave pairing (Arai et al., 2000) that is regarded as a potential candidate for the order parameter in this material.

The possibility of an anisotropic gap in $(ET)_2$ salts has, however, been seriously contested by heat capacity measurements on κ-$(ET)_2$ Cu[N(CN)$_2$]Br (Elsinger et al., 2000), which revealed a clear exponential temperature dependence of the electronic specific heat below T_c. Similar behaviour was noted also for κ-$(ET)_2$Cu(NCS)$_2$ (Müller et al., 2002; Wosnitza et al., 2003), pointing towards conventional BCS pairing. To check the role of phonons in the pairing mechanism, extensive isotope effect studies were performed by several workers (Auban-Senzier et al., 1993; Carlson et al., 1992; Kini et al., 1996, 1997, 2001) by replacing the isotopes ^{1}H, ^{12}C, and ^{32}S by the corresponding heavier ones at different sites of either the donor molecules, their counter-anions, or both. Surprisingly, these efforts revealed either a very large or often a negative shift (*inverse isotope effect*). These are ascribed to the so-called *geometrical* isotope effect resulting from changes in the unit cell volume or, equivalently, the internal pressure caused by the isotope substitutions (Lang, 1996). By replacing only the central atoms of the donor (i.e. four ^{12}C and eight ^{32}S), the observed 1% decrease (Pedron et al., 1997) in T_c was, however, consistent with the BCS prediction and suggested that at least partial involvement of phonons cannot be ruled out from the superconductivity mechanism.

By far, the strongest support for an unconventional pairing in these compounds comes from the pressure–temperature phase diagram and the pressure-effect studies on $(ET)_2$ salts discussed earlier. Superconductivity appears in proximity to the insulating AFM state, the latter giving way to the former as the applied pressure is increased. The origin of superconductivity is ascribed to AFM fluctuations. In the case of the mixed compound λ-$(BETS)_2Fe_xGa_{1-x}Cl_4$, there exists a common boundary between the superconducting and AFM insulating phases for $0.35 < x < 0.5$ (Figure 14.17) (Uji et al., 2002), which makes a strong case for unconventional superconductivity. The presence of non-magnetic disorder and impurities has a deleterious effect on the T_c of $(ET)_2X$ salts, which is an accepted feature of unconventional pairing. An example of this is the way in which ethylene disorder, frozen in during fast cooling, adversely affects superconductivity. Further, although individually β-$(ET)_2I_3$ and β-$(ET)_2IBr_2$ are found to be superconducting, when they are alloyed to form β-$(ET)_2(I_3)_{1-x}(IBr_2)_x$, the phenomenon is quenched (Tokumoto et al., 1987) because of disorder. Such disorder, however, does not necessarily occur every time through anion mixing. Mutual ordering can take place, as with the example of the BETS salt already mentioned. Similarly, for the mixed salt κ-$(ET)_2Cu[N(CN)_2]Cl_{1-x}Br_x$, the ordering of Cl and Br keeps the superconductivity intact (Kushch et al., 1993; Shibaeva et al., 1993).

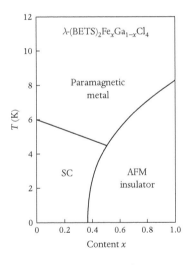

Figure 14.17 Phase diagram of λ-$(BETS)_2F_x$ $Ga_{1-x}Cl_4$ manifesting a common boundary between the superconducting and AFM phases (after Uji et al., 2002).

14.4 Superconducting fullerides

The discovery of C_{60} fullerene (Kroto et al., 1985) as the third stable form of carbon led to the advent in 1991 of the alkali-metal doped fullerides (the solid form of fullerene-based compounds), which were found to be conducting (Haddon et al., 1991) and also superconducting (Hebard et al., 1991; Rosseinsky et al., 1991). The potassium-doped fulleride K_3C_{60}, the first member of the fulleride family to be discovered, showed superconductivity below 19.5 K. Since then, about 30 fulleride compounds with the general formula M_3C_{60}, with M an alkali (A), alkaline earth (AE), or rare-earth (RE) metal or their combinations in the required stoichiometry, have been identified as superconductors, with some possessing the highest T_c values among the organic class (Table 14.6). At ambient pressure, the optimum T_c reported is 33 K (Tanigaki et al., 1991) for the ternary fulleride $RbCs_2C_{60}$, while the binary compound Cs_3C_{60} exhibits a still higher $T_c = 40$ K, albeit at an imposed pressure of 1.2 GPa (Palstra et al., 1995).

Surprisingly, the fullerides possess a dominant Coulomb repulsion and at low temperatures are expected to be a Mott insulators, and yet they exhibit a relatively high T_c exceeding the generally accepted BCS limit and their superconductivity seems mostly conventional. There are many detailed topical reviews and monographs on these novel materials (Gunnarsson, 1997, 2004; Saito and Yoshida, 2000; Han et al., 2005; Ott, 2008).

14.4.1 C_{60} fullerene, fullerite, and doped fullerides

A C_{60} molecule comprises a cluster of 60 carbon atoms forming a nearly spherical closed shell of 0.71 nm diameter. The hollow molecule possesses a truncated icosahedral structure and closely resembles a soccer ball whose surface consists of 12 pentagons and 20 hexagons of carbon atoms, as shown schematically in Figure 14.18(a). Figure 14.18(b) depicts a real-time scanning tunnelling microscope (STM) image (Narlikar et al., 1994) of an isolated molecule. It possesses the highest possible molecular symmetry, with 30 carbon double bonds of length 0.14 nm between the neighbouring hexagons and 60 single bonds of length 0.145 nm separating the neighbouring pentagons and hexagons. At ambient temperature, the C_{60} molecules bind together with a weak van der Waals force, to form a molecular solid, termed *fullerite*, with a close-packed fcc lattice (Figure 14.18(c)) having $a_0 = 1.417$ nm, with two neighbouring molecules 1 nm apart. Since the molecules are soluble in benzene, a fullerite single crystal can be readily grown by slow evaporation of a benzene solution. The pristine sample can be doped with an alkali metal such as potassium to form the potassium *fulleride* K_3C_{60}, simply

Table 14.6 C_{60} fulleride superconductors

Fulleride	Structure	Category	Lattice parameters (nm)	T_c (K)
K_3C_{60}	fcc	I	1.4240	19.5
K_2RbC_{60}	fcc	I	1.4267	23
K_2CsC_{60}	fcc	I	1.4292	24
KRb_2C_{60}	fcc	I	1.4337	27
Rb_3C_{60}	fcc	I	1.4384	29.5
Rb_2CsC_{60}	fcc	I	1.4431	31.3
$RbCs_2C_{60}$	fcc	I	1.4555	33
Na_2CsC_{60}	fcc	I	1.4473	29.6
Li_2CsC_{60}	fcc	I	1.4080	10.5
$(NH_3)_xNaK_2C_{60}$	fcc	I	1.435–1.440	8–13
$(NH_3)_xNaRb_2C_{60}$	fcc	I	1.450–1.453	8.5–17
$(NH_3)_xK_3C_{60}$	fcc	I	1.432	8.5
$(NH_3)_4Na_2CsC_{60}$	fcc	I	1.4473	30
$(NH_3)K_3C_{60}$	fco (face-centred orthorhombic)	IV	$a_0 = 1.4917, b_0 = 1.4971, c_0 = 1.3692$	28[+]
Cs_3C_{60}	bco (body-centred orthorhombic) $\rightarrow$ bcc[+]	IV	$a_0 = 1.1843, b_0 = 1.2220, c_0 = 1.1464$	40[+]
Na_2KC_{60}	Simple cubic	II	1.4122	2.5
Na_2RbC_{60}	Simple cubic	II	1.4092	3.5
Ca_5C_{60}	Simple cubic	II	1.401	8.4
$Na_xN_yC_{60}$	Simple cubic	II	1.4204	12
$Na_xH_yC_{60}$	Simple cubic	II	1.4356	15
$K_xH_yC_{60}$	fcc	I	1.4351	19.5
$K_{2.75}Ba_{0.25}C_{60}$	fcc	III	—	15
Ba_4C_{60}	bco	IV	$a_0 = 1.1610, b_0 = 1.1235, c_0 = 1.0883$	6.7
Sr_4C_{60}	bco	IV	—	4.4
$Yb_{2.75}C_{60}$	Simple orthorhombic	V	$a_0 = 2.7874, b_0 = 2.7980, c_0 = 2.7873$	6
Sm_xC_{60}	Simple orthorhombic	V	$a_0 = 2.817, b_0 = 2.807, c_0 = 2.7873$	8
La_xC_{60}	—	V	—	12.5
$K_3Ba_3C_{60}$	bcc	IV	1.1245	5.6
$Rb_3Ba_3C_{60}$	bcc	IV	1.1338	3–5

Data from Saito and Yoshida (2000) and Mourachkine (2008).

by heating them together in a evacuated tube at 400°C. The potassium vapour diffuses into the empty space between C_{60} molecules of the FCC lattice. Potassium atoms in the fulleride become ionised to form the positive ion K^+ while each C_{60} accepts three electrons:

$$3K + C_{60} \rightarrow 3K^+ + [C_{60}]^{3-}$$

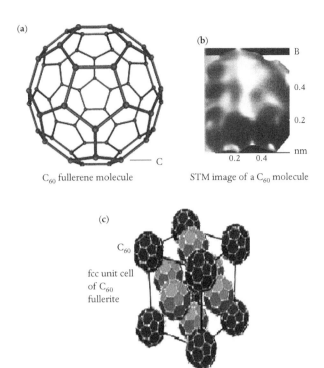

(a)

(b)

C_{60} fullerene molecule

STM image of a C_{60} molecule

(c)

C_{60}

fcc unit cell of C_{60} fullerite

Figure 14.18 (a) Schematic structure of a C_{60} fullerene molecule (buckyball). (b) STM image of an individual buckyball showing the characteristic arrangement of carbon atoms in the form of hexagons and pentagons (after Narlikar et al., 1994). (c) The fcc structure of C_{60} fullerite.

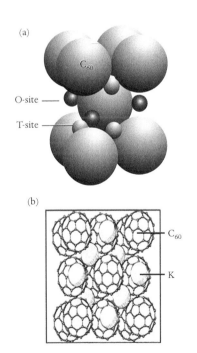

(a)

C_{60}

O-site

T-site

(b)

C_{60}

K

Figure 14.19 (a) Schematic fcc structure of M_3C_{60} depicting the O- and T-sites. (b) Structure of superconducting K_3C_{60}.

The three electrons on each fullerene molecule are free to move across to the neighbouring molecules, and this is what gives rise to its electrical conductivity.

In most cases, the original fcc structure of the host lattice is preserved, with only a slight increase in the lattice parameter a_0, depending on the ionic size of the inserted ions. In the case of M_3C_{60}, to accommodate the M atoms, there is a larger octahedral site (the O-site) and two smaller tetrahedral sites (the T-sites) per C_{60} molecule available in the fcc lattice (Figure 14.19(a)). While the O-site is large enough for any alkali ion, the T-sites have smaller radii than K^+, Rb^+, and Cs^+ but larger than Na^+. The structure of superconducting K_3C_{60} is shown in Figure 14.19(b). The structure can be modified if larger-sized or a greater number of M atoms are introduced into the lattice. For example, bct (body-centred tetragonal) or bcc lattices result in M_4C_{60} and M_6C_{60}, respectively, both being insulators. Similarly, the larger ionic size of Cs is responsible for the bcc structure of Cs_3C_{60}.

An interesting feature of the fullerite structure is that at room temperature the fullerene molecules are orientationally disordered and are rapidly rotating, which makes all four atoms of the unit cell equivalent, and the structure, as described above, is fcc. On cooling below 260 K, a first-order phase transition occurs to an orientationally ordered (frozen) structure of simple cubic type having a space group $Pa3$, with the neighbouring C_{60} molecules having different orientations relative to each other. In doped fullerides, however, the presence of the inserted atoms often tends to inhibit the free rotation of the molecules at room temperature and also influences the ground-state behaviour by preventing neighbouring molecules from optimising their interaction. For example, in the presence of large alkali ions such as K and Rb, lowering of the temperature does not lead to any phase transition, except that the initial fully ordered fcc structure of space group *Fm3* changes to a slightly disordered fcc structure of space group *Fm3m* through what is known as *merohedral disordering* (Stephens et al., 1991). However, in the presence of smaller Na ions at the T-sites in the $Na_2M_xC_{60}$ family, inter-C_{60} interaction is allowed, which gives rise to an fcc $\rightarrow$ simple cubic transition to a *Pa3* phase (Tanigaki et al., 1994), with all the four atoms of the unit cell having different orientations, analogous to the case of the undoped state.

14.4.2   Electronic features of fullerides

The HOMO (highest occupied molecular orbital) of a fullerene molecule possesses five degenerate levels of h_u symmetry, while its LUMO (lowest unoccupied molecular orbital) is triply degenerate, with t_{1u} symmetry. In the undoped state, the h_u is fully occupied while t_{1u} is empty, which makes a fullerite insulating/semiconducting with a band gap of 1.5 eV. In M_3C_{60}, which is metallic and shows superconductivity, the band is half-filled, while in M_6C_{60} it is fully occupied, which makes it an insulator. In the remaining compounds, such as MC_{60}, M_2C_{60}, M_4C_{60}, and M_5C_{60}, the band is partially filled and, as per rigid band model, they should have been metallic. The breakdown of the rigid band picture indicates the presence of strong electron correlations. Since the electrons in the conduction band are concentrated on the C_{60} molecules, the electron screening becomes ineffective, resulting in a strong on-site Coulomb repulsion with energy U ($\approx$ 1–1.25 eV) $>> W$. This suggests that this half-filled fulleride should have been a Mott–Hubbard insulator and not a metal. This puzzle has been successfully addressed theoretically by Lu (1994) and Gunnarsson et al. (1996), who pointed out that the orbital degeneracy L at each site markedly raises the U/W criticality limit from 1 to 2.5 and thus prevents the insulating transition in M_3C_{60}.

In this situation, how does a very weak electron–phonon interaction of energy of order 0.1 eV overcome the large U and also survive to produce appreciably high T_c values in doped fullerides? In conventional superconductors, the *Migdal approximation* takes into account very different energy scales for phonons and electrons and markedly simplifies the interaction by ignoring the complicated vertex corrections (Schrieffer, 1994). This brings in retardation effects that substantially reduce the strength of Coulomb interaction in favour of superconductivity. In M_3C_{60} compounds, however, the energy scales of electrons and

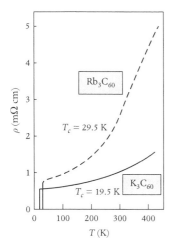

Figure 14.20 Resistive superconducting transition curves for K_3C_{60} and Rb_3C_{60} (after Baumgartner, 1998).

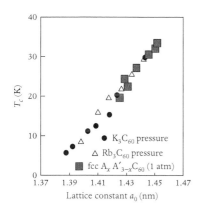

Figure 14.21 The T_c dependence of a_0 for doped fullerides (after Saito and Yoshida, 2000).

phonons are close to each other and the energy hierarchy on which the Migdal approximation is based is not present to supply a retarded electron–phonon interaction. Because of this, phonons alone have been considered inadequate to explain the high T_c values of doped fullerides, and alternative mechanisms have been proposed. But, at the same time, it has been argued that the phonon mechanism could be sufficient if the Coulomb interaction were efficiently screened (Gunnarsson and Zwicknagl, 1992).

The above features are manifested in unusual properties of doped fullerides. At room temperature, the resistivity exhibits characteristic metallic behaviour with typical values of 1.5 and 2.5 mΩ cm, respectively, for K_3C_{60} and Rb_3C_{60} (Hou et al., 1995). These correspond to an electron mean free path of a few angstroms, i.e., fractions of nanometres and less than the separation of the neighbouring C_{60} molecules. Consequently, the observed metallic conduction is unexpected, and the R–T behaviour observed (Figure 14.20) is strikingly different from that of a conventional metal (Baumgartner, 1998).

14.4.3 Classification of fulleride superconductors and T_c–lattice-parameter behaviour

The structural characteristics and T_c values of various fulleride superconductors are listed in Table 14.6. Following Gunnarson (1997), these superconductors may be classified into five broad categories I to V listed in the table, based on their structure and C_{60} valence.

Several authors (Tanigaki et al., 1991, 1992; Sparn et al., 1991; Fleming et al., 1991; Yildirim et al., 1995) have studied the interrelation between the lattice parameter a_0 and T_c for the doped fullerides of the first two categories, which include the largest number of compounds. T_c is found to increase monotonically with the lattice parameter a_0, as depicted in Figure 14.21 (Saito and Yoshida, 2000). The rate of increase is, however, found to be different for the two categories, as may be seen from the figure. The increase in the lattice parameter is dictated by the ionic size of the dopant. With increasing a_0, the conduction band narrows, which increases the DOS at E_F and thereby enhances T_c. Interestingly, this behaviour for compounds of one category is independent of the type of dopant introduced and in this correlation, for achieving a particular T_c, what really matters is the size of a_0. If essentially the same a_0 could be achieved through various means, such as changing the alkali dopant, incorporation of neutral ammonia, or imposed pressure, the ensuing T_c values would not be too different. The T_c behaviour of the compounds of the first two categories is mostly consistent with this contention, although some exceptions have been noted for Rb_3C_{60} and Rb_2CsC_{60} (Schirber et al., 1993; Diederichs et al., 1997).

14.4.4 Superconducting properties and the nature of the superconducting state

Among fulleride superconductors, K- and Rb-doped compounds have been more extensively studied for their various superconducting parameters, and the compiled data for these two fullerides are listed in Table 14.7. The gap coefficients in the BCS weak-coupling limit have been found from both μSR and optical reflectivity studies (Ramirez, 1994). The large λ and small coherence length ξ of these fullerides make them extreme type II superconductors, with large Ginzburg–Landau parameter κ (>130), which is corroborated by their very large upper critical fields and equally small lower critical fields.

In order to check whether the superconductivity is conventional phononic-type or otherwise, the fulleride superconductors have been studied for their isotope-effect behaviour. For full substitution of ^{13}C for ^{12}C in K_3C_{60} and Rb_3C_{60}, the value of the carbon isotope-effect exponent α, in $T_c \propto m^{-\alpha}\kappa$ is found to be 0.30 ± 0.06 and 0.30 ± 0.05, respectively (Chen and Lieber, 1992, 1993; Lieber and Zhang, 1994). Surprisingly, however, for incomplete substitution of ^{13}C, unexpectedly larger values of $\alpha > 0.5$ have been encountered that are yet to be understood (Ebbesen et al., 1992; Zakhidov et al., 1992; Ramirez et al., 1992). Nevertheless, the values for full substitution are a strong indication of a conventional phonon mechanism. On the other hand, no isotope effect is observed for the alkali metals dopants (Burk et al., 1994). This suggests that alkali phonons are not involved in superconductivity. This is perhaps consistent with the observed a_0–T_c behaviour being independent of the intercalated alkali metal. The phonons that are considered to play the major part in superconductivity belong to the high-frequency intramolecular H_g mode, with the low-energy intermolecular C_{60}–C_{60} and alkali–C_{60} vibrations being less important. The light mass of C atoms and the stiff bonding from the curved surface of C_{60} result in high-frequency phonons and a strong electron–phonon coupling to enhance T_c.

The conventional s-wave superconductivity of alkali-doped fullerides is corroborated by both NMR (Stenger et al., 1995) and μSR (Kiefl et al., 1993) studies of Rb_2CsC_{60} showing the characteristic Hebel–Slichter peak below T_c. Also, the Knight shift in NMR measurements

Table 14.7 Various superconducting parameters of K_3C_{60} and Rb_3C_{60}

Fulleride	T_c (K)	$H_{c1}(0)$ (mT)	$H_{c2}(0)$ (T)	$\lambda(0)$ (nm)	$\xi(0)$ (nm)	κ	$\Delta C/T_c$ (mJ mol^{-1}K^{-2})	$2\Delta/k_B T_c$
K_3C_{60}	19.5	13	49	480–600	2.6	130–170	64 ± 12	3–5.3 ± 0.2
Rb_3C_{60}	29.5	12	78	420–460	2.0	140–160	75 ± 14	4.1–5.2 ± 0.2

From Gunnarson (1997) and Saito and Yoshida (2000).

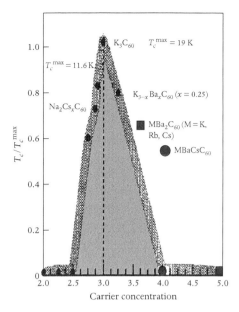

Figure 14.22 T_c as a function of carrier concentration per C_{60} molecule, manifesting a sharp peak at the C_{60} valence $m = 3$ (after Yildirim et al., 1996).

exhibits a decrease on cooling across T_c, which supports the contention of s-wave superconductivity (Tyco et al., 1992; Stenger et al., 1995). However, there is some evidence also of an unconventional mechanism. For example, T_c manifests a sharp dependence on the concentration of alkali atoms, showing a pronounced peak for the stoichiometry M_3C_{60}, namely, at three M-atoms per molecule, which corresponds to exact half-filling of the t_{1u} orbital by electrons. As one moves away from three atoms in either direction, T_c decreases rapidly (Yildirim et al., 1996; Meddeb et al., 2004). The results of Yildirim et al. are shown in Figure 14.22. Such a decrease is, however, unexpected in the conventional theory, where the DOS decreases rather smoothly. Further, the doped fullerides possess a low superfluid density and a small E_F, features that are common to unconventional superconductors. As a result, the possibility of unconventional electronic pairing mechanisms has not been ruled out (Baskaran and Tosatti, 1991; Friedberg et al., 1992; Chakravarti et al., 1991; Chakravarty and Kivelson, 1999). However, Han et al. (2003) have found that it is not necessary to invoke an electronic mechanism to explain the carrier-concentration dependence of T_c in doped fullerides, since the electron–phonon coupling symmetry of local Jahn–Teller phonons in C_{60} can account for the observed dependence.

14.5 Graphite intercalation compounds (GICs)

Superconducting GICs, intercalated with alkali metals, have been in existence since 1965, when they were discovered by Matthias's group (Hannay et al., 1965) and further studied by others (Koike et al., 1978;

(a)

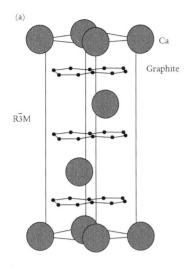

Ca

Graphite

R̄3M

(b)

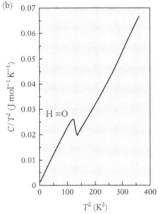

H = 0

Figure 14.23 (a) Rhombohedral unit cell and corresponding hexagonal structure of the graphite intercalation compound CaC_6. (b) Characteristic specific heat anomaly at the superconducting transition (in zero field) at $T_c = 11.4$ K (after Kim et al., 2006).

Kobayashi and Tsujikawa, 1981; Iye and Tanuma 1982a, b). They were found to have $T_c < 1$ K, but, by intercalating a greater concentration of atoms in the graphite layers by applying pressure, this could be markedly raised, although the highest T_c achieved in this way was just 5 K for C_2Na (Belash et al., 1987a, b, 1989a–c, 1990). Broadly, GICs are essentially graphite analogues of fullerites, and having been discovered earlier, they provided a useful stimulus to the development of the latter. As with fullerides, the intercalants make the host material electron-doped. However, in contrast to fullerides, which are isotropic and exhibit three-dimensional behaviour, the GICs, owing to their layered structure, are anisotropic. Strictly, however, they are not quasi-two-dimensional, since, owing to their low T_c, their coherence length ξ is large and generally exceeds the interlayer distance. The stiff bonding of the curved surface of the carbon cage of the fullerene molecule is partly responsible for the observed higher T_c of fullerides than of GICs formed with planar graphene sheets.

Until recently, because of their low T_c values, GICs did not attract much attention, although they displayed interesting anisotropy that decreased with increasing dopant content, and their H_{c2} values exhibited an unusual linear temperature dependence down to low temperatures (Belash et al., 1989a, 1990). These features were explained theoretically by Jishi et al. (1991) and Jishi and Dresselhaus (1992). However, the unexpected discovery (Weller et al., 2005; Emery et al., 2005) of relatively high T_c values in YbC_6 and CaC_6 of 6.5 K and 11.5 K (which is enhanced to 15.1 K under a pressure of 10 GPa: Debessai et al., 2010), respectively, has resulted in considerable experimental and theoretical attention being directed at this class of materials (Mazin, 2005; Calandra and Mauri, 2005; Lamura et al., 2006; Upton et al., 2007, 2010; Hinks et al., 2008). Figure 14.23(a) shows the crystal structure of CaC_6, which exhibits a pronounced specific heat jump at $T_c = 11.4$ K (Figure 14.23(b)) (Kim et al., 2006). The pressure dependence of T_c for this compound is depicted in Figure 14.24 (Debessai et al., 2010). Table 14.8 presents the T_c values of some representative members of the GIC family.

The materials are commonly synthesised by subjecting a highly ordered pyrolytic graphite (HOPG) sample to the vapour or a solution of the atoms that are to be intercalated. For example, to synthesise CaC_6, Weller et al. (2005) used Ca vapour transport, while Emery et al. (2005) immersed HOPG in a Li–Ca solution (75 at% Li) at 350°C for

Table 14.8 Representative superconductors of the GIC family with their T_c values

GIC	CaC_6	$Li_3Ca_2C_6$	YbC_6	NaC_2^*	KC_3^*	NaC_3^*	LiC_2^*	SrC_6	KC_6^*	$KHgC_8$	KC_8	RbC_8
T_c (K)	11.5	11.15	6.5	5	3	2.3-3.8	1.9	1.65	1.5	1.4	0.14	0.025

*Intercalation was carried out under pressure.
Data from Emery et al. (2008) and Belash et al. (1989a).

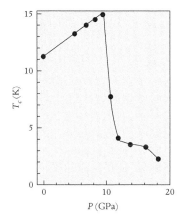

Figure 14.24 Pressure dependence of T_c for CaC_6 (after Debessai et al., 2010). For 20 GPa $< P <$ 32 GPa, no superconductivity was detected at $T_c > 2$ K.

several days, in a process known as *flux intercalation*. Li is the first to be intercalated into graphite layers, followed by a slow exchange of Ca for Li. In the case of isostructural CaC_6 and YbC_6, the intercalation leads to an ordered structure ($P6_3/mmc$) in which the intercalants form a triangular array sandwiched between graphene layers (Weller et al., 2005). This structure resembles that of MgB_2, a potential superconductor for technological applications with a T_c of 39 K, which forms the topic of Chapter 15.

With regard to the origin of superconductivity in GICs, both unconventional electronic (Csányi et al. 2005) and conventional phononic (Mazin, 2005; Calandra and Mauri, 2005, 2007; Upton et al., 2010) mechanisms have been considered, with particular attention being paid to CaC_6 and YbC_6. As with the superconducting fullerides, the dominant mechanism driving superconductivity in these materials seems the conventional type, although some of its key aspects remain unresolved. Both penetration depth (Lamura et al., 2006) and thermal conductivity (Sutherland et al., 2007) measurements on these two materials are in full accord with isotropic *s*-wave pairing with no multiple gaps. Surface impedance studies of CaC_6 confirm that it is a fully gapped weakly coupled superconductor free from nodes. Isotope-effect studies by Hinks et al. (2008) corroborated Mazin's (2005) finding that the Ca-isotope effect coefficient α was significantly more dominant than that of C, suggesting that superconductivity is determined mainly by Ca vibrations parallel to the graphene layers. On the other hand, Calandra and Mauri (2005) theoretically found $\alpha(Ca) \approx \alpha(C)$, indicating that C vibrations transverse to the layers are equally important. Consequently, the phonons of either Ca alone or of both Ca and C become coupled to the Fermi surface of the intercalated Ca, which results in conventional singlet-state pairing. On the other hand, Upton et al. (2010) have suggested that the phonon coupling occurs primarily with the electron-doped graphene Fermi surface rather than with the intercalate Fermi surface. Clearly, further experiments are needed to resolve these uncertainties.

14.6 Summary

Three categories of organic superconductors formed the mainstay of this chapter. Outside these categories, superconductivity of boron-doped diamond (see Chapter 7) and carbon nanotubes (CNTs) were also mentioned. While the former is fully established, the situation remains inconclusive with the latter. There are unconfirmed reports (Tang et al., 2001; Tsebro et al., 1999; Zhao and Wang, 2003) of superconductivity in single-walled CNTs with T_c ranging from 20 K to room temperature even, which still remains to be experimentally confirmed. Also, surprisingly, HOPG has been found to exhibit superconductivity above 200 K (Kopelevich et al., 2003), but these observations presently lack the reproducibility to be scientifically credible. As to the three categories of organic superconductors presented here, clearly they constitute a rich field in which many of the key issues remain open for further investigation. One of the most pursued problems with all of these materials is the nature of their superconducting state and its associated driving mechanism for electron pairing. In quasi-one-dimensional Bechgaard salts, triplet pairing resulting from magnetic fluctuations seems to be responsible, while for the quasi-two-dimensional ET salts, a spin-singlet state with d-wave symmetry is favoured. The strongest support for an unconventional superconducting state in these salts stems from their proximity to the tunable magnetic state as revealed by complex phase diagrams, and the way superconductivity is adversely affected by non-magnetic impurities and disorder. In sharp contrast to these low-dimensional organic salts, both the doped fullerides and GICs appear definitely to belong to the conventional superconducting class, with some form of electron–phonon interaction as its driving mechanism.

CHAPTER 15

Superconducting magnesium diboride

While discussing superconducting graphite intercalated compounds (GICs) in Chapter 14, we pointed out their structural similarity to magnesium diboride (MgB_2), which comprises alternate graphene-like layers of B between which a triangular Mg layer is sandwiched. Superconductivity of MgB_2, with a remarkably high $T_c = 39$ K, was discovered relatively recently in early 2001 (Nagamatsu et al., 2001), and the material possesses many striking features.

1. It is the only member of the AlB_2 family (hexagonal space group $P6/mmm$) possessing such a high T_c of 39 K; the other members are either non-superconducting (ReB_2) or possess $T_c < 10$ K (e.g. MB_2 with M = Nb, Ta, Zr, or Be) (Yamamoto et al., 2002; Gasparov et al., 2001; Young et al., 2002; Felner, 2001). Interestingly, superconductivity in some of these materials was known even before the discovery of MgB_2, although their stoichiometry and crystal structure were slightly different (Cooper et al., 1970; Leyarovska and Leyarovska 1979; Higashi et al., 1986).

2. The 39 K T_c of MgB_2 is surpassed to date only by HTS cuprates, some of the ruthene-cuprates, and recently discovered pnictide and chalcogenide superconductors.

3. Despite its large T_c, which goes beyond the generally accepted BCS limit (McMillan, 1968), the driving mechanism for superconductivity of MgB_2, surprisingly, is still the conventional phonon-mediated one.

4. It possesses a remarkably low normal-state resistivity $\rho(42\ \text{K}) \approx 0.3\mu\Omega$ cm.

5. T_c is insensitive to the residual normal-state resistivity; that is, various defects and disorders have little impact on T_c.

6. MgB_2 is the first established example of multiband superconductivity, possessing two distinct energy gaps of about 2 and 7 meV.

7. It is an extreme type II superconductor whose upper critical field can vary from 14 T for sintered bulk samples to 74 T for thin films.

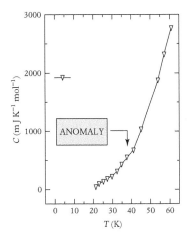

Figure 15.1 Specific heat of MgB_2 over a range of temperatures, showing an anomaly near 40 K (after Swift and White 1957).

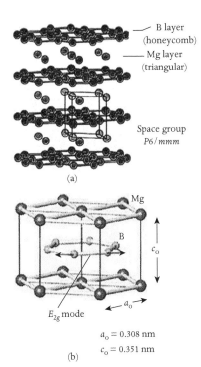

$a_o = 0.308$ nm
$c_o = 0.351$ nm

Figure 15.2 (a) Crystal structure of MgB_2. (b) Lattice parameters and the relevant phonon mode.

8. Polycrystalline samples of MgB_2 show transparency of electric current to grain boundaries; that is, the latter do not manifest *weak-link* effects to degrade the critical current.

9. The lighter constituent elements, namely B and Mg, make this material particularly light in weight, suitable for applications and large installations.

10. The raw material is cheap and readily available: at the time of discovery of its superconductivity in 2001, the material was commercially available from suppliers in quantities of metric tons as a chemical reagent used for making other borides.

11. MgB_2 stands today as a promising contender to replace the Nb–Ti and Nb_3Sn superconducting wires and cables that for the last 40 years have been the backbone of superconductor technology.

It seems almost inexplicable that this material missed the attention of superconductivity chasers until 2001, by which time some of the other members of the same family were already known for their superconductivity. One possible explanation could be that MgB_2 is a simple s, p-metal compound that does not contain any constituent elements from either the transition metal or lanthanide groups, which were generally considered necessary for high T_c. Probably, no-one thought that this simple intermetallic compound would be superconducting, let alone with such a high T_c. Interestingly, however, heat capacity measurements made on MgB_2 in 1957 (Swift and White, 1957) had, in fact, revealed a feeble signature of the characteristic anomaly (Figure 15.1) at about 40 K, which then remained unnoticed. It would be interesting to speculate how the existing road map of superconductivity would have changed if the 39 K T_c of MgB_2 had been discovered in the late 1950s when a good many of the potential A-15 superconductors were yet to be discovered. There exist many topical reviews on superconducting MgB_2 (Narlikar, 2002; Muranaka et al., 2005; Dahm, 2005; Dou et al., 2005; Goldacker and Schlachter, 2005; Xi et al., 2005).

15.1 Crystal structure and T_c

The crystal structure (hexagonal) of MgB_2 (of space group $P6/mmm$) is shown in Figure 15.2(a). It comprises alternate layers of B and Mg atoms, the former is graphene-like with a honeycomb pattern while the latter forms a triangular arrangement halfway between the neighbouring B-layers, or vice versa. The Mg atoms are intercalated at the centres of the boron hexagons. Each B atom is surrounded by three other equidistant (0.178 nm) B atoms, while the neighbouring atoms in the Mg plane are separated by the lattice parameter $a_0 = 0.308$ nm. The planar layers are separated by $c_0 = 0.351$ nm (Figure 15.2(b)). The

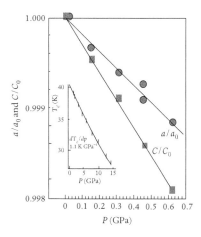

Figure 15.3 Variation of lattice parameters a and c (both normalized with respect to their corresponding ambient-pressure values) with applied pressure (after Schilling et al., 2002). The decrease of c/c_0 is about twice as fast as that of a/a_0. The inset shows T_c depression due to applied pressure (after Goncharov et al., 2002).

structure is anisotropic, and the anisotropy factor is found to vary from 2 to 5. Its coherence length is around 4–5 nm (Junod et al., 2002), which is, however, sufficiently large in comparison with interplanar distances. The material is therefore just anisotropic and not strictly quasi-two-dimensional.

The thermal expansion coefficient of MgB_2 along the c-axis is approximately double that along the basal plane, that is, $\alpha_c \approx 11.4 \times 10^{-6} K^{-1}$ and $\alpha_a \approx 5.4 \times 10^{-6} K^{-1}$ at 11–297 K (Jorgensen et al., 2001a), which is corroborated by others (Margadonna et al., 2001; Oikawa et al., 2002). As one might understand, this anisotropy is very important in MgB_2 conductor development, where the strains due to anisotropic thermal contraction during cooling need to be kept under control (Goldacker et al., 2003; Goldacker and Schlachter, 2005). Similarly, there is anisotropy in the lattice compressibility under imposed pressure (Schilling et al., 2002), with the c-parameter decreasing much more (Figure 15.3) than the a-parameter, implying that the in-plane Mg–Mg bonds (or B–B bonds) are stronger than the out-of-plane Mg–B bonds. As well as decreasing c_0, imposed pressure causes a lowering of T_c at varying rates with $dT_c/dp = -0.35$ to -1.6 K GPa^{-1} (Bordet et al., 2001; Lorenz et al., 2001b; Tissen et al., 2001; Tomita et al., 2001; Goncharov et al., 2001; Deemyad et al., 2001; Schlachter et al., 2002). The results of Goncharov et al. (2001) are shown in the inset of Figure 15.3. The decrease in T_c is in accord with conventional BCS theory and is due to a reduction in the density of states (DOS) at E_F resulting from the contraction of the Mg–B bonds.

The optimum $T_c \approx 39$ K of MgB_2 corresponds to the maximum value of the c_0 parameter, and the changes in the c-axis length are found to strongly influence the relevant E_{2g} phonon modes of the B plane, responsible for superconductivity. The presence of internal strains can increase c_0 and thereby enhance T_c of MgB_2 to 41.8 K (Muranaka et al., 2005). Systematic studies of $Mg_{1-x}Al_xB_2$ (Slusky et al., 2001) have revealed an intimate correlation between increasing Al content, depression in T_c, and a decrease in c_0. With Al substitution, for $0.1 < x < 0.25$, there exist two phases, however, which complicates the superconducting behaviour. A host of substitutional studies with Mn, Co, Fe, Cu, Zn, Al, Li, and so on at the Mg site have shown decrease in T_c (Felner, 2002), as depicted for some of these dopants in Figure 15.4. Substitution at the Mg^{2+} site is expected to introduce holes (for monovalent ions such as Li^{1+}) or electrons (for trivalent and higher-valent dopants). Both experimental data (Masui et al., 2002; Kim et al., 2003; Fisher et al., 2003) and band structure calculations (Rosner et al., 2002; Antropov et al., 2002) corroborate a hole-type superconductivity of MgB_2 and, therefore, from the conventional BCS mechanism, in general one expects its T_c to degrade with electron doping and rise with hole doping. This is found to be true with most of the above substitutions at

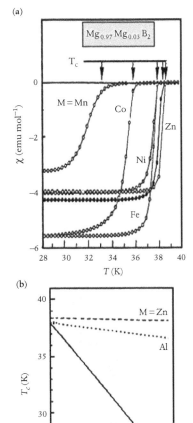

the Mg site. Surprisingly, the Li^{1+} substitution that introduces holes is also found to suppress T_c (Felner, 2002).

Since the B planes, as mentioned, are important for superconductivity, the T_c depression is expected to be more pronounced when dopants such as C, Si, O, and F are partially substituted at B sites. Because of its higher valence than B, C substitution in B planes is expected to promote electron doping. Magnetic susceptibility data (Felner, 2002) on C-doped samples are shown in Figure 15.5(a). The observed (Figure 15.5(b)) rate of T_c depression in $MgB_{2-x}C_x$ is much higher than in $Mg_{1-x}Al_xB_x$ (Figure 15.4(b)), although both these substitutions lead to electron doping respectively in the B and Mg planes (Felner, 2002). In all such substitutional studies carried out with any host material, in general, there is a discrepancy between the nominal and actual concentrations of dopants present in the sample. Because of this, the data reported by different authors often tend to differ. In the case of C-substituted MgB_2 samples, the inconsistency is greater for polycrystalline samples, where the dopant atoms, instead of occupying the B sites, may form clusters within the grain boundaries. Also, heat treatment may give rise to various impurity phases such as Mg_2C_3 and MgB_2C_2 (Muranaka et al., 2005), because of which the properties of MgB_2 may be greatly affected. C substitution at B sites markedly lowers T_c, but, due to the ensuing disorder, H_{c2} may be substantially raised and the high-field J_c at $T << T_c$ significantly improved (Muranaka et al., 2005). Similarly, O and Si substitution at B sites can dramatically enhance J_c at $T << T_c$ in high magnetic fields, although T_c is reduced (Eom et al., 2001). In contrast to this, the dopants Ti, Zr, and Al at Mg sites help to improve J_c in low magnetic fields (Zhao et al., 2003; Okabe, 2003). It is, however, found

Figure 15.4 (a) Magnetic susceptibility curves showing the effect of 3% substitution of Mg by Mn, Co, Ni, Fe, or Zn. (b) T_c depression resulting from substituting Zn, Al, or Mn at the Mg-site. (After Felner (2002).)

Figure 15.5 (a) Magnetic susceptibility curves of C-substituted (at B sites) MgB_2. (b) T_c depression of MgB_2 with C content. (After Felner (2002).)

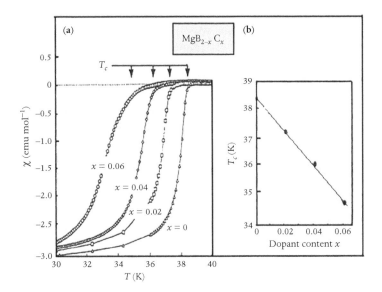

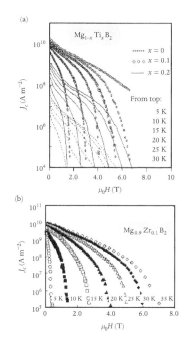

Figure 15.6 Critical current density in a magnetic field of MgB$_2$ samples doped at Mg sites with (a) Ti and (b) Zr (after Zhao et al., 2003).

that Ti and Zr give rise to particles of second phase, which promote strong flux pinning in low fields. The J_c increases resulting from Ti and Zr substitution (Zhao et al., 2003) are illustrated in Figures 15.6(a) and (b), respectively. Zhao et al. have found these substitutions to markedly increase the tolerance of J_c to water and humidity.

15.2 Conventional superconductivity of MgB$_2$

15.2.1 Isotope effect

Early support for a conventional phonon mechanism of superconductivity in MgB$_2$ came from measurements of the isotope effect (Bud'ko et al., 2001; Canfield et al., 2002). On replacing ^{11}B by ^{10}B, the critical temperature, as determined by resistivity, magnetic susceptibility, and specific heat measurements, increased by 1.0 K, giving a B-isotope effect coefficient $\alpha_B \approx 0.26$. The pertinent resistivity and magnetic susceptibility curves are shown in Figures 15.7(a) and (b), respectively. Similar studies by Hinks and co-workers (Hinks et al., 2001; Hinks and Jorgensen, 2003) found $\alpha_B \approx 0.30$ and $\alpha_{Mg} \approx 0.02$. The large value of α_B/α_{Mg} (≈ 15) indicates that superconductivity of MgB$_2$ is primarily driven by phonons from B planes, with Mg planes not being so relevant. Various considerations (Liu et al., 2001) showed that phonons having a pronounced coupling with the charge carriers at E_F belonged to the anharmonic planar B-stretching optic phonon mode (Figure 15.2) E_{2g} with an energy range of 58–82 meV. Choi et al. (2002a, b) pointed out that the lower value of $\alpha_B < 0.5$ than for a conventional superconductor was theoretically fully consistent with the BCS mechanism if the anharmonicity of E_{2g} phonons was taken into account. The large phonon frequency associated with the small B mass is therefore mainly responsible for the observed $T_c \approx 40$ K of MgB$_2$. Choi et al. further showed that, while the phonon anharmonicity brings down the T_c of MgB$_2$ from 55 K to 39 K, the anisotropy enhances it from 19 K to 39 K. Consequently, the observed high T_c of MgB$_2$ is possibly also a consequence of its anisotropy, and it could have been even higher if the phonon anharmonicity had been absent.

15.2.2 Support for conventional pairing

The conventional superconductivity of MgB$_2$, in the form of isotropic s-wave pairing, is further substantiated by a host of other experimental studies, such as ^{11}B-NMR (Kotegawa et al., 2001), photoemission spectroscopy (Takahashi et al., 2001), inelastic neutron scattering (Osborn et al., 2001), pressure effects on T_c (Schilling et al., 2002), tunnelling spectroscopy (Karapetrov et al., 2001; Sharoni et al., 2001), and penetration depth measurements (Niedermayer et al., 2002, Manzano et al.,

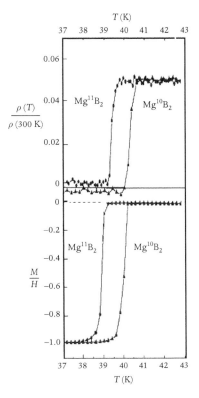

Figure 15.7 Boron-isotope effect of T_c manifested by resistivity and magnetic measurements at the superconducting transition (after Canfield et al., 2002).

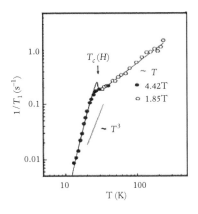

Figure 15.8 Temperature dependence of ^{11}B-NMR spin–lattice relaxation rate $1/T_1$ in MgB$_2$ measured in $\mu_0 H = 4.42$ and 1.85 T, showing a Hebel–Slichter resonance peak just below T_c (after Kotegawa et al., 2001).

2002; Ohishi et al., 2003). ^{11}B-NMR measurements (Kotegawa et al., 2001) reveal the presence of a Hebel–Slichter peak (Hebel and Slichter, 1959) just below T_c, followed by an exponential temperature dependence of the spin–lattice relaxation time (Figure 15.8). These are the characteristic features of the conventional s-wave superconductivity. Knight shift data from NMR experiments (Jung et al., 2001) showed a decrease in spin susceptibility below T_c consistent with the spin-singlet state. To check the role of phonons in the superconductivity of MgB$_2$, several authors (Osborn et al., 2001; Hinks et al., 2001; Muranaka et al., 2002) studied the phonon DOS using inelastic neutron scattering (INS), which was consistent with the calculated values. They could further predict theoretically that E_{2g} in-plane modes were strongly coupled to the planar B bands at E_F, and thus a strong connection between the phonon states and the ensuing superconductivity was indirectly evidenced.

High resolution photoemission spectroscopy (PES) of MgB$_2$ also corroborates the s-wave superconductivity (Takahashi et al., 2001). The leading edge of the PES spectrum of MgB$_2$ in the normal state ($T = 50$ K $> T_c = 39$ K) coincides with that of Au, used as a reference material, indicating the absence of a gap (Figure 15.9). On cooling to 15 K, in the superconducting state, the spectrum shifts to the higher-binding side, consistent with the opening up of a gap.

The leading edge of the superconducting MgB$_2$ spectrum, however, remains parallel to that of Au. This indicates the gap symmetry to be of isotropic s-wave type with a magnitude of about a few millielectronvolts. This is further confirmed by a numerical fit that corroborated the gap size of about 4.5 meV.

The pressure studies on MgB$_2$ showing a decrease in T_c under pressure were mentioned in Section 15.1. These measurements allow an accurate determination of the change in T_c with unit-cell volume V. Schilling et al. (2002) found their experimental pressure data of T_c to fit very well with the predictions of the McMillan formula, which shows that superconductivity in MgB$_2$ originates from the conventional BCS mechanism.

Although measurements of the penetration depth $\lambda(T)$ in MgB$_2$ initially yielded a power-law temperature dependence in the form T^n, with $n \approx 2$ or 2.7 (Chen et al., 2001b; Promin et al., 2001), contrary to the conventional behaviour, subsequent studies all yielded an exponential dependence, characteristic of conventional s-wave superconductivity (Niedermayer et al., 2002; Ohishi et al., 2003; Manzano et al., 2002; Carrington and Manzano, 2003). Specific heat data on MgB$_2$ are generally found to be complicated by the presence of multibands at E_F (see Section 15.3). Although some early reports (Wälti et al., 2001) found the measured data at $T << T_c$ to be governed by conventional isotropic s-wave superconductivity, it is now generally understood that a more involved analysis is needed that considers formation of two distinct

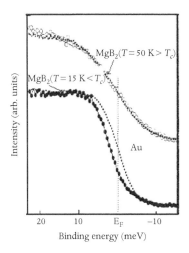

Figure 15.9 High-resolution photoemission spectroscopy (PES) studies of MgB_2 (after Takahashi et al., 2001). PES spectra of the sample in superconducting and normal states are compared with those of Au (see the text).

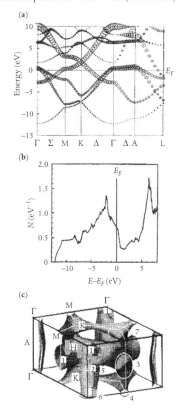

Figure 15.10 (a) Energy bands, (b) total DOS (after Antropov et al., 2002), and (c) Fermi-surface topology (after Kortus et al., 2001) of MgB_2.

energy gaps below T_c (Bouquet et al., 2001; Junod et al., 2002; Fisher et al., 2002b). The presence of two gaps was also observed in MgB_2 by scanning tunnelling spectroscopy, where the spectrum revealed two overlapping gaps (Karapetrov et al., 2002). Extensive tunnelling studies of MgB_2 made using point contact junctions with Cu (Szabó et al., 2001), Au (Gonnelli et al., 2002; Kohen and Deutscher, 2001, Schmidt et al., 2002), Pt (Gonnelli et al., 2002), Nb (Li et al., 2001), and In (Gonnelli et al., 2002) reveal the presence of two gaps with temperature dependence closely following the BCS theory.

15.3 Band structure and two superconducting gaps

Since the discovery of high T_c in magnesium diboride, its electronic band structure has been examined by many authors (An and Pickett, 2001; Rosner et al., 2002; Kortus et al. 2001; Antropov et al., 2002; Profita et al., 2002; Choi et al., 2002a, b), which underscores MgB_2 as perhaps the first confirmed example of a multiband superconductor. The honeycomb-like B planes that we have seen to be very important in the superconductivity of MgB_2 also dominate in their p-orbitals in its electronic structure. It has in-plane σ-bonding derived from B-p_x, p_y orbitals, and out-of-plane π-bonding from B-p_z orbitals. Calculations find that there are two incompletely filled 2D-σ-bands, with carriers as holes, and two 3D-π-bands, carrying electrons and holes, crossing the Fermi energy of MgB_2. The energy bands, the total DOS, and the ensuing Fermi surface topology are shown in Figures 15.10(a), (b), and (c), respectively. The Fermi surface is a complicated one in which the 3D-π-bands form tubular networks while the 2D-σ-bands give rise to 2D-cylindrical surfaces around the Γ–point. These two types of bands, σ and π, are found to nearly equally share the total electron DOS at E_F. As already mentioned, the phonons that couple strongly with the carriers at E_F are those belonging to the E_{2g} optic phonon mode with longitudinal bond stretching vibrations confined to the B plane. These in-plane vibrations are coupled much more strongly to the planar σ-bands than to 3D-π-bands. As a result of this highly anisotropic electron–phonon interaction resulting from the E_{2g} phonon mode, the former gives rise to a much larger superconducting gap of $2\Delta_{0\sigma} \approx 7$ meV, while the latter gives a smaller gap $2\Delta_{0\pi} \approx 2$ meV; both gaps appear (vanish) together below (above) $T_c \approx 39$ K (Figure 15.11), which indicates strong interband coupling. Both gaps obey the conventional BCS temperature dependence. The measured data using tunnelling, Raman spectroscopy, specific heat, etc. (Bouquet et al., 2001; Bohnen et al., 2001; Karapetrov et al., 2001; Laube et al., 2001; Schmidt et al., 2001; Szabó et al., 2001; Iavarone et al., 2002; Takasaki et al., 2002; Fisher et al., 2002a) are in full accord with the presence of a two-gap structure, giving $2\Delta_{0\sigma} / \kappa_B T_c \approx 3.8$–5.3 and $2\Delta_{0\pi} / \kappa_B T_c \approx 1.2$–1.5.

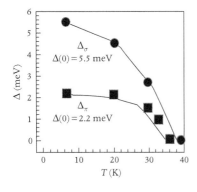

Figure 15.11 The two energy gaps in MgB$_2$ for σ- and π-bands, showing BCS-like temperature variation (after Tsuda et al., 2003).

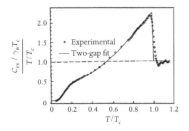

Figure 15.12 Specific heat of superconducting MgB$_2$ with a two-gap fit (after Fisher et al., 2002b).

Figure 15.12 shows (Fisher et al., 2002b) the specific heat behaviour in the two-gap structure. The transition to the superconducting state occurs in two stages, with the lower exponential decrease corresponding to the smaller gap.

Interestingly, this gap structure in MgB$_2$ is stable against interband impurity scattering, and the gaps will merge only when the impurity level or disorder is excessively large. Mazin et al. (2002) have argued that the 3D-π-bands are antisymmetric with respect to the B-plane, while the 2D-σ-bands are symmetric, with the result that the interband impurity scattering rate is substantially lower than the intraband scattering rate. What this means is that the impurities can influence the individual gaps, but nothing much will happen between them and the two gaps will remain intact.

15.4   Implications of two gaps

One of the prominent shortcomings of MgB$_2$ is that bulk samples possess a relatively low upper critical field of about 14 T, which is even smaller than of Nb–Ti alloys and approximately one-half that of of Nb$_3$Sn. To become competitive with these two technologically long-established superconductors one of the challenging problems is to enhance H_{c2} of MgB$_2$. As we saw in Chapter 5, in a conventional single-band superconductor, both $H_{c2}(0)/T_c$ (equation (5.19)) and the slope dH_{c2}/dT_c (close to T_c) scale with the normal-state resistivity ρ_n, and consequently the upper critical field can be increased by raising the normal-state resistivity, provided T_c and the coefficient of electronic specific heat γ do not decrease appreciably. This approach has proved useful for enhancing H_{c2}. For MgB$_2$, having two bands σ and π, some modification in the above criterion is to be expected. The conduction process in MgB$_2$, in the two-band model, is accordingly governed by the electron (or hole) diffusivities D_σ and D_π. By achieving pronounced decreases in D_σ and D_π, one may hope to realise a significant enhancement in the upper critical field.

Clearly, MgB$_2$ presents three distinct impurity scattering channels for enhancing resistivity: two individual intraband scattering channels in the σ- and π-bands and their interband scattering. The existence of these multiple scattering channels offers the possibility to enhance H_{c2} of MgB$_2$ far beyond the reach of the usual one-gap superconductors. For thin films of MgB$_2$, H_{c2} values in excess of 40 T (even 70 T) have already been reported (Liu et al., 2001; Wang et al., 2001; Gurevich et al., 2004), which is considered to be a consequence of its two-gap superconductivity. Through appropriate cationic substitutions for Mg and B, the diffusivities D_σ and D_π can be suitably increased to achieve a much larger upper critical field in MgB$_2$.

Another interesting effect arising from the two-gap structure in MgB_2 is that the anisotropy factor for H_{c2}, given by $\Lambda_{c2} = H_{c2}^{ab}/H_{c2}^{c}$, decreases from 5 at low temperatures to about 2 near T_c, while this change is usually less than 20% for a conventional single-gap superconductor. On the other hand, the anisotropy factor for H_{c1}, given by $\Lambda_{c1} = H_{c1}^{ab}/H_{c1}^{c}$, decreases with increasing temperature, with the result that the prediction of the Ginzburg–Landau theory, namely, $H_{c2}^{ab}/H_{c2}^{c} = H_{c1}^{c}/H_{c1}^{ab}$, no longer holds for MgB_2 (Cubitt et al., 2003; Xu et al., 2001). Finally, close to the T_c of MgB_2, H_{c2}^{c} varies linearly with temperature, while H_{c2}^{ab} shows a pronounced upward curvature, which is not observed in a single-gap superconductor. All these unusual features of MgB_2 can be explained in terms of its two-gap structure, the widely different anisotropies of its two Fermi surfaces, and its interband coupling interaction of moderate strength (Dahm and Schopohl 2003).

The influence of the presence of two energy gaps in MgB_2 is revealed also in an anomalous coherence peak (Figure 15.13) in the temperature dependence of the real part of the microwave conductivity σ_s^{MW}/σ_N (normalised to its value just above T_c) of thin films at $T \approx 0.5T_c$ (Jin et al., 2003). In conventional superconductors, such a peak is analogous to the Hebel–Slichter peak in the temperature dependence of the NMR spin relaxation rate and is normally observed near $0.9T_c$, as shown for Nb in Figure 15.13. In MgB_2, the presence of the lower gap is believed to be responsible for shifting the peak downwards on the temperature scale (Jin et al., 2003).

Recent observations (Moshchalkov et al., 2009) on MgB_2 made using the Bitter pattern technique have revealed an unconventional vortex structure that appears as a combination of the intermediate-state structure of a type I superconductor and the characteristic Abrikosov mixed-state structure of a type II superconductor. The observed patterns are stripe- and gossamer-like, and are attributed to the simultaneous presence of type I superconductivity arising from π-bands and type II superconductivity from σ-bands. Because of such an unusual superconductivity, MgB_2 has been suggested to be a *type-1.5 superconductor* (Moshchalkov et al., 2009), which is neither fully type I nor fully type II, but in-between the two categories.

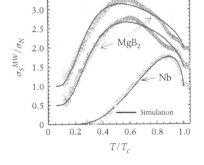

Figure 15.13 Normalised microwave conductivity of two typical thin-film samples of MgB_2 as a function of reduced temperature (after Jin et al., 2003). Also shown for comparison are the data for Nb.

15.5 MgB_2 for practical applications

MgB_2 possesses numerous attractive features for technological use. Among intermetallic compounds, it possesses the highest $T_c = 39$ K, which is sufficiently large to achieve an acceptable performance at $T \approx 20$ K using either a liquid hydrogen bath cryostat, which is much cheaper than using liquid helium, or any conventional closed-cycle refrigerator. Because of its large T_c and high H_{c2} (in thin films), which exceed those of both Nb_3Sn ($T_c = 18.5$ K, $\mu_0 H_{c2} = 24$ T) and Nb–Ti alloys ($T_c \approx 10$ K, $\mu_0 H_{c2} = 14$ T), MgB_2 is a serious competitor for these two superconductors that have dominated technological applications for more than 40 years. Also, the material is much lighter in weight, which carries special advantages for its use in large superconducting installations. But when we compare MgB_2 with the above two Nb-based materials, we realise also its prominent drawbacks. MgB_2 is a brittle intermetallic, whereas Nb–Ti forms ductile alloys that are mechanically compatible with Cu for their co-processing. This allows ready fabrication of Cu-stabilized multifilamentary conductors of Nb–Ti. Such is not the case with MgB_2. In the case of Nb_3Sn, which is also a brittle intermetallic, there exist many ingenuous fabrication routes such as the bronze process, internal Sn diffusion, and the *in situ* alloy process that allow easy fabrication of stabilized multifilamentary conductors in which several thousands of very fine Nb_3Sn filaments are embedded in bronze wires and tapes. There are no similar fabrication routes for making long lengths of MgB_2 conductors, and instead one is mostly required to follow the powder-in-tube process, which has many limitations. Because of these fabrication-related hurdles, MgB_2 has so far not been able to supersede Nb–Ti and Nb_3Sn.

A strong aspect of MgB_2 is that, unlike HTS cuprates, it exhibits no grain-boundary weak-link effect. The weak-link behaviour is prevented in MgB_2 by its much greater coherence length ($\xi \approx 5$ nm) and a significantly lower electrical resistance of the grain boundaries. Figure 15.14 depicts a number of normalised conductance spectra at room temperature obtained at the centres of differently located grain boundaries (Narlikar et al., 2002), along with a typical spectrum observed in the middle of one of the adjoining grains. All the spectra observed are characteristic V-shaped, typical metallic type, and the normalised conductance near $V = 0$ for the boundaries does not noticeably vary among them and is only marginally lower than for the grain interior. This suggests that the amorphous grain-boundary region in MgB_2 is of low-resistivity metallic type and the grains are mutually coupled through S–N–S-type junctions having only a marginal effect in suppressing J_c. The reason for this may be traced to the high normal-state electrical conductivity of MgB_2 and also to the semimetallic type of band structure that it possesses. In the case of semimetals such as Bi, Si, and Ge, for example, the amorphous state exhibits a metallic character with enhanced conductivity. As a result, the grain boundaries in MgB_2, having width greater than $\xi \approx 5$ nm, continue to remain transparent to the flow of electrical current, which is an important factor for realising high J_c. The role of grain boundaries in MgB_2 is therefore primarily to serve as pinning entities for enhancing J_c, which is realised by synthesising samples having very small grain size. Other potential defects present in MgB_2, serving as pinning centres, are fine precipitates of

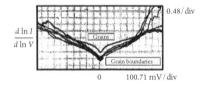

Figure 15.14 Conductance spectra obtained within the amorphous grain boundary regions and also within the crystalline grain, all showing typical metallic behaviour (after Narlikar et al., 2002).

Table 15.1 Comparison of the characteristic parameters of Nb_3Sn, YBCO, and MgB_2

Superconductor	T_c (K)	θ_D (K)	$2\Delta_0/k_BT_c$	μ_0H_c (T)	μ_0H_{c2} (T)	μ_0H_{c1} (T)	$-\mu_0(dH_{c2}/dT)_{T_c}$ (T K^{-1})	$\lambda(0)$ (nm)	$\xi(0)$ (nm)	κ
Nb_3Sn	18.5	230	4.8	0.52	24	0.13	1.6	39	11.5	3.5
YBCO	91	425	5.0	1.0	150	0.03	2.3	150	0.3–2.0	> 100
MgB_2	39	850–1050	1.2–4.8	0.26	14–16	0.018	0.56	185	4.9	38

Data based on Junod et al. (2002) and Muranaka et al. (2005).

MgO. Stacking faults and crystal dislocations have also been seen in thin films of MgB_2 by TEM, which can provide further pinning (Zhu et al., 2002). Additional advantages of MgB_2 are its high carrier density and simple binary composition free from costly or rare ingredients. Further, being an intermetallic, MgB_2 is free from non-stoichiometry, which results in readily reproducible T_c and comparative insensitivity to residual normal-state resistivity. These factors are favourable for developing MgB_2 as a practical conductor for technological applications. Some of the characteristic parameters of MgB_2 are compared with those of Nb_3Sn and YBCO in Table 15.1.

15.6   Material synthesis

MgB_2 in bulk form is synthesised by two separate approaches called *ex situ* and *in situ*. The *ex situ* method uses commercially produced MgB_2 powder as the starting material, while the *in situ* method starts with unreacted Mg and B. In the former, sintering has to be carried out under high pressures of several gigapascals to form strong bonds between MgB_2 particles. The density of the sintered product is almost ideal and the material is mostly free from pores and voids. The grain size of such samples generally varies from 1 to 5 μm in diameter and the zero field J_c from 10^6 to 10^9 A m^{-2} at T = 20 K and 5 K, respectively. In the presence of a magnetic field, J_c is typically around 10^8 A m^{-2} at 5 T, 5 K and at 3 T, 20 K. However, the commercial MgB_2 powder used may contain impurities that can degrade T_c below 39 K and thereby adversely affect also other properties.

With the *in situ* method, by using B and Mg of higher purity, one may produce much purer MgB_2 than is available from the *ex situ* method. Also, the grain size is generally smaller than 0.5 μm. Pressed powder mixtures of Mg and amorphous (microcrystalline) B, with Mg in excess of the required stoichiometry of MgB_2, sealed inside a Ta tube, are made to react in the temperature range 550–1200°C. The excess Mg ensures that there is adequate Mg available to allow the complete conversion of B into MgB_2. The melting temperature of Mg being 650°C, the reaction at lower temperatures requires longer heating times. In the low-temperature reaction, the grain size of the reacted material formed

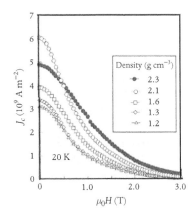

Figure 15.15 $J_c(H)$ of bulk MgB$_2$ samples at 20 K possessing varying densities (after Muranaka et al., 2005).

is determined primarily by the microcrystalline grain size of the B powder used, while, above 650°C, both temperature and reaction time are relevant. For preparing bulk samples, the reaction temperature commonly used is 950°C, with a heating time of 2–5 hours. A drawback of *in situ* processed samples is that their density is almost half of the ideal density realised with the *ex situ* method under pressure. The reason for this is that MgB$_2$ is much denser than Mg and B, and formation of MgB$_2$ by the reaction of Mg + 2B is therefore accompanied by about 27% volume contraction. The actual sample density is only about 73% of the theoretical density, and such materials contain a large concentration of voids and wide grain boundaries with poor intergrain connectivity. These are serious issues that limit J_c, as seen in Figure 15.15 (Muranaka et al., 2005). Consequently, much effort has been directed at obtaining dense samples of MgB$_2$ with strong intergrain connectivity.

A successful *in situ* approach to obtain dense samples of MgB$_2$, without the application of external pressure, known as the *reactive Mg-liquid infiltration (Mg-RLI) process*, is due to Giunchi et al. (2006). The process involves filling a tubular stainless steel container with a solid Mg rod at its centre, surrounded by microcrystalline B powder in the ratio of 1:2. After welding the container, it is annealed in a conventional furnace at 950°C for 3 hours to form MgB$_2$ by infiltration of molten Mg into the surrounding B powder and their mutual reaction. When the reaction is complete, the B space is completely occupied by MgB$_2$ with the original Mg space being vacant. The sample density of MgB$_2$ realised this way is 2.4 g cm^{-3}, as against the theoretical value of 2.6 g cm^{-3}. The process shown to be compatible with fabrication of MgB$_2$ cables.

15.7   Nanoparticle doping for enhancing J_c

The relatively large $\xi \approx 5$ nm of MgB$_2$ allows optimum pinning to be achieved by adding artificial pinning centres (APCs) in the form of nanoparticles of comparable size for realising improved $J_c(H)$ performance at a more convenient temperature of $T \approx 20$ K. Naturally formed crystal defects in pure MgB$_2$, such as grain boundaries, dislocations, and stacking faults, are inadequate for meeting the desired level of $J_c(H)$ performance, and consequently nanoparticles of numerous materials have been introduced in MgB$_2$ as APCs (Dou et al., 2002; Rui et al., 2004; Sumption et al., 2005; Lezza et al., 2006; Yeoh et al., 2006a) for achieving $J_c(H)$ of up to 10^9 and 10^8 A/m^2 at 10 and 20 K, respectively, in an applied field of 5 T (Kim et al., 2006; Yamamoto et al., 2005). The various APCs tried include nano(n)-SiC (Dou et al., 2002; Soltanian et al., 2003), n-SiO$_2$ (Rui et al., 2004), n-TiO$_2$ (Xu et al., 2004), ZrO$_2$ (Chen et al., 2004), Y$_2$O$_3$ (Wang et al., 2002), n-Co$_3$O$_4$ (Awana et al., 2006), n-carbon (Yeoh et al., 2006a), carbon nanotubes (CNTs) (Yeoh et al., 2006b), and

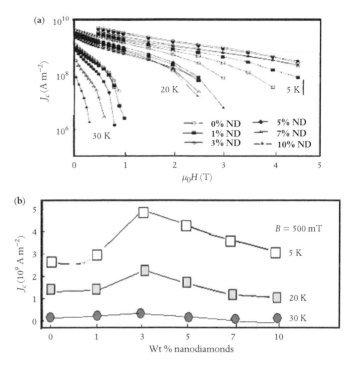

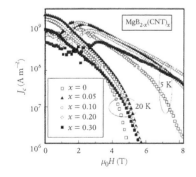

Figure 15.16 (a) $J_c(H)$ behaviour of MgB$_2$ doped with nanodiamonds (ND). (b) In low fields, the optimum J_c is observed at 3% doping (after Vajpayee et al., 2007).

Figure 15.17 Critical current curves at 5 K and 20 K for MgB$_2$ samples formed by doping with carbon nanotubes at varying amounts x at the B site (after Dou et al., 2003). As may be seen, less than 20% doping improves the critical current performance.

n-diamond (nD) (Cheng et al., 2003; Vajpayee et al., 2007). Figure 15.16(a) illustrates the $J_c(H)$ behaviour (Vajpayee et al., 2007) of nD-doped MgB$_2$ samples: MgB$_2$–nD$_x$, with the concentration x of nD varying from 0 to 10 wt%. J_c values at 5 K, 20 K, and 30 K were estimated through magnetisation using Bean's critical-state model. Interestingly, $J_c(H)$ plots at 5 K show the pristine sample to carry the lowest current at all applied fields, while at 20 and 30 K the samples with nD concentration greater than 3% exhibit an inferior performance, possibly due to their lower T_c. As may be seen in Figure 15.16(b), at all temperatures and at a fixed low magnetic field (500 mT), the optimum J_c is realised for 3% nD. At both 5 K and 20 K, J_c of the 3% nD sample at low fields is nearly twice that of pristine MgB$_2$ which indicates the importance of incorporating nanodiamonds for enhancing J_c. Similar behaviour is observed for CNT-doped MgB$_{2-x}$(CNT)$_x$ samples depicted in Figure 15.17 (Dou et al., 2003). The results show a broad range of CNT concentration between 5 and 20 wt % for optimum J_c. As may be seen, the maximum J_c observed with this doping exceeds 10^8 A m^{-2} at 20 K and 4 T or at 5 K and 8 T.

Nanoparticles of various silicides such as SiO$_2$, MgSi$_2$, ZrSi$_2$, and WSi$_2$ have been found to serve as effective APCs in MgB$_2$ (Ma et al., 2003) to enhance J_c. They have, however, little chemical affinity with Mg or B, and these particles therefore serve more as additives. Similarly, nanoparticles of Y$_2$O$_3$ have been found to increase H_{irr} at 4.2 K, although the effect is much less at 20 K (Wang et al., 2002). Dou et al. (2005) found n-BN to be a successful APC for J_c enhancement of MgB$_2$.

Extensive work by Dou and his collaborators, reported in some of the references cited above, has established n-SiC as the most successful dopant for improving both H_{irr} and $J_c(H)$ of bulk MgB$_2$, a contention that has since been corroborated by other authors (Matsumoto et al., 2003; Sumption et al., 2004). The significant features of SiC doping are, first, that it leads to J_c enhancement at both low and high fields, with the increase being by more than an order of magnitude at certain fields, and, second, that the J_c enhancement takes place at all temperatures up to T_c. These special features have been related to interband scattering effects due to the presence of two gaps as well as much improved flux pinning (Wang et al., 2001; Gurevich et al., 2004). During sintering, SiC particles partially decompose to form Mg$_2$Si, which provides additional pinning entities, while some of the C atoms released substitute at the B sites, which enhances H_{c2}. In addition, the SiC particles suppress grain growth and at their interface with the matrix give rise to a large dislocation density, which can provide strong pinning (Mikheenko et al., 2007). The $J_c(H)$ behaviour of SiC-doped MgB$_2$ is depicted in Figure 15.18 (Dou et al., 2002). As may be seen, doping with 10 wt % SiC raised J_c to 10^9 A m^{-2} at 3 T and 20 K and $H_{irr} \approx 8$ T, although the density of the sample was just 50% of the theoretical density. By decreasing the particle size of the precursor powder of SiC, the performance is significantly improved and the optimum behaviour is realised when the particle size used is least (about 20 nm diameter). However, agglomeration of such fine particles at grain boundaries can significantly lower J_c by the weak-link effect. In Figure 15.19, the $J_c(H)$ performance of 10 wt% SiC-doped MgB$_2$ (Dou et al., 2002) in the form of nanoparticles is compared with that of Ti-doped (Zhao et al., 2001) and Y$_2$O$_3$-doped (Wang et al., 2002) MgB$_2$, as well as with MgB$_2$ thin film (Eom et al., 2001) and Fe-clad MgB$_2$ tape

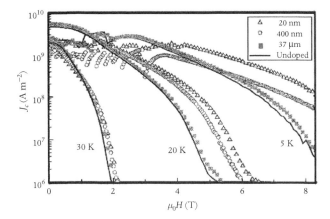

Figure 15.18 Critical current curves for MgB$_2$ samples doped with 10 wt% SiC particles of size ranging from 20 nm to 37 µm, measured at 5, 20, and 30 K (after Dou et al., 2005). Also shown are the curves for undoped MgB$_2$.

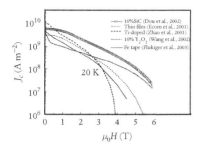

Figure 15.19 Critical current curves for various MgB$_2$ samples studied by different workers (after Dou et al., 2005). The general superiority of the SiC-doped sample may be noted.

(Flükiger et al., 2003). The overall superiority of SiC-doped MgB$_2$ above 1 T field is clearly corroborated.

15.8 Conductor development: wires and tapes of MgB$_2$

Soon after the discovery of superconductivity in MgB$_2$ in 2001, the material was prepared (Canfield et al., 2002) in a wire form of short segments by transforming B fibres into MgB$_2$ by subjecting them to Mg vapour at 950°C. However, for fabricating continuous long lengths of stabilized MgB$_2$ conductors for practical applications, there have been mainly two approaches that have been followed: the *powder-in-tube (PIT)* process and the *coated-conductor* process. The PIT process for MgB$_2$ has been developed for both *ex situ* and *in situ* approaches (Glowacki and Majoros, 2002). The highest-performing MgB$_2$ wires result from the *in situ* approach, but for the fabrication of superconducting magnet prototypes, one has to pursue the operationally difficult *wind-and-react technology*. On the other hand, the *ex situ* approach is found to be more convenient for manufacturing long lengths of MgB$_2$, albeit their performance is not as good. But the conductors fabricated using this approach have demonstrated their suitability for winding magnets through more friendly *react-and-wind technology* (Braccini et al., 2007).

There are a number of approaches for producing coated conductors, including the doctor blade technique (Chen et al., 1995), screen printing (Kang et al., 1996), spraying (Baker et al., 1994), and ink-jet printing (Glowacki, 2000), suitably adapted for MgB$_2$.

The *ex situ* and *in situ* PIT processes are schematically illustrated in Figures 15.20(a) and (b), respectively. In the former, the starting material is a compacted powder of commercial MgB$_2$, while with the latter it is a compacted powder of Mg + 2B. The tube used for the PIT is of Nb or Ta, which is inserted into the outer Cu tube to prevent reaction of the powder with Cu. The composite tube is swaged and drawn down to around 2 mm diameter with an intermediate annealing. The wire is next inserted in a stainless-steel jacket (for mechanical reinforcement) before being deformed to a final diameter of about 0.8–1.5 mm. The wire is then sintered at 875–950°C in an Ar/H$_2$(5%) atmosphere for 2–3 hours. Instead of Nb or Ta along with Cu, one may use Fe. Tubes of Ag and of Cu–Ni have also been used in the PIT process (Glowacki and Majoros, 2002). Since Fe is much cheaper and reacts only marginally with Mg/MgB$_2$, it is an attractive sheath material. By packing together single-core wires in a separate billet, followed by swaging and drawing, multifilamentary wires of MgB$_2$ have also been produced (Goldacker et al., 2003), but performance-wise they are inferior.

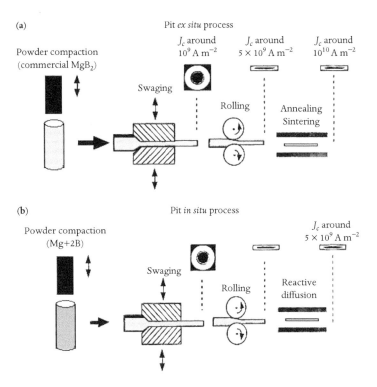

Figure 15.20 PIT process for MgB_2 conductor development: (a) *ex situ*; (b) *in situ* (after Glowacki and Majoros, 2002).

Interestingly, the presence of a ferromagnetic Fe sheath around the MgB_2 core is found to promote a significant enhancement of dissipation-free supercurrent densities that exceed that of the bare wire (with iron sheath denuded) by more than an order of magnitude (Pan et al., 2003). This phenomenon, which is observed in transverse magnetic fields, has been termed an *overcritical state*, and was first theoretically predicted by Genenko et al. (2000) for a thin superconducting strip in a ferromagnetic environment. The observed behaviour occurs owing to the presence of the magnetic environment, causing a redistribution of supercurrents from the sample edge to its interior, which allows an enhanced current flow in the sample, giving an *overcritical current density* $J_c^{oc} > J_c$.

To make coated conductors of MgB_2 with the *in situ* approach, a suspension of mixed Mg + 2B powders is prepared using acetone. By spreading the suspension on a stainless-steel strip and allowing it to evaporate, a uniform coating of powder mixture is formed on the stainless-steel surface. The deposited layer is made denser by applying pressure on its surface and is covered by another stainless-steel strip of the same size. The sandwich thus formed is rolled down (Figure 15.21) to the required thickness and is then subjected to reactive sintering to form a composite conductor of stainless steel/MgB_2/stainless steel, as may be seen at the bottom right of Figure 15.21. In the *ex situ* approach,

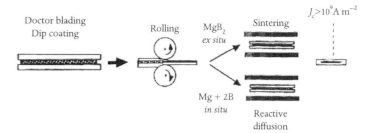

Figure 15.21 Coated-conductor route for MgB_2 (after Glowacki and Majoros, 2002).

the suspension used for coating the stainless-steel strip is of commercial MgB_2 powder and the sandwich containing the MgB_2 is subjected to sintering after the final rolling is complete, as shown in the upper right of Figure 15.21. Thick films (thickness > 1 μm) of 90% MgB_2 + 10% MgO with a good intergrain connectivity are speedily formed at $T = 660°C$, followed by rapid cooling to room temperature, and their $J_c \approx 8 \times 10^8$ A m^{-2} at 5 K and 1 T (Yao et al., 2004).

15.9 Summary

We have reviewed the scientific advances and technological progress achieved with MgB_2 since the discovery of its superconductivity in 2001. Its T_c of 39 K stands at the peak among intermetallics and exceeds the generally accepted BCS limit, and yet the weight of experimental evidence presented here corroborates its superconductivity as being driven by the conventional phonon mechanism. MgB_2 represents the first known example of a multiband superconductor possessing two energy gaps, as experimentally confirmed through several spectroscopic techniques. Its higher critical temperature and upper critical field have made MgB_2 a serious competitor for both Nb–Ti and Nb_3Sn, which have been the backbone of the superconductor technology for the last four decades. Although commendable progress has been made with conductor fabrication using MgB_2, further attention needs to be directed at developing conductors with c-axis texturing and at producing multifilamentary configurations having a large number of fine, uniform, and very dense filaments of MgB_2 embedded in a metallic matrix. The recent development of cables formed by the reactive Mg-liquid infiltration process, briefly described in this chapter, might provide one such approach.

CHAPTER 16

High-temperature cuprate superconductors

The HTS cuprates arrived most unexpectedly in 1986. Until this time, the highest T_c of 23.2 K had been held for the preceding 12 years by the A-15 superconductor Nb_3Ge. Although during this period many novel superconductors had evolved with unusual properties, none possessed a higher T_c. In 1986, what emerged as a most unexpected break-through in T_c was the extraordinary discovery of Bednorz and Müller (1986) announcing superconductivity in a mixed metal copper oxide $La_{2-x}Ba_xCuO_{4-\delta}$ ($x = 0.15$) (in short La-214) at $T_c \approx 30$ K. Hitherto, no one had thought a ceramic cuprate formed with a lanthanide element would turn out to be a superconductor, let alone one possessing such a high T_c. The discovery set off an unprecedented explosion of research on the new type of such materials on a global scale that led to a dramatic further increase in T_c. More than 100 HTS cuprate systems have since been discovered with T_c exceeding 30 K and rising up to 138 K for $Hg_{0.8}Tl_{0.2}Ba_2Ca_2Cu_3O_8$ (in short Hg-1223) (Sun et al., 1994; Lokshin et al., 2001; Putilin et al., 2001), which is the highest-T_c ambient-pressure superconductor known to date. The T_c of Hg-1223 itself got pushed up from 134 K to 164 K under a pressure of about 30 GPa (Putilin et al., 1993; Schilling et al., 1993; Chu et al., 1993; Nuñez-Regueiro et al., 1993). This is already more than halfway to room temperature, and there seem no reasons to preclude superconductivity occurring up to yet higher temperatures, even at room temperature, in the near future. Although the overall progress made with HTS cuprates since 1986 has been enormous, their fundamental understanding and wide-scale technological exploitation are still formidable challenges. Various factors such as their anisotropy, grain-boundary weak-link effects, and inadequate flux-pinning centres have proved to be major constraints in realising their practical application. Nevertheless these problems are being surmounted with novel fabrication approaches that form the main subject of Chapter 17.

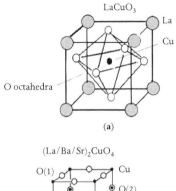

LaCuO$_3$

O octahedra

(a)

(La/Ba/Sr)$_2$CuO$_4$

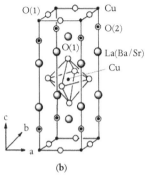

(b)

Figure 16.1 Crystal structures: (a) perovskite LaCuO$_3$; (b) perovskite-related K$_2$NiF$_4$ structure of La$_2$CuO$_4$.

16.1   Genesis of HTS cuprates

Bednorz and Müller's (1986) discovery of high-T_c cuprates was a result of applying chemical considerations and cationic substitutions rather than seriously invoking any complex physical theory. Taking full advantage of the mixed valence of copper, their reasoning led them to a partial substitution of La^{3+} by Ba^{2+} in La$_2$CuO$_4$, whereby they could readily vary the ratio of Cu^{3+}/Cu^{2+} in a controlled fashion. Accordingly, in contrast to the non-superconducting metallic perovskite LaCuO$_3$ (ABO$_3$ type; Figure 16.1(a)) with Cu^{3+}, and the insulating La$_2$CuO$_4$, possessing the perovskite-related K$_2$NiF$_4$ structure (Figure 16.1(b)) with Cu^{2+}, the substituted compound La$_{1.84}$Ba$_{0.16}$CuO$_4$ (in short LBCO) with Cu$^{2.15+}$ became metallic and superconducting at $T_c \approx 30$ K. The T_c of La$_{1.85}$M$_{0.17}$CuO$_{4-\delta}$ was raised to 38 K (Takagi et al., 1987; Cava et al., 1987) when M = Ba^{2+} was replaced by another alkaline earth, Sr^{2+}, of slightly smaller radius (in short LSCO), but the T_c fell to below 20 K for M = Ca^{2+} (Kishio et al., 1987) and Na$^+$ (Markert et al., 1988; Subramanian et al., 1988). The general formula of these materials is La$_{2-x}$M$_x$CuO$_4$ (M = Ba, Sr, Ca, or Na; $x = 0.15$–0.17). In all of the above instances with M having a smaller valence than of La, the cation substitution gave rise to *hole-doped* La-214 superconductors. The first *electron-doped* analogue of this cuprate system (Tokura et al., 1989; Markert and Maple, 1989; Markert et al., 1989) was R$_{2-x}$M$_x$CuO$_{4-\delta}$ (R = rare earth Pr^{3+}, Nd^{3+}, Sm^{3+}, or Eu^{3+}; M = Ce^{4+} or Th^{4+}; $x \approx 0.1 - 0.18$ and $\delta \approx 0.02$), with the highest T_c of about 25 K for NCCO (i.e. Nd$_{2-x}$Ce$_x$CuO$_{4-\delta}$). Cava et al. (1990a) were able to add another Cu–O plane to LSCO by synthesising La$_{2-x}$Sr$_x$CaCu$_2$O$_6$, with $x = 0.35$, exhibiting hole superconductivity at 60 K.

Successful clues to aid the search for new HTS cuprate systems came from pressure studies on LBCO. Chu et al. (1987a) noted a large and positive pressure coefficient of T_c for LBCO, where the onset of superconductivity under external pressure occurred at an exceptionally high temperature of 52.5 K (Chu et al., 1987b). They simulated this chemically (internal pressure) in LBCO by replacing La^{3+} with the smaller Y^{3+} ion. This way, they succeeded in achieving a dramatic leap, raising T_c to about 92 K for the YBCO cuprate (Wu et al., 1987) which became the first superconductor at liquid nitrogen temperature. The high-T_c phase was identified as a triple-layer perovskite, namely BaCuO$_3$YCuO$_3$BaCuO$_3$, but possessing some oxygen deficiency, with its composition being expressed by YBa$_2$Cu$_3$O$_{7-\delta}$ (called Y-123 or YBCO). Its layered crystal structure, depicted in Figure 16.2(a), comprises double Cu–O planes and a single plane carrying Cu–O chains. A relatively marginal decrease in oxygen content from O$_7$ to O$_{6.3}$ completely quenches superconductivity, making the material antiferromagnetic and changing the structure from orthorhombic to

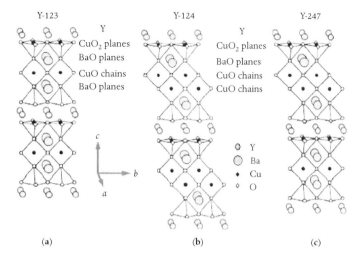

Figure 16.2 Crystal structures of (a) Y-123, (b)Y-124, and (c) Y-247.

tetragonal. It was soon realised that the formation of the high T_c phase was not confined to Y^{3+} alone, but, in fact, superconducting R-123 readily formed (Hor et al., 1991) with all lanthanides R, except Ce and Tb, which resulted in non-superconducting multiphases. This is generally attributed to their relatively bigger ion size and the possibility of their valence state being 4+. In the case of Pr-123 (PBCO), except for a rare observation of high-temperature unstable superconductivity that gradually disappeared with time (Zou et al., 1998), in the majority of instances the synthesised material has been found to be insulating. $RBa_2Cu_4O_8$ (i.e. R-124, where R is any lanthanide other than Ce, Tb, or Pr) with T_c ≈ 80 K was subsequently discovered, which is an analogue of R-123 possessing an additional plane containing Cu–O chains (Tallon et al., 1990), while a combination of 123 and 124 systems evolved as an ordered intergrowth in the form of the Y-247 system with $T_c \approx 95$K (Genoud et al., 1994). The crystal structures of Y-124 and Y-247 are shown in Figure 16.2(b) and (c), respectively.

The exciting development of the lanthanide-based compounds was shortly followed by the advent of analogous HTS cuprate systems that were synthesised free from any lanthanide ions.

Michel et al. (1987) had found T_c of Bi–Sr–Cu–O compounds to range from 7 K to 22 K (later it was raised to 34 K), which was suggestive of replacement of La by Bi in LSCO, although the resulting crystal structure was different from the original K_2NiF_4 structure. The structure of the single-CuO-layer ($n = 1$) compound $Bi_2Sr_2CuO_6$ (i.e. Bi-2201) is illustrated in Figure 16.3(a). Subsequently Maeda et al. (1988) found a substantially enhanced $T_c \approx 80$–110 K for Bi–Sr–Ca–Cu–O (in short BSCCO) with the generic formula $Bi_2Sr_2Ca_{n-1}Cu_nO_{2n+4}$. For $Bi_2Sr_2CaCu_2O_8$ (i.e. Bi-2212), having two ($n = 2$) Cu–O layers (Figure 16.3(b)), T_c ranged from 90 K to 95 K, while for $Bi_2Sr_2Ca_2Cu_3O_{10}$

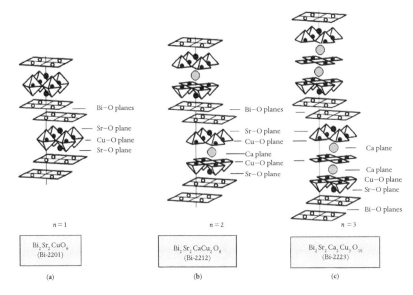

Figure 16.3 Crystal structures of BSCCO superconductors: (a) Bi-2201, (b) Bi-2212, and (c) Bi-2223.

(i.e. Bi-2223), containing three ($n = 3$) Cu–O planes (Figure 16.3(c)), T_c was 110 K (Zandbergen et al., 1988). Partial substitution of Bi by Pb, i.e., $(BiPb)_2Sr_2Ca_2Cu_3O_{10}$, did not affect T_c, but helped its phase stability in processing (Sunshine et al., 1988). By completely replacing Bi by Tl and Sr by Ba, that is, Tl–Ba–Ca–Cu–O (in short TBCCO), one may have single- or double-Tl-layered compounds (Sheng and Hermann, 1988) with generic chemical compositions $TlBa_2Ca_{n-1}Cu_nO_{2n+3}$ and

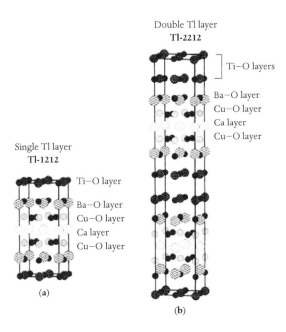

Figure 16.4 Crystal structure of (a) the single-Tl-layer compound Tl-1212 and (b) the double-Tl-layer compound Tl-2212.

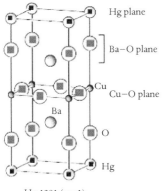

Hg plane

Ba–O plane

Cu–O plane

Cu

O

Hg

Ba

Hg-1201 ($n = 1$)

(a)

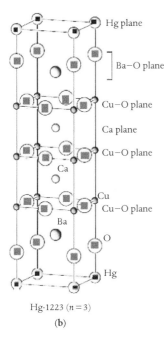

Hg plane

Ba–O plane

Cu–O plane

Ca plane

Cu–O plane

Cu–O plane

Ca

Cu

Ba

O

Hg

Hg-1223 ($n = 3$)

(b)

Figure 16.5 Crystal structure of (a) Hg-1201 with a single Cu–O layer and (b) Hg-1223 with three Cu–O layers and possessing the optimum $T_c = 134$ K.

$Tl_2Ba_2Ca_{n-1}Cu_nO_{2n+4}$, respectively. As with the BSCCO system, the nominal composition and processing schedule determine the T_c. and structural details of the synthesised phases of the TBCCO system. Among single-Tl-layer compounds, Tl-1223 and Tl-1234 are noted for their high T_c values of 133 K (Iyo et al., 2001a) and 127 K (Iyo et al., 2001b), respectively, while, in the double-Tl-layer class, Tl-2223 and Tl-2234 possess T_c of 128 K (Tristan Jover et al., 1996) and 119 K (Chen et al., 1993), respectively. The crystal structures of the single- and double-layered compounds Tl-1212 and Tl-2212 are depicted in Figure 16.4(a) and (b), respectively.

The search for new HTS materials with even higher T_c values culminated in 1993, when Tl was replaced by Hg, leading to single- and double-layered Hg systems $HgBa_2Ca_{n-1}Cu_nO_{2n+3}$ and $Hg_2Ba_2Ca_{n-1}Cu_nO_{2n+4}$ (Putilin et al., 1993; Schilling et al., 1993). As already mentioned, the single-Hg-layered compound Hg-1223 formed with three Cu–O planes ($n = 3$), which otherwise exhibits superconductivity at 134 K, has so far manifested the highest $T_c = 138$ K when Hg is marginally doped with 0.2% Tl. The crystal structures of Hg-1201 and Hg-1223 are presented in Figure 16.5(a) and (b), respectively. Although many high-temperature superconductors have since been discovered, for example $AuBa_2CaCu_2O_7$ ($T_c = 82$K; Bordet et al., 1997), $Pb_2Sr_2(R, Ca)Cu_3O_8$ (R = Y, Eu, Sm, or Pr, $T_c = 77$ K; Cava et al., 1988), $PbSr_2Ca_2Cu_3O_9$ ($T_c = 122$ K; Tamura et al., 1997), and $CuBa_2Ca_3Cu_4O_{11}$ ($T_c = 117$; Ihara, 2001a), $Cu_2Ba_2Ca_3Cu_4O_{12}$ ($T_c = 113$; Chu, 1997), so far none has surpassed the T_c of Hg-1223. Prominent HTS systems discovered to date are listed in Table 16.1.

16.2 General features of HTS cuprates

16.2.1 Structure

As indicated in Table 16.1, the HTS cuprates are either tetragonal or orthorhombic with their unit cells formed by stacking various layers along the c-direction which makes the c-parameter much longer than the other two. One or more Cu–O planes (CuO_2) that are present are particularly important, as they constitute the *superconducting layers* where the mobile charge carriers (electrons or holes) reside and that undergo superconducting condensation when cooled below T_c. Other intervening metal oxide and pure metal layers present, such as Tl–O, Bi–O, Hg–O, Ba–O, Sr–O, and Cu–O chains, and Y and Ca layers, provide the necessary structural framework for the pertinent number of CuO_2 planes to exist. The mixed-valence character of some of these oxide layers (e.g. Tl–O, Bi–O, Hg–O, and Cu–O chains) makes them serve as *charge reservoir layers* supplying mobile carriers to superconducting CuO_2 planes. Simultaneously, these layers render a low dimensionality to their

Table 16.1 Critical temperature of some prominent HTS cuprates with their maximum T_c values

Material	Short form	T_c (K) (max)	Cu–O planes per unit cell	Crystal structure
La/Nd-214 and La/Nd-2126:				
$La_{1.84}Sr_{0.16}CuO_4$	LSCO or La-214	38	1	Tetragonal (T-phase)
$Nd_{1.85}Ce_{0.15}CuO_4$	NCCO or Nd-214	25	1	Tetragonal (T′-phase)
$SmLa_{0.8}Sr_{0.2}CuO_4$	SmLSCO	24	1	Tetragonal (T*-phase)
$La_{1.65}Sr_{0.35}CaCu_2O_6$	La-2126	60	2	Tetragonal
R-123 and related systems:				
$YBa_2Cu_3O_7$	YBCO or Y-123	92	2 + 1 Cu–O chains	Orthorhombic
$YbBa_2Cu_3O_7$	Yb-123	89	2 + 1 Cu–O chains	Orthorhombic
$GdBa_2Cu_3O_7$	Gd-123	94	2 + 1 Cu–O chains	Orthorhombic
$NdBa_2Cu_3O_7$	NBCO or Nd-123	95	2 + 1 Cu–O chains	Orthorhombic
$YBa_2Cu_4O_8$	Y-124	80	2 + 2 Cu–O chains	Orthorhombic
$Y_2Ba_4Cu_7O_{15}$	Y-247	95	Both planes and chains	Orthorhombic
$LaBaCaCu_3O_7$	La-1113	80	3	Tetragonal
BSCCO system:				
$Bi_1Sr_2Ca\,Cu_2O_7$	Bi-1212	102	2	Tetragonal
$Bi_2Sr_2CuO_6$	Bi-2201	34	1	Tetragonal
$Bi_2Sr_2CaCu_2O_8$	Bi-2212	95	2	Tetragonal
$Bi_2Sr_2Ca_2Cu_3O_{10}$	Bi-2223	110	3	Tetragonal
$Bi_2Sr_2Ca_3Cu_4O_{12}$	Bi-2234	110	4	Tetragonal
$(BiPb)_2Sr_2Ca_2Cu_3O_{10}$	(Bi-Pb)-2223	110	3	Tetragonal
TBCCO:				
$TlBa_2CuO_5$	Tl-1201	50	1	Tetragonal
$TlBa_2CaCu_2O_7$	Tl-1212	82	2	Tetragonal
$TlBa_2Ca_2Cu_3O_9$	Tl-1223	133	3	Tetragonal
$TlBa_2Ca_3Cu_4O_{11}$	Tl-1234	127	4	Tetragonal
$Tl_2Ba_2CuO_6$	Tl-2201	90	1	Tetragonal
$Tl_2Ba_2CaCu_2O_8$	Tl-2212	110	2	Tetragonal
$Tl_2Ba_2Ca_2Cu_3O_{10}$	Tl-2223	128	3	Tetragonal
$Tl_2Ba_2Ca_3Cu_4O_{12}$	Tl-2234	119	4	Tetragonal
HBCCO:				
$HgBa_2Cu_1O_5$	Hg-1201	97	1	Tetragonal
$HgBa_2CaCu_2O_7$	Hg-1212	128	2	Tetragonal
$HgBa_2Ca_2Cu_3O_9$	Hg-1223	134	3	Tetragonal

(Continued)

Table 16.1 *(Continued)*

Material	Short form	T_c (K) (max)	Cu–O planes per unit cell	Crystal structure
$HgBa_2Ca_3Cu_4O_{11}$	Hg-1234	127	4	Tetragonal
$HgBa_2Ca_4Cu_5O_{13}$	Hg-1245	110	5	Tetragonal
$HgBa_2Ca_5Cu_6O_{15}$	Hg-1256	107	6	Tetragonal
$Hg_2Ba_2CaCu_2O_8$	Hg-2212	44	2	Tetragonal
$Hg_2Ba_2Ca_2Cu_3O_{10}$	Hg-2223	45	3	Tetragonal
$Hg_2Ba_2Ca_3Cu_4O_{12}$	Hg-2234	114	4	Tetragonal
$HgBa_{1.8}\,Tl_{0.2}Ca_2Cu_3O_{10}$	Hg-1223 with Tl	138	3	Tetragonal

Data taken from Hott et al. (2005).

electronic structure by serving as *blocking* or *spacer layers* between the superconducting layers. They weaken their electronic coupling in the *c*-direction to make them quasi-two-dimensional, and promote a significant anisotropy in both the superconducting and normal states. The ratio of the normal-state resistivities along and perpendicular to the *c*-direction, ρ_c / ρ_{ab}, ranges from the order of 10^2 to the order of 10^5 and the critical current across the layers in the superconducting state can be orders of magnitude lower than in the *ab*-plane. This poses a serious constraint and challenge for HTS conductors to achieve high J_c for practical applications.

Looking at Table 16.1, a common trend readily visible with BSCCO, TBCCO, and HBCCO systems is that T_c initially goes up with *n* (the number of Cu–O planes contained in the unit cell), reaches a maximum, and then, with further increase of *n*, it either decreases or shows saturation (Figure 16.6). The reason for such behaviour is not known, but introducing more coupled planes beyond $n = 3$ may lead to an increase in non-uniform doping, with an adverse effect on T_c.

16.2.2 Generic phase diagram and pseudogap

The parent or undoped HTS cuprates are all antiferromagnetic (AFM) Mott–Hubbard insulators. Typically, such a compound with Cu in the d^9 state has a half-filled $d_{x^2-y^2}$ band and should normally be metallic. This, however, is prevented by strong electron–electron interaction, which causes an on-site Coulomb repulsion (a large Hubbard energy U), leading to a splitting of the band into lower (filled) and upper (empty) Hubbard bands. The on-site repulsion along with Pauli's exclusion principle disfavour hopping of charge carriers between the neighbouring Cu sites, making the material insulating. However, a small amount of hopping is facilitated by the spins on the adjacent Cu ions of CuO_2 planes becoming antiferromagnetically oriented (to satisfy the exclusion

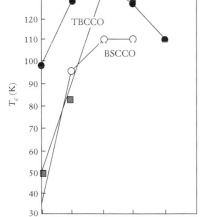

Figure 16.6 T_c as a function of the number of Cu–O planes per unit cell for three HTS systems (Table 16.1).

279

principle) in the form of a Néel lattice with characteristic temperature T_N, with the AFM order being mediated by O-2p orbitals (Anderson, 1987).

The parent compound, with charge-carrier density (number of holes per Cu) $p = 0$, with 300 K $< T_N <$ 400 K, can be chemically doped with holes either by partial substitution of trivalent cations (such as La^{3+}) by divalent ones (such as Sr^{2+}) as in LSCO or simply by increasing the oxygen content x, as in La_2CuO_{4+x}, $YBa_2Cu_3O_{6+x}$ (alternatively $YBa_2Cu_3O_{7-\delta}$, where $x = 1 - \delta$), etc. In the case of LSCO, the dopant (Sr) concentration $x = p$. But this is not so straightforward with other cuprates having many Cu–O planes and also chains, where the planar hole concentration p can be estimated through *bond valence sums* (Brown, 1989) and titration measurements. Electrons can be doped by partial substitution of trivalent ions by tetravalent ions, as in NCCO. As charge carriers are progressively introduced in the parent compound, a series of new phases of different physical properties evolve as a function of temperature and hole concentration, leading to the generic phase diagram of Figure 16.7 (Timusk, 1999; Orenstein and Millis, 2000; Tallon and Loram, 2001; Varma, 1997; Alff et al., 2003; Norman et al., 2005; Millis, 2006; Cho, 2006), which holds for the majority of HTS cuprates.

As depicted in Figure 16.7, the long-range AFM order is speedily destroyed when the carrier content rises from $p = 0$ to $p \approx 0.03$ holes/Cu, above which a spin-glass state with short-range order is formed. Superconductivity spans the range 0.05 $< p <$ 0.27 in the shape of an inverted parabola exhibiting an optimum T_c at $p_{opt} = 0.16$, the temperature scale being material-dependent. For $p > 0.27$ there is no superconductivity and the compound is n-type metallic. This parabolic superconducting behaviour roughly follows the equation $T_c = T_{c,opt}[1 - 82.6(p - 0.16)^2]$ (Presland et al., 1991). With Sr dopant, near $p = 0.125$ (i.e. 1/8 holes/Cu), the parabolic behaviour exhibits an anomaly in the form of a tiny dip (Figure 16.7), which is far more pronounced in LBCO. It is attributed to *spin-charge stripe* fluctuations. The charges introduced by doping into the AFM lattice are aligned parallel to form equidistant charge and spin stripes (Fujita et al., 2004, Abbamonte et al., 2005; Kim et al., 2008; Dunsiger et al., 2008) arising out of competition between phase separation (the AFM insulator expelling the doped holes) and long-range Coulomb repulsion. The stripes formed are either statistically ordered (LBCO) or slowly fluctuating (LSCO) and are related to a low-temperature tetragonal structure.

Clearly, the range 0.05 $< p <$ 0.19 corresponds to the region where short-range AFM correlations coexist with superconductivity, and they, along with phonons, probably mediate Cooper pairing. The material with $p < p_{opt}$ is called *underdoped*, while for $p > p_{opt}$, it is called *overdoped*. The underdoped regime has a lower superfluid density, a greater

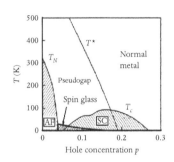

Figure 16.7 Generic phase diagram for hole-doped HTS (after Tallon, 2005). There is AFM ordering at low p (holes per Cu) and superconductivity at a higher p. There is a spin-glass state at intermediate doping coexisting with superconductivity. The pseudogap energy gradually falls to zero at a critical doping of $p = 0.19$. With increasing p, the system moves towards the normal Fermi-liquid metallic state.

anisotropy, and stronger fluctuations. It shows a poor metallicity, with the normal-state resistivities $\rho_{ab}(T)$ in the basal plane and $\rho_c(T)$ along the c-direction both exhibiting a semiconductor-like activated temperature dependence. The charge transport along the c-direction takes place via tunnelling across the CuO_2 planes. In the optimally doped system, $d\rho_{ab}(T)/dT$ is positive but $d\rho_c(T)/dT$ is very often negative, which is indicative of two-dimensional behaviour. However, the latter derivative can also be positive, depending on the material. In the overdoped state, both $d\rho_{ab}(T)/dT$ and $d\rho_c(T)/dT$ are positive, which is a characteristic of a three-dimensional metallic state.

The most interesting feature of the HTS cuprates, as revealed by host of techniques, including resistivity (Bucher et al., 1993; Batlogg et al., 1994; Ito et al., 1993), thermopower (Tallon, 1995), magnetic suscept-ibility (Hwang et al., 1994), NMR (Warren et al., 1989; Takigawa et al., 1991; Alloul et al., 1993), neutron scattering (Rossat-Mignod et al., 1991), heat capacity (Loram et al., 1993), infrared spectroscopy (Homes et al., 1993; Basov et al., 1994), and angle-resolved photoemission spectroscopy (ARPES) (Ding et al., 1996; Loeser et al., 1996), is the formation of a pseudogap in the normal state when the compound is underdoped. In the underdoped regime, at $T >> T_c$, there is an unusual redistribu-tion of electronic states that occurs below a temperature T^* near E_F and resembles a superconducting energy gap (Timusk and Statt, 1999; Buchanan, 2001). This is termed a *pseudogap*, and its energy or T^*, in contrast to T_c, monotonically decrease as p increases (Figure 16.7) in the underdoped regime.

There are different interpretations of the PG and its interplay with superconductivity, which may be briefly mentioned (Hüfner et al., 2008; Norman et al., 2005; Timusk and Statt, 1999). Emery and Kivelson (1995) describe the PG below T^* in terms of uncorrelated (incoher-ent) electron pairs (*preformed Cooper pairs*) existing in the normal state, as against the coherent pairs of the superconducting state, the former serving as a precursor to the latter. Some of the theories based on *res-onating valence bonds* (RVBs), which invoke spin–charge separation into *spinons* (spin $^1/_2$, charge 0) and *holons* (spin 0, charge $+e$), relate the spin PG to pairing of spinons (Anderson, 1987; Randeria et al., 1992; Zhang, 1997). In this scenario, the T^* line (Figure 16.7) merges gradually with the superconductivity dome in the strongly overdoped region (not shown) of the phase diagram such that, for all dopings, $T^* > T_c$ (Millis et al., 2006). There are also two alternative scenarios. In one of them, the T^* line ends on the superconducting dome close to $T_{c,\mathrm{opt}}$ and thus there is no PG within the dome (Cho, 2006). On the other hand, several experimental techniques, including NMR, Raman scattering, infrared spectroscopy, electrical resistivity, and ARPES, have indicated (Loram et al., 1994, Tallon et al., 1999) that the decrease of T^* with dopant con-centration, as illustrated in Figure 16.7, might in fact continue inside the superconducting dome after its intersection and T^* would vanish at $T = 0$ at the critical doping $p_{crit} = 0.19$ holes/Cu in a lightly overdoped regime. Such behaviour is suggestive of this point being a quantum critical point, and at p_{crit} superconductivity is most robust, with maximum condensation energy, and is least sensitive to impurities (Tallon et al., 1995, 1999; Tallon and Loram, 2000).

In this model (Loram et al., 1994), the pseudogap represents a second-order parameter in competition with the superconducting order parameter. The jump in the electronic specific heat at T_c shows a decrease as the pseudogap is formed (Loram et al., 1994; Ghiron et al., 1993). The pseudogap is a precursor to the superconducting gap and both possess the same d-wave symmetry (Loesser et al., 1996; Ding et al., 1996). Since both T^* and short-range AFM order vanish together at $p_{crit} = 0.19$, the formation of the pseudogap seems closely associated with the latter. Interestingly, the pseudogap phase is observed (Lee, 2000; Iguchi et al., 2001) to carry vortex-like excitations at $T >> T_c$, indicating the presence of superconducting fluctuations (Meingast et al., 2001; Lortz et al., 2003) and giving credence to the existence of preformed Cooper pairs in the normal state.

Another interesting possibility of the pseudogap being a competing second type of order envisages it as a *hidden order* (Chakravarty et al., 2001) in the form of a $d_{x^2-y^2}$-*density-wave (DDW) state* that involves alternating orbital currents from one CuO_2 plane to the next (Figure 16.8). In this framework, a conventional charge-density wave (CDW) is conceived of as an s-density-wave state. The contention that the pseudogap is a hidden order and that it competes with superconductivity is supported by the transport studies of Watanabe et al. (2004) on BSCCO single crystals. In yet another version of the preformed pair model, T^* is interpreted as the temperature for charge stripes to develop superconducting correlations, and at a lower temperature these

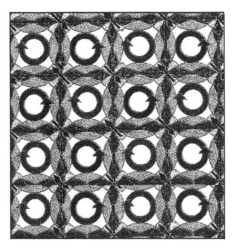

Figure 16.8 Bond currents of alternating orientations (Chakravarty et al., 2001) giving rise to a DDW state forming a hidden order associated with the pseudogap (after Hott et al., 2005).

stripes become Josephson-coupled to realise bulk superconductivity (Tranquada et al., 1995; Emery and Kivelson, 1998; Yamada et al., 1998).

The phase diagram of electron-doped HTS is analogous to what we have discussed above for a hole-doped system, although it has not been studied in such depth, since the synthesis of electron-doped HTS is much more demanding than that of hole-doped HTS. In NCCO and PCCO, for example, superconductivity occurs in rather narrow ranges of $0.14 < x < 0.17$ and $0.13 < x < 0.2$, respectively (Tsuei and Kirtley, 2000a, 2002), which makes the task of preparing reproducible samples more difficult.

16.2.3 Strange normal state

In contrast to conventional superconductivity, which evolves out of a simple Fermi-liquid (FL) metallic state, the high-temperature superconductivity of cuprates stems from a strange normal state having anomalous properties. Consequently, the unusual characteristics of the normal state are considered relevant to high-T_c phenomenon. Table 16.2 compares some of the normal-state features of conventional and cuprate superconductors. The origin of the anomalous characteristics of cuprates clearly lies in strong correlation effects involving spins of the Cu-d^9 ions and the doped holes in their CuO_2 planes. The most prominent anomalous normal-state properties of HTS cuprates are the linear temperature dependence of the resistivity ρ, the power-law temperature dependence (commonly T^3) of the NMR spin–lattice relaxation rate

Table 16.2 Normal states of conventional superconductors and HTS cuprates compared

Normal-state characteristics	Conventional superconductors	HTS cuprates
Resistivity $\rho(T)$	$\sim T^2$	$\sim T$
Hall effect	R_H T-independent $\cot\theta_H = R_H/\rho \sim T^{-2}$	RH $\sim T^{-1}$ and $\sim T$ (at low T) $\cot\theta_H \sim T^2$
Magnetic susceptibility	T-independent	T-dependent
Thermoelectric power	Sensitive to Fermi surface (FS) and average v_F	Insensitive to FS and influenced by hole density
AFM correlations	Absent	Strong
Pseudogap	Absent	Present
Quasiparticle lifetime $1/\tau(T,\omega)$	$\sim a\,T^2 + b\omega^2$	$\sim aT + b\omega$
Spin excitation spectrum	T-independent	Peaked at $\sim (\pi/a, \pi/a)$
Maximum excitation strength	~ 1 state eV^{-1}	~ 20–300 states eV^{-1}
Spin excitation energy	$\sim E_F$	$\omega_{sf} \sim T \ll E_F$

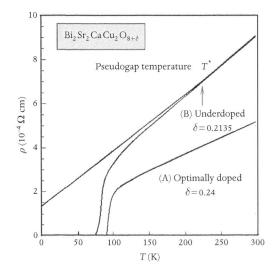

Figure 16.9 Curve (A) shows the typical linear R–T behaviour of optimally doped HTS. The depicted curve is for Bi-2212 with an optimal oxygen concentration of 8.24. Curve (B) is for the same material when underdoped with a Bi concentration of 8.2135. Around 200 K, the curve shows a small downward curvature, which is believed to be the pseudogap temperature T^*.

of Cu spins, a Hall angle that diverges as T^2 over a wide temperature region, an optical conductivity that shows a Drude peak with a width $1/\tau$ varying linearly with T (consistent with the linear behaviour of ρ), and a pronounced mid-infrared peak above the Drude peak.

The resistivity $\rho_{ab}(T)$ of an optimally doped compound, instead of exhibiting the characteristic T^2 behaviour of the FL theory, is found to be linear in T from T_c to about 500–1000 K. Curve (A) in Figure 16.9 (Watanabe et al., 2004c) is a typical example of such behaviour manifested by an optimally doped Bi-2212 sample. Interestingly, the slope of this linear behaviour is found to be nearly uniform for all such optimally doped compounds, despite differences in their phonon spectrum and spin fluctuations. Further, resistivity in these materials does not seem to be affected by impurity scattering and is of the order of 100 $\mu\Omega$ cm for most of them. The resistivity of all HTS cuprates is highly anisotropic, with ρ_c/ρ_{ab} ranging from the order of 10 to the order of 10^5, which is in strong contrast to the isotropic conductivity of conventional systems. In the overdoped region, the resistivity varies as T^2, in accordance with three-dimensional FL behaviour.

The Hall effect in HTS in the normal state is again anomalous and shows a strong temperature dependence, and, further, it is sensitive to impurities and disorder. The relaxation time τ estimated from the Hall angle θ_H has a T^{-2} temperature dependence, in contrast to the T^{-1} dependence deduced from resistivity. The anomalous temperature dependence of the Hall angle is explained in terms of anisotropic scattering at different regions of the Fermi surface (Stojkovic and Pines, 1996).

The normal-state properties of HTS, particularly in the underdoped regime, are affected by pseudogap formation (Timusk and Statt,

1999). The normal-state resistivity $\rho_{ab}(T)$ in the marginally underdoped regime shows an anomaly in the form of a downward curvature in the linear behaviour at T^*, as depicted by curve (B) in Figure 16.9. The thermoelectric power S in the conventional normal state, which is sensitive to the average Fermi velocity v_F, becomes independent of the individual Fermi surface and is influenced mainly by hole concentration. Also, the absolute value of S is controlled by the carrier concentration rather than by other differences that may exist in similar compounds (Liang, 1998).

16.2.4 Pairing mechanism and unusual isotope effects

While the mechanism that binds the charge carriers into Cooper pairs still remains to be established for HTS cuprates, the dominant role of magnetic interactions seems to be favoured in the discussed scenario. The proximity of both the pseudogap and the superconducting dome to the AFM phase in the T–p phase diagram (Figure 16.7) provides a strong indication of a spin-related mechanism. Also, short-range spin fluctuations are present well within the superconducting dome until the critical concentration p_c at which superconductivity is most robust. Inelastic neutron scattering experiments (Aeppli et al., 1999) also provide credence to the presence of magnetic fluctuations.

To ascertain the possible involvement of phonons in pairing mechanism, the HTS cuprates have been subjected to isotope-effect studies. The early observations (Bourne et al., 1987; Batlogg et al., 1987) of the oxygen isotope effect (OIE) on optimum-T_c cuprate samples made by changing ^{16}O to ^{18}O had depicted only a feeble isotope-effect exponent $\alpha = 0.02$–0.1, as against the value of 0.5 expected for conventional phonon mediated superconductivity. Phonons were therefore regarded as being redundant in high-temperature superconductivity. The situation has changed little as far as optimally doped HTS are concerned, where the measured exponents have remained insignificant as before. However, more interestingly, recent studies (Zhao et al., 2001; Khasanov et al., 2004; Keller, 2005; Keller et al., 2005, Chen et al., 2007; Keller and Bussmann-Holder, 2010) have revealed the unexpected results that the exponent α for these materials can be very large when the samples are underdoped and also that it increases as the number of Cu–O layers in the unit cell is reduced. Typical illustrations of these are shown respectively in Figure 16.10(a) (Keller and Bussmann-Holder, 2010) and (b) (Chen et al., 2007). As may be seen from the former, *site-selective oxygen isotope effect (SOIE)* studies reveal that the planar oxygen ions in $Y_{1-x}Pr_xCu_3O_{7-\delta}$ primarily contribute almost 90% of the total OIE due to apex, chain, and planar oxygen at all doping levels, which corroborates the greater importance of in-plane oxygen ions in comparison with out-of-plane ones. Second, as T_c goes down with underdoping, α increases even beyond 0.5. This increase is generally attributed to the

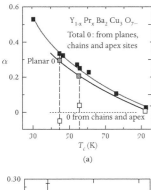

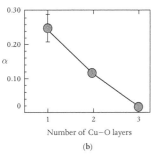

Figure 16.10 (a) Site selective oxygen isotope effect in (Y/Pr)-123 (after Keller and Bussmann-Holder, 2010). (b) Oxygen isotope effect in BSCCO as a function of the number of Cu–O layers per unit cell of the material (after Chen et al., 2007).

formation of a pseudogap (Tallon, 2005), which derives some support from the fact that no such increase is found for overdoped samples free from pseudogap formation (Keller, 2005; Keller et al., 2005). Although these findings do not necessarily support the role of lattice-mediated pairing, they do emphasise the non-trivial role of phonons in HTS. The results of Chen et al. (2007) in Figure 16.10(b) show that in optimally doped samples of the BSCCO series, $Bi_2Sr_2Ca_{n-1}Cu_nO_{2n+4+\delta}$ (for $n = 1, 2, 3$), the OIE exponent α gradually increases towards 0.5 as the number of CuO_2 layers in the unit cell decreases from $n = 3$ to $n = 1$, accompanied by a reduction in the optimum T_c. Clearly, in this situation, there is no pseudogap formation to justify the increase in α. These findings nevertheless underscore the important role of phonons and interlayer coupling in HTS cuprates, but, at the same time, it is evident that the phonons are not fully responsible for the high T_c. As discussed in Chapter 15, in the case of MgB_2 ($T_c = 39$ K), which is considered to be a phonon-mediated superconductor, the value of $\alpha \approx 0.25 < 0.5$, suggesting the involvement of phonons in T_c to be partial.

In various HTS, unusual oxygen and copper isotope effects have been observed for a host of other quantities besides T_c, such as the superconducting gap 2Δ, λ_{ab}, T^*, the spin-glass temperature T_g, the effective Cooper pair mass m^*, and the EPR line width (Furrer, 2005; Keller, 2005; Khasanov et al., 2008a, b). Interestingly, substantial isotope effects have been observed for these quantities in several families of HTS cuprates at varying doping levels using different measuring techniques (Zhao and Morris, 1995a, b; Zhao et al., 1998; Keller, 2005; Bussmann-Holder et al., 2005). Clearly, purely electronic mechanisms for HTS would be inadequate in explaining these phenomena, and there is a need to introduce lattice contributions as well.

16.2.5 Symmetry of the order parameter

Shortly after the discovery of HTS cuprates, experimental investigations involving, among others, flux quantisation, NMR, Knight shift, and Josephson tunnelling showed that the current carriers in these compounds were paired in a singlet state emerging from simple s-wave (BCS-type), extended s-wave, or d-wave pairing. Simple s-wave pairing indicates a homogeneous and isotropic superconducting energy gap 2Δ at the Fermi surface, with various superconducting parameters, such as electronic specific heat C_e, thermal conductivity K_e, ultrasonic coefficients, $\lambda_{ab}(T)$, and NMR relaxation time, all exhibiting an exponential temperature dependence, that is, $\exp(-2\Delta/k_BT)$, below T_c. For both the extended s-wave and d-wave states, the energy gap is anisotropic, having line nodes on the Fermi surface, which, in the former case, has the same rotational symmetry as the crystal and, in the latter case, a lower symmetry.

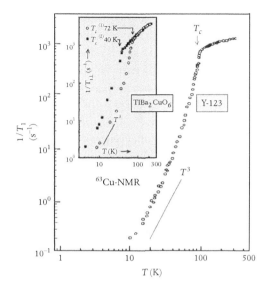

Figure 16.11 Temperature dependence of the spin-lattice relaxation rate in ^{63}Cu-NMR of YBa$_2$Cu$_3$O$_7$ (after Kitaoka et al., 1988). The inset shows similar results on single-crystal samples of TlBa$_2$CuO$_6$ with the applied field normal to the c-axis (after Fujiwara et al., 1990). Both plots show the absence of a coherence peak below T_c and the T^3 dependence of the relaxation rate at low temperatures below T_c, consistent with unconventional superconductivity.

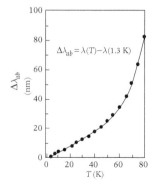

Figure 16.12 Penetration depth, measured at low temperatures below T_c parallel to the basal plane of Y-123 single crystal, showing a linear temperature dependence instead of the exponential dependence followed by conventional superconductors (after Hardy et al., 1993).

The Knight shift of ^{63}Cu in nuclear resonance studies of most cuprate superconductors (reviewed by Asayama et al., 1996a) reveals that the spin component of the susceptibility vanishes below T_c, in accordance with a spin-singlet pairing state, which may occur in both conventional s-wave and unconventional d-wave orbital channels. However, the relaxation rate $1/T_1$ exhibits unconventional behaviour in two respects. First, the relaxation rate in the normal state is not proportional to T and its absolute value is much larger than that estimated from the band structure. Second, the relaxation rate, in contrast to its conventional exponential behaviour, decreases rapidly below T_c without manifesting a coherence peak. These features are well exhibited by NMR–NQR (nuclear quadrupole resonance) studies of various HTS cuprates reviewed by Asayama et al. (1996a, b) and Rigamonti et al. (1998). The results obtained on Y-123 (Kitaoka et al., 1988) and Tl-2201 (Fujiwara et al., 1990) single crystals, for example, are depicted in Figure 16.11. The absence of a Hebel–Slichter coherence peak and the observed T^3 behaviour below T_c are considered strong arguments for the presence of unconventional pairing with the order parameter having line nodes at the Fermi surface.

Most of the above mentioned experiments, which constitute a *non-phase-sensitive approach*, have been found to be consistent with the d-wave symmetry $d_{x^2-y^2}$ for HTS cuprates, with their various superconducting parameters, instead of an exponential temperature dependence, manifesting power-law behaviour (which holds also for extended s-wave pairing) of the form T^n, with n being an integer. For example, the electronic specific heat C_e varies as T^2 (Moler et al., 1994), the penetration depth (Figure 16.12) λ_{ab} as T (Hardy et al., 1993), and the NMR relaxation time $1/T_1$ as T^3, as already discussed.

To fully confirm the symmetry of the order parameter, however, one has to determine both the magnitude and the phase of the gap function, and in this respect the Josephson junction experiments are particularly powerful (Wollman et al., 1995; Tsuei et al., 1994; Tsuei and Kirtley, 2000a, 2002). Wollman et al. were able to confirm the d-wave superconductivity of YBCO by studying YBCO–Au–Pb Josephson junctions in a SQUID configuration where the junctions were fabricated on the a- and b-edges of a YBCO single crystal. Equally convincing support for d-wave pairing stemmed from tunnelling studies on tricrystal thin films (Figure 16.13) of YBCO and TBCCO, as discussed by Tsuei and Kirtley (2000a, 2002). The experiment was essentially based on the original prediction of Geshkenbein et al. (1987) that if the superconductivity across the grain boundaries were of d-wave type, it would give rise to a π-shift and spontaneous generation of a half-flux quantum $\phi_0/2$ on any path encircling the node. In the case of simple s-wave pairing, this would be a full flux quantum ϕ_0 instead. Using a high-resolution scanning SQUID microscope, Tsuei and Kirtley (2000a) did indeed detect the magnetic flux threading through the superconducting circular path around the node to be $\phi_0/2$, in strong confirmation of pure d-wave symmetry.

On the other hand, the c-axis Josephson tunnelling between Pb–YBCO junctions, both twinned (Sun et al., 1994) and untwinned (Kouznetsov et al., 1997), revealed a substantial s-wave component together with the d-wave pairing. These experiments showing the presence of significant s-wave pairing along with d-wave, that is, $s + d$, seem to be consistent with the unusual isotope effects discussed earlier.

The situation is, however, debatable with electron superconductors such as NCCO and PCCO. This is surprising, especially as they have very similar crystal structures and phase diagram as of hole doped ones. Both $\lambda_{ab}(T)$ (Alff et al., 1999; Kim et al., 2003; Skinta et al., 2002) and the tunnelling spectra of microbridges in the ab-plane of c-axis-oriented thin films (Kashiwaya et al., 1996; Kleefisch et al., 2001) of these

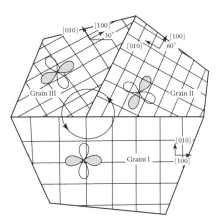

Figure 16.13 Josephson tunnelling experiments on tricrystal thin films (after Tsuei and Kirtley, 2000, 2002).

electron compounds favour *s*-wave pairing. However, tricrystal measurements on electron compounds have revealed a magnetic flux of $\phi_0/2$ around the grain-boundary node, in accord with $d_{x^2-y^2}$ pairing (Tsuei and Kirtley, 2000a, b, 2002). Also, photoelectron spectroscopy (Sato et al., 2001) and π-SQUID experiments (Chesca et al., 2003) favour the same contention for NCCO and $La_{2-x}Ce_xCuO_4$. Tunnelling spectra of hole-doped HTS generally show zero bias anomaly (ZBA) (Iguchi et al., 2000) with an amplitude that varies with the in-plane misorientation angle. The electron-doped cuprates do not, however, manifest a ZBA (Kashiwaya et al., 1996; Kleefisch et al., 2001). Clearly, the electron-doped cuprates behave differently from their hole-doped counterparts, and an acceptable explanation of this seems still missing.

16.3 Prominent HTS cuprate systems

16.3.1 La-214

This is the first family of HTS superconductors discovered, possessing the highest $T_c = 38$ K for $La_{1.84}Sr_{0.16}CuO_4$ and with a unit cell having a single CuO_2 plane. As mentioned previously, their superconductivity stems from an otherwise-insulating compound La_2CuO_4 having a perovskite-related tetragonal K_2NiF_4 structure when La^{3+} is partly substituted by Sr^{2+}, that is, $La_{2-x}Sr_xCuO_4$ (LSCO) with $0.05 < x < 0.27$. The lattice parameters of $La_{2-x}Sr_xCuO_4$ are $a = 0.3773$ nm and $c = 1.3166$ nm. Partial substitution of La (which is the largest lanthanide ion) by other rare-earth ions in LSCO gives rise to a systematic decrease in both *a* and *c* lattice parameters of the doped compound. This is known as *lanthanide contraction*, which is also noted in RE-123 compounds.

The advantages of the La-214 family of superconductors are their simple crystal structure, the ready feasibility of growing good-quality single crystals free from twinning, the presence of a single CuO_2 plane with no ambiguities introduced by chains or modulated structures, and finally the fact that both *p*- and *n*-type superconductivity can be conveniently realised. As a result, this family has been much exploited for basic research on HTS, although these materials do not offer any special features to make them commercially attractive for practical use.

The materials belonging to the La-214 system are noted for three tetragonal phases (see Table 16.1): (1) the T-phase of the hole compounds such as LSCO and LBCO; (2) the T′-phase of the electron compound NCCO; and (3) the T*-phase belonging to SmLSCO. Their crystal structures are illustrated in Figure 16.14. In the T-phase (Figure 16.14(a)), the four planar Cu atoms at the centre of the unit cell form an octahedron with two out-of-plane oxygen atoms (apical oxygen), a feature that is missing in the structure of the T′-phase (Figure 16.14(b)). The separation of Cu and O atoms is smallest in the CuO_2 planes. The T*-phase

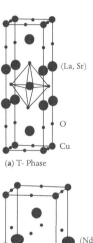

(a) T- Phase

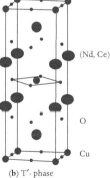

(b) T′- phase

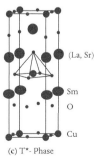

(c) T*- Phase

Figure 16.14 Crystal structures of (a) T-phase, (b) T′-phase, and (c) T*-phase.

(Figure 16.14(c)) has its central Cu atom encased by five oxygen atoms in a pyramidal configuration. Its structure is essentially a combination of a NaCl-type block layer of T-phase and a CaF_2 type block layer of T'-phase. In contrast to the other two phases, its structure lacks a centre of symmetry, with the result that the T^*-phase possesses properties quite different from those of the T- and T'-phases. However, the properties of the T^*-phase have been much less investigated owing to difficulties in synthesising pure samples. On the other hand, the properties of T-phase compounds have been more extensively studied because of their higher T_c.

It is interesting to note that the parent insulating compound, stoichiometric La_2CuO_4, can be made metallic and even superconducting by a small change in oxygen stoichiometry in either direction, that is, $O_{4+\delta}$ or $O_{4-\delta}$ (Schirber et al., 1988; Beille et al., 1987; Yoshizaki et al., 1988). The former can be achieved by oxygenation under pressure and the latter by argon annealing or low-pressure oxygenation. The same effects may also be realised respectively through La deficiency (La_{2-x}) or La excess (La_{2+x}) and maintaining O_4.

16.3.2   R-123

R-123 is the first family of HTS cuprates discovered with T_c exceeding the liquid-nitrogen temperature of 77.4 K. The crystal structure of Y-123 (R = Y) or YBCO, as mentioned in Section 16.1 (Figure 16.2(a)), comprises triple layers of oxygen-deficient perovskite blocks, shown in Figure 16.15 (where the various Cu and O sites are indicated). The fully oxygenated sample with the optimum $T_c \approx 90$ K having the stoichiometry $YBa_2Cu_3O_{7-\delta}$ with $\delta = 0$ is orthorhombic with $a \approx 0.383$ nm, $b \approx 0.389$, and $c \approx 1.166$ nm. As mentioned earlier, leaving out Ce and Tb (which form multiphases) and Pr-123 (PBCO, which is an insulator), all other lanthanides from La to Yb form isostructural superconductors with $T_c \approx 90$–96 K and, similar to La-214, exhibit the lanthanide contraction.

The unit cell (Figure 16.15) of the compound includes double layers of CuO_2 that are slightly buckled and a single planar layer carrying linear Cu–O chains running along the b-axis. The former are separated by a pure Y plane, while a Ba–O plane separates the CuO_2 plane from the outermost layer of chains. Clearly, there are two crystallographically distinct sites for Cu, Cu(1) and Cu(2). Depending on the oxygen content, the former can have 2-, 4- or even 6-fold coordination, while the latter can exist only in the square pyramidal 5-fold coordination. The O_7 compound gives a 4-fold coordination to the Cu(1) site, leading to CuO chains (or CuO_4 ribbons) and Ba acquires a 10-fold coordination. The structure and the existence of superconductivity are both sensitive to O stoichiometry: for $\delta = 1$ (i.e. O_6), the compound is fully

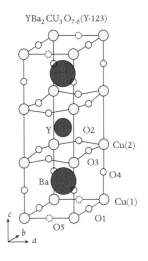

$YBa_2CU_3O_{7-\delta}$(Y-123)

Figure 16.15 Triple oxygen-deficient perovskite structure of Y-123.

tetragonal (with $a = b$) and an AFM insulator, while for $\delta = 0$ (i.e. O_7), the orthorhombic distortion is optimum and T_c is maximum. In the former situation, the O1 site (see Figure 16.15) is totally unoccupied and the structure is devoid of chains.

The effect of oxygenation is to cause the O1 site to progressively fill, and the material becomes doped with holes and simultaneously starts to become orthorhombic, with $b > a$. For $O_{\sim 6.4}$, the magnetic order is quenched and superconductivity appears. Thus, the Cu–O chains from which the oxygen content of the compound can be reversibly changed simply by pumping it in and out, serve as charge reservoirs and also the centres of *O–T* (orthorhombic-to-tetragonal) *transformation*. As may be seen from Figure 16.16, T_c responds in a nonlinear way to δ manifesting a plateau region between $O_{6.7}$ and $O_{6.5}$ with $T_c \approx 60$ K (Cava et al., 1988a, 1990b). This region, known as *ortho-II*, is of altered orthorhombicity in which the alternate chains are removed from the unit cell, which amounts to the doubling of its *a*-parameter. This situation corresponds to the underdoped regime accompanied by a large pseudogap (Alloul et al., 2009).

Above 750°C, YBCO is tetragonal, with the dissolved O distributed uniformly between O1 and O5 sites. During sample processing, when it is cooled to room temperature, it is transformed to an orthorhombic structure with O1 sites filled and O5 empty. The shear strains associated with the transformation are relieved by twinning. Formation of twin boundaries poses a serious problem in growing good-quality single crystals for basic studies. However, single crystals of YBCO have been grown without twinning by applying a uniaxial stress at 970°C during the annealing stage (Lin et al., 1992).

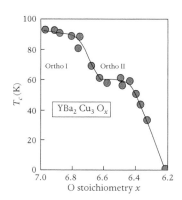

Figure 16.16 T_c of Y-123 as a function of oxygen stoichiometry (after Cava et al., 2008).

16.3.3 BSCCO, TBCCO, and HBCCO systems

The Bi-based cuprates BSCCO were the first HTS system not to contain any rare-earth element, appearing soon after LSCO and RBCO. The three most important compounds of this family are Bi-2212 (i.e. $Bi_2Sr_2CaCu_2O_{8-\delta}$, optimum $T_c = 95$ K), Bi-2223 (i.e. $Bi_2Sr_2Ca_2Cu_3O_{10-\delta}$, optimum $T_c = 110$ K), and BiPb-2223 (i.e. $(Bi_{0.85}Pb_{0.15})_2Sr_2Ca_2Cu_3O_{10-\delta}$, optimum $T_c = 110$ K) Figure 16.3(b) presents the tetragonal unit cell of Bi-2212 with lattice parameters $a = b \approx 0.54$ nm and $c = 3.089$ nm (Giannini et al., 2005). It comprises two Bi–O layers that may be conducting or semiconducting, depending on the content of intercalated O, two SrO layers, and a single Ca layer sandwiched between two conducting CuO_2 layers. The two Bi–O layers are weakly bonded, which allows the material to be easily cleaved across the two layers, leaving a Bi–O layer at its surface. The material exhibits the phenomenon of intergrowth, which can result in a local disorder in the sequence of layers and a change in the *c*-parameter. Further, there is a small lattice mismatch between the Bi–O and Sr–O layers, which gives rise to a structural modulation along the b-axis with a period of 4.76, although this has little impact on superconductivity.

The Bi-2223 compound with three CuO_2 layers sandwiching two pure Ca layers possesses a comparatively high T_c of 110 K, which makes it more attractive for practical applications. Its unit cell with a large c-parameter (≈ 3.7009 nm) is depicted in Figure 16.3(c) (Giannini et al., 2005). However, its synthesis in phase-pure form has proved problematic, and therefore Bi-2212, which is easy to synthesise, is often preferred for applications despite its slightly lower T_c. However, a partial substitution of Pb for Bi in $(Bi_{0.85}Pb_{0.15})_2Sr_2Ca_2Cu_3O_{10-\delta}$ dramatically improved the stability of the 2223 phase (Sunshine et al., 1988; Mizuno et al., 1988), which made Pb-substituted 2223 a most attractive choice for fabrication of the first generation of wires and tapes of HTS cuprates. A major problem with BSCCO materials, however, is their large anisotropy, exceeding 100, which imposes grain alignment of conductors as a mandatory condition for achieving high J_c for magnet applications.

In these compounds, Bi can be replaced by Tl and Hg to yield TBCCO and HBCCO with even higher T_c, as shown in Table 16.1. In Tl-based compounds, Sr can completely replace Ba to form TSCCO. But these are difficult to synthesise unless Tl is 50% substituted by Pb or Bi (Liu and Edwards, 1991, Izumi et al., 1991; Ganguli and Subramanian, 1991), and further their $T_c < 100$K make them less attractive. TBCCO with higher T_c can be more easily formed with either single or double Tl layers (Table 16.1). The unit cells of typical single- and double-layer compounds are shown in Figure 16.4. In both Bi- and Tl-based cuprates, the optimum T_c corresponds to a hole doping of 0.17 holes per Cu, which is good agreement with the generic phase diagram discussed earlier. As regards Hg-based cuprates, their representative structure is shown

Table 16.3 Characteristic parameters of prominent optimally doped cuprates

Cuprate	T_c (K)	ξ_{ab} (nm)	ξ_c (nm)	λ_{ab} (nm)	λ_c (nm)	$\mu_0 H_{c2}^{ab}$(T)	$\mu_0 H_{c2}^{c}$(T)	κ_{ab}
LSCO	38	3.3	0.25	200	2000	80	15	61
NCCO	25	7–8	0.15	120	2600	7	—	15
YBCO	92	1.3	0.2	145	600	150	40	115
Bi-2212	95	1.5	0.1	180	700	120	30	120
Bi-2223	110	1.3	0.1	200	1000	250	30	150
Tl-1223	133	1.4	0.1	150	—	160	—	110
Hg-1223	134	1.3	0.2	177	3000	190	—	136

Data taken mostly from Mourachkine, A. *High Temperature Superconductivity in Cuprates: The Non-Linear Mechanism and Tunneling Measurements.* Kluwer, London (2002), p. 60.

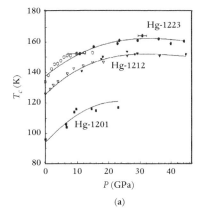

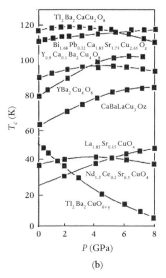

Figure 16.17 (a) Pressure dependence of T_c of some of the HBCCO materials (after Gao et al., 1994). Note that T_c of Hg-1223 is raised to more than 160 K. (b) Pressure effects on T_c of other HTS systems (after Mori et al., 1991).

in Figure 16.5 and Hg-1223 has the record T_c of 138 K (Table 16.1), and with pressure (Figure 16.17(a) and the figure (b)) its highest T_c reached exceeds 160 K.

16.3.4 Copper-based HTS

Another important class of superconductors, possessing $T_c > 100$ K, are copper-based (with Cu additional to that in the cuprate component of the compound) with the general formula $CuBa_2Ca_{n-1}Cu_nO_{2n+2}$ (Jin et al., 1994; Wu et al., 1994; Ihara, 1995, 2001b). They are analogous to single-layer Tl compounds Tl-12$(n-1)n$, where Tl is replaced by Cu. As one can see, these materials are free from toxic and volatile components and also there are no lanthanide elements present. The respective T_c values (Ihara, 2001a) of Cu-1223, Cu-1234, and Cu-1245 are 120 K, 118 K, and 98 K. At the time of their discovery, these materials were believed to possess very low anisotropy, but subsequently this was challenged. Nevertheless, besides their high T_c, these materials are noted also for their high J_c and high H_{irr}. However, the phases are stable only under pressure, which necessitates their synthesis using a diamond anvil cell (about 35 GPa), which imposes constraints on the production of long lengths as wires and tapes for practical applications.

16.4 Substitution studies in HTS

The HTS systems that have been discussed here have been subjected to extensive substitution studies (Narlikar et al., 1989a, b; Agarwal and Narlikar, 1994; Gupta et al., 2005). The significance of such studies can be understood from the fact that in most instances the discovery of these materials has come about through different substitutions and dopings carried out on the related low-T_c or even insulating systems. Partial or complete substitutions of any of the constituent cations or anions of the parent system are in general attempted with four broad objectives:

1. New phases may be formed with higher T_c.
2. Hazardous, less abundant, and more expensive components may be replaced with safer, abundant, and cheaper ones.
3. Anomalies or unforeseen results may lead to new physics or insights into the high-T_c phenomenon;
4. Cation substitutions may lead to the formation of nano-sized regions where T_c is locally reduced. Such regions, through ΔT_c pinning, can effectively restrict the movement of highly mobile pancake-type flux vortices (Zhao et al., 2002b, 2003; Muralidhar et al., 2002b; Cheng et al., 2005; Murakami et al., 2005) and effectively promote $J_c(H)$. While a continuous flux line in a three-dimensional superconductor

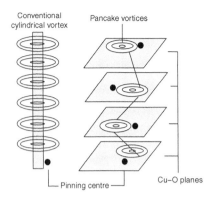

Figure 16.18 A conventional long cylindrical vortex can be pinned by one or a limited number of pinning centres (left), but in the case of highly mobile pancake vortices formed in various Cu–O planes of layered superconductors such as HTS, each such vortex needs pinning centres to pin them individually.

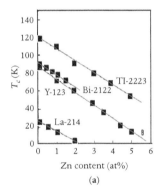

(a)

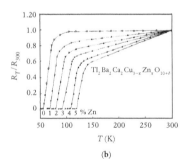

(b)

Figure 16.19 (a) Effect of Zn substitution at the Cu site in four HTS systems. (b) Resistive transitions of Tl-2223 samples containing 0–5 at% Zn in place of Cu. (After Agarwal and Narlikar (1994).)

can be readily pinned by a single or few pins, the separate pancakes formed in different Cu–O layers of a quasi-two-dimensional superconductor require a much larger number of such pins to prevent them from moving (Figure 16.18). The intrinsically very short coherence length $\xi \approx 1$ nm of HTS necessitates having nano-sized artificial pinning centres (APCs) to realise acceptably high transport current densities for large-current applications.

As discussed earlier, the important common feature of all the HTS cuprates is the presence of one or more two-dimensional Cu–O networks where superconductivity resides, and consequently any substitution that directly or indirectly interferes with the Cu–O networks tends to have the largest, and generally deleterious, effect on superconductivity by creating disorder in the networks. This is manifested by a host of Cu-site substitutions that have been studied in the above systems. In the case of the R-123 system, the relevant Cu site for superconductivity is Cu(2) (Figure 16.15). The most widely investigated dopant at this site is Zn, which is found to drastically suppress T_c. The results of Zn substitution (Agarwal and Narlikar, 1994) for four cuprate systems are summarised in Figure 16.19(a). As may be seen, the rate of T_c suppression for all the systems is nearly the same, about 10–12 K per %Zn, which appears to be greater than for any other dopant at Cu or non-Cu sites. For example, Ni has less than half the effect of Zn in lowering T_c. R–T curves for Zn-doped Tl-2223 are depicted in Figure 16.19(b). The fact that Zn produces near-identical effects in suppressing T_c in all four HTS systems indicates that, irrespective of their rather widely varying T_c values, the basic mechanism of superconductivity is similar in the four systems. The large T_c depression due to Zn has been attributed to its fixed valence, which at the Cu sites inhibits charge transport in the Cu–O layers, thus destroying both normal conductivity and superconductivity. Alternatively, it has been argued that both Zn and Ni induce magnetic moments in their vicinity that destroy superconductivity (Fukuzumi et al., 1996).

The effects of partial substitution of Fe, Co, Ga, and Ni at the Cu sites of Bi-2212 are depicted in Figure 16.20, which shows that they degrade T_c respectively at rates of 7, 5, 3, and 2.5 K per % of the dopant. In the R-123 system, Fe produces a comparatively smaller T_c depression, possibly because it prefers the Cu(1) chain site to the Cu(2) plane site

Non-Cu-site substitutions may also cause significant T_c degradation by indirectly producing disorder in the Cu–O planes. Examples of this are Pr (Narlikar et al., 1999) and Ca substitutions at the R (or Y) sites of R(Y)-123 (Awana and Narlikar, 1994; Awana et al., 1994a). In fully oxygenated samples of $Y_{1-x}M_xBa_2Cu_3O_{7-\delta}$ (M = Pr or Ca), Pr and Ca give rise to O loss from the Cu–O planes, which is responsible for the T_c decrease of the substituted samples (Figure 16.21). The O-depleted

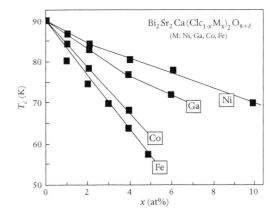

Figure 16.20 Suppression of T_c of Bi-2212 with Cu partially substituted by Ni, Ga, Co, or Fe (after Agarwal and Narlikar, 1994).

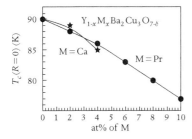

Figure 16.21 T_c suppression in $Y_{1-x}M_xBa_2Cu_3O_{7-\delta}$ due to substitution of M(Pr, Ca) at the Y site (Narlikar et al., 1999; Awana and Narlikar, 1994; Awana et al., 1994).

regions in the Cu–O planes have their T_c locally reduced and they serve as effective pinning centres (Huhtinen et al., 2007) to enhance J_c (Figure 16.22) through ΔT_c pinning (Chapter 6). It is further argued that both Pr and Ca substantially help in marginalising the adverse effects of grain boundaries as weak links, either by lowering their resistance or making them weakly superconducting through a supply of extra charge carriers or O. Ca substitution has therefore been pursued in the conductor development of YBCO.

16.5 Summary

In this chapter, we have discussed the basic features of the prominent high-T_c cuprate superconductors in terms of their structure and properties. Although these are complex materials, they have several common features that help us to understand them. Structure-wise, all are layered perovskite or related compounds with long orthorhombic or tetragonal

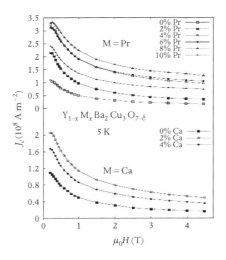

Figure 16.22 J_c enhancement of YBCO resulting from partial replacement of Y by Pr (upper curves) and Ca (lower curves) (after Huhtinen et al., 2007).

unit cells containing superconducting layers, spacer or blocking layers, and charge-reservoir layers, each having a distinct and important role in their superconductivity. This kind of structural arrangement makes the materials quasi-two-dimensional and their superconductivity markedly anisotropic. The highest T_c value of 138 K is realised when Hg-1223 ($T_c = 134$ K) is doped with Tl, and it is increased to 164 K under pressure. In principle, even higher T_c values are not ruled out for new systems still to be discovered.

The HTS cuprates are mostly AFM Mott insulators with low carrier concentrations and $T_N \approx 300$–400 K, the latter speedily decreasing with hole doping. The materials exhibit many unusual features, such as the presence of a pseudogap, stripe phases, and coexistence of short-range magnetic fluctuations with superconductivity, which are manifested in a generic T–p phase diagram discussed in the present chapter. The HTS cuprates are noted for their strange normal state and an unconventional superconductivity with possible d-wave or $d + s$ pairing. These materials are extreme type II superconductors possessing huge critical fields, but their characteristic features such as anisotropy, grain-boundary weak links, and hard and brittle nature are posing technological hurdles for practical use. In Chapter 17, we will see the ingenious approaches followed to overcome these problems.

Thin-film technology and conductor development of HTS cuprates

Although HTS cuprates are type II superconductors possessing a huge H_{c2} and large T_c (Table 16.1), some of their characteristic features, as mentioned in Chapter 16, impose serious constraints on their practical use. First, because of their ceramic character, these oxide superconductors are intrinsically hard and brittle, which makes fabrication of wires and tapes in long lengths a formidable task. Besides, one is confronted with a number of factors that adversely affect their performance as practical electrical conductors, bulk materials for levitation, and in related applications, as well as thin films as components of sensitive superconducting electronic devices. In this chapter, we take stock of the progress made with HTS cuprates in the form of wires, tapes, and thin films, while the development of bulk HTS materials is considered separately in Chapter 18.

17.1   Microstructural aspects

17.1.1   Short coherence length, granularity, and anisotropy problems

17.1.1.1 *Pinning centres*

The very short coherence length ξ (a few nanometres), which is the outcome of their higher T_c (as $\xi = 0.18\,hv_F/k_B T_c$), necessitates the presence of pinning centres of a few nanometres diameter to cause effective flux pinning in HTS. Normally, naturally occurring crystal lattice defects, which provide strong pinning in conventional low-temperature superconductors possessing a much longer ξ, turn out to be too coarse in size to become effective pins in HTS. Nanosized defects are simply absent or just not there in sufficient concentration to meet the strong pinning requirement in these materials. This problem has been successfully tackled by introducing (*ex situ*) the required concentration of artificial nanosized pinning entities, such as carbon nanotubes, nanodiamonds, and nano-SiC particles, as described in Chapter 15 on MgB_2 Also, as discussed in Chapter 16, atomic- or nano-sized disordered regions in Cu–O planes resulting directly or indirectly from substituting Pr, Ca, Zn, Ni, etc. at Cu or rare-earth sites of YBCO can enhance J_c through ΔT_c pinning.

17.1.1.2 Pancake vortices

In HTS cuprates, because $\xi_c < \xi_{ab}$ (Table 16.3) there is poor super-conducting coupling between multilayers along the c-direction in comparison with that in the ab-plane. This leads to a partial breaking up of cylindrical vortices into stacks of highly flexible *pancake vortices* that are highly mobile and more difficult to pin, as mentioned in Chapter 16 (Figure 16.18). Consequently, stronger small-sized pinning centres in larger numbers are needed for flux pinning in HTS, otherwise the flux creep becomes dominant and degrades J_c at elevated temperatures $T < T_c$.

17.1.1.3 Weak links

Defects such as grain boundaries in HTS, having their boundary width $w \geq \xi$, serve as *weak links* of S–N–S type, with *Josephson-like* low current density J_c, which becomes further degraded in the presence of small applied magnetic fields. The mechanism of grain growth is such that impurities that cannot be fitted into the intragrain lattice structure of the growing grains are pushed aside into the growth front to accumulate in the grain-boundary regions. The weak-link effect becomes significant when the boundary regions contain insulating impurity phases or oxygen non-stoichiometry (O vacancies). The boundaries are mechanically strained regions from where the oxygen is readily squeezed out. The resulting grain-boundary Josephson junction is of S–I–S type, with a very low J_c. Such materials in which superconducting grains are interconnected by a network of weak links are known as *granular superconductors*. Polycrystalline HTS cuprates are intrinsic granular superconductors, which explains why most of the initial reports of J_c and $J_c(H)$ measurements on them had yielded disappointing values of J_c (less than about 100 A cm^{-2}).

17.1.1.4 Grain-boundary misorientation

Another grain-boundary problem is that J_c decreases exponentially with the misorientation angle θ between the neighbouring grains, following the relation $J_c(\theta) = J_c(0)\,\exp(-\theta/\theta_c)$, where $\theta_c \approx 5°$ is the angle above which J_c decreases abruptly (Figure 17.1) (Dimos et al., 1988). Chaudhari et al. (1990) ascribed this to the relative orientation of the d-wave order parameter, but the observed suppression of J_c was too large to agree with this contention (Sigrist and Rice, 1994; Hilgenkamp and Mannhart, 2002; Yokoyama et al., 2007). As discussed in Chapter 16, in the case of R-123, a remedial measure for this is partial substitution of

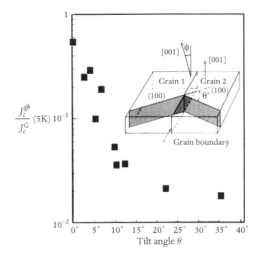

Figure 17.1 The ratio of the grain-boundary critical current density to the average intra-grain critical current density, as measured for a YBCO bicrystal, shows a strong correlation with the tilt angle between the grains, and the ratio decreases exponentially with the misorientation angle θ (after Dimos et al., 1988).

R^{3+} by $C\,a^{2+}$, which, besides providing additional holes to the boundary regions, is found to enhance the intragrain pinning (Zhao et al., 2002a; Huhtinen et al., 2007).

17.1.1.5 *Anisotropy aspect*

All the HTS superconductors of the cuprate family are anisotropic. In the *ab*-plane, the normal state is metallic, while it is semiconductor-like in the *c*-direction. In the superconducting state, J_c in the *ab*-plane is optimum and can often exceed the value along the *c*-direction by more than a few orders of magnitude. Clearly, for achieving optimum current flow, a current-carrying tape must be textured such that grains parallel to its surface are all aligned in the *ab*-plane and the tape surface throughout remains normal to the *c*-direction. Such *c-axis texturing* is a *uniaxial texturing*, which is mandatory with BSCCO-based conductors possessing a large anisotropy ($\gamma >> 25$). Since $\xi_{ab} > \xi_c$, with this texturing, the grain-boundary weak-link effect for the current flow in the *ab*-plane is significantly reduced. On the other hand, with Y-123 and Tl-1223, a substantial improvement in J_c can occur only if the orientation of the *c*-axis and the in-plane orientations of the *a*- and *b*-axes are all correct (Dimos et al., 1988), that is, through *biaxial texturing*. This is a significant hurdle in fabricating long lengths of the conductor, as the processes used for biaxial texturing are time-consuming and expensive. For this, one has to achieve first a similar texturing in a polycrystalline Ni, Ni-alloy, or Ag substrate, or in the buffer layer used, which in turn serves as a template to form a biaxially textured HTS layer. The material thus produced has properties nearly as good as those of epitaxial HTS films deposited on single crystals, which exhibit the greatest J_c and the sharpest transition at T_c. However, for long HTS, conductors, this approach is impractical owing to its high cost, limitations on size, and poor mechanical stability.

17.1.1.6 *Grain alignment in HTS conductors*

Broadly, there are three main approaches followed to achieve grain alignment: (i) applying a mechanical force to compress grains in the desired direction; (ii) applying strong magnetic field gradients to paramagnetic grains embedded in a resin; and (iii) reforming the molten sample in a furnace having the required thermal gradient (i.e. melt texturing). Of these, (i) and (iii) have been found more popular with HTS cuprates. The HTS systems developed for wires and tapes are mainly BSCCO (the *first generation of HTS conductors*, or 1G) and TBCCO and YBCO (the *second generation of HTS conductors*, or 2G), although the toxicity of Tl has been a deterrent factor with Tl-based systems. For BSCCO superconductors, approach (i) is used, and for YBCO approach (iii). Weak van der Waals bonding between the adjacent Bi–O layers of BSCCO compounds allows grain alignment via mechanical processing steps such as rolling or pressing. Although BSCCO compounds possess a large anisotropy factor $\gamma \approx 25$–165 as against $\gamma \approx 5$ for the other two systems, their simplicity in fabrication led them to become the 1G HTS conductors. On the other hand, YBCO and TBCCO, which, with their higher J_c form the 2G HTS, are fabricated in the form of coated conductors using more complicated fabrication routes, to be described later.

The grain-boundary problems are, however, much more complex and not fully resolved. It is not feasible to avoid grain boundaries in kilometre-long wires and tapes, and the realistic approach being followed (Zhao et al., 2002a) is one directed at making the conductor *weak-link free* rather than *grain-boundary free* by suitable doping (e.g. Ca at Y sites). Although the results are encouraging, they are still far from the optimum solution. In particular, the grain-boundary current density $J_{GB} << J_c$ is still low for applications. Second, although J_{GB} is improved by doping, this simultaneously lowers the intragrain J_c (Zhao, 2002a). Significant improvement is, however, expected if the boundary regions can be kept free from insulating impurities and phases so that the S–I–S kind of weak links, which are more deleterious for J_c, are avoided. The presence of conducting additives such as particles of Ag is, however, helpful. In practice, all these approaches are followed to overcome grain-boundary problems.

17.1.2 Critical current behaviour of single crystals

We now examine how the intrinsic anisotropy influences J_c in HTS single crystals. In these extreme type II superconductors, as discussed in Chapter 6, flux pinning through the magnetic interaction is expected to be weak and the core interaction is more relevant. For the optimum core interaction, the pinning entities are normal particles of radius ξ, the range of coherence. The core energy per unit length of the flux line is given approximately by $\frac{1}{2}\mu_0 \pi H_c^2 \xi^2$. The force needed to

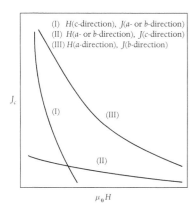

Figure 17.2 Expected $J_c(H)$ behaviour for an anisotropic superconductor with the field and the current along different directions.

displace the core by a distance ξ from the pinning centre is $\frac{1}{2}\mu_0\pi H_c^2\xi^2$, and the corresponding J_c in an arbitrary external field $\mu_0 H$ would be $(\frac{1}{2}\mu_0\pi H_c^2\xi/\phi_0)(1-h)$, where the factor $1-h$ represents some arbitrary field dependence of J_c, with $h = H/H_{c2}$ being the reduced field. For an orthorhombic HTS single crystal, such as R-123, one can have the following three different orientations of mutually perpendicular field and current, as depicted in Figure 17.2:

(I) $H(c\text{-axis})$, with the corresponding

$$J_c(a\text{- or }b\text{-axis}) = \frac{\frac{1}{2}\mu_0\pi H_c^2\xi^{\parallel}}{\phi_0}\left(1 - \frac{H}{H_{c2}^{c\text{-axis}}}\right);$$

(II) $H(a\text{- or }b\text{-axis})$, with the corresponding

$$J_c(c\text{-axis}) = \frac{\frac{1}{2}\mu_0\pi H_c^2\xi^{\perp}}{\phi_0}\left(1 - \frac{H}{H_{c2}^{ab\text{-plane}}}\right);$$

(III) $H(a\text{-axis})$, with the corresponding

$$J_c(b\text{-axis}) = \frac{\frac{1}{2}\mu_0\pi H_c^2\xi^{\parallel}}{\phi_0}\left(1 - \frac{H}{H_{c2}^{a\text{-axis}}}\right).$$

In situation (I), $H_{c2}^{c\text{-axis}}$ has the lowest value, while J_c is lowest in situation (II), where the current flow is across the layers. The optimum situation with respect to current and field is represented by (III) in Figure 17.2, where both H and J are confined to the ab-plane. This situation further offers constraints to flux motion (along the c-direction) and thereby favours a higher J_c. As discussed in Chapter 6, the flux line lattice (FLL) moves via movement of its dislocations. Because of the very short ξ_c of these materials, smaller than the lattice spacing, a force equivalent to the *Peirls–Nabarro force*, which is known to oppose movement of crystal dislocations in solids, is expected to be dominant in the FLL to suppress easy motion of flux line dislocations. Situation (III) is what is recreated as best as possible through grain alignment in the ab-plane of polycrystalline wires, tapes, or deposited thin films. Such grain-aligned samples cannot fully replace single crystals, but they are far superior to powders of randomly oriented grains, and for long lengths of wires and tapes for which single crystals are irrelevant, they offer the only alternative to realise acceptable technical compatibility for practical applications.

17.2 Prominent techniques for depositing HTS films

The techniques followed may be classified as physical and chemical processing routes and are listed in Figure 17.3. The physical routes are mostly vacuum-based and, in general, provide high-quality films needed for research or small electronic devices. The vacuum requirement,

Physical routes	Chemical routes
Pulsed laser deposition (PLD)	Chemical vapour deposition (CVD)
Electron-beam (EB)/thermal co-evaporation	Metal−organic (MOCVD)
Sputtering (single- or multiple-target magnetron)	Trifluoroacetae (TFA-MOD)
	Ex situ BAF2
	Liquid-phase epitaxy (EPE)
Molecular-beam epitaxy (MBE)	Sol−gel, spin, spray or dip coating, screen printing etc.

Figure 17.3 Prominent physical and chemical techniques for depositing HTS films.

however, adds to the cost of the deposition process, and these routes therefore are generally not the preferred options for coating long lengths of substrates. The chemical deposition routes are based on chemical reactions between inorganic or organic precursors containing the pertinent metallic species, resulting in the formation of the required layer on a suitable substrate. The chemical routes are either one-step or two-step processes. The former include CVD (chemical vapour deposition) and MOCVD (metal–organic chemical vapour deposition) in which the superconducting oxide phase is formed directly after chemical reactions in the vapour phase on the substrate. The two-step processes include spray and various dip-coating types of methods in which the first step is to coat the substrate with an unreacted precursor, followed by a step that transforms it into the compound layer of HTS.

17.2.1   Substrate requirement

For depositing epitaxial HTS cuprate films, the choice of a suitable substrate carries significant weight, irrespective of the deposition route followed. The basic requirements of the substrate are as follows:

1. single-crystalline, having a good lattice match with the HTS cuprate;
2. freedom from any mutual chemical interaction;
3. thermal stability under oxygen pressure during deposition;
4. suitable hardness to realise a mirror-like surface finish;
5. non-magnetic or only weakly magnetic;
6. thermal coefficient of expansion/contraction matching that of the HTS cuprate;
7. for practical applications, easy availability at a reasonable cost.

For deposition over several-kilometre lengths, such as for coated conductors, the substrate should, in addition, be metallic and thin (about

20–50 μm thick), flexible, and sufficiently strong for winding and bending.

Some of the commonly used substrate materials are single crystals of yttrium-stabilised zirconia (YSZ), MgO, $SrTiO_3$ (STO), $LaAlO_3$ (LAO), $LaGaO_3$ (LGO), and sapphire (α-Al_2O_3). Because these materials generally do not react chemically with HTS, they are also used as buffer layers for metallic substrates such as Ni, and they effectively serve as a diffusion barrier between the deposited HTS layer and the metallic tape. The buffer performance of these materials is further improved by depositing a secondary buffer layer of R_2CuO_4 (Gao, 2002).

17.2.2   Physical routes for depositing HTS films

17.2.2.1   *Pulsed laser deposition (PLD)*

This technique of laser ablation works for depositing stoichiometric superconducting thin films, or buffer layers, from stoichiometric targets (Witanachchi et al., 1988; Wu et al., 1989; Koren et al., 1989; Gupta et al., 1989). A pulsed laser beam of energy density 1–5 J cm^{-2} is focused on a dense stoichiometric target of HTS cuprate placed inside a vacuum chamber (about 10^{-7} Torr) to generate a plume of ejected atoms or molecules in a direction transverse to the target surface. The target is rotated to prevent it from getting pierced by the laser beam. The commonly used sources for ablation are a pulsed KrF excimer laser ($\lambda \approx$ 248 nm) or the third-harmonic pulses of a Nd-YAG laser ($\lambda \approx$ 355 nm). The substrate, heated to 700–725°C, is located about 5 cm away from the target. During ablation, the oxygen partial pressure is set at about 200 mTorr, while after deposition the substrate is allowed to cool slowly in about 700 Torr of O_2 with a dwell time of 30 minutes at 400°C. No post-deposition heat treatment is needed, and the $T_c = 92$ K of the deposited YBCO film (thickness $\approx$ 150 nm) on STO substrate is very sharp, having a width $\Delta T_c < 1$ K and a commendable current-carrying capacity $J_c(77$ K, 0 T$) \approx (1–5) \times 10^{10}$ A m^{-2} (Wu et al., 1989; Koren et al., 1989; Gupta et al., 1989).

17.2.2.2   *Electron-beam/thermal evaporation*

In this technique (with vacuum better than 10^{-7} Torr), electron guns are used for evaporating the constituent elements (or their oxides) with low vapour pressure such as Y and other rare earths. Resistively heated boats of transition metals such as W, Ta, and Mo (called *effusion cells*) may alternatively be used for thermal evaporation. The vapours resulting from co-evaporation are made to condense as a film on a heated (600–700°C) substrate. The mean free path of evaporating species inside the high vacuum chamber is considerably greater than the distance between the evaporating material and the substrate. The deposited HTS film is oxygenated at 900°C to become superconducting. When the co-evaporation is carried out in the presence of ozone (Bozovic et al., 1990) or of atomic or activated oxygen, the approach is known as *reactive evaporation* and is used for *in situ* growth of the superconducting film.

17.2.2.3 *Sputtering*

Inert-gas ions are used to sputter-eject clusters of atoms or molecules from an HTS target with uniform compositions. The deposition of the ejected material takes place on a heated substrate at about 600–800°C, located close to the target. The deposited film is required to be separately annealed in an oxygen atmosphere to make it superconducting. For *in situ* superconductivity, the oxygen is injected through a nozzle placed close to the substrate during sputter deposition.

To compensate for the deviations observed in stoichiometry, suitable off-stoichiometric targets have been used in conjunction with radiofrequency magnetron sputtering (Adachi et al., 1987; Kamada et al., 1988). However, better results are obtained with off-axis sputtering, where the substrate is located adjacent to the magnetron gun with its planar surface held along the axial direction of the sputter source (Gao et al., 1992). Alternatively, co-sputtering is performed using three separate targets respectively for Y, Ba, and Cu (Silver et al., 1987) to achieve better control over the composition of the consolidated deposit.

17.2.2.4 *Molecular-beam epitaxy (MBE)*

Typically, an MBE system consists of three ultrahigh-vacuum chambers respectively for loading/unloading, substrate outgassing, and MBE growth, and the chances of any contamination of the deposited film are significantly marginalised. In sequential deposition, the constituent species are deposited layer-by-layer without breaking the vacuum. Compound formation takes place by solid state reaction between the layers in the post-deposition treatment (Tsaur et al., 1987; Bao et al., 1987). For *in situ* growth of epitaxial films of YBCO using *reactive MBE*, Kwo et al. (1988) used co-evaporation from two electron-beam sources for Y and Cu along with an effusion cell heater for Ba. To achieve the oxidation of the deposit, a beam of reactive O_2 was directed onto the heated substrate. A detailed account of thin-film processing of HTS cuprates may be found in the review by Rijnders and Blank (2005).

17.2.3   Chemical routes for depositing HTS films

17.2.3.1 *CVD and MOCVD*

Chemical vapour deposition (CVD) and *metal–organic chemical vapour deposition (MOCVD)* present many advantages: (i) they allow the growth of films possessing high melting points at relatively low temperatures; (ii) they are compatible with various processing environments from low-pressure, to atmospheric-pressure, to plasma, (iii) there is faster epitaxial growth than with physical routes; (iv) they are suitable for coating homogeneous layers on large dimensions; and (v) they are relatively cheaper than physical routes. In CVD or MOCVD, the gaseous precursor molecules are transported to the reactor chamber containing the substrate, where the vapours chemically react to form the required thin film. The pressure inside the chamber is around a few torr and the use of costly vacuum equipment is avoided. For depositing films of HTS cuprates as well as various buffer layers, many metal–organic precursors are available for MOCVD, including metal alkoxides $M(OR)_n$, β-diketonates $M(RCOCHCOR')_n$, and β-diketonatoalkoxides $M(OR)_{n-1}(RCOCHCOR')_n$ (with R and R$'$ = Me, Et, or Bu), and alkyl or cyclopentadienyl derivatives. These are in solid form and are vaporised by sublimation at 100–300°C.

The growth process in MOCVD involves these precursors, sublimed into vapour form, being transported using an inert carrier gas into the growth chamber of the CVD reactor, where the required compound is formed by chemical reaction of vapours in proximity to the heated substrate. For the synthesis of BSCCO, temperatures from 140 to 230°C have been used (Yamane et al., 1989; Zhang et al., 1989) for the transport of volatile Bi, Sr, Ca, and Cu precursors into the reaction chamber with argon as carrier gas. Superconducting coatings 1–3 μm in thickness are commonly realised on MgO substrates at 890°C. For depositing 1–10 μm thick YBCO films, β-diketonate chelates of Y, Ba, and Cu have been used as precursors, along with STO or YSZ substrates at 600–850°C) (Watanabe et al., 1989; Zhang et al., 1989).

In many instances, the different precursors used for the constituent atoms in MOCVD differ significantly in their volatility and sublimation temperatures, which makes it desirable to have all the precursors in the form of a single solution. The problem has been successfully tackled in different ways, such as *aerosol-assisted CVD (AACVD)* (Weiss et al., 1993; Meffre et al., 1999), *band-evaporation MOCVD (BEMOCVD)* (Wahl et al., 1968), and *pulsed-injection MOCVD (PIMOCVD)* (Felten et al., 1995; Senateur et al., 1997). A consolidated account of progress in MOCVD for HTS deposition may be found elsewhere (Beauquis et al., 2004).

17.2.3.2 *MOD–TFA and* ex situ *BaF₂ processes*

These two processes are essentially similar except in their precursors, and in the final stage they merge with each other. *Metallorganic decomposition (MOD)* involves a combination of precursors of the desired species of atoms linked to suitable organic groups in the appropriate composition ratio, which are subsequently allowed to decompose on a substrate at an elevated temperature, a process named *pyrolysis*. In conventional MOD, octyl salts are used as starting materials. These are dissolved in methanol and the solution is coated onto the substrate and heat-treated. Initially, formation of $BaCO_3$ was problematic, but

McIntyre and co-workers (McIntyre et al., 1990; McIntyre and Cima, 1994) demonstrated $J_c(77K) > 10^{10} A\ m^{-2}$ when the film was deposited (on STO and $LaAlO_3$ substrates) by the MOD process using Y, Ba, and Cu trifluoroacetates (TFAs) as starting materials, which did not form $BaCO_3$. To start with, solutions of Y, Ba, and Cu acetates in stoichiometric quantities are reacted with trifluoroacetic acid, followed by drying, which results in a blue powder. The precursor solution or gel is obtained by mixing the powder with methanol. The substrate is then coated with the precursor by spraying, spinning, or dipping techniques. The thermal processing, as per heating schedule of Figure 17.4 (Falter et al., 2002), involves two chemical reactions respectively at 400 and 775–800°C, leading to (Smith et al., 1999)

$$1/2\ Y_2Cu_2O_5 + 2BaF_2 + 2CuO + 2H_2O \rightarrow YBa_2Cu_3O_{6.5} + 4HF$$

$$(17.1)$$

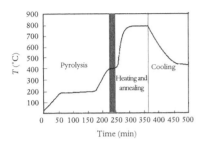

Figure 17.4 Heating schedule in the MOD process (after Falter et al., 2002).

The oxygen-deficient YBCO film on the substrate is oxygen-annealed for several hours at 400°C to achieve high J_c and T_c for optimum superconductivity performance. It is important that BaF_2 reacts with water vapour and is fully transformed to BaO ($BaF_2 + H_2O \rightarrow BaO + 2HF$) to promote YBCO formation. The reaction (17.1) has to be precisely controlled to get epitaxial nucleation of $YBa_2Cu_3O_{6.5}$ grains and achieve their textured growth (Castaño et al., 2003) for high J_c. Among the shortcomings of the process are the extended time schedule followed to remove organic ingredients and to avoid stress-induced cracking during calcining/pyrolysis, the porosity of the deposited films, and the evolution of toxic HF gas.

Some of the above drawbacks are circumvented in *ex situ BaF₂ process*. In this approach, which is again based on reaction (17.1), the precursor is a mixture of Y and Cu compounds and BaF_2, subjected to electron-beam evaporation on a substrate, followed by firing at 775–800°C. The advantage of the method is that YBCO films of 5 μm thickness can be deposited rapidly (Solovyov et al., 1997) which makes it a potential route for producing long conductors.

17.2.3.3 *Miscellaneous wet chemical methods*

These essentially include coating methods for realising thick films (thickness > 10 μm) such as *spray pyrolysis*, the *sol–gel process*, and various forms of dip coating such as the *doctor blade method* and *screen printing*. In spray pyrolysis of BSCCO films, an aqueous solution of Bi, Sr, Ca, and Cu nitrates in the required stoichiometric ratios is sprayed through a nozzle over a heated MgO substrate at 400°C, using compressed oxygen as the carrier gas. After depositing 40–50 μm thickness, the substrate is slowly cooled to room temperature. The sample is further heated to 845°C for 15–20 hours in air, which results in $T_c > 100\ K$

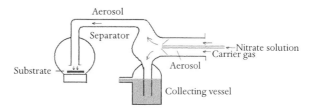

Figure 17.5 Spray pyrolysis (aerosol deposition) technique for depositing thick films of Y-123, Bi-2223, and Tl-2223 (after Jergel, 1995).

without formation of any low-T_c phases (Nobumasa et al., 1988; Hsu et al., 1989). A similar approach of spraying a nitrate solution of Y, Ba, and Cu was followed by Saxena et al. (1988) for depositing 5 μm films of YBCO on LiNbO$_3$ substrates. However, the disadvantage of nitrate decomposition approaches is that the removal of nitrogen during heat treatment involves an exothermic reaction, which affects the morphology of the film and poses problems if the film is too thick (Glowacki, 2005). In the aerosol deposition route (Kodas et al., 1989; Jergel et al., 1992; Jergel, 1995), nitrate salts of Y, Ba, and Cu are again used for the starting solution, which is passed through an aerosol generator to form micrometre-size droplets. These are next transported through a furnace at about 1000°C, where the droplets react with O$_2$ carrier gas to form a fine superconducting powder free from contamination, possessing an optimum T_c. This powder can be used to make thick films or sintered pellets. The experimental set-up of Jergel et al. (1992) is shown schematically in Figure 17.5 By using metal–organic solutions instead of nitrates, and also keeping the processing temperatures below 600°C, Jergel (1995) could obtain good-quality thick films also of Bi-2223 and Tl-2223, with J_c(10 T, 4.2 K) $> 10^9$ A m^{-2}.

In the sol–gel method (Barboux et al., 1988), a solution is formed by mixing Y hydroxide with Ba and Cu acetates and converted via complex chemical reactions to a colloidal suspension of fine particles of size 1 nm–1 μm, called a *sol*. This is followed by transformation of the sol into a highly viscous *gel* through successive hydrolysis and condensation reactions (Figure 17.6). The gel is applied to the substrate surface by dip-coating or spin-coating techniques, followed by heating at 800–920°C for a few hours and oxygenation at 400°C for 12 hours to induce superconductivity. The film thickness is controlled by the concentration and viscosity of the gel and the number of times it has been applied (Mutlu et al., 2000). Paranthaman et al. (2001) have reported J_c(77 K) $> 10^{10}$ A m^{-2} for sol–gel samples of YBCO.

Doctor blading and *screen printing* are essentially tape-casting methods that are commonly applied in the semiconductor and ceramic industry. In the case of HTS, they have proved economical for making homogeneous coatings over large surface areas (Riemer, 1988). The doctor blade method involves printing, coating, or spreading the precursor paste or a slurry, generally prepared with organic compounds, on a substrate

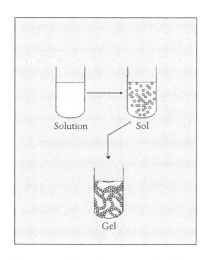

Figure 17.6 Schematic representation of transformation of solution to sol and then to gel.

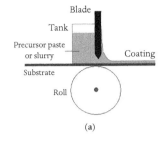

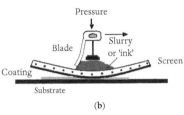

Figure 17.7 Techniques for making thick HTS films: (a) doctor blade technique; (b) screen printing.

or tape using a sharp surgical blade (Figure 17.7(a)). The organic compounds in the slurry are removed by pyrolysis at 600°C for a few hours. This is followed by oxygenation at 800 to 930°C for several hours to make the film superconducting.

Screen printing is carried out (Figure 17.7(b)) by making the slurry, or so-called *ink*, pass through the open areas of a patterned screen by applying pressure so as to *print* the required pattern on the substrate. In the case of YBCO, the ink for printing (Bailey et al., 1991) is prepared using pre-synthesised YBCO by first crushing it into fine powder, possibly with a ball mill. It is then mixed with ethylcellulose to make a paste or slurry. The process is repeated a few times until a homogeneous and sufficiently thick coating is achieved. This is followed by annealing at about 950°C. A modified version of this approach is the method of *ink-jet printing* due to Glowacki (2000). Using a nozzle, well-defined fine droplets of the precursor solution or sol–gel, serving as *ink*, are fired onto the substrate, to print the desired pattern on its surface. The wetting angles of the droplets on the surface, determined by the surface tension of the ink with respect to the substrate, play an important part in achieving resolution of the print. With the right viscosity of the ink used, one can achieve precision printing.

17.2.3.4 *Conventional and hybrid liquid-phase epitaxy*

In the method of *liquid-phase epitaxy (LPE)*, film growth occurs from a liquid flux by having the material in the molten state in direct contact with the substrate surface. The growth therefore occurs near thermodynamic equilibrium, which favours better crystal quality in terms of homogeneity and crystal lattice, accompanied by formation of very flat surfaces (Scheel et al., 1991, 1994; Klemenz and Scheel, 1993, 1996; Scheel, 1994; Yoshida et al., 1994; Yamada, 2000; Yamada and Hirabayashi, 2001) compared with many of the other deposition techniques already discussed. The growth occurs at low supersaturation, which results in layer-by-layer growth, albeit at an emerging screw dislocation site at the surface, giving rise to characteristic growth spirals. In the case of LPE-YBCO film, the seed layer, for instance, may be of Y- or Nd-123 or of $NdGaO_3$, $LaGaO_3$, and $LaSrGaO_4$ (Klemenz and Scheel, 1993; Yamada and Hirabayashi, 2001). It is further necessary to have the melting temperature of the seed greater than, that of the melt which is achieved by adding BaF_2 or Ag to the melt to lower the melting temperature to below 880°C. An obvious drawback of LPE-grown YBCO film is that, although its T_c is high (about 90 K), it generally possesses a much lower J_c(77 K) $< 10^8 A/m^2$, due to lack of sufficient density of flux-pinning centres. However, J_c, for example, can be markedly increased by two orders of magnitude with proton irradiation (Vostner et al., 2003). The success achieved with conventional LPE-grown films of RBCO and BSCCO has been promising with respect to superconducting properties

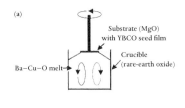

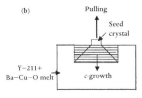

Figure 17.8 Schematic illustration of (a) liquid-phase epitaxy (LPE), which is very similar to (b) Czochralski's top-seeded melt growth (TSMG) process.

(Balestrino et al., 1989; Takeya and Takei, 1989; Scheel et al., 1991; Klemenze and Scheel, 1996; Yamada, 2000; Yamada and Hirabayashi, 2001).

The experimental arrangement for LPE (Figure 17.8(a)) is similar to the *top-seeded melt growth (TSMG)* method (Figure 17.8(b)) used for crystal growing. When a single-crystalline substrate with matching lattice parameters contacts a supersaturated melt, an epitaxial film starts growing on its surface. The thickness of the layer formed is controlled by the melt temperature (greater than about 1000°C) and the duration of their contact. LPE is a complex process and, in the case of YBCO for example, during cooling of the melt, a solid Y-211 phase (Y_2BaCuO_5) crystallises out, which peritectically reacts with the liquid to form Y-123 particles around Y-211. When all the 211 phase is used up, Y-123 continues to crystallise from the molten film, forming Y-123, $BaCuO_2$, and CuO phases through the eutectic reaction (Glowacki, 2005).

In this conventional LPE process, it is problematic to maintain the required supersaturation at the growth interface. This is firstly because of limited solubility of the rare earth R (or Y in YBCO) in the molten flux matrix and secondly because the diffusion of R to the growth front across the stagnant liquid layer is retarded (Kursumovic et al., 2000; Qi and MacManus-Driscoll, 2001). Crystal growth from a thin liquid layer has long been known for whisker and dendrite growth, and such a quasi-liquid layer is found to exist on the growth surface even below the bulk melting temperature. These considerations led to the concept of the vapour–liquid–solid (VLS) growth mode, which was extended to so-called tri-phase epitaxy. This epitaxy mode is known as *hybrid LPE*, or *HLPE*, and was first applied by Yoshida et al. (1996) to a thin liquid layer on a growing YBCO film surface. Ohnishi et al. (2004) found a high rate of YBCO film growth in the HLPE mode when the liquid layer was fed by vapour produced by electron-beam co-evaporation, which allowed the required level of supersaturation to be maintained at the growth interface.

17.2.3.5 *Films of Tl- and Hg-based HTS cuprates*

Much of what we have discussed so far has been for BSCCO and RBCO systems, which do not contain any volatile components. The fabrication of thin films of Tl- and Hg-based HTS cuprates in a conventional way calls for a compromise to be drawn between stoichiometry and volatility of the constituent species, namely Tl and Hg. The problem is circumvented by following a two-step process: (I) deposition of the precursor layer devoid of Tl or Hg, that is, containing only non-volatile species such as Ba, Ca, and Cu, followed by (II) *ex situ* thallination anneal or Hg-vapour annealing (Li et al., 1999; Klimonsky et al., 2002). This step is carried out in a sealed tube under vacuum to prevent evaporative loss which further protects the environment from toxic Tl and Hg fumes. Both

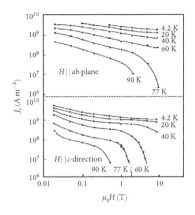

Figure 17.9 Critical current curves at various temperatures for textured thin-film short samples of Tl-1223 formed by the spray pyrolysis technique on a YSZ substrate, measured along and perpendicular to the basal plane (after deLuca et al., 1993). Commendably high values of the critical current density are noted.

Tl- and Hg-based superconductors, such as Tl-1223, Tl-2223, Tl-2212, Hg-1223, and Hg-1212, in thin film form have been successfully synthesised by various processing routes such as spray pyrolysis, aerosol, and MOCVD (deLuca et al., 1993; Adachi et al., 1997; Bramley et al., 1999; Jorda et al., 2004, 2005; Malandrino and Fragalá, 2004; Inoue et al., 2004), and the reported J_c(77 K) values are around $5 \times 10^9 A\ m^{-2}$. Critical current curves with magnetic field along and transverse to the basal plane of a textured Tl-1223 film on a YSZ substrate produced by the spray pyrolysis technique (deLuca et al., 1993) are shown in Figure 17.9.

In the MOCVD technique for coating Tl cuprates (Hitchman and Jensen, 1993), the precursors used for realising the Ba–Ca–Cu matrix, as the first step, are metal–organic complexes, namely tetraglyme-based for Ba and Ca and acetylacetone-based for Cu, and the preferred substrates are YSZ, LaAlO$_3$, and MgO. This is followed by a Tl-vapour diffusion step for realising either single- or double-sided superconducting films of TBCCO. Tl cuprates, owing to their much higher T_c and promising J_c(77 K), remain as major contenders of the second generation of YBCO-coated conductors.

17.2.4 Practical status of YBCO thin films

During the last few years, short samples of YBCO films, grown on different substrates, have been commercially offered for research and applications by a number of vendors in the USA, Japan, Germany, and South Korea. These epitaxial (biaxially textured) films of thickness 0.2–2.0 μm, deposited on single crystals of sapphire, LAO, or MgO, can sustain J_c(77 K, 0T) $> 10^{10} A\ m^{-2}$. The wafers, typically available, are circular (10 or 20 cm diameter) or rectangular (10cm × 20cm) in shape, carrying a uniform layer of HTS, which in turn is coated with a thin Au film to help the user make current contacts for measurements. By replacing expensive conventionally grown single crystals with much cheaper IBAD, ISD, or RABiTS (to be described in Section 17.3) synthesised quasi-single-crystalline substrates, it has been possible to bring down the overall costs of the commercially prepared thin films without making many compromises compared with previously available single-crystal architectures for electronic device applications.

17.3 Conductor development

The various techniques of depositing films and coatings of HTS discussed so far have had a key role in the fabrication of thin-film electronic devices and in conductor development. Conductor technologies for long lengths of wires and tapes have progressed mainly for three systems:

1. BSCCO: Bi-2212, Bi-2223 and (Bb, Pb)-2223;
2. R-123: Y-123 (YBCO);
3. TBCCO: Tl-1223, (Tl, Bi)-1223, and (Tl, Pb)-1223.

The typical short-sample critical current curves (Glowacki, 2004) of these three HTS materials at 77 K are shown in Figure 17.10. Clearly, YBCO stands out as having the best performance. However, for this, the conductor needs biaxial texturing. Although, compared with the other two materials, the performance of BSCCO at 77 K is poor, at lower temperatures it becomes quite impressive. Figure 17.11 compares the $J_c(H)$ behaviour of Bi-2212 with those of the established low-temperature superconductors Nb–Ti and Nb$_3$Sn at 4.2 K (Schwartz et al., 2008), showing the general superiority of BSCCO.

In the case of BSCCO, the popular fabrication techniques used for industrial production of wires and tapes, which are uniaxially textured (along the c-axis), are the *powder-in-tube (PIT)* process and various *dip-coating* methods. For Y-123 and Tl-1223, belonging to the second generation of HTS conductors, the *coated-conductor routes* are followed. These yield biaxially textured thin or thick HTS films deposited on pretextured metallic substrates or suitable pretextured buffer layers using the various physical or chemical deposition routes discussed earlier. The pretexturing of the substrate and the buffer layer are carried out through separate processes. During the last 15–20 years, BSCCO/Ag conductors have been produced in lengths of several kilometres and have attained a practical level of development for technological applications. Ag is the only matrix material that does not react chemically with HTS and allows for adequate oxygen diffusion. The close melting temperatures of BSCCO and Ag, at about 850°C, allow the close electrical and mechanical contacts between the two materials that needed for stable performance of the composite conductor. Coated conductors of Y-123 are in a stage of rapid development, while progress with Tl compounds is promising. In addition, Hg-1212 has also been under consideration as a promising material to form coated conductors (Wu et al., 2002). Some of the characteristic features of these cuprates are

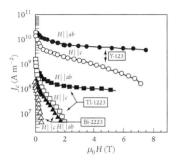

Figure 17.10 Transport $J_c(H)$ behaviour at 77 K of three HTS thin films in parallel and transverse magnetic fields, showing the superior performance of Y-123 (after Glowacki, 2004).

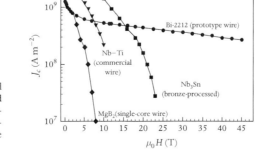

Figure 17.11 Low-temperature (4.2 K) critical current performance of Bi-2212 compared with that of low-temperature superconductors, namely commercial Nb–Ti and bronze-processed Nb$_3$Sn along with that of single-core MgB$_2$ (after Schwartz et al., 2008).

Table 17.1 Some characteristic parameters of HTS conductors

Material	Bi-2223	Bi-2212	Y-123	Tl-1223	(Hg, Re)-1212
Fabrication process	PIT	PIT (partial melting)	Coated-conductor	Coated thin films	Coated thin films
Structure	Ag/Bi2223 MF tape	Ag/Bi2212 MF tape	Hastelloy with textured buffer layer	Substrate LaAlO$_3$	Substrate SrTiO$_3$
T_c(K)	110	95	92	116	122
J_c(A m^{-2})	$(2-3.5) \times 10^8$ (0 T, 77 K)	$> 10^9$ (20 T, 4.2 K)	$(1-2) \times 10^{10}$ (0 T, 77 K)	7×10^9 (0 T, 77 K)	4.9×10^{10} (0 T, 77 K)
I_c	> 100 (0 T, 77 K)	> 300 (20 T, 4.2 K)	> 100 (0 T, 77 K)	> 100 (0 T, 77 K)	> 100 (0 T, 77 K)

Data from Narlikar, A.V. (ed.). *High Temperature Superconductivity 1: Materials*. Springer, Heidelberg (2004).

compared in Table 17.1. The various approaches followed for conductor development of HTS have been extensively reviewed by several authors (Beauquis et al., 2004; Glowacki, 2004; 2005; Hott, 2004; Inoue et al., 2004; Jorda et al., 2004; Jorda, 2005; Su et al., 2004; Yamada and Shiohara, 2004).

17.3.1 BSCCO conductors

17.3.1.1 *Powder-in-tube (PIT) process*

All three BSCCO compounds have been commercially fabricated using the PIT process. Typically, for multifilamentary wires and tapes of Bi-2223, the calcined precursor powders are packed into an Ag tube (Figure 17.12) and drawn into a wire of hexagonal cross-section. This is then cut into short lengths and stacked into an assembly of 7, 19, 37,

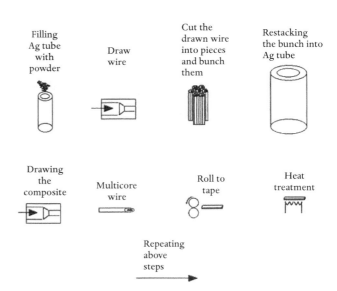

Figure 17.12 Sequential steps of the powder-in-tube (PIT) process for HTS.

55, 61, 85, or more strands inside another Ag tube. The composite is drawn or rolled into a multifilamentary wire or tape form, which is subjected to thermal processing at 822–835°C for 100 hours, which leads to formation of the 2223 phase (Tanaka et al., 1997). This is followed by one or two repetitions of room-temperature intermediate pressing or rolling and then heating for 50 hours each time. This is called thermomechanical processing and gives rise to grain alignment and improved intergrain connectivity. The final thickness of the tape is about 0.2–0.4 mm. The respective J_c and I_c values realised are $(2-3) \times 10^8 \mathrm{A\ m^{-2}}$ and more than 100 A at 77 K and 0 T (Hayashi et al., 2001; Kobayashi et al., 2001). The procedure followed for fabricating $(\mathrm{Bi, Pb})_2\mathrm{Sr}_2\mathrm{Ca}_2\mathrm{Cu}_3\mathrm{O}_{10}$ tapes is similar. It is worth pointing out that in both materials, with and without Pb, the calcined powders that are cold-rolled contain various secondary phases, such as Bi-2212, $\mathrm{Ca}_2\mathrm{CuO}_3$, $\mathrm{CaSrCuO}_2$, CaO, and CuO, as well as (in the Pb-doped compound) $\mathrm{Ca}_2\mathrm{PbO}_4$. This is a serious drawback with the 2223 system in comparison with Bi-2212, where the precursor powders are mostly single-phase and there are no complications arising from secondary phases. This makes Bi-2212 a much cleaner material suitable for high-J_c applications. The formation kinetics of both Pb-doped and Pb-free Bi-2223 phases have been discussed by Grivel and Flukiger (1996, 1998) together with the Bi-2212 phase.

In the case of Bi-2212 (Sato et al., 2001; Hasegawa et al., 2001), the PIT process followed varies slightly from that just described, in that the heat treatments involve the partial melting of the 2212 precursor powders. The material, after cold-working, is thermally processed with the following two alternative heating schedules. In the *slow-cooling (SC)* approach (Kumakura, 1996) depicted in Figure 17.13, the maximum temperature reached is 880°C, which slightly exceeds the melting temperature, after which the rolled conductor is allowed to cool slowly $(5-10°\mathrm{C\ h^{-1}})$ to room temperature, which results in *c*-axis-aligned grains. In the second approach, called *isothermal partial melting (IPM)*, advantage is taken of the fact that Bi-2212 can be crystallised, without increasing the temperature, simply by raising the oxygen partial pressure $P(\mathrm{O}_2)$, which essentially lowers its melting temperature (Strobel et al., 1991; McManus-Driscoll et al., 1994; Funahashi et al., 1999). Both these approaches result in Bi-2212/Ag tapes with high $J_c(4.2 \mathrm{\ K, \ 20 \ T}) \approx 5 \times 10^9 \mathrm{A\ m^{-2}}$, while the engineering critical current density J_c^{Eng}, as determined by the overall cross-section of the conductor, is an order of magnitude lower, namely $J_c^{\mathrm{Eng}} \approx (5-6) \times 10^8 \mathrm{A\ m^{-2}}$ (Kitaguchi et al., 1999a, b; Kitaguchi, 2000). These values are commendably high, making Bi-2212 an ideal choice for conductors in insert magnets to raise the optimum steady magnetic field of $\mathrm{Nb}_3\mathrm{Sn}/\mathrm{NbTi}$ hybrid high-field magnets towards 25–30 T at $T \approx 2.2$–4.2 K. Because Bi-2212 has a lower $T_c \approx 91$–,95 K it can carry an acceptable J_c only below 20 K and is not suitable to function at 77 K. On the other hand, Bi- or (Bi, Pb)-based

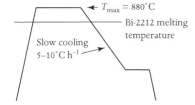

Figure 17.13 Heating schedule for Bi-2212 wire/tape, which involves partial melting of the compound layer (after Kumakura et al., 1996).

2223 conductors, possessing a higher $T_c = 110\,\text{K}$, exhibit flux-pinning behaviour inferior to that of Bi-2212, and to date they have not been considered for magnet applications. However, they are well suited for power transmission cables, functioning at 77 K.

17.3.1.2 *Dip-coating route for BSCCO*

An alternative process that has become popular for the commercial fabrication of BSCCO conductors is the dip-coating route, which, as discussed earlier, includes various approaches such as the doctor blade method (Togano et al., 1992; Shimoyama et al., 1993), screen printing (Glowacki, 2000), spin coating (dos Santos et al., 1991), and sol–gel (Okano et al., 1992). A continuous dip-coating method (Kumakura et al., 1997) that can be used also to fabricate pancake coils of BSCCO for high magnetic fields is schematically illustrated in Figure 17.14.

17.3.1.3 *Status of BSCCO as practical conductors*

Bi-based 2223 and 2212 phases, as 1G conductors, have both already reached a practical level of technical compatibility and lengths of several kilometres are routinely available. (Bi, Pb)-2223 forms the core of today's commercial HTS wire. The composite structure of the conductor generally carries about 30–40% of 2223 phase embedded in an Ag-alloy matrix. The commercially developed 2223 conductors are generally in flat tape form (0.2 mm × 4 mm), with their critical stress at room temperature varying from 65 to 265 MPa, critical tensile strain (77 K) from 0.3% to 0.4%, critical compressive strain (77 K) from 0.15% to 0.25%, and minimum bend diameter from 70 to 100 mm. Long conductors can

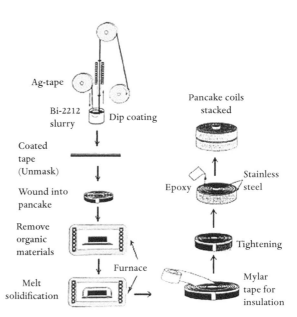

Figure 17.14 Dip-coating process for pancake coils of Bi-2212 (after Kumakura et al., 1997).

sustain an average of 150 A in end-to-end I_c at 77 K. This corresponds to an engineering critical current density of about 1.8×10^8 A m^{-2} and a $J_c \approx 4.5 \times 10^8$ A m^{-2}, while the conductor performance is much improved on lowering the temperature. At 30 K and 2 T, for example, $J_c \approx 1 \times 10^9$ A m^{-2}.

Commercial Bi-2212 conductors are commonly round-shaped wires that, at 4.2 K, can typically sustain current densities of 2.0×10^9 A m^{-2} in 10 T and 1.8×10^9 A m^{-2} in 20 T magnetic fields. They have been used as insert magnets to increase the maximum field strength of conventional LTS magnets in the range of 22–30 T.

17.3.2   Coated conductors of YBCO

For more than two decades, epitaxial thin films of Y-123 on single-crystalline substrates deposited using various physical and chemical techniques have demonstrated an exceptionally sharp superconducting transition at $T_c = 92$K, a very high critical current density $J_c > 10^{10}$ A m^{-2} at liquid nitrogen temperature of 77 K, and an equally commendable irreversibility field H_{irr} (at 77 K). All these features are significantly superior to those of the BSCCO system, and this is what has led the YBCO system to become the second generation of HTS conductors. In this orthorhombic system, with $a \neq b$, the intergrain mismatch in the ab-plane, through which the current flows, gives rise to pronounced weak-link problems that degrade J_c, for which uniaxial texturing of the c-axis alone is not enough. Consequently, the simple approaches followed for fabricating BSCCO conductors have never worked satisfactorily for YBCO and other systems such as Tl-1223 and Hg(Re)-1212, and the conductors so produced have failed to sustain an acceptable J_c(Bellingeri et al., 1998; Yamada and Shiohara, 2004).

The coated conductors possess a complex architecture of stacked functional multilayers as depicted in Figure 17.15(a) and (b). To begin with, there is a metallic substrate tape, which can be made from a range of materials (Glowacki, 2005), including Ni, Ni-based Hastelloy, Ni–Fe, Ni–Cu, Ni–V, Ni–W, Ni–Cr, stainless steel, Ag, and Ag alloys. Of these Hastelloy, stainless steel, Ag, and Ag alloys are used as untextured substrates, while the other materials provide textured tapes. Ni and Ni–Fe are magnetic and are not suitable for superconducting AC applications, but they work satisfactorily for DC applications. Between the substrate and the superconducting layer, there are one or more buffer layers whose purpose is first to prevent any contamination of the superconducting layer from the substrate. Additionally, the buffer layers, if they are properly textured, help to compensate the lattice mismatch of the HTS layer and lead to an improved J_c. The topmost buffer layer, serving as the template for YBCO, is called the *cap layer*. Since the buffer layers are often of insulating oxide materials, the top surface of YBCO is

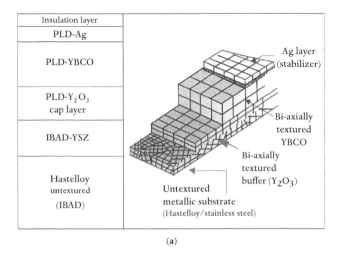

| Insulation layer |
| PLD-Ag |
| PLD-YBCO |
| PLD-Y₂O₃ cap layer |
| IBAD-YSZ |
| Hastelloy untextured (IBAD) |

(a)

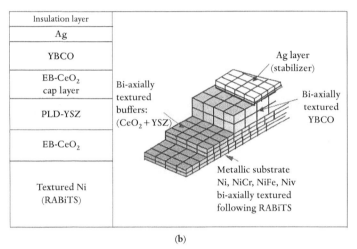

| Insulation layer |
| Ag |
| YBCO |
| EB-CeO₂ cap layer |
| PLD-YSZ |
| EB-CeO₂ |
| Textured Ni (RABiTS) |

(b)

Figure 17.15 Typical schematic illustrations of multilayer architectures formed using (a) an untextured Hastelloy/stainless steel substrate (in the IBAD process) and (b) a textured Ni/Ni-alloy substrate (in the RABiTS process).

coated with a conducting layer such as Ag, which is in turn covered by electrical insulation (not shown in Figure 17.15). The function of the Ag layer is to provide stabilisation and protection against accidental quenching of the conductor. Engineering considerations necessitate that all the functional layers, including the HTS coating and the substrate, should be optimally thin, otherwise J_c^{Eng} is markedly lowered. Ideally, one would like to have a substrate thickness of about 20 μm with a buffer layer less than about 0.5 μm thick and an HTS coating of about 5 μm on both sides of the substrate.

To realise biaxial texturing of YBCO, a variety of processes have been developed, which fall into two groups, based on their texture controlling layer:

(I) Textured buffer-layer-based coated conductors can be formed on untextured substrates, such as Ni-based Hastelloy or stainless steel, using *ion-beam-assisted deposition (IBAD)* (Iijima et al., 1992, 1993) or *inclined substrate deposition (ISD)* (Hasegawa et al., 1997).

(II) For textured metallic-substrate-based coated conductors, including those formed on substrate tapes of various cubic metals such as Ni, Ag, Cu, and their cubic alloys (already mentioned), the template texturing can be achieved through rolling and recrystallisation, primarily as what are known as *rolling-assisted biaxially textured substrates (RABiTS)* (Goyal et al., 1997).

Commonly used techniques for depositing HTS layers are pulsed-laser deposition (PLD), metal–organic deposition (MOD), including trifluoroacetate (TFA)-MOD, chemical vapour deposition (CVD), electron-beam (EB) deposition, thermal evaporation, liquid-phase epitaxy (LPE), and the *ex situ* BaF_2 process. Some of these routes are followed also for depositing buffer layers.

In approach (I), the commonly used substrate is polycrystalline untextured (Ni-based) Hastelloy, on which is deposited a biaxially textured buffer layer (generally an oxide such as YSZ, GZO, or CeO_2), fabricated using IBAD or ISD (Figure 17.15(a)). The textured buffer layer serves as a template for the YBCO layer. In approach (II), the metallic substrate of cubic structure (commonly Ni or a Ni alloy) itself is first biaxially textured to serve as a template for subsequent layers of the architecture (Figure 17.15b).

Some of the prominent cubic buffer materials suitable for biaxial texturing are listed in Table 17.2 along with the deposition techniques

Table 17.2 Various buffer layers, their lattice mismatch with YBCO and Ni, and the deposition techniques used to fabricate them

Buffer	Cubic lattice parameter (nm)	% mismatch with YBCO	% mismatch with Ni	Technique
MgO	0.4210	9.67	17.74	EB or TE
NiO	0.4177	8.89	16.96	SOE, MOCVD
$BaZrO_3$	0.4193	9.27	17.34	MOD, PLD
$SrRuO_3$	0.5573	3.08	11.17	Sputtering
$SrTiO_3$	0.3905	2.16	10.26	MOD
$LaMnO_3$ (LMO)	0.3880	1.60	9.70	Sputtering
$LaNiO_3$	0.5457	0.98	9.07	Sputtering
Eu_2O_3	1.0868	0.54	8.64	MOD
CeO_2	0.5411	0.12	8.22	EB/TE, ED
Gd_2O_3	1.0813	0.07	8.17	MOD, EB
$Gd_2Zr_2O_7$ (GZO)	0.5264	−2.64	5.47	MOD, ED
YSZ	0.5139	−5.03	3.07	Sputtering, PLD

EB, electron-beam evaporation, TE, thermal evaporation, ED, electrodeposition, PLD, pulsed-laser deposition, MOD, metal–organic deposition; MOCVD, metal–organic chemical vapour deposition.
Data based on Glowacki (2005) and Paranthaman (2010).

used. The table also gives information about their lattice mismatch with respect to the Ni substrate (in the RABiTS process) and YBCO. As mentioned earlier, it is quite common to have more than one buffer layer (Beauquis et al., 2004; Glowacki, 2004, 2005; Yamada and Shiohara, 2004; Paranthaman, 2010) deposited by PLD or other deposition techniques to achieve improved texturing and a greater purity of HTS. Some of the commonly used multiple layered buffer architectures that have yielded high $J_c > 10^{10}$ A m^{-2} include Y_2O_3/YSZ, $CeO_2/YSZ/CeO_2$, $CeO_2/Y_2O_3/CeO_2$, $LaNiO_3/YSZ/CeO_2$, $CeO_2,/MgO$, $NiO/YSZ/Y_2O_3$, and $Y_2O_3/MgO/LMO$ (Aytug et al., 2000; Knauf et al., 2001; Nemetschek et al., 2002; Beaquis et al., 2004; Paranthman, 2010). A typical multilayer buffer architecture of a RABiTS-based YBCO conductor is depicted in Figure 17.16. As may be seen from Table 17.2, these materials, in general, possess a relatively smaller lattice mismatch with both YBCO and Ni that favours a better texturing with RABiTS conductors. Also, different deposition techniques, mentioned in Table 17.2, are used to fabricate other functional layers of the architecture.

17.3.2.1 *Texturing of buffer layer*

Ion-beam-assisted deposition (IBAD)

IBAD is the most popular technique for producing a biaxially textured oxide buffer layer template (mainly of YSZ, GZO, CeO$_2$, or MgO) on an untextured polycrystalline metallic substrate, depicted in Figure 17.17. Two sputtering sources used, one for deposition of the buffer material and the other that operates simultaneously to assist in growing a biaxial texture on the deposited layer. In this process, the $\langle 111 \rangle$ axis of YSZ becomes aligned in the direction of the Ar$^+$ beam. Studies have revealed that the ion-beam texturing is most pronounced when the incident ion beam from the second ion source is directed at an angle of 55° to the substrate normal (Iijima et al., 1993). The deposition-texturing process in IBAD takes place at a rate of about 70 nm/h, which is very slow. To circumvent this problem, the IBAD sputtering process is used only to deposit a very thin seed template layer of CeO$_2$, on which YSZ is deposited by the much faster PLD technique (Yamada and Shiohara,

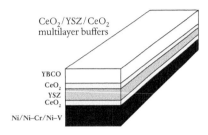

Figure 17.16 Typical RABiTS-based YBCO conductor architecture containing multilayer buffers.

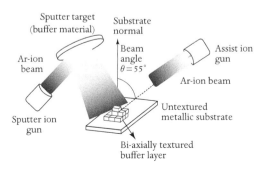

Figure 17.17 Schematic illustration of an ion-beam-assisted deposition (IBAD) system used for making multilayered YBCO conductors.

2004). In terms of the texturing achieved, a 10 nm thick MgO layer grown using IBAD is as good as a 1 μm thick YSZ layer grown using the same technique, which means a substantial time saving is achieved by having a very thin template of MgO (Paranthaman, 2010).

Inclined substrate deposition (ISD)

This technique allows fabrication of textured YSZ (or CeO_2 or MgO) on a Hastelloy substrate using PLD. The in-plane ordering of YSZ is achieved through a shadowing effect by inclining the substrate during deposition, and in ISD the $\langle 110 \rangle$ axis of the buffer layer becomes aligned with the ablation direction. The technique is shown schematically in Figure 17.18. This approach is faster and simpler than IBAD, although the J_c values reported are marginally lower, $8 \times 10^9 A\,m^{-2}$ (Bauer et al., 1999).

17.3.2.2 Metallic substrate and texturing

Rolling-assisted biaxially textured substrates (RABiTS)

This technique (Goyal et al., 1997) exploits the fact that when polycrystalline fcc metals such as Cu, Au, Ag, and Ni and some of their fcc alloys are mechanically rolled and suitably annealed for recrystallisation, the emerging texture is the characteristic cubic one, described by $\{100\}\,\langle 001 \rangle$ (i.e. the rolling plane is $\{100\}$ and the rolling direction $\langle 001 \rangle$). Some of these materials, for example Ag, Ni, and Ni alloys, when rolled and annealed, meet the requirement of biaxially textured substrates for coated conductors, since their lattice constants match closely the a- and b-parameters of the YBCO compound. Moreover, this approach allows

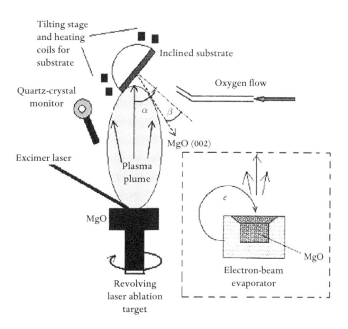

Figure 17.18 Schematic arrangement of inclined substrate deposition (ISD) using a PLD system. The inset shows the EB evaporation system as a common alternative to PLD.

fast fabrication of long lengths of the textured substrate free from the experimental constraints of the IBAD and PLD techniques, which makes RABiTS simple and cost-effective. However, a buffer layer is still needed between the Ni/Ni-alloy substrate and YBCO, since direct deposition of the superconductor on the textured substrate results in its chemical contamination and degradation of superconducting properties. In practice, this problem has been overcome with the multilayer buffers already mentioned (Figure 17.16). The alloying of Ni with solutes such as Cr, V, Cu, and Fe improves its yield strength and does not affect texturing, while Mo, W, Nb and Ta markedly improve texturing, besides adding to its strength (Eickmeyer et al., 2001), and $J_c(77\,K)$ values for such conductors remain quite high, of the order of $10^{10}\,A\,m^{-2}$.

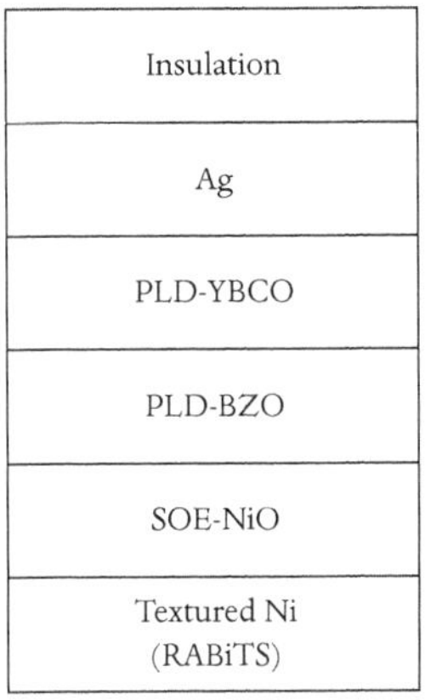

Surface oxidation epitaxy (SOE)

One of the problems encountered in RABiTS is the formation of a non-epitaxial polycrystalline NiO film during buffer deposition. In their approach, called *surface oxidation epitaxy (SOE)*, as a modification of RABiTS, Matsumoto et al. (1999) could control the oxidation of Ni and epitaxially grow NiO as a biaxially textured buffer layer on the Ni tape. Before depositing YBCO, SOE tape was further deposited with a second buffer layer of $BaZrO_3$ (BZO) using PLD (Watanabe et al., 2002), and such a conductor (Figure 17.19) sustained a $J_c(77\,K)$ of $1 \times 10^{10}\,A\,m^{-2}$.

| Insulation |
| Ag |
| PLD-YBCO |
| PLD-BZO |
| SOE-NiO |
| Textured Ni (RABiTS) |

Figure 17.19 Architecture of SOE–YBCO tape.

17.3.2.3 *Practical status of YBCO-coated conductors*

The YBCO-coated tapes, as 2G conductors, developed to date are still to reach the same commercial status that we find for the BSCCO 1G conductors. YBCO-coated conductors have been under intense development internationally for more than a decade, especially in France, Germany, Japan, China, South Korea, and the USA. The central idea is to grow an epitaxial layer of HTS with biaxial texturing on several kilometres' length of a metallic substrate, with buffer and other functional layers introduced to improve the overall texture quality and purity of the coated superconductor. The advantages of very high current density at 77 K and vastly reduced operational costs at liquid nitrogen temperature are expected to make the 2G superconductors cost-effective, and, in fact, manufacturers generally believe that this will bring down the overall cost of 2G superconductors by a factor of 2–5 in comparison with the existing 1G superconductors and lead to their widespread adoption for a variety of technological applications. Progress achieved up to 2010 is indicated in Table 17.3. As may be seen, lengths of more than a kilometre of conductor with a high critical current at 77 K in its self-field has been demonstrated, and it looks promising that longer lengths of 2G superconductors will become readily available in the coming years, allowing widespread exploitation of superconductivity at 77 K. Presently,

Table 17.3 Year-wise progress in length and critical current sustained at liquid nitrogen temperature and under self-field for commercially manufactured YBCO coated conductors

	January 2002	July 2002	July 2003	October 2004	January 2005	July 2005	July 2006	January 2007	January 2008	January 2009	January 2010
Length (m)	—	1.0	18	62	97	158	427	595	795	1030	1065
Critical current I_c (A) at 77 K/0 T	—	< 90	~100	103	105	115	185	185	190	230	285

the coated conductors are being offered commercially by two manufacturers in the USA at a price (2010) of around $40 per metre (Bomberg et al., 2010).

There appears, however, to be a disparity between the level of general technological exploitation of 2G superconductors and the advances in basic research into the properties of these materials that were discussed earlier, one of the prime reasons being that the research studies have led to a rather wide choice of processing techniques and conductor architectures, all yielding high-quality samples of coated conductors. This has led to indecision among potential manufacturers regarding their optimal production strategies and processing routes for long lengths of 2G superconductors. We have discussed many processing routes suitable for industrial production, including RABiTS, IBAD, and ISD for textured buffered substrates. For depositing HTS, the focus is on MOD, MOCVD, PLD, and electron-beam deposition, although the last two, requiring costly vacuum equipment, are options that manufacturers are generally reluctant to adopt. The cheaper production routes are expected to substantially lower the current prices of these conductors in the near future.

17.3.3 Tl-based coated conductors and thin films for applications

Like YBCO, the Tl-based HTS cuprates belong to the second generation (2G) of materials, still under development for applications. Among the four prominent HTS families, Tl cuprates possess a higher T_c than RBCO and BSCCO, a greater oxygen stability than YBCO, a much smaller anisotropy than BSCCO, and a lower toxicity than Hg-based cuprates. Although both single- and double-Tl–O-layer phases (Tl-1223 -1212, -2122, and -2223) have been chosen for magnet and power applications, because of their high critical magnetic field, J_c, and T_c, the single-layer compounds have the general advantage that their J_c is less affected by magnetic fields. This is attributed to the stronger coupling of the superconducting wave functions between Cu–O planes in the single-layer Tl compounds. On the other hand, because of their comparatively lower microwave surface resistance R_s, the double-layer phases are better suited for use in passive microwave devices (Dew-Hughes, 1997). The higher-T_c phase ($T_c = 125$ K) Tl-2223 is attractive for devices functioning at a higher temperature. In conductor development, one aims to achieve a high J_c, while for device applications R_s should be small. There is no correlation between J_c and R_s. The former is determined by the current path and the flux-pinning centres and the latter by intrinsic material characteristics and the surface conditions.

Viable techniques for growing thick and thin films of Tl-based cuprates were discussed in Section 17.2.3.5 and they form the backbone for developing both conductor architecture and thin-film electronic devices. In the case of the former, a convenient conductor architecture consists of LAO, YSZ, and CeO_2-capped YSZ, prepared using the PLD method, which forms the substrate for a Tl-cuprate layer (Ren et al., 1998; O'Connor et al., 1999). In this way, (Tl, Bi)-1212 samples with $T_c = 94-100$ K and $J_c(77$ K$) = 6 \times 10^9 - 1 \times 10^{10}$A m^{-2} grown by PLD and post-deposition annealing on LAO/CeO/YSZ have been reported (Lao et al., 2000). Epitaxial (Tl, Bi)-1212 films have also been grown on YSZ and CeO_2-capped RABiTS metallic substrates (Goyal et al., 1996; Norton et al., 1996) with overall $J_c(77$ K$) > 5 \times 10^9$A m^{-2}. Similarly, other Tl compositions, such as Tl-1223, Tl-2212, and Tl-2223, have manifested promising J_c and T_c values that have made Tl-based HTS strong candidates to serve as 2G HTS conductors for power and high-field magnet applications (Goyal, 2005; Jorda, 2005; Bhattacharya and Paranthaman, 2010).

As already mentioned, the double-layer Tl compounds, synthesised in single-crystal or quasi-single-crystal form, with c-axis texturing, are strong contenders for electronic microwave devices, such as antennae and mixers, microstrip and cavity resonators, oscillators, and filters. Their substrates, besides providing a good structural match with the deposited film, form part of the microwave circuitry, and as mentioned earlier, they are required to fulfil some important criteria that are not very easy to meet: (1) a dielectric loss tangent less than 10^{-5}; (2) a relative dielectric constant of about 10; and (3) isotropy and invariance with respect to temperature and electric and magnetic fields. STO, for instance, provides an excellent epitaxial match, but is ferromagnetic, its relative dielectric constant is more than 1000 at 80 K, and also it is lossy. Further, its growth suits single-layer Tl compounds. Sr from the substrate tends to diffuse at the Ba sites, which leads to deterioration of the Tl-1223 compound. LaO, which is pseudocubic, has loss tangent less than 10^{-5} at 77 K, and does not react with the deposited film, provides a convenient substrate for realising optimum superconducting properties. Its drawback is that it undergoes a structural transition at about 500°C that causes twinning during cooling and anisotropy in its relatively large dielectric constant of about 24. MgO is cubic, its dielectric constant is about 10, and its loss tangent is also less than 10^{-5} at 77 K, and it is ideal from a dielectric point of view. Although its structural match is not ideal for epitaxy, MgO has been found satisfactory for growing device-quality films. A still better substrate for Tl compounds is, however, sapphire, α-Al$_2$O$_3$, which has a loss tangent of 10^{-7} at 77 K and provides a better epitaxial match. However, a thin buffer of CeO$_2$, which provides a structural match for epitaxy, is needed to prevent Al diffusion to the deposited compound layer during the thallination reaction at 850°C. To prevent the formation of barium cerate, this reaction is carried out at a lower temperature of 750°C under a reduced oxygen partial pressure. The optimum thickness for low microwave surface resistance R_s for the device is about 700 nm, since with greater thicknesses the epitaxy deteriorates, with lowering of T_c. Typically, $R_s = 130~\mu\Omega$ for Tl-2212, while it is less than 200 $\mu\Omega$ for Tl-2223 at 10 GHz (Holstein et al., 1993) and 145 $\mu\Omega$ (at 77 K, 10 GHz) for (Tl, Pb)–1223 on an LAO substrate (Lauder et al., 1993).

17.3.3.1 *Practical status of Tl-based conductors and thin films*

Against the background of the experience gained with the fabrication of YBCO coated conductors, the studies described here have confirmed that there are no pertinent practical hurdles in realising long lengths of Tl-based HTS conductors. But, despite their larger T_c than YBCO, coated conductors of Tl-based HTS presently do not seem to offer any distinct advantages in respect of critical current at 77 K. This is because their J_c values are lower than those of YBCO, and further the toxicity and volatility of Tl and the need for two-step processes tend to deter manufacturers from opting for these materials for long conductors. Some of these drawbacks seem to hold also for thin films for fabrication of passive microwave devices, although the practicality and technical advantages of these devices over conventional ones have now been fully established.

17.4   Summary

The focus of this chapter has been on technological compatibility of HTS cuprates and progress made in their thin-film technology and conductor development to meet the challenges posed by practical applications. The complications arising from grain boundaries due to the short coherence length of these materials have been discussed and the importance of texturing has been emphasised as a means to combat them. In order to realise a high J_c(77 K) of the order of 10^{10} A m^{-2} or more for electronic devices or electrical conductors for power engineering applications, deposited HTS films have to be grown with biaxial texturing on suitable single-crystalline substrates. Such films of YBCO are now commercially available for research and electronic applications. In the case of conductor development, there is the seemingly impossible task of achieving the required epitaxial growth of HTS films over several kilometre lengths of the metallic substrate. We have seen in this chapter a variety of ingenious approaches that have been followed to effectively meet this challenge with remarkable success. The state of the art of synthesising HTS films using physical and chemical deposition methods has been presented, and novel techniques for producing complex multilayer architectures of long HTS conductors have been described. The development and the present status of 1G BSCCO conductors have been discussed, and novel strategies pursued with 2G coated conductors of YBCO and other potential HTS have been presented. The 2G YBCO conductors have already reached a practical level of technological development, with manufacturers offering 1 km length with acceptable J_c, albeit at a high cost. With the vast variety of available fabrication routes discussed in this chapter, there is every hope that the coming years will find further improvements in quality and significantly reductions in prices.

CHAPTER 18

Bulk HTS cuprates

During the last 15 years, there have been worldwide efforts directed to develop *bulk HTS cuprates*, especially of the RBCO type 123 system, for both active and passive applications. With ingenious approaches followed in their processing, RBCO pellets can be successfully synthesised possessing the optimum $T_c(>94$ K) and large current densities such as $J_c(77$ K, 1 T) $> 10^8$ A m^{-2} and $J_c(50$ K, 10 T) $> 10^9$ A m^{-2} (Hott et al., 2005). Pronounced flux pinning in bulk material allows *trapping* of a large magnetic induction in the bulk, and thereby such material can serve as a high-field permanent magnet of exceptional stability and homogeneity that far surpass the capabilities of conventional permanent magnets. The only constraint on the functioning of such permanent magnets is that their temperature has to be maintained below T_c. Bulk superconductors are attractive choices for a host of passive applications, including maglev vehicles, magnetically levitated conveyor systems, low-loss magnetic bearings, and flywheels for energy storage. They hold potential for applications in superconducting brushless synchronous motors, magnetic resonance imaging, fusion experiments, and purification of polluted water. Besides, bulk HTS cuprates are being used as high-current leads for energising superconducting electromagnets and various superconducting power engineering devices. Currently, the production cost of these materials exceeds 1000 Euros per kg, which is greater by an order of magnitude than the market price of conventional permanent magnets, such as NdFeB. However, the price of the HTS cuprate materials is expected to fall as their overall production rises to several tons per annum and their applications become wider and more popular. This chapter focuses on the progress made with the fabrication and properties of bulk RBCO and related superconductors. Several reviews that should be useful for supplementary reading have been written on this topic (Jin et al., 1992; Hull, 2001, 2004; Ohara et al., 2001; Tanaka, 2003; Ikuta 2004; Murakami, 1992, 1996, 2005; Wang and Wang, 2005; Muralidhar et al., 2006).

18.1   General considerations

In Chapter 17, we saw that in conductor development and thin-film processing of HTS, T_c and $J_c(H)$ are the most relevant superconducting parameters that are required to be optimised. These are, however, not the only requirements with bulk superconductors. The magnetisation M of a bulk superconductor is given by $M = AJ_cR$, where R is the radius of the domain or grain where the magnetisation current loop is formed and A is a geometrical constant. When the grain size is large, the current loop is bigger and so is its magnetisation. Thus, apart from high J_c, the fabrication process for bulk superconductors must result in large-sized c-axis textured grains. A large J_c along with a bigger grain size R together promote an enhanced M and trapped flux. With the growth techniques presented in this chapter, it is readily possible to fabricate bulk materials with large grains of $R > 5$ cm (Murakami, 2005).

The value of J_c is commonly estimated from the magnetisation of a small sample cut out of a large grain and therefore is not fully representative of the J_c of the entire grain. In the case of bulk superconductors, the essential requirement is that the entire sample or a large-sized grain, and not merely a small part of it, must possess an appreciable J_c. In view of this, besides high J_c and T_c, the most relevant parameter for bulk HTS is the *trapped* (or *remanent*) *field* (or *flux*), to be discussed in Section 18.5. To retain a large magnetic flux, the material is required to withstand a high magnetic pressure due to its trapped flux, under which it can readily break. It should therefore have adequate mechanical strength and structural reinforcement to serve as a high-field permanent magnet. Strategies followed for achieving this are discussed in Section 18.6.

18.2   Melt processing of bulk YBCO samples

Various forms of *melt processing* techniques have been invented to realise textured bulk samples of YBCO and related systems of optimum T_c, possessing large-sized grains with high J_c and trapped field. There are two objectives to be met. First, the texturing must ensure a higher intergrain $J_c(H)$ by overcoming the grain-boundary weak-link and anisotropy problems. More specifically, the number of high-angle grain boundaries formed must be minimised and the grains formed must be textured with c-axis aligned perpendicular to their surface. Second, the techniques must be directed to produce the optimum size and distribution of flux-pinning centres and a high intragrain $J_c(H)$. The most prominent processes pursued are melt-textured growth (MTG) (Jin et al., 1988), quench–melt growth (QMG) (Murakami et al., 1989), melt–powder–melt growth (MPMG) (Murakami et al., 1991), top-seeded melt growth (TSMG) (Lee et al., 1994; Cardwell, 1998), and oxygen-controlled melt growth (OCMG) (Murakami et al., 1994). Of these, MTG was the first to be developed and established the importance of texturing in improving J_c. It did not, however, introduce pinning centres in the material. As a result, the behaviour of MTG samples was essentially that of a good-quality single crystal devoid of defect structure. Consequently, $J_c(H)$ was not sufficiently high. In the other four processes, a high intragrain J_c is achieved by introducing normal particles of Y-211 (i.e. Y_2BaCuO_5) phase as part of the melt processing technique, leading to a considerably enhanced J_c. For achieving the required texturing, all these processes follow a common crucial strategy of cooling the melt very slowly under a temperature gradient, leading to grain alignment and reducing the formation of high-angle grain boundaries. These techniques differ in their ways of varying the size and distribution of 211-phase particles.

The general principle of the melt process for YBCO is that above the reaction temperature of about 1010°C in air (the peritectic temperature), Y-123 decomposes into Y-211 and a liquid phase by peritectic reaction: Y-123→Y-211 + liquid Ba–Cu–O + xO_2. Solidification takes place by the same reaction in reverse, where Y-211 particles react with the liquid phase to produce the stoichiometric Y-123. The latter forms at the expense of the former; that is, the particle size of the 211 phase decreases as the 123 phase grows. If the process were to continue for long, the whole reaction product would turn into Y-123 phase. However, the reaction is stopped at a stage where small unreacted 211 particles are trapped in a Y-123 matrix. The interfaces between 123 and 211 phases serve as effective pinning centres for vortices to enhance the intragrain J_c. By using a starting composition having excess 211 phase, the number density of 211 particles can be increased. $J_c \propto V_f/d$, where V_f is the volume fraction and d the average diameter of 211 particles; this is well corroborated experimentally (Figure 18.1) (Murakami, 1996).

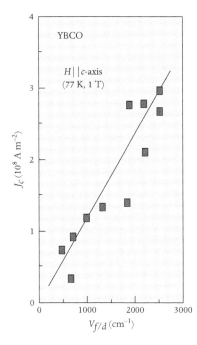

Figure 18.1 Plot showing $J_c \propto V_f/d$, where V_f is the volume fraction and d the average diameter of 211-phase particles (after Murakami, 1996).

18.2.1 Melt-textured growth (MTG)

Melt texturing (Jin et al., 1988, 1989) is essentially a method of achieving a directional solidification of YBCO from its molten state. It involves heating the YBCO compound to a temperature of 1030–1100°C (which is above the peritectic temperature of about 1010°C at ambient pressure) in an O_2 atmosphere for its peritectic decomposition, followed by slow successive cooling (2–10°C h^{-1}) in different temperature regions as shown in Figure 18.2(b); the sectional phase diagram is depicted in Figure 18.2(a). As the natural preference of growth of YBCO is along the basal plane directions in comparison with the c-direction, the molten Y-123, even in the absence of temperature gradients, tends to solidify as large parallel plates if the cooling rates are maintained low. The role of the temperature gradient is mainly to avoid multiple nucleation of these parallel plates at different sites during crystallization, and this promotes perfect alignment and stacking of plates separated by

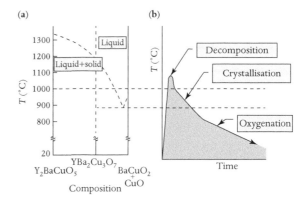

Figure 18.2 (a) Sectional phase diagram of the YBCO system and (b) heating schedule for melt texturing (after Jin et al., 1988, 1989).

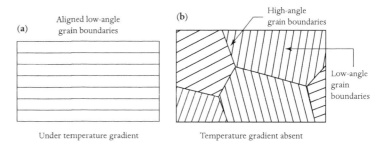

Figure 18.3 (a) There is grain-boundary alignment under a temperature gradient, but (b) no such alignment takes place in the absence of a temperature gradient.

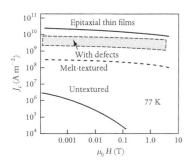

Figure 18.4 Typical J_c values of a melt-textured Y-123 sample (with and without defects) compared with those of an untextured bulk sample and an epitaxial thin film (after Jin et al., 1992). The performance of the bulk-textured sample is second only to that of the thin-film sample.

low-angle grain boundaries, as depicted schematically in Figure 18.3(a). The structure is devoid of high-angle boundaries. In the absence of a temperature gradient, however, multiple nucleation takes place, favouring a grain structure with multiple high-angle grain boundaries that are randomly distributed as shown in Figure 18.3(b), although within an individual grain the plates are uniformly stacked. The overall J_c and $J_c(H)$ of such material, in which the current path has to cross multiple high-angle grain boundaries, are vastly reduced owing to the weak-link and anisotropy effects.

The MTG process, as it did not introduce pinning centres, led to low J_c (77 K, 0 T) $= 1.7 \times 10^8 \mathrm{A \ m^{-2}}$ and J_c (77 K, 1 T) $= 4 \times 10^7 \mathrm{A \ m^{-2}}$ (Jin et al., 1988). But, for the first time, MTG brought out the advantages of texturing. Figure 18.4 shows J_c of differently processed YBCO samples. As may be seen, the MTG grain-aligned sample has a much improved $J_c(H)$ compared with the random-grain sample. The figure further illustrates the importance of defects as pinning centres and epitaxial deposition in further enhancing the critical current behaviour.

18.2.2 Quench–melt growth (QMG)

The peritectic decomposition of 123 phase is just one way of realising 211-phase particles during cooling of the melt. Murakami et al. (1989) noted that during solidification of the melt, in the reverse reaction, the particles formed were too coarse, with a rather uneven size distribution that did not favour optimum pinning. They instead followed an alternative approach of producing 211-phase particles by peritectic decomposition of Y_2O_3 through the reaction Y_2O_3 + liquid phase $\rightarrow$ Y-211, occurring above 1200°C. This led to a more even distribution of the finer particles of 211 phase in the melt. In their QMG process, the sample was quickly heated to above 1200°C and splat-quenched to achieve a finer distribution of 211-phase particles. The quenched sample was subsequently remelted in the 211 + liquid-phase region and allowed to solidify very slowly to yield a highly textured 123 phase with very fine whiskers of 211 phase for improved pinning. The sample thus produced, after oxygenation at about 500°C, exhibited J_c (77 K, 1 T) $> 10^8$ A m^{-2}.

18.2.3 Melt–powder–melt growth (MPMG)

In their QMG approach, discussed above, where 211 phase formed was nucleated at Y_2O_3, Murakami et al. (1991) subsequently discovered that agglomeration of the latter caused a non-uniform distribution of 211-phase particles in different regions of the sample, which resulted in a scatter in the measured J_c values and trapped fields. To overcome this problem, they powdered the quenched material, mixed and compacted it, and, after remelting, left it for very slow cooling in a thermal gradient, which caused texturing. The sample is required to be oxygenated for a few hours at about 500°C for optimum superconducting properties. When the pristine powder composition is 211-rich, the cooling results in 211-phase particles dispersed in a 123-phase matrix, which contributes to enhanced flux pinning as well as to mechanical strengthening. Murakami et al. realised that one could dispense with the quenching step of QMG and instead simply use the melt–powder (M–P) step, in a process called MPMG. The thermal processing schedule of MPMG is illustrated in Figure 18.5. The observed J_c (77 K, 1 T) $> 3 \times 10^8$ A m^{-2} obtained by this method, together with a commendable trapped flux observed in magnetisation studies, made the material attractive for levitation and magnetic-bearing devices. This process had five advantages:

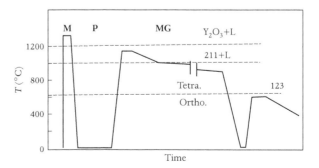

Figure 18.5 Heating schedule of the MPMG process (after Murakami et al., 1991; Murakami, 1992).

the size of the Y_2O_3 can be varied by mechanical powdering; the distribution of Y_2O_3 is uniform, free from the agglomeration found in QMG; rapid heating and quenching is not required; samples of various sizes and shapes can be realised; and additives can be introduced during solid state reaction. For example, extra preformed fine particles of 211 phase can be added to enhance the number density of pinning centres.

18.2.4 Top-seeded melt growth (TSMG)

This process involves (Murakami, 2005) seeding a thin crystal on top of the molten sample of YBCO (Figure 18.6) with matching lattice parameters so as to produce oriented grain nucleation during solidification. In order to induce c-axis texturing, the most suitable seed crystal for YBCO is Nd-123 or Sm-123 having the same structure and near-matching lattice parameters as Y-123. Also, both these compounds have a higher melting temperature than the peritectic decomposition temperature of YBCO and are therefore stable under thermal processing. In this situation, called *cold seeding*, the seed crystal can be mounted on top of the pellet before the melting process begins. A new cold seeding approach and its heating schedule are depicted in Figure 18.7 (Wu et al., 2009). If the melting and peritectic temperatures are close, the seed crystal is introduced only after the peritectic temperature has been crossed and the melt cooling has begun, which is called *warm seeding*. But, in any case, the decomposition temperature of the seed has to be sufficiently high to prevent its chemical reaction with the growing material. Alternatively, one must add some Ag, Ag_2O, or Au powder to the material to bring down its peritectic temperature to below that of the seed crystal. Interestingly, these additives that lower the decomposition temperature also help in refining the 211-phase particles and simultaneously make the material mechanically stronger against cracking (Ikuta, 2004;, Murakami 2005). Addition of Ag, for example, improves the fracture strength of large-grain YBCO from 1 MPa to nearly 10 MPa (Mityamoto et al., 1999). Much effort has been directed at reducing the size of 211-phase (or Nd-422) particles for enhancing flux pinning. In general, the size varies from 1 to 20 μm, while the lowest size measured is confined to the submicrometre rage for particles of Gd-211. For achieving texturing, as with the other approaches discussed previously, the molten material is allowed to cool very slowly in the presence of a thermal gradient, with, in the final stage, oxygenation at 500°C to optimise T_c and other superconducting properties.

18.2.5 Oxygen-controlled melt growth (OCMG)

In the case of melt growth of R-123 containing lighter rare-earths (LREs), such as R = Nd, Sm, and Gd, the 123 phase formed is found to

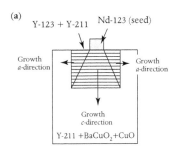

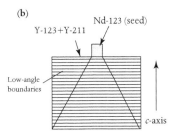

Figure 18.6 (a, b) Schematic illustration of the grain growth of Y-123 in the TSMG process with Nd-123 as the seed.

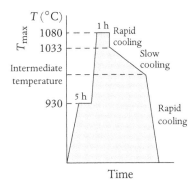

Figure 18.7 Thermal schedule for cold seeding in TSMG (after Wu et al., 2009).

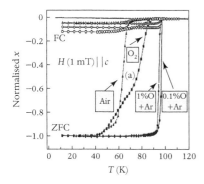

Figure 18.8 Superconducting transition curves (variation of magnetic susceptibility with temperature) of bulk Nd-123 samples synthesised under reduced oxygen partial pressure (after Murakami, 1996).

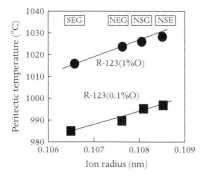

Figure 18.9 Processing under a reduced oxygen atmosphere lowers the peritectic temperature (after Murakami, 2000). SEG, NEG, NSG, and NSE represent RBCO compounds where a single rare earth R is replaced by three light rare earths in equal proportion (33%) from the group Nd, Sm, Eu, and Gd.

contain anti-site defects where the LRE, having a trivalent 3+ state, tends to occupy Ba^{2+} sites, leading to an increased rare-earth stoichiometry (e.g. $Nd_{1+x}Ba_{2-x}Cu_3O_7$). This results in a lower carrier concentration and reduced T_c. However, the formation of such anti-site defects becomes unfavourable if the melt growth is performed under a reduced O_2 atmosphere (Murakami et al., 1994; Murakami, 1996) and, as may be seen in Figure 18.8, this enhances T_c. Processing under a reduced atmosphere (in Ar with 1% partial pressure of O_2) essentially lowers the peritectic temperature of $Nd_{1+x}Ba_{2-x}Cu_3O_7$, which prevents formation of the anti-site phase. In Figure 18.9, this effect is seen for the triple-rare-earth-based R-123 compounds, to be described later. This processing favours the formation of smaller submicrometre-size Nd-422 particles for higher J_c. Although the bulk of the synthesised material in a reduced oxygen atmosphere is stoichiometric, possessing the highest $T_c = 96$ K, it nevertheless carries a distribution of small local regions of 3–5 nm size (Murakami, 2005) where the Nd concentration is larger and where T_c is locally suppressed. Such regions give rise to the previously described ΔT_c *pinning*, which is essentially part of a more general κ-*pinning*, discussed in Chapter 6. These regions serve as weakly superconducting pins for flux lines and cause J_c and magnetisation to increase with magnetic field and temperature. The ensuing *peak effect* (or *fish-tail effect*) is illustrated in Figures 18.10 and 18.11, showing secondary peaks in the magnetisation and J_c–H curves (Murakami, 1996). Figure 18.12 schematically presents the distribution of such regions, which, under low applied magnetic fields, are superconducting and exhibit low flux pinning. Under high magnetic fields, when these regions become normal, although the bulk of the matrix is still superconducting, they provide a much stronger pinning and a higher J_c. A similar effect results with increasing temperature. An increase in J_c with H is the characteristic feature of the OCMG-processed 123 compounds formed with LREs, and is generally not manifested by 123 systems based on Y or heavier rare earths. This has made the LRE-123 systems particularly important for fabricating bulk superconductors for applications.

Figure 18.10 Nd-rich 123-phase regions formed in stoichiometric Nd-123 serve as weakly superconducting pinning centres that become normal in high field and exhibit enhanced pinning and larger magnetisation and J_c (after Murakami, 1996, 2005). This gives rise to a peak effect in the $J_c(H)$ and magnetisation curves (see Figures 18.11 and 18.12).

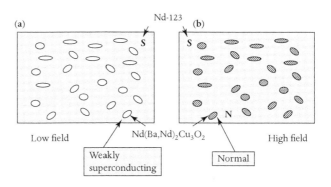

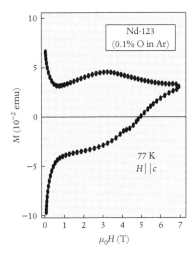

Figure 18.11 Peak effect in magnetisation of Nd-123 (after Murakami, 1996).

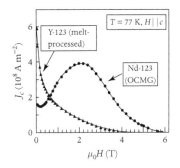

Figure 18.12 Peak effect in $J_c(H)$ behaviour of Nd-123 (after Muralidhar et al., 2000).

18.3   Effective pinning centres in bulk HTS

18.3.1   The 211 phase

Melt-processed samples, as revealed by transmission electron microscopy (TEM) (Murakami, 1992, 1997) may possess a highly complex defect structure that contains a variety of intrinsic crystal defects such as twin boundaries, stacking faults, dislocation networks, microcracks, micropores, oxygen defects, and composition variations that, in principle, can serve as pinning centres for flux vortices in superconductors. However, since the coherence length in YBCO is small, $\xi \approx 5-15$ nm, most of these structural features are believed to be too coarse to cause dominant pinning. Pinning in bulk HTS is basically of two kinds: *background pinning* due to fine normal particles of Y-211 phase and ΔT_c *pinning* resulting from composition fluctuation. As discussed earlier (Figure 18.1) for 211-phase particles, $J_c \propto V_f/d$, which means that the volume fraction of 211-phase particles needs to be large and their coarsening (i.e. the increase in their diameter d) should be prevented to achieve a higher J_c. The common steps for realising this during sample processing are (i) rapid heating to the decomposition temperature, (ii) lowering of the peritectic temperature by suitable additions (e.g. Ag, Ag_2O, or Au), and (iii) reducing the hold time of the sample at the peritectic temperature. Additives such Ce, Pt, CeO_2, and PtO_2 have an important role in two ways. First, they serve as nucleation sites for 211 phase. Second, their presence enhances the boundary energy between the 211 phase and the liquid phase, which reduces the coarsening of the particles by Ostwald ripening (Ikuta, 2004). Generally 0.5% of additives are adequate for an optimum outcome, since their effect in refining the 211-phase particles becomes saturated at higher concentrations. For optimum pinning, the pin size has to be twice the coherence length (i.e. it should be in the nanometre range). The smaller peritectic temperature of Gd-123 favours refinement of the particle size of Gd-211 during sample processing, and these secondary-phase particles are therefore preferred for improved pinning. The smallest size of a few tens of nanometres has been realised for Gd-211 by using the ball-milling approach in conjunction with Ag addition (Nariki et al., 2002a). Ball milling with Y_2O_3–ZrO_2 balls has been used to reduce the size of commercial Gd-211 particles for incorporation in different 123-type superconductors. In this way, the particle size of Gd-211 has gone down from 20 μm to about 70 nm. For improved flux pinning, it is found effective to supplement up to 40 mol% of extra Gd-211 particles besides those resulting from peritectic decomposition. The 211-phase particles yield high J_c values in low magnetic fields, whereas more effective ΔT_c pinning, discussed in Section 18.3.2, provides high $J_c(H)$ values.

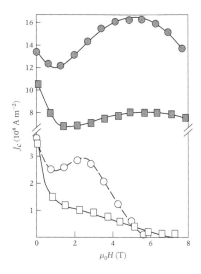

Figure 18.13 Both undoped (rectangles) and Zn-doped (circles) YBCO samples exhibit the peak effect at 55 K (filled symbols) and 75 K (unfilled symbols) (after Krabbes et al., 2000).

18.3.2 Pinning centres with ΔT_c pinning

Although the problem of anti-site defect formation is largely overcome in bulk samples of Nd-123 by application of the OCMG process, there are fine regions of 30–50 nm size still present in its solid solution, where T_c is locally reduced owing to some excess Nd occupying the Ba sites of the 123 crystal structure. A local variation in T_c in these nano regions results in strong ΔT_c *pinning* and the peak effect as mentioned above. Such pinning centres are intrinsic to 123-type compounds formed with LREs such as Nd, Sm, Eu, and Gd, although their effect decreases in that order as their atomic weight rises and ionic size decreases. Surprisingly, however, there are reports of secondary peaks in $J_c(H)$ behaviour in 123-type compounds of the heavier rare earths Dy and Y, which normally do not form anti-site defects (Nariki and Murakami, 2002; Ullrich et al., 1993; Matsushita et al., 2001; Krabbes et al., 2000, 2002). The results of Krabbes et al. (2000, 2002) showing a peak effect in undoped Y-123 at 55 and 75 K are shown in Figure 18.13. Their origin has been attributed to oxygen-depleted clusters present in their solid solution (Nariki and Murakami, 2002). Around such clusters, T_c is locally suppressed and they effectively provide ΔT_c pinning.

In the heavier-rare-earth-based 123 compounds, the peak effect can result also from Cu-site substitutions. As discussed in Chapter 16, local disorder in Cu–O planes of Y-123, such as produced by Zn substitution in small concentration (about 0.1%), can effectively pin the pancake-type vortices formed in these planes (Krabbes et al., 2000, 2002). This creates small local regions of suppressed T_c in CuO_2 planes, which cause ΔT_c pinning, leading to the peak effect. In this way, with Zn substitution, the trapped field can be significantly enhanced, much beyond that of the pristine sample (Krabbes et al., 2002). The peak effect at 55 and 75 K of Zn-doped Y-123 is displayed in Figure 18.13.

18.3.3 Artificial pinning centres

Nanosized defects can be artificially introduced into superconductors, in the form of columnar defects or local disordered regions, by means of heavy ions (Civale, 1997) and neutron irradiation (Sauerzopf, 1991; Wacenovsky et al., 1993; Aleksa et al., 1998), which results in a dramatic increase in J_c. In addition, by doping the material with fissionable U, followed by neutron irradiation, one can cause internal damage through fission fragments (Thompson et al., 1997), resulting in marked rise in J_c. These approaches are, however incompatible, with mass production for large-scale applications. A more feasible approach is to introduce artificial pinning centres in the form of MgO nanorods (Yang and Lieber, 1996), carbon nanotubes, nanodiamonds (Fossheim et al., 1995; Huang et al., 1997), etc., as discussed, for example, in Chapter 15 for MgB_2.

18.4   Ternary 123 bulk compounds

The 123-type HTS compounds, when formed using the OCMG process, with more than one LRE element at the rare-earth sites of the 123 structure, are noted for their exceptionally high J_c and strong flux pinning (Muralidhar et al., 2000). These are $(NdEuGd)_3Ba_2Cu_3O_7$ (for short NEG-123), NdSmGd-123 (for short NSG-123), and SmEuGd-123 (for short SEG-123). Here, the properties can be tailored by suitably changing the ratios of the three rare-earth elements at the rare-earth sites. The secondary 211-phase particles of approximately micrometre size produced during peritectic decomposition of these systems comprise the corresponding three mixed rare earths at the rare-earth sites of the 211 phase. The way in which their peritectic temperature varies with changing ion size and oxygen partial pressure is shown in Figure 18.9. For these materials, TEM has further revealed submicrometre-sized particles, identified as Gd-211 (Koblischka et al., 2000). Each of these LREs has a distinct role in superconductivity: Nd enhances flux pinning at low fields, Gd enhances flux pinning at intermediate and high fields, and Eu controls the position of the secondary peak in the $J_c(H)$ curve (Muralidhar et al., 1998, 2001). Commonly, each LRE is at an equal concentration of 0.33. The ball-milling approach, using Y_2O_3–ZrO_2 balls, has led to the formation of two new types of Zr-rich nanosized precipitates caused by Zr diffusion. Of these, (LRE) Ba_2CuZrO_y of 40–60 nm size and the still smaller (LRE, Zr)$BaCuO_y$ (Muralidhar et al., 2003, 2006) of less than 30 nm are responsible for exceptionally large pinning observed even at a liquid oxygen temperature of 90.4 K in the close vicinity of T_c.

As an illustration of the high J_c values observed, Figure 18.14 shows the field dependence (with H parallel to the c-axis) of the critical current density (at 77 K) of NEG-123 samples containing 30 and 40 mol% Gd-211 particles refined to 200, 100, and 70 nm size by ball milling for 0.3, 2, and 4 hours. While the commercial Gd-211 powder yields much lower values, the presence of refined 211-phase particles results in remarkably high optimum $J_c \approx 1.40 \times 10^9$ and 1.92×10^9 A m^{-2} respectively for 30% and 40% samples in their remanent fields at 77 K. The 40% sample manifests a high value of 1.10×10^9 A m^{-2} at 77 K and 3 T. For the first time, these materials, containing the newly discovered Zr-rich phases previously mentioned, revealed equally commendable J_c values at liquid oxygen temperature of 90.4 K, as depicted in the inset of Figure 18.14. Similar large $J_c(H)$ values were obtained also for other triple LRE-123 compounds (Muralidhar et al., 2004a, b, 2006), and have been improved still further (Muralidhar et al., 2005, 2006). The superior electromagnetic properties of these triple LRE-123 superconductors are believed to be due primarily to a uniform distribution of Gd-211 particles of relatively small size (100 nm), much smaller

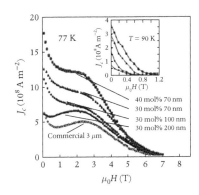

Figure 18.14 Very high $J_c(H)$ of NEG-123 containing 30–40 mol% of Gd-211 particles of 70–200 nm size. Also shown is the inferior performance resulting from coarse commercially produced Gd-211 particles. The inset shows an equally commendable performance at liquid oxygen temperature of 90 K. (After Muralidhar et al. (2006).)

(<60 to 30 nm) precipitates of Zr-rich phases previously described, and the presence of anti-site disorder in the mixed LRE solid solution.

18.5 Trapped field

The trapped (remanent) magnetic field (flux) directly manifests the capability of the entire sample to pin the magnetic flux when the external magnetic field is switched off after high-field magnetisation. In order to evaluate the optimum performance of the bulk sample, it is necessary to magnetise it in a sufficiently high magnetic field exceeding the expected trapped-field limit. In the *quasi-static magnetisation* method (Figure 18.15(a)), the trapped field can be recorded with the sample in either the *field-cooled (FC)* or *zero-field-cooled (ZFC)* state. In the former case, the sample is first magnetised just above T_c in sufficiently high magnetic field and then cooled to the required temperature in the superconducting state when the applied field is removed. In the ZFC case, the sample is first cooled to the required temperature in the absence of an external field. A high magnetic field is subsequently applied and removed to trap the field in the sample. The external field required in the FC case is half that in the ZFC case, and the former method of magnetisation is therefore preferred for realising a trapped field. Very large trapped fields generally call for *pulsed-field magnetisation (PFM)*, which exceeds the present limit of DC fields (Figure 18.15(b)). With PFM, there is an unavoidable movement of flux vortices past pinning centres, which leads to heating of the sample and dissipation of the trapped field. This problem can be overcome by lowering the frequency and amplitude of the pulses applied and allowing sufficient cooling time between successive pulses.

Experimentally, the trapped field is measured by mounting a Hall sensor on the sample surface and mapping the field distribution by scanning the sensor over the surface. However, the trapped field thus

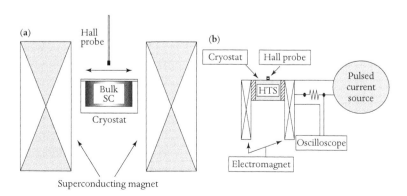

Figure 18.15 Methods followed to produce and measure trapped flux: (a) DC magnetisation and (b) pulsed-field magnetisation.

measured at the surface is less than the actual value in the sample interior. Similarly, the net field present between two closely placed samples, each having a trapped field, may exceed their individual trapped field values, and in this way higher magnetic fields can be produced by placing many samples close together. A cylindrical sample with an axial bore containing a trapped field may serve as a solenoid having a magnetic field in the open space of its bore. Trapped fluxes reported in different 123-type bulk HTS are given in Table 18.1. The trapped field decreases rapidly as the temperature is increased, which, for illustration, is shown in Figure 18.16 for 30 mm diameter NdBCO and SmBCO samples (Ikuta et al., 2000). If the sample contains cracks, macroscopic inhomogeneities, or weak links, or is of poor quality, having a low density of flux-pinning centres, then (i) the trapped field is markedly reduced and (ii) it decays with time owing to enhanced flux creep effects. In the ideal case, the trapped flux density varies uniformly, with a maximum at the centre of the material (Figure 18.17(a)). The presence of cracks causes the trapped flux to escape non-uniformly, leaving many smaller peaks as shown in Figure 18.17(b). During the last 10 years, with improved fabrication techniques, trapped fields in YBCO and related bulk materials have reached very high values, well beyond the requirements for many applications.

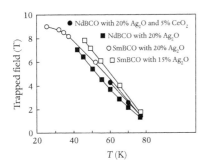

Figure 18.16 Trapped field as a function of temperature for 30 mm diameter NdBCO and SmBCO samples (after Ikuta et al., 2000). The additives in the starting powder are indicated for each sample.

18.6 Mechanical strengthening

The intrinsic mechanical strength of the material turns out to be the limiting factor holding the trapped magnetic field. This calls for structural reinforcement of the material for practical use. The presence of fine 211-phase particles, of course, helps to withstand large magnetic pressures, but that alone is not adequate. Addition of Ag or AgO_2, which helps in refining 211-particle size, also prevents cracking of the sample. Clamping of the sample using a metallic ring or encasing it in a metallic jacket are other reinforcement strategies that have been followed (Fuchs et al., 2000). In addition, an impregnation approach, with phenol–formaldehyde resin plus polyamide-imide as hardener, has been used to fill up cracked regions at the sample surface (Tomita and Murakami, 2000, 2003). This also protects the surface against corrosion due to moisture. When the bulk sample is vacuum-impregnated with epoxy resin, the latter permeates into the bulk through surface cracks to fill internal voids and cracks, and this can lead to a tenfold increase in mechanical strength. The impregnation technique has been further improved by using various 123-type compounds as fillers added to the resin for filling up voids and cracks in the sample (Tomita and Murakami, 2001).

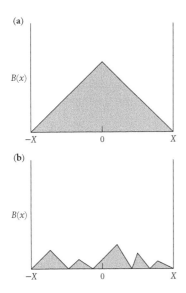

Figure 18.17 Schematic illustration of the trapped field densities of (a) an ideal and (b) a cracked cylindrical bulk sample.

Table 18.1 Trapped field in various bulk superconductors of 123 type

Material	Sample diameter (mm)	Temperature (K)	Distance of probe sample surface (mm)	Hall from Trapped field (T)	Reference
Sm-123 (10% Ag$_2$O)	36	77	0.64	1.7	Ikuta et al. (1998)
Sm-123 (10% Ag$_2$O)	30, 36	77	2.0	2.1	Ikuta et al. (1998, 1999); Mizutani et al. (1999)
Sm-123 (10% Ag$_2$O)	30, 36	25	2.0	9.0	Ikuta et al. (1998,1999); Mizutani et al. (1999)
Sm-123 (10% Ag$_2$O)	45	77	2.0	1.3	Kohayashi et al. (2000)
Sm-123 (10% Ag$_2$O)	60	77	Contact	2.0	Nariki et al. (2003)
Sm-123	24	47	Contact	13.69	Tomita and Murakami (2003)
Nd-123 (15–20% Ag$_2$O)	30	77	Contact	1.4	Ikuta et al. (2000)
Nd-123 (15–20% Ag$_2$O)	30	40	Contact	8.0	Ikuta et al. (2000)
Nd-123 (15–20% Ag$_2$O)	30	77	Contact	1.8	Murakami (2005)
Gd-123 (Ag-doped)	32	77	0.5	1.5	Nariki et al. (2000)
Gd-123 (20% Ag$_2$O)	32	77	1.2	2.0	Nariki et al. (2002)
Gd-123 (20% Ag$_2$O)	32	55	0.7	6.7	Nariki et al. (2002)
Gd-123 (20% Ag$_2$O)	32	77	1.2	2.05	Nariki et al. (2002a)
Gd-123 (20% Ag$_2$O)	50	77	1.2	2.54	Nariki et al. (2002a)
Gd-123 (20% Ag$_2$O)	50	77	0.7	2.7	Nariki et al. (2002a)
Gd-123 (20% Ag$_2$O)	48	77	0.7	3.34	Nariki et al. (2002b)
Gd-123	50	77	Contact	2.6T	Nariki et al. (2002b)
Gd-123	65	77	Contact	3.05	Nariki et al. (2003)
Gd-123 (two pellets)	65	77	Contact	4.3	Nariki et al. (2003)
Dy-123 (10% Ag$_2$O)	32	77	1.2	1.4	Nariki et al. (2001)
Dy-123 (10% Ag$_2$O)	48	77	1 ?	1.7	Nariki et al. (2001)
Y-123	45	77	Contact	0.72	Fuchs et al. (1997)
Y-123 (Ag-doped)	26	22.5	1.0	14.35	Fuchs et al. (1997)
Y-123 (Ag-doped)	26	51.5	1.0	8.5	Fuchs et al. (1997)
Y-123 with 1.12% Zn	25	77	Contact	1.12	Krabbes et al. (2000)
Y-123 without Zn	25			0.75	
Y-123 (two pellets)	26.5	29	Contact	17.24	Tomita and Murakami (2003)
Y-123 (four pellets)	20	77	Contact	3.1	Weinstein et al. (1996)
Y-123 (four pellets)	20	42	Contact	10.1	Weinstein et al. (1996)
Y-123 (four pellets)	20	62	Contact	8.1	Ren et al. (1997)
NEG-123	22	77	Contact	0.7T	Muralidhar et al. (2002)
NSG-123	30	77	Contact	1.2	Muralidhar et al. (2002a)

18.7   Summary

In this chapter, we have reviewed the spectacular developments in bulk HTS superconductors for high-magnetic-field applications and have seen that the values of trapped magnetic field realised to date have surpassed the requirements for many applications. Some commercial products utilising bulk HTS as a new type of permanent magnet are already on the market, including high-field generators for laboratory experiments, magnetic separators for water purification, superconducting motors, and magnetron sputtering devices. With the advent of triple-LRE-based HTS, as a new class of superconductors possessing large $J_c(H)$ values in the liquid oxygen temperature range, it has been possible to perform levitation applications at these temperatures, which until recently was considered unfeasible. Various ingenious approaches invented for fabrication of bulk HTS for practical use have been discussed in this chapter, including numerous strategies to optimise their superconducting and mechanical properties. Presently, the latter seem to set a limit to the optimum magnetic field that can be trapped without the material becoming cracked. Here the resin impregnation approach seems to hold significant promise.

CHAPTER 19

Ruthenates and ruthenocuprates

Ruthenates and ruthenocuprates form an intriguing class of superconductors that were discovered in the mid 1990s. The former emerged as a result of extensive search to find an alternative transition metal oxide to fully replace the copper oxide of HTS to yield a possible new series of superconductors. Surprisingly, of all transition metals, only Ru could replace Cu in LSCO to exhibit superconductivity, albeit with a low $T_c \approx$ 1.5 K. But, interestingly, this ruthenate provides an example of unconventional spin-triplet pairing. Another novel class of superconductors to be discussed in this chapter are the *ruthenocuprates*, the offshoots of ruthenates and cuprates. We have come across many materials depicting an interesting interplay of low temperature superconductivity and magnetism. In ruthenocuprates, we revisit this exciting scenario displaying an unusual interplay of high-temperature superconductivity with weak ferromagnetic (W-FM) order.

This chapter focuses on the novel features of superconducting ruthenates and ruthenocuprates, which in recent years have become front runners in basic superconductivity research. Several extended papers and review articles have been published on them that the reader should find stimulating (Lorenz et al., 2003; Felner, 2003, 2006; Ovchinnikov, 2003; 2003a, b; Braun, 2005; Chu et al., 2005; Nachtrab et al., 2006; Klamut, 2008).

19.1 A superconductor in the ruthenate family: Sr_2RuO_4

The members of the ruthenate family possess the generic formula $(Sr/Ca)_{n+1}Ru_nO_{3n+1}$ (Ruddlesden and Propper, 1958), where n is the number of Ru–O sheets forming the layered structure. Interestingly, each member of the family, based on either Sr or Ca, having different n values is found to possess different properties, ranging from ferromagnetic (FM), antiferromagnetic (AFM) to superconducting (Table 19.1) for $n = 1, 2$, and infinity. The materials with $n > 2$ are hard to synthesise, although $Sr_4Ru_3O_{10}$ ($n = 3$) has been reported to be an itinerant ferromagnet (Ovchinnikov, 2003a).

Table 19.1 Principal features of Sr- and Ca-based ruthenates with single, double, and infinite Ru–O layers

	Number of Ru–O layers		
	$n = 1$: single layer	$n = 2$: double layer	$n = \infty$: infinite layer
Sr-based material	Sr_2RuO_4	$Sr_3Ru_2O_7$	$SrRuO_3$
Characteristics	p-wave superconductor ($T_c \approx 1.5$ K), Fermi-liquid metal, exchange-enhanced	Non-superconducting, Non-Fermi-liquid metal, quantum metamagnet	Non-superconducting, non-Fermi-liquid metal, ferromagnet ($T_M = 165$ K)
Ca-based material	Ca_2RuO_4	$Ca_3Ru_2O_7$	$CaRuO_3$
Characteristics	Non-superconducting, Mott insulator, antiferromagnet ($T_N = 113$ K)	Non-superconducting, poor metal, antiferromagnet ($T_N = 56$ K)	Non-superconducting, poor metal, strong paramagnet

Sr_2RuO_4 is the only confirmed ruthenate exhibiting superconductivity, although $Sr_3Ru_2O_7$ has been predicted to be a superconductor, but only if it could be synthesised as a high-purity single crystal (Ovchinnikov, 2003a). Sr_2RuO_4 with $T_c = 0.93$ K was first reported by Maeno et al. (1994). It was soon realised that its T_c was adversely affected by the presence of non-magnetic impurities and crystal defects, as single-crystal samples of higher purity yielded the highest $T_c = 1.489$ K (Mao et al., 1999, 2000). Bulk superconductivity was soon confirmed through specific heat measurements (Maeno et al., 1996) and the material was found to be well within the clean limit as its $l/\xi \approx 30$, where l being the electron-mean-free path and ξ, the range of coherence (Maeno et al., 1994, 1996).

Sr_2RuO_4 is clearly analogous in chemical formula and isostructural to the well-known HTS cuprate LSCO (i.e. $(La, Sr)_2CuO_4$) of Chapter 16, possessing a body-centred tetragonal (K_2NiF_4) structure in which there is no La and Ru^{4+} fully replaces Cu^{2+}, as shown in Figure 19.1(a). Ru^{4+}

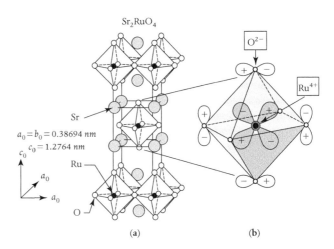

$a_0 = b_0 = 0.38694$ nm
$c_0 = 1.2764$ nm

Figure 19.1 (a) The crystal structure of Sr_2RuO_4 and (b) the octahedral arrangement of oxygen atoms. Note that the orbital lobes are directed not at the O^{2-} ions along the x-, y-, and z-axes but in between them.

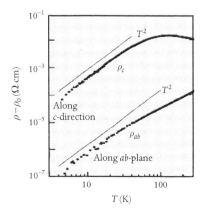

Figure 19.2 Resistivity along the *c*-axis is about 100 times greater than along the basal plane and, furthermore, above T_c, the normal state is a Fermi-liquid type manifesting T^2 behaviour (after Maeno et al., 1996).

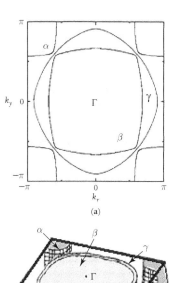

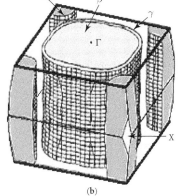

Figure 19.3 (a) The band structure and (b) the Fermi surface formed with three cylindrical sheets (after Oguchi, 1995; Mackenzie et al., 1996a, b; Singh, 1995).

atoms in the compound are placed at the centres of octahedra formed by oxygen atoms (Figure 19.1(b)), responsible for controlling its properties. Conduction occurs in partially filled *d*-bands of Ru that are hybridised with *p*-orbitals of O in Ru–O planes. Hybridisation is much less in the *z*-direction, which gives the compound a two-dimensionality, similar to cuprates. As with LSCO, the material exhibits anisotropy (Figure 19.2), with $\rho_c/\rho_{ab} \approx 100$ (Maeno et al., 1996). In the normal state, the resistivity along both directions manifests T^2 behaviour. Further, superconductivity is type II (Mao et al., 1999, 2000), with $H_{c2}//ab$-plane$(0) = 1.5$ T and $H_{c2}//c$-axis $(0) = 0.07$ T.

19.1.1 Synthesis

As with HTS cuprates, a polycrystalline sample of Sr_2RuO_4 is commonly synthesised by solid state diffusion using the constituent oxides, nitrates, or carbonates. However, for achieving an optimum T_c, a high-purity single crystal is needed. For this the floating-zone method has been commonly used (Mao et al., 1999, 2000). Rods prepared with 15% excess Ru as flux are melted in an Ar atmosphere containing 10% O_2 under a pressure of 3 bar. In general, smaller feed rates (around 4 or 4.5 cm h^{-1}) result in a low defect density with an optimum T_c of 1.5 K.

19.1.2 Band structure and Fermi surface

Band-structure calculations of Sr_2RuO_4 by the local-density approximation (LDA) method (Oguchi, 1995; Singh, 1995; Mackenzie et al., 1996a) revealed three bands crossing the Fermi level: α (hole band) and β and γ (electron bands). The bands correspond respectively to d_{xz}-, d_{yz}-, and d_{xy}-orbitals. These orbitals have lobes that are directed not at the O^{2-} ions along the *x*-, *y*-, and *z*-axes, but in between them (Figure 19.1(b)), and the electrons in these orbitals form the Fermi surface (Figure 19.3(a) and (b)) as three cylindrical sheets. The band structure is highly anisotropic, with the dispersion along the *c*-axis being the least. This is indicated by their Fermi velocities $V_{Fz} = 1.4 \times 10^6$ cm s^{-1} and $V_{Fx} = V_{Fy} = 2.4 \times 10^7$ cm s^{-1}. The anisotropy causes the Fermi surfaces to look like corrugated cylinders, which is corroborated by Shubnikov–de Haas and de Haas–van Alphen (dHvA) measurements on high-purity single crystals (Mackenzie et al., 1996a, b; Yoshida et al., 1999). For superconductivity, the active band is the γ-band, for which the spin–fluctuation interaction is strongest, giving the largest effective mass of the electron. It is the γ-band where superconductivity is first to occur, which, on further cooling, spreads to the other two bands. In Sr_2RuO_4, the valence state of Ru is 4+, which leaves four electrons in the 4*d*-shell. These four electrons are distributed nearly equally among the three orbitals, which means each filling up to 4/3 (Oguchi, 1995; Singh, 1995). This promotes the material acquiring a metallic state without any additional doping; that is, it is *self-doped*.

19.1.3 Differences from LSCO and unusual features

1. As discussed in Chapter 16 on HTS cuprates, the parent compound $LaCuO_4$ is an antiferromagnetic insulator that needs to be adequately externally doped with Ba or Sr before it becomes conducting and displays high-temperature superconductivity with optimum $T_c \approx$ 30−36 K. Its Ru analogue, Sr_2RuO_4, as already mentioned, is self-doped, but with a much lower optimum $T_c \approx 1.5$ K. While the T_c and the phase diagram of HTS cuprates are controlled mainly through doping, in Sr_2RuO_4 it is the angular displacements and distortion of the RuO_6 octahedra of Figure 19.1 that have this role (Ovchinnikov, 2003a).

2. In Sr_2RuO_4, superconductivity is detected only when the sample purity is high. Even a small concentration of non-magnetic impurities (Mackenzie et al., 1998) or crystal defects (Mao et al., 1999) seriously suppresses superconductivity (Figure 19.4(a)). Similarly, T_c is found to be sensitive to residual normal-state resistivity or impurity content (Figure 19.4(b)) (Mao et al., 1999). In sharp contrast, superconductivity of HTS cuprates is substantially robust against inhomogeneities and defects. The high sensitivity of T_c to non-magnetic impurities gave the first indication of spin-triplet superconductivity in Sr_2RuO_4.

3. The normal state of the ruthenate is of Fermi-liquid (FL) type. Despite the large anisotropy in resistivity, both $\rho_c(T)$ and $\rho_{ab}(T)$, as mentioned earlier, at low temperatures follow the characteristic FL T^2 law (Figure 19.2) (Maeno et al., 1996), although their resistivity coefficients are quite different. Above 100 K, $\rho_c(T)$ displays semiconductor-like behaviour analogous to some of the cuprate superconductors.

4. Both systems are electron-correlated, although the correlations seem stronger in HTS cuprates. As such, Sr_2RuO_4 is generally considered as moderately electron-correlated (Pérez-Navarro et al., 2000). Strong electron correlation (SEC) effects become relevant when the on-site Coulomb repulsion $U >> W$, where W is the bandwidth. $U_{3dCu} > U_{4dRu}$, which makes the correlation effects comparatively less pronounced in Sr_2RuO_4, as $U \geq W$. The correlations have been seen in the ruthenate through various studies. For example, the valence band photoemission spectra of Sr_2RuO_4 (Yokoya et al., 1996) showed the density of states (DOS) at E_F to be about three times smaller than theoretically estimated (Schmidt et al., 1996). Similarly, angle-resolved photoemission spectroscopy (ARPES) studies of Sr_2RuO_4 by Inoue et al. (1996) found $U/W \approx 1.7$, which favoured SEC, although these measurements left some uncertainty owing to possible surface

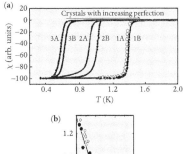

(a)

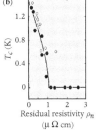

(b)

Figure 19.4 The T_c of Sr_2RuO_4 (single crystals) (a) shows an increase with crystal perfection and (b) is sensitive to the residual resistivity (after Mao et al., 1999).

contamination of the sample. More convincingly, dHvA oscillations (Mackenzie et al., 1996a) showed the effective electron mass in Sr_2RuO_4 to be larger by a factor of 3–12 than its free-electron value, when measured along different directions. The largest value of 12 corresponds to the γ-band mentioned above. These observations drew support from specific heat (Maeno et al., 1996) and electron photoemission (Puchkov et al., 1998) studies.

5. In the cuprate superconductors, the Cooper pairs are of spin-singlet type in the d-wave orbital channel (Bickers et al., 1987). In the ruthenate, the pairs are believed to be of spin-triplet type, having been formed out of the FL state in the p-wave orbital channel (Rice and Sigrist, 1995). Interestingly, in this respect, the ruthenate is similar to superfluid ^{3}He, which possesses (i) the p-wave symmetry of the paired-electron condensate emerging from a well formed FL state and (ii) a relatively low superfluid transition temperature of about 1 mK. This makes Sr_2RuO_4 essentially a solid state analogue of superfluid ^{3}He.

6. Sr_2RuO_4 exhibits no long-range magnetic order, but its magnetic susceptibility is more than seven times larger than the free-electron value (Maeno et al., 1994). Its Stoner factor $IN(0)$, where $N(0)$ is the DOS at E_F and I is the exchange parameter, lies close to 1 (Oguchi, 1995; Singh, 1995), which makes the material unstable with respect to transition from a paramagnetic to an FM state. FM fluctuations are therefore expected to be strong in Sr_2RuO_4. This leads to one of the prominent differences between the cuprate and ruthenate superconductors. In the latter, the carrier motion occurs against a background mainly of FM fluctuations, while in the former the fluctuations involved are AFM and they are believed to mediate Cooper pairing in the d-wave channel. In contrast, the FM fluctuations in Sr_2RuO_4 are what seem primarily responsible for its spin-triplet pairing (Mazin and Singh, 1997), but the ensuing T_c is much lower than in the AFM-mediated HTS cuprates. The experimental evidence for such a pairing mechanism involving FM fluctuations stems from NMR studies (Imai et al., 1998). Also, as may be seen from Table 19.1, the Sr-based members of the ruthenate family, such as $SrRuO_3$, are either FM or exchange-enhanced paramagnetic, which are suggestive of FM correlations. According to the theoretical considerations of Mazin and Singh (1999), AFM fluctuations are also present, as revealed by elastic neutron scattering data (Sidis et al., 1999), but their role in triplet superconductivity remains questionable (Kikugawa and Maeno, 2002). Clearly, as in cuprates, where there is competition between AFM fluctuations and d-wave superconductivity, in the ruthenate there is a similar competition between both FM and AFM fluctuations and p-wave superconductivity, in which, below 1.5 K, the latter finally dominates.

7. Finally, it is natural to ask why LSCO, and HTS cuprates in general, possess a much higher T_c than Sr_2RuO_4. The reason for this is believed to be the opposite types of Cooper pairing interactions, namely the AFM fluctuations in the former and the FM fluctuations in the latter. The triplet superconductivity resulting from the latter is not so robust. In cuprates, which are more strongly correlated, the singlet pairing in the d-wave channel as induced by AFM fluctuations is much more pronounced, yielding a distinctly higher T_c. Second, it is argued (Kuz'min et al., 1999; Ovchinnikov 2003a) that the higher T_c of cuprates is supported by the presence of a *van Hove anomaly*, which, however, does not occur in the ruthenate.

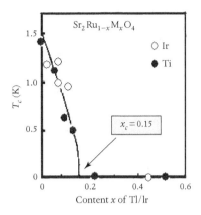

Figure 19.5 Effect on T_c of in-plane partial substitution of magnetic Ir and non-magnetic Ti at Ru sites (after Kikugawa et al., 2002).

19.2 Unconventional superconductivity

19.2.1 Impurity scattering

Conventional superconductivity is mostly immune to scattering from non-magnetic impurities and defects, which is not the case with the ruthenate superconductor. Mackenzie et al. (1998) observed that Al and Si in minute concentrations of 100 ppm could quench superconductivity of Sr_2RuO_4. Drastic suppression of superconductivity due to non-magnetic impurities is a pointer to the triplet pairing in Sr_2RuO_4. Figure 19.5 shows that both magnetic Ir^{4+} and non-magnetic Ti^{4+} have near-identical effects in rapidly suppressing superconductivity of $Sr_2Ru_{1-x}M_xO_4$ (where M = Ir or Ti) for a critical concentration $x_c = 0.15$ (Kikugawa et al., 2002). This is suggestive of p-wave pairing.

19.2.2 NQR/NMR studies

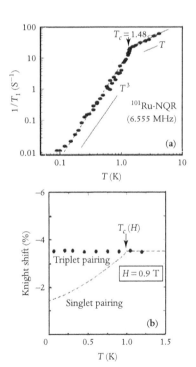

Figure 19.6 (a) Nuclear spin–lattice relaxation rate of ^{101}Ru-NQR as a function of temperature, and (b) Knight shift in Sr_2RuO_4 (after Ishida et al., 1998, 2000b).

Nuclear quadrupole resonance (NQR) experiments on Sr_2RuO_4 failed to show the characteristic Hebel–Slichter coherence peak in the nuclear spin–lattice relaxation rate at T_c (Ishida et al., 1997). Also, as may be seen in Figure 19.6(a), its temperature dependence at low temperatures is non-exponential. As we previously discussed in Chapters 13 and 14 on heavy fermion and organic superconductors, these features constitute important indications of unconventional superconductivity.

Perhaps the strongest support for the spin-triplet pairing in Sr_2RuO_4 came from the NMR measurements of ^{17}O by Ishida et al. (1997, 1998). For superconductors with antiparallel (singlet) pairing, the spin susceptibility, as measured through the Knight shift, tends to vanish as T is lowered much below T_c. Ishida et al. found the Knight shift to be invariant within the experimental error (Figure 19.6(b)), as was to be expected with the triplet state. This conclusion received credence also from the scattering measurements of polarised neutrons by Duffy et al. (2000).

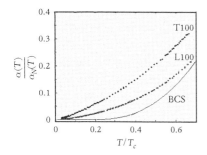

Figure 19.7 Variation of normalised attenuation coefficient with reduced temperature (after Lubien et al. 2001). Deviation of both transverse and longitudinal modes from the conventional exponential (BCS) behaviour is noted.

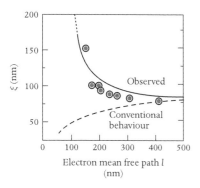

Figure 19.8 Observed dependence of the range of coherence for Sr_2RuO_4 on electron mean free path compared with the conventional behaviour (after Mao et al., 1999).

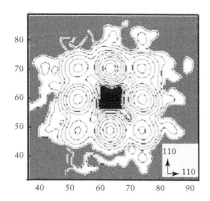

Figure 19.9 Contour plot of FLL neutron diffraction pattern (magnetic field 20 mT), manifesting a square vortex lattice (after Kealey et al., 2000). Axes give pixel numbers.

19.2.3 Various other properties

In conventional s-wave superconductors with isotropic gap, various superconducting properties below T_c vary exponentially as $\exp(-\Delta_0/k_BT)$. On the other hand, if the gap has nodes and superconductivity is unconventional, such as with some of the heavy fermion and organic superconductors, the quasiparticle density at $T << T_c$ varies as a power law rather than exponentially. In Sr_2RuO_4, at low temperatures, various parameters, such as specific heat (Nishizaki et al., 1999, 2000), thermal conductivity (Tanatar et al., 2001; Izawa et al., 2001), penetration depth (Bonalde et al., 2000), and NQR relaxation rate (Ishida et al., 2000a), all exhibit power-law temperature behaviour, indicating the presence of line nodes. Similarly, for ruthenate samples, ultrasonic attenuation studies by Lupien et al. (2001) also found a $T^{1.8}$ power-law behaviour at temperatures down to $T_c/30$ (Figure 19.7), supporting the existence of an unconventional gap structure. The $H_{c2}(T)$ behaviour of Sr_2RuO_4 deviates markedly from the conventional BCS theory and, interestingly, the range of coherence $\xi_{ab}(0)$, contrary to the conventional behaviour, increases with decreasing electron mean free path l (Mao et al., 1999) (Figure 19.8).

19.2.4 Vortex lattice

It was predicted theoretically by Agterberg (1998) that for p-wave pairing the vortex lattice formed with H parallel to the c-axis would be square instead of the triangular shape formed in conventional superconductors. Such a square vortex lattice was confirmed through neutron diffraction studies of Sr_2RuO_4 (Figure 19.9) (Riseman et al., 1998; Kealey et al., 2000), which gave credence to p-wave superconductivity

19.2.5 Tunnelling measurements

Nelson et al. (2003, 2004) confirmed the unconventional superconductivity of Sr_2RuO_4 by studying the tunnelling behaviour of junctions formed between the ruthenate and $Au_{0.5}-In_{0.5}$ in DC-SQUID configuration and showed that the phase of the order parameter in Sr_2RuO_4 changes by π under inversion. The experiment thus provided the first definitive confirmation that Sr_2RuO_4 is an odd-parity superconductor. More recently, the spin-triplet pairing in Sr_2RuO_4 has been supported by tunnelling measurements between a sample (with $T_c = 1.5$ K) and a triple-layer Ag–Pb–Ag sandwich with $T_c > 1.5$ K (Myers et al., 2009).

19.2.6 Multiple phases

As discussed earlier for heavy fermion superconductors, formation of multiple phases is taken as an indication of unconventional superconductivity. In triplet superconductors, the possible states are often

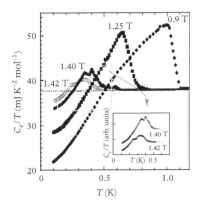

Figure 19.10 Occurrence of multiple phases in Sr_2RuO_4 (after Deguchi et al., 2004).

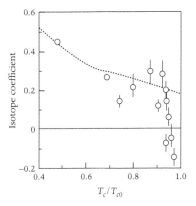

Figure 19.11 Variation of oxygen isotope effect coefficient for samples of different T_c (after Mao et al., 2001). T_{c0} represents the critical temperature of a pure sample.

degenerate, which allows creation of many phases by changing the external parameters. The possibility of the presence of another phase in Sr_2RuO_4 is revealed at high fields and low temperature and confirmed through high-resolution specific heat and thermal conductivity measurements (Deguchi et al., 2004). The specific heat studies showed that in a well-aligned magnetic field, the single specific heat jump of Sr_2RuO_4 splits into two, indicating the formation of another phase as seen in Figure 19.10 (and more clearly in the inset).

19.2.7 Isotope effect

The oxygen isotope effect has been examined on changing from ^{16}O to ^{18}O in Sr_2RuO_4 (Mao et al., 1999), and the observed behaviour is found to be unusual. The results of Mao et al. (2001) are depicted in Figure 19.11. As may be seen, the isotope-effect exponent for impure samples with low T_c is positive and increases towards 0.5 as T_c is lowered. With increasing sample purity, as T_c approaches the optimum value, the exponent abruptly deviates from the curve by changing its sign to become negative, giving rise to an *inverse isotope effect*. For an optimum-T_c sample ($T_c = 1.5$ K), the exponent $\alpha \approx -0.15$. Mao et al. (1999) suggested that this unusual $\alpha(T_c)$ behaviour could be the outcome of anharmonic phonon effects. Similarly, the observed impurity effect was theoretically accounted for by a phonon mechanism (Kuwabara and Ogata, 2000; Sato and Kohmoto, 2000; Zhitomirsky and Rice, 2001). According to Shimahara (2003), the inverse isotope effect is indicative of phonons having a non-trivial role in superconductivity, while Schnell et al. (2006), have argued that both phonons and spin fluctuations are equally responsible for superconductivity, although neither alone can cause the superconductivity of Sr_2RuO_4.

19.3 Summary of the current status of ruthenate superconductors

Experimental indications seem to favour both a node-less gap and the presence of line nodes. Also, one needs to fully differentiate nodes from locations of deep minima. Although the possibility of triplet superconductivity looks strong, better-quality samples, which deviate from the conventional BCS theory, do not fully follow the expected behaviour of p-wave superconductors. Second, definitive experimental evidence of FM fluctuations mediating pairing is lacking, although there seems every possibility of this happening. The possible role of AFM fluctuations, theoretically predicted and experimentally observed, in superconductivity of Sr_2RuO_4 remains to be understood. According to theoretical considerations of Kikugawa and Maeno (2002), AFM fluctuations have no constructive role in the superconductivity of Sr_2RuO_4.

It is worth mentioning that, even besides unconventional superconductivity, Sr_2RuO_4 is otherwise an unusual material manifesting interesting effects. For instance, as with some of the organic superconductors discussed earlier, it is predicted to display the *spontaneous quantum Hall effect* (QHE), and that too in the absence of an applied field (Volovik, 1988). The origin of spontaneous QHE is attributed to supercurrents. The effect is expected to be rather feeble and poses experimental challenges for its detection.

19.4   Superconducting ruthenocuprates

This class of materials, which emerged soon after the discovery of superconducting Sr_2RuO_4, are an interesting combination of ruthenate and HTS cuprate, being, as the name suggests, formed with both Ru and Cu oxides. Extraordinarily, in ruthenocuprates, ferromagnetism, although weak, which normally excludes superconductivity, seems to coexist with superconductivity. There is a weight of evidence gathered through tunnelling spectroscopy (Felner et al., 1997, 2000), muon-spin rotation (Shengelaya et al., 2004), and magneto-optic and Raman studies (Williams and Ryan, 2001) that indicates that in these materials both superconductivity and magnetic states coexist within the same crystalline grain. However, as far as their understanding goes, the situation today with this class of superconductors is even worse than with Sr_2RuO_4. For instance, although there is little doubt about the issue of coexistence, the origin of weak ferromagnetic(WFM) ordering itself remains debated. Besides AFM and FM, there are other intriguing magnetic phases observed in these systems at different temperatures and magnetic fields, and it is not clear whether these are of intrinsic or extrinsic origin. These magnetosuperconductors, with $T_M > T_c$, are termed *superconducting ferromagnets* (Felner, 1997) to distinguish them from the previously discovered *ferromagnetic superconductors* such as $ErRh_4B_4$ and $HoMo_6S_8$, with $T_M < T_c$.

19.4.1   Two families and their crystal structures

There exist two families of ruthenocuprates, but with broadly similar characteristics and discovered around the same time in the mid 1990s. These are (I) Ru-1212, that is, $RuSr_2RCu_2O_8$ (Bauernfeind et al., 1995a, 1996, 1999), and (II) Ru-1222, that is, $RuSr_2R_{2-x}Ce_xCu_2O_{10-\delta}$ (Bauernfeind et al., 1995a, b; Felner et al., 1997), where, in both cases, the rare earth R = Gd, Eu, Sm, or Y. In the literature, Ru-1222 is also sometimes written as Ru-2122, where the R-site occupancy is mentioned first. The rare earths R = Y, Er, Ho, and Dy can be incorporated only by high-pressure, high-temperature (HPHT) processing and are therefore comparatively less studied. Er, Ho, and Dy seem to form only the FM state in Ru-1212 at $T_M = 150$–170 K, with no convincing indication of superconductivity (Kawashima and Takayama-Muromachi, 2003). In the case of the Er-based compound, although diamagnetism was observed below 43 K, a zero-resistance state was not detected. Both compounds formed with Y, however, manifest T_M as well as T_c, but very often T_c fails to show up resistively, although the resistive T_c has also been reported (Felner et al., 2007). This has been ascribed to formation of cracks (Awana and Takayama-Muromachi, 2003), which inhibit the sample reaching the zero-resistance state. Table 19.2 gives the T_c and T_M values for both families (with Ce content in Ru-1222 of $x \approx 0.25$–0.5) containing different R. The compounds mentioned in the table are only those manifesting both superconductivity and magnetism. In the case of the Ru-1222 compounds, the table also shows T_{2M} (also called T_{irr} in the literature), which represents the Curie temperature T_C below which a W-FM phase is formed. Below this temperature, both field-cooled (FC) and zero-field-cooled (ZFC) samples exhibit hysteresis in magnetic susceptibility with temperature. All the values given in the table are meant to serve just as indications, since they tend to vary significantly owing to small differences in sample processing. Both systems display an interesting interplay and coexistence of superconductivity with magnetic order, in similar temperature ranges, with $T_c \approx 10-70$ K and $T_M \approx 130-220$ K. As may be seen from the chemical formulae of the two systems, Ru-1222 manifests oxygen non-stoichiometry while Ru-1212 does not. That is, the oxygen content $10 - \delta$ in Ru-1222 can be altered by high-pressure oxygenation or nitrogenation, which respectively enhance or lower the carrier concentration.

Ruthenocuprates have given rise to excitement for four reasons: (i) for the first time, the coexistence of superconductivity and magnetism has been associated with a high or moderately high T_c; (ii) the pertinent

Table 19.2 T_c and T_M for Ru-1212 and Ru-1222 compounds with different rare-earths R manifesting both superconductivity and magnetism

Ru-1212			Ru-1222			
R	T_M (K)	T_c (K)	R	T_M (K)	T_{2M} (K)	T_c (K)
Gd	~132	~45	Gd	150–160	80–100	~42
Eu	~132	~32	Eu	150–160	80–100	~32
Sm	~146	~12	Sm	~220	~140	~35
Y	~150	~45	Y	~152	100	30–39

Table 19.3 Representative materials with different values of T_M and T_c

Material	T_c(K)	T_M $(= T_N$ or $T_C)$ (K)	T_M/T_c
Ruthenocuprates	20–50	130–180	~5
R-123 (HTS)	~90	0.15–2.2	~0.001–0.025
Quaternary borocarbide	8–22	1.5–11	~0.1–0.5
Heavy fermion ($CeCu_2Si_2$)	0.7	0.8K	~1.1
$ErRh_4B_4$	~9	~0.93	~0.1
$HoMo_6S_8$	~2	~0.7	~0.035
Y_9Co_7	~2	~4.5	~2.25

magnetic state seems to be FM; (iii) the magnetic ordering temperature $T_M \gg T_c$, which is uncommon in the other known systems manifesting the interplay phenomenon (Table 19.3); and (iv) the internal magnetic field due to FM order, if it lies between the H_{c1} and H_{c2} values of the material, presents the interesting possibility of a *spontaneous vortex phase (SVP)* being formed in the absence of an external field (Sonin, 1998; Sonin and Felner, 1998).

Both families of ruthenocuprates stem (Figure 19.12) from Y-123, that is, $YBa_2Cu_3O_{7-\delta}$ (Figure 19.12(a)), which, for convenience, can alternatively be written as $Cu^{(1)}Ba_2YCu^{(2)}_2O_{7-\delta}$. Here $Cu^{(1)}$ represents the Cu located in chains and $Cu^{(2)}$ that in planes. Ru-1212 (Figure 19.12(b)) is formed by replacing Y by Eu for example (or R = Gd, Sm, or Y), CuO chains by RuO_2 planes, and BaO planes by SrO planes. In the absence of chains, the structure is tetragonal, with space group $P4/mmm$, and, for the Eu-containing compound, lattice parameters $a_0 = 0.3835$ nm and $c_0 \approx 3a_0 = 1.1573$ nm. The almost perfect match of $a_0 \approx c_0/3$ promotes the formation of microdomains with the c-axis of neighbouring domains aligned along different principal directions. In the new structure thus formed, the RuO_2 planes (or, more correctly, RuO_6 octahedra) are the source of magnetism, resulting from strong Ru-$4d_{xy,yz,zx}$–O$2p_{x,y,z}$ hybridisation with a magnetic moment $\mu(Ru) \approx 1\mu_B$, and they also serve as the charge reservoir, the role originally played by CuO chains in the Y-123 structure. Interestingly, in the superconducting ruthenate, RuO_2 planes are responsible for superconductivity, while the same planes in ruthenocuprates provide magnetism instead. This is because a slight decrease in the carrier concentration in these planes stabilises magnetism and such a decrease in Ru-1212 is governed by charge transfer between CuO_2 and RuO_2 layers. In contrast to Y-123, this structure carries only one Cu site. The Ru ion is octahedrally coordinated and RuO_6 octahedra deviate from their ideal positions by a small rotation of about $14°$ about the c-axis, accompanied by a small tilting that lowers the Cu–O–Ru angle to $173°$ (McLaughlin et al., 1999). The reason for

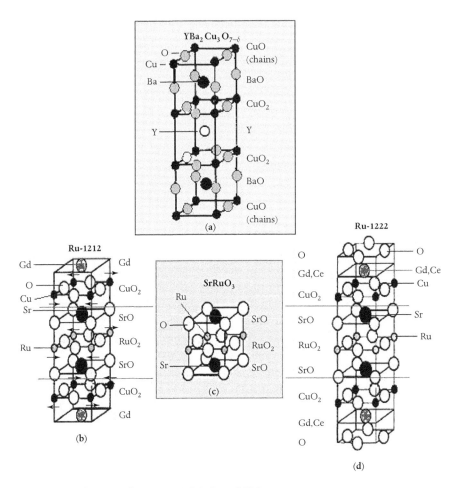

Figure 19.12 Crystal structures of (a) Y-123, (b) Ru-1212, (c) $SrRuO_3$, and (d) Ru-1212.

this is the mismatch between the larger Ru–O and smaller Cu–O bond lengths, which slightly changes the space group to $P4/mbm$.

The Ru-1222 lattice (Figure 19.12(d)) is formed by replacing Eu (or, more generally, R) in the Ru-1212 structure by a fluorite block, $(Eu, Ce)_2O_2$. This insertion results in a shift in the next perovskite block along the $\langle 110 \rangle$ direction, which doubles the unit cell along the c-axis. The structure continues to be tetragonal with space group $I4/mmm$, possessing the lattice parameters (for R = Eu) $a_0 = 0.3846$ nm and $c_0 = 2.872$ nm. Owing to the closeness of the ionic radii of Eu^{3+} and Ce^{4+}, the lattice parameters of Ru-1222Eu (i.e. Ru-1222 containing Eu: $RuSr_2Eu_{2-x}Ce_xCu_2O_{10-\delta}$) remain invariant with x. An interesting feature of both Ru-1212 and Ru-1222 structures is the presence of layer blocks of $SrRuO_3$ (Figure 19.12(c)) and CuO_2–R–CuO_2. The former is known to be an itinerant FM with Curie temperature $T_C = 160$ K, while the latter constitutes the most relevant component for high T_c

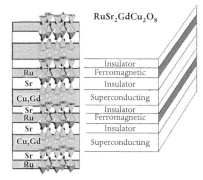

$RuSr_2GdCu_2O_8$

Ru	Insulator
	Ferromagnetic
Sr	Insulator
Cu,Gd	Superconducting
Sr	Insulator
Ru	Ferromagnetic
Sr	Insulator
Cu,Gd	Superconducting
Sr	
Ru	

Figure 19.13 Ruthenocuprates as natural multilayers.

in the R-123 system. These features give credence to the occurrence of both FM and high-T_c superconductivity in ruthenocuprate families. The spatial separation of the layer blocks prevent the superconducting state being destroyed by FM and allows their coexistence. The structure of Ru-1222 can alternatively be viewed as a naturally occurring multilayer stacking of FM–insulator–superconductor–insulator–FM blocks (Figure 19.13).

Band-structure calculations by Pickett et al. (1999) for Ru-1212Gd found that the layered structure and ensuing bond separation favour the coexistence of two antagonistic phenomena. Interestingly, the large magnetic moment of Gd^{3+} ($\approx 7\mu_B$) has no effect on the electronic structure of the CuO_2 and RuO_2 planes because of insufficient overlap of their respective wavefunctions. The electronic structure of the CuO_2 planes is independent of the magnetic state of the RuO_2 planes and there is very little overlap between them that allows coexistence of magnetism with superconductivity. The internal field of the FM-ordered state leads to a Zeeman splitting of the Fermi surface that normally excludes the formation of a spin-singlet state. Allowing for the finite angular momentum of the electron pairs, the calculations indicated that the coupling between the FM and superconducting layers was neither too strong nor too weak, but just of adequate strength to allow the formation of a Fulde–Ferrell–Larkin–Ovchinnikov (FFLO)-type superconducting state. The reader may recall that we had discussed the formation of such a state in Chapters 13 and 14 on heavy fermion and organic superconductors. In this situation, a suitable spatial modulation of the superconducting or the FM order parameter, or both, may lead to their coexistence.

19.4.2 Synthesis of ruthenocuprate samples

Polycrystalline samples of Ru-1212 and Ru-1222 containing Gd or Eu are synthesised under ambient pressure by routine solid state reaction routes. With other rare earths, such as Y, Dy, Ho, and Er, one has to follow a high-pressure, high-temperature (HPHT) route. The exact procedures followed tend to vary among different research groups, with the result that the measured properties show a significant scatter. In the case of Gd-based samples, for example (Bauernfeind et al., 1995a; Lorenz et al., 2001a, 2003), the starting materials (RuO_2, Gd_2O, $SrCO_3$, CuO, and CeO_2 (the latter being required for Ru-1222) in powder form are preheated for 12 hours at 600–800°C and subsequently thoroughly mixed in the appropriate ratios of 1:2:1:2 or 1:2:2:2 and calcined for 16 hours at 960°C. This is followed by grinding, compacting, and further sintering in progressively increasing temperatures for 10 to 24 hours. In the final stage, the sintering of Ru-1212 samples at 1065°C is extended up to 2 weeks in an oxygen atmosphere, and this particular long heat treatment is found to be crucial in the sample synthesis. To improve the carrier density of Ru-1222, which exhibits oxygen non-stoichiometry, the synthesised samples are further annealed (for about 100 hours) at 420°C under an oxygen pressure of about 50–200 atm. Alternatively, if one wants to have oxygen depletion, the heating is performed using nitrogen, instead of oxygen (Awana 2005).

Single crystals and thin films of Ru-1212Gd and Ru-1222Gd have been grown by self-flux (Lin et al., 2001; Watanabe et al., 2004) and laser ablation (Lebedev et al., 2005; Matveev et al., 2004) methods, respectively, but their quality still needs to be significantly improved. The best thin films of Ru-1222 have been grown by the flux-assisted solid phase epitaxy technique (Lu et al., 2005).

19.5 Superconductivity, general features

19.5.1 Valence state and superconductivity

The valence states of different constituent ions forming the two ruthenocuprate systems have been investigated using various techniques (Felner, 2006), such as Mössbauer spectroscopy (MS), X-ray absorption spectroscopy (XAS), NMR, and X-ray absorption near-edge structure (XANES), which confirm that the R ions are trivalent, Ce is tetravalent, and Ru is pentavalent. In the case of Ru-1212, XANES studies indicated the presence of Ru^{4+} and Ru^{5+} in a proportion of 40:60 (Liu et al., 2001), where the mixed valence is possibly caused by charge transfer from CuO_2 planes to RuO_2 planes (Mandal et al., 2002). However, Mössbauer spectroscopy yielded the Ru valence in Ru-1212Gd as 5+ (deMarco et al., 2002). In Ru-1222, the average Ru valence stood close to or marginally lower than 5+, which increased with oxygenation (Felner, 2006).

As the two ruthenocuprate systems have evolved out of the HTS system, it is only natural that their superconductivity is governed by the same considerations as for HTS cuprates. The charge state of Y-123 is expressed by

$$Y^{3+} + Ba_2{}^{2+} + Cu_2{}^{2.2+} + Cu^{2.6+} = O_7{}^{2-}.$$

If p is the hole concentration per CuO_2 layer, its charge state is $2 + p$, with $p \approx 0.05-0.25$. The corresponding equation for Ru-1212, for example, can be written as

$$Ru^{(5-2p)} + Sr_2{}^{2+} + Gd^{3+} + Cu_2{}^{2+p} = O_8{}^{2-},$$

giving $p \approx 0.08$. Clearly, this value is much smaller than $p \approx 0.16$, the value corresponding to the optimum T_c in the phase diagram of HTS cuprates (Chapter 16). The underdoped nature of the Ru-1212 system is corroborated by the temperature-dependent resistivity $\rho(T)$, displayed

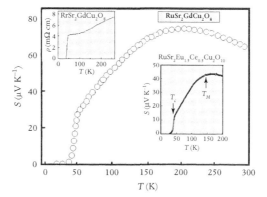

Figure 19.14 Thermopower as a function of temperature for Ru-1212Gd (Lorenz et al., 2003). The top-left inset shows the resistivity behaviour (after Lorenz et al., 2003). The bottom-right inset shows the thermopower behaviour of Ru-1222Eu/Ce (after Felner, 2006).

in the top-left inset of Figure 19.14 for a Ru-1212 sample, while the main figure presents its thermoelectric power $S(T)$ (Lorenz et al., 2003). The bottom-right inset shows $S(T)$ for Ru-1222 (Felner, 2006). The resistivity behaviour shown manifests a smooth curvature below room temperature, which indicates the opening up of the pseudogap that generally signifies an underdoped regime. The thermoelectric power remains positive throughout, which is indicative of holes being the charge carriers. The relatively large values of $S \approx 50-100\,\mu V\,K^{-1}$ observed at room temperature are also an indication of the underdoped state of the material. With the $S(T)$ data, one may apply an empirical relation, valid for HTS cuprates, for estimating p (Obertelli et al., 1992), namely $S(290\ K) = 992\ \exp(-38.1p)$. The estimated p-values, 0.06–0.1, for Ru-1212 and Ru-1222 corroborate the underdoped nature of both systems. It is therefore not surprising that the experimentally observed T_c values in the ruthenocuprates (10–50 K) are much lower than the $T_c \approx 90\ K$ of optimally doped R-123.

In the case of Ru-1222Eu, $RuSr_2Eu_{2-x}Ce_xCu_2O_{10-\delta}$, the parent compound corresponding to $x = 1$ (i.e. $RuSr_2EuCeCu_2O_{10}$) is an insulator like La_2CuO_4 or $YBa_2Cu_3O_6$. To make it conducting and superconducting, it needs hole doping in its CuO_2 planes, which is achieved by a suitable variation of the R^{3+}/Ce^{4+} ratio and/or by increasing the oxygen content through more (or high-pressure) oxygenation. Superconductivity is observed only for $0.4 < x < 0.8$, with the optimum value corresponding to a Ce concentration $x = 0.6$ (Figure 19.15(a)). Similarly, as may be seen in Figure 19.15(b), the T_c of a synthesised Ru-1222Eu sample with composition $RuSr_2Eu_{1.5}Ce_{0.5}Cu_2O_{10-\delta}$ shows a gradual rise with oxygen pressure. However, in the case of Ru-1212, having a fixed oxygen stoichiometry, oxygenation is of little help to enhance the carrier density. Here the related compound synthesised by a HPHT route, namely $Ru_{1-x}Sr_2GdCu_{2+x}O_{8+\delta}$, is found more suitable (Figure 19.15(c)), as also is the Ce-substituted composition $RuSr_2Gd_{1-y}Ce_yCu_2O_8$ (Klamut et al., 2001a, b, 2003). In the former

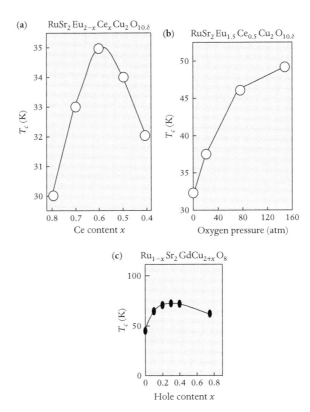

Figure 19.15 T_c as a function of (a) Ce content for Ru-1222, (b) oxygen pressure in Ru-1222 (after Felner et al., 2003, 2006) and (c) T_c as a function of hole concentration in Ru-1212 (after Klamut et al., 2003).

case, there is an additional hole doping of Cu and also non-stoichiometry of oxygen that allows increase in hole density through oxygenation. This way, for $x = 0.4$, with oxygenation, T_c can be enhanced to 72 K while the magnetic state is quenched. The inverted parabolic type of T_c variation with x in $Ru_{1-x}Sr_2GdCu_{2+x}O_{8+\delta}$ is depicted in Figure 19.15(c) (Klamut et al., 2001a). In the latter compound, the hole density can be tuned by varying the Ce concentration y, with an increase in y raising T_M but lowering T_c. Thus, the charge state of both types of ruthenocuprate systems can be varied by cationic aliovalent substitutions, leading to changes in Cu valence and/or oxygen content.

Having a close structural similarity with Y-123, it is not surprising that the ensuing superconductivity of ruthenocuprates is not much different either. Point-contact tunnelling spectroscopy studies of Ru-1212Gd by several authors (Giubileo et al., 2003; Ummarino et al., 2003; Piano et al., 2005) confirmed the presence of an order parameter with characteristic nodes in the d-wave channel, as previously observed for HTS cuprates. The superconducting gap 2Δ varied from 2.7 to 6 mV (Ummarino et al., 2003; Piano et al., 2005), giving a gap parameter $2\Delta/k_BT_c \leq 4$, that is, of weak-coupling type.

19.5.2 Unusual superconductivity

While Ru-1222 seemed to pose no serious problems, there were serious difficulties, at least at the beginning, in establishing superconductivity in the Ru-1212 family. The materials of this class did not readily manifest the Meissner effect to confirm their bulk superconductivity, which led to serious doubts about the existence of a superconducting state in these compounds. Subsequently, following experiments on samples of improved quality, it was finally accepted that Ru-1212 compounds do indeed exhibit superconductivity, albeit possessing some exotic or uncommon characteristics. These include (i) a very feeble diamagnetic signal and weak or often missing bulk Meissner effect, (ii) a huge magnetic penetration depth, (iii) an unusual magnetic field dependence of T_c, and (iv) novel pressure effects on superconducting and magnetic states. Some of these features are comparatively less distinct in Ru-1222 compounds. Nevertheless, because of these unusual traits, one of the most controversial questions about both systems has been whether their superconducting state is really microscopically homogeneous. At the same time, the bulk superconductivity of both Ru-1212 (Tallon et al., 2000) and Ru-1222 (Chen et al., 2001a) has been substantiated by the observations of sizable specific heat jumps $\Delta C/T$ at T_c. Here, yet another atypical behaviour reported has been that the observed specific heat anomaly is independent of or even increases with the applied field (Tallon et al., 2000; Felner, 2003). However, the latter observation was not corroborated by Chen et al. (2001a).

19.5.2.1 *Weak Meissner effect*

The diamagnetic signal associated with the Meissner effect in ruthenocuprates is generally detected in a low magnetic field and at a temperature T_d much lower than the onset T_c (Bernhard et al., 2000). This and sometimes even the absence of the Meissner effect in these compounds can be readily understood. First, the weak Meissner effect or its absence is ascribed to the internal magnetic field or the ensuing spontaneous vortex phase (SVP) due to the coexisting FM state. The presence of a SVP may also account for $T_d < T_c^{\text{onset}}$. Between these two temperatures, the vortex state is believed to be present even in the absence of an external magnetic field. Below T_d, flux is expelled from some magnetic domains. This argument derives some credence from the fact that in Nb-1212 and Nb-1222, which are free from FM order, owing to the absence of Ru, the diamagnetic signal is strong and the Meissner state is readily observed (Felner et al., 2000b). The occurrence of SVP in both Ru-1222 (Sonin and Felner, 1998; Felner et al., 2000a) and Ru-1212 (Bernhard et al., 2000; Tokunaga et al., 2001) has been discussed and is supported by NMR studies.

19.5.2.2 Huge magnetic penetration depth λ

The above mentioned uncommon features could also result from the size of the magnetic penetration depth λ observed in ruthenocuprate compounds (Xue et al., 2002; Lorenz et al., 2003). The estimated value of the intragrain penetration depth in Ru-1212Gd is $\lambda(0 \, \text{K}) \approx 2\text{–}3 \, \mu\text{m}$ ($T_c \approx 40 \, \text{K}$) rising to more than $6 \, \mu\text{m}$ ($T_c \approx 30 \, \text{K}$). These are huge values in comparison with other superconductors and are of the same order as the average grain diameter of the synthesised compounds. Large λ corresponds to a low superconducting condensate density and it occurs when superconducting grains contain non-superconducting inhomogeneities. This would account for the observed very feeble diamagnetic signal and frequently missing Meissner effect. These observations, along with a few other considerations, led the research group at Houston (Lorenz et al., 2003; Chu et al., 2005) to propose an *electronic phase separation model* for these materials to explain the coexistence of superconductivity with W-FM. Accordingly, at $T < T_M$, intragrain AFM domains are formed that are separated by very fine FM domains on a nanoscale. The ensuing intragrain superconductivity at $T < T_c$ exhibits the exotic and unusual properties mentioned above. Such a phase separation essentially makes the compound inhomogeneous on a nanoscale, with T_c only representing the phase-lock temperature of a Josephson junction array (JJA) coupled through weak links across the FM domains. Clearly, in this situation, the Meissner phase will not be observed unless a macroscopic phase coherence across the FM nano regions is established. On the other hand, these intragrain weak links could be associated with c-axis Josephson coupling.

19.5.2.3 Granular nature

Similar to HTS cuprates, ruthenocuprates are anisotropic type II superconductors, possessing a large upper critical field H_{c2} and small coherence length ξ. For instance, Ru-1222Gd possesses a large upper critical field $\mu_0 H_{c2}^{ab}(0) = 39 \, \text{T}$ and $\mu_0 H_{c2}^{c}(0) = 8 \, \text{T}$, with corresponding coherence lengths $\xi^{ab}(0) = 14 \, \text{nm}$ and $\xi^{c}(0) = 2.8 \, \text{nm}$ (Escote et al., 2002). Despite the fact that the material contains a metallic RuO_2 layer, at 5 K, 0 T it carries (Felner et al., 2003) a very low critical current density of $2 \times 10^5 \, \text{A m}^{-2}$, which is at least two orders of magnitude lower than for Y-123. This could be the outcome of the internal magnetic field or the SVP state, coexisting with superconductivity. Furthermore, $J_c(H)$ shows an extreme sensitivity to applied magnetic field, and a field of just 10 Oe suppresses J_c by an order of magnitude. This is indicative of the current carriers being driven through Josephson weak links, which is a characteristic feature of granular structure. In the case of Nb-1222, which is free from any magnetic order, the observed $J_c(5 \, \text{K}, 0 \, \text{T})$ is comparatively large, $1.575 \times 10^7 \, \text{A m}^{-2}$ (Felner, 2006).

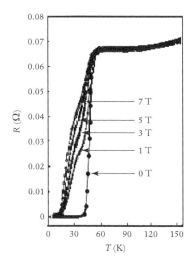

Figure 19.16 $R(T, H)$ curves for Ru-1222 Gd/Ce, showing step-like resistive transitions (after Awana, 2005).

Small coherence lengths make the grain boundaries in polycrystalline samples serve as weak links, and the materials exhibit granular characteristics. For instance, $R(T, H)$ measurements show characteristic step like behaviour (Figure 19.16) (Awana, 2005). The initial partial resistance drop at T_c^{onset}, due to an intragrain effect, is sharp, and, at lower temperatures, is followed by an intergrain weak-link contribution that shows a significant broadening under an applied magnetic field. T_c^{onset} is, however, not very sensitive to H.

Particularly interesting are the two critical temperatures T_c and T_p, corresponding respectively to intra- and intergrain regions of the sample, with $T_p < T_c$. In ruthenocuprate superconductors, T_c and T_p have been readily detected both resistively and magnetically (Figure 19.17) (Lorenz et al., 2003). Both T_c and T_p are suppressed by an applied field and, as to be expected, the effect is more pronounced with T_p, which is sensitive to weak links. Surprisingly, however, in Ru-1212 (Chu et al., 2005), even in a low magnetic field, the intragrain T_c exhibits an anomalously large suppression behaviour, as shown in Figure 19.18. The curve exhibits a curvature over the entire temperature range that is opposite to that displayed by a conventional bulk superconductor (inset of Figure 19.18). This suggests that even the intragrain T_c in ruthenocuprates is really a phase-locked Josephson junction array coupled temperature between the phase-separated AFM domains formed within individual grains of the material. This is consistent with the phase-separation model of the Houston group.

19.5.2.4 *Pressure effects on* T_c, T_p, *and* T_M

Ruthenocuprates exhibit unusual effects under imposed mechanical pressure P. Both T_c and T_p, measured resistively and inductively for Ru-1212, are found to increase linearly with pressure (Figure 19.19(a)), but the effect is more pronounced with T_p (Lorenz et al., 2003). The corresponding dT_c/dP and dT_p/dP values for the two critical temperatures measured for Ru-1212Gd are respectively 1.06 K GPa^{-1} and 1.8 K

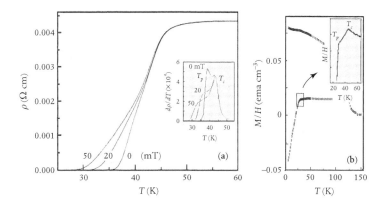

Figure 19.17 Intra- and intergrain critical temperatures T_c and T_P in Ru-1212Gd, as observed (a) resistively and (b) magnetically (Lorenz et al., 2003).

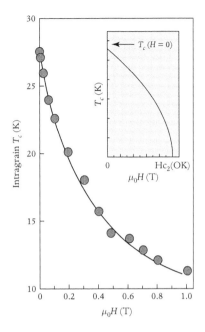

Figure 19.18 Anomalous $T_c(H)$ behaviour in Ru-1212Eu (after Lorenz, 2003). The inset shows the conventional $T_c(H)$ behaviour.

GPa^{-1} (Lorenz et al., 2003). The larger value observed for T_p is due to improved intergrain connectivity caused by the imposed pressure. The pressure effect on the magnetic ordering temperature T_M is, however, much greater (Figure 19.19(b)), represented by $dT_M/dP = 6.7$ K GPa^{-1} (Lorenz et al., 2003). Comparing the pressure coefficient of T_c with those measured for various HTS cuprates (with no competing magnetic order, however), the former is 3–4 times smaller. This indicates that in Ru-1212 the pressure primarily affects the FM order coexisting with superconductivity. Because of their mutual competition, a stronger effect on T_M of FM leaves a lesser effect on T_c of superconductivity. Interestingly, this is again in line with the electronic phase-separation model referred to earlier. Contrary to the above observations, the Ru-1222Sm compound (Oomi et al., 2002), however, shows a much larger pressure coefficient of 4.7 K GPa^{-1} for T_c and a negative pressure coefficient of -12.9 K GPa^{-1} for its T_M (Figure 19.19(c)). This suggests that the orders mutually compete.

19.6 Magnetic states and coexistence of T_M and T_c

Although various techniques such as neutron powder diffraction (NPD), NMR, μSR, and Mössbauer spectroscopy have exhibited mutual contradictions in the general characteristics of Ru-1212 and Ru-1222 systems, the magnetic structure of Ru-1212, as indicated by the arrows shown in its crystal structure (Figure 19.12(b)), is described as AFM planar Ru order. There is, however, no similar consensus existing on the magnetic

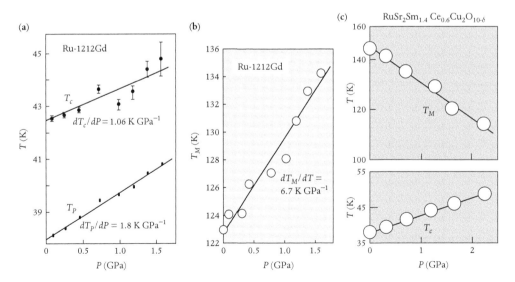

Figure 19.19 Pressure dependences of (a) T_c and T_P and (b) T_M in Ru-1212Gd (Lorenz et al., 2003) and (c) T_c and T_M in Ru-1222Sm/Ce (after Oomi et al., 2002).

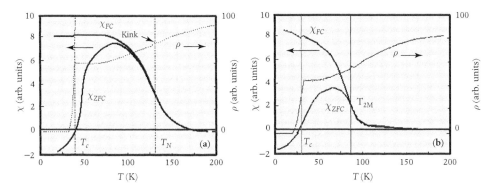

Figure 19.20 Schematically illustrated susceptibility and resistivity behaviours of (Fig.(a)) after Ru-1212 and (Fig.(b)) after Ru-1222.

structure of Ru-1222, and a good deal of current understanding of this family has evolved through studies of its magnetic state behaviour.

Figure 19.20 schematically depicts the characteristic temperature dependence of the magnetic susceptibilities of the two systems below 200 K. Also shown for comparison are the corresponding resistivity curves (schematic) in the same temperature range. As shown in Figure 19.20(a), for a Ru-1212 sample, cooled in a small field (5 Oe), upon lowering the temperature from 200 K, the field-cooled (FC) DC susceptibility χ_{FC} rises sharply at $T_M(T_N) \approx 132$ K (for a Ru-1212Gd sample, for example), suggesting a magnetic transition. It continues to increase slowly with further cooling, until it starts saturating, and finally flattens by exhibiting a kink at T_c (~30 K for Ru-1212Gd), but without diamagnetism, which corresponds to the onset of superconductivity. On warming the sample, after zero-field cooling (ZFC), the susceptibility χ_{ZFC}, first displaying a diamagnetic signal at 4 K due to a Meissner state, starts rising and exhibits a bend at T_c (≈ 30K), and then slowly increases with temperature to finally merge with the earlier χ_{FC} curve at a temperature slightly lower than T_M. Between T_c and T_M, the FC and ZFC parts of the curve show a magnetic hysteresis, which diminishes with increasing field and, with cooling under a larger field of a few tenths of a tesla, wipes out the ZFC curve. An abrupt rise of χ_{FC} at T_M and the occurrence of hysteresis are possible signatures of an incipient W-FM state below T_M. The resistivity curve in Figure 19.20(a) shows the superconducting drop leading to a zero-resistance state at T_c (≈ 30 K). A small kink in the resistivity behaviour commonly shows up at the magnetic ordering temperature T_M. Above T_M, the material is paramagnetic, while below this temperature it is AFM, albeit with a component that is W-FM and remains unaffected below T_c.

The magnetic behaviour of Ru-1222 (Figure 19.20(b) is similar, but significantly more complicated. The major magnetic order in both compounds is AFM in nature, with a small FM component detectable at a lower temperature in Ru-1222 instead of the higher temperature T_M in Ru-1212. In fact, Ru-1222 displays two or three magnetic transitions above T_c, their origin, whether intrinsic or extrinsic, still remaining uncertain. Most commonly, the magnetic characteristics of Ru-1222 exhibit two transitions: one at T_{2M} (alternatively called T_{irr}) $\approx 60-120$ K, depending on the strength of the magnetic field, and the other at a higher temperature $T_M \approx 130-220$ K, as schematically depicted in Figure 19.20(b). T_{2M} represents the temperature where the ZFC curve merges with the FC curve. Below this temperature, Ru-1222 exhibits hysteresis in the FC and ZFC curves (in low fields of about 5 Oe), which becomes narrower with increasing temperature and finally disappears at $T = T_{2M}$. The effect of increasing the magnetic field is to lower the temperature T_{2M}, and in a relatively large magnetic field of a few tenth of a tesla the ZFC curve is no longer observed. Below T_{2M}, the compound is weakly FM. On lowering the temperature to $T = T_c(10\text{--}50$ K$) < T_{2M}$, where the ZFC curve enters the diamagnetic region, the material becomes superconducting without affecting the W-FM state. The higher magnetic ordering temperature T_M is associated with the AFM state, above which the material is paramagnetic. The origin of T_M in Ru-1222 is believed to be related to the formation of small clusters of impurity phases exhibiting a pronounced $Ru^{4+}-Ru^{4+}$ exchange interaction. Both T_M and T_{2M} have been observed in all Ru-1222 compounds, and in that sense they are intrinsic to the system. Besides these commonly observed magnetic transitions, there is yet one more AFM-like transition reported in between T_M and T_{2M} at $T \approx 120$ K (Felner, 2006), the origin of which, however, remains debated. When the two compounds contain Gd, both of them exhibit an AFM order due to Gd ions (magnetic moment $7\mu_B$) formed at about 2.2–2.6 K, coexisting with superconductivity. This is analogous to the AFM order observed in Gd-123.

The contention of the W-FM ordering of Ru moments in both Ru-1212 (Bernhard et al., 1999) and Ru-1222 (Felner et al., 2003) compounds is firmly supported by the observations of the DC magnetisation curves measured below T_M (Ru-1212) and T_{2M} (Ru-1222) in low magnetic fields, which display characteristic FM-like hysteresis loops. A typical FM type of low-field hysteresis curve observed in Ru-1222Eu at $T = 5$ K $< T_{2M}$ (Felner, 2006) is depicted in Figure 19.21. It shows the two characteristic parameters, the remanent moment M_{rem} and the coercive field H_C, to be respectively $0.41\mu_B/$Ru and 190 Oe (Felner, 2003). The relatively large M_{rem}/M_{sat} ratio is consistent with FM-like order in Eu-containing Ru-1222. Both M_{rem} and $H_C(T)$ vanish at $T_{2M} = 80$ K. In the case of Ru-1212Gd, the similar hysteresis curve observed (Awana, 2005) is illustrated in the inset of Figure 19.21, where the W-FM characteristics observed at low temperatures disappear at the higher ordering temperature T_M. At $T > T_M$ (i.e. in the paramagnetic state),

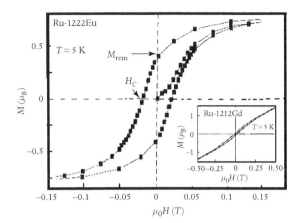

Figure 19.21 Isothermal magnetisation (at 5 K) of Ru-1222Eu (main figure, after Felner, 2006) and of Ru-1212Gd (inset, after Awana et al., 2005).

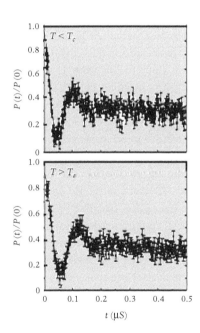

Figure 19.22 Zero-field μSR of Ru-1212Gd (after Bernhard et al., 1999).

the Curie–Weiss behaviour leads to a positive Curie temperature θ_c, which is indicative of the FM character of the Ru spins and $\mu_{\text{eff}} \approx 3.2\mu_B$ (Williams and Krämer, 2000). The presence of W-FM was further substantiated by zero-field muon-spin rotation measurements on Ru-1212Gd below T_M, which detected microscopically homogeneous internal magnetic fields over a sample volume of more than 80% and confirmed the bulk nature of the magnetic state. Further, the oscillatory component of muon-spin polarisation remained invariant across the superconducting transition (Figure 19.22) (Bernhard et al., 1999), indicating the magnetic state to be unaffected by superconductivity. Similar observations were made for Ru-1222 compounds below $T < T_{2M}$ (Shengelaya et al., 2004). ESR (Fainstein et al., 1999) and zero-field NMR (ZFNMR) experiments (Tokunaga et al., 2001) on both systems indicated the exchange coupling between the Ru ions to be of FM type, which gave further credence to the earlier findings. On the basis of magnetisation and μSR measurements, it was suggested that the FM ordering of the Ru sublattice occurs with the easy axis directed along the ab-plane (Bernhard et al., 1999).

Unexpectedly, the results of NPD studies were not in accord with these other findings. Surprisingly, for Ru-1212Gd samples, NPD revealed no FM phase, but showed (Lynn et al., 2000) the Ru sublattice of the above compounds to order in an AFM manner along all crystallographic directions (i.e. G-type AFM), with the easy axis along the c-direction. Modified band-structure calculations incorporating a 14° rotation of the RuO_6 octahedra also seemed to favour AFM, rather than FM or W-FM, ordering of the Ru (Nakamura et al., 2001). If there was any in-plane FM component at all, it could not possibly have an overall (Ru plus Gd) value exceeding $0.1\mu_B$ (Lynn et al., 2000). In this situation, a spin-flop transition and a sudden rise in the FM signal were predicted as resulting from the application of an imposed magnetic field directed along the *c*-direction. Interestingly, these predictions were found to be consistent with the NPD measurements by Lynn et al.

These contradictory outcomes posed an important question as to how the presence of W-FM order substantiated through major microscopic probes such as magnetisation, μSR, NMR, and EPR could be reconciled with the observations of AFM as confirmed by NPD. Broadly, this challenge has been effectively met by two mutually exclusive and independent models that are respectively based on whether the magnetic state formed is homogeneous or inhomogeneous. NPD (Shibata, 2002), ZFNMR (Tokunaga et al., 2001), and μSR (Bernhard et al., 1999), for example, indicate the magnetic structure to be homogeneous and bulk. On the other hand, an inhomogeneous structure of the magnetic order is suggested the specific heat measurements, which manifest only a broad hump at the magnetic phase transition (Klamut, 2008). There are indications (Lorenz et al., 2003; Chu et al., 2005) that the magnetic state formed is really inhomogeneous on the microscopic scale but that this does not readily show up owing to the inadequate resolution of the experimental techniques used. In the situation where the magnetic order is assumed homogeneous, a possible explanation of W-FM order has been derived from the *canted spin model*. Accordingly, while the dominant ordering of Ru ions, as revealed by NPD, is bulk AFM with the easy axis perpendicular to the layers, the whole subsystem is slightly canted so as to yield a net W-FM component along the *ab*-plane (Lynn et al., 2000; Jorgensen et al., 2001a). On applying a magnetic field, the Ru spins cant further away from the G-type AFM, and at 7 T the spins are almost fully FM. Nakamura and Freeman (2002), taking into account this canting, showed that the magnetic moment of Ru ions projected along the AFM *c*-axis was $1.16\mu_B$, while along the FM axis in the *ab*-plane it was $0.99\mu_B$, which is much greater than the suggested upper limit of $0.1\mu_B$. Modified calculations by Butera et al. (2001) found the Ru ordering in the RuO_2 sheets to be FM and the inter-plane coupling to be AFM. The canting is possibly a consequence of antisymmetric exchange coupling of Dzyaloshinsky–Moriya (Dzyaloshinsky, 1958; Moriya, 1960a, b) type between neighbouring Ru moments (Felner et al., 1997), caused by tilting of the RuO_6 octahedra from the *c*-axis (Knee et al., 2000).

The Ru-1222 compounds have posed intriguing controversies related to the many different magnetic orders that are observed, which remain unresolved. One of the much debated issues has been whether the magnetic ordering is long-range or short-range. While a number of reports (Bernhard et al., 1999; Felner et al., 2004b, 2005; Nigam et al., 2007; Xue et al., 2003) suggest the occurrence of long-range AFM and FM order at different temperatures as discussed already, NPD measurements (Lynn et al., 2007) did not reveal any long-range order. Corroborating this, both thermo-remanent magnetisation and frequency-dependent AC susceptibility measurements in

Ru-1222 showed the formation of a short range spin-glass (SG) state (Cardosa et al., 2003). Other reports (Felner, 2006; Klamut, 2008; Nigam, 2010; Awana et al., 2011) are in accord with this. For Ru-1222Eu, observations by Awana et al. (2011) show the formation of a SG state at the spin-freezing temperature $T_f = 82\,\text{K}$ and, on further cooling, inhomogeneous FM clusters are formed, embedded in an SG matrix. FM order and the SG state coexist below T_f, and superconductivity occurring at $T_c < 30\,\text{K}$ does not affect or get affected by the above coexisting features of the magnetic state. The origin of the *cluster spin-glass state* is attributed to the randomly occurring exchange interactions between Ru ions in penta- and tetravalent states. The state that exists in low DC magnetic fields is affected when the field is increased, and for fields beyond 5 T there is a changeover from the cluster SG state to a superparamagnetic state (Nigam, 2010).

The Houston group (Chu et al., 2000, 2005; Lorenz et al., 2002, 2003; Xue et al., 2000, 2002) has argued that the magnetic order formed is microscopically inhomogeneous and the supposition of canted spins does not seem to hold. Their phase-separation model was mentioned in Section 19.5.2.2 in connection with the unusual superconductivity manifested by ruthenocuprates. In this model, intragrain AFM domains form as a result of phase separation and, in the boundaries between the domains, an FM (but not superconducting) state appears on a nanoscale that matches the size of the coherence length. This explains both the AFM order observed by NPD and the W-FM detected through magnetisation and other probes. This makes the intragrain structure inhomogeneous on a nanoscale, serving as phase-locked Josephson junction arrays. Superconductivity develops, with T_c corresponding to the phase-locked transition, as discussed earlier. The observation of FM clusters in Ru-1222Eu (Xue et al., 2003) is consistent with the phase-separation model.

19.7 Cationic substitutions in Ru-1212 and Ru-1222, effect on T_c and T_M

Similar to HTS cuprates, cationic substitutions at the Cu^{2+} site have the largest impact in depressing T_c of ruthenocuprates. The isovalent Zn^{2+} and Ni^{2+} are the most extensively studied cations, and with Zn both systems fast lose their superconductivity upon 2.5–3% of substitution (Felner et al., 2004a), but with little impact on T_M (Bernhard et al., 1999, 2000; Felner et al., 2004a). The fact that T_M remains invariant with these dopants indicates that the cationic disorder ensuing from the substitution does not affect the RuO_2 planes where the magnetic state originates. Similarly, complete substitution of pentavalent Nb for Ru^{5+} retains superconductivity of Ru-1222 (Cava et al., 1992) albeit at a lower temperature, but more effectively suppresses its magnetism. Up to a 50% level of substitution, however, the original W-FM continues to coexist with superconductivity (Felner et al., 2000a). This effect is larger in Ru-1212, which shows a greater suppression of both T_c and T_M (Feng et al., 2001).

With Al substituted at Ru sites, superconductivity is detected only up to a 5% level of substitution, beyond which it is quenched. This is ascribed to oxygen loss and the lattice strains that develop as a result of the mismatch caused by differing ion sizes. Samples with 7% and 10% substitution exhibited only magnetic order. The temperature of W-FM order, which occurred at 82 K for a pristine sample, was reduced to 67 K with 10% Al substitution, owing to the ensuing dilution of magnetic interactions (Felner et al., 2004a). With Sn, the effect on magnetic ordering was much more pronounced (Figure 19.23), with the temperature being reduced from 91 K to about 40 K for a 5% Sn-substituted sample, the reason for which remains to be understood (Felner et al., 2004b). In the case of Ru-1212 having a fixed O_8 stoichiometry, substituting Sn^{4+} for Ru ($4+/5+$ state) mainly contributes towards enhancing the hole density, which is manifested by an increase in T_c^{onset} from 36 K to 48 K after 2% Sn substitution, but there is a simultaneous decrease in T_M (McLaughlin and Attfield, 1999). With 4% Sn, T_c^{onset} is pushed up to 78 K and T_M is fully suppressed. Substitution effects of non-magnetic Sb in Ru-1222 are similar to those of Al (Shi et al., 2003), while the presence of Fe in both Ru-1212 and Ru-1222 compounds drastically lowers the magnetic order as well as superconductivity (Felner and Asaf, 1997). Fe tends to occupy both Ru and Cu sites and simultaneously lower both T_c and T_M. Similarly, superconductivity of Ru-1212 is fully quenched by substituting Ru by 5% Co (Escamilla et al., 2004). Substitutional effects with Ti, Ir, and Rh at the Ru sites in Ru-1212 are again very similar, with both T_c and T_M decreasing rapidly on 10–20% substitution (Hassen et al., 2003).

Aliovalent substitutions at Sr^{2+} and R^{3+} sites in both Ru-1212 and Ru-1222 compounds influence T_c or T_M either by changing the hole concentration p or by creating disorder through ion-size mismatch. With

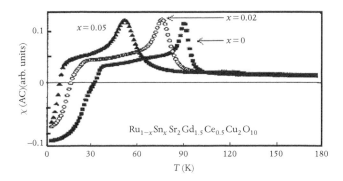

Figure 19.23 Effect of Sn substitution at Ru sites in Eu-1222 on T_c and T_{2M} (after Felner et al., 2004, 2004a, b).

10% Ca substitution for Gd in Ru-1212, T_c is raised to 61 K (Klamut et al., 2001b) owing to an increase in hole density. On the other hand, replacing Gd^{3+} in Ru-1212Gd by a much larger ion, such as La (both being 3+), creates local disorder, with about 9% substitution completely suppressing T_c. However, T_M is affected only after 20% substitution (Liu et al., 2005). Substitution of 5% Ce at the Gd sites enhances T_M from 132 K to 145 K, which is ascribed to small structural alterations involving tilting of the RuO_6 octahedra (Klamut et al., 2001a). In Ru-1212R, T_c increases as the ionic size of R is reduced. For instance, T_c of Ru-1212Eu ($T_c \approx 35$ K) increases to about 48 K on replacing Eu by Gd, and a partial replacement of Gd by Dy raises T_c to 58 K (Liu et al., 2001).

The effect of hydrogenation was investigated by Felner et al. (1998), who found that it lowered the T_c of Ru-1222EuH$_{0.07}$, but markedly enhanced the W-FM characteristics, with T_M being pushed up to a maximum reported value of 225 K. The latter is probably caused by a transfer of electrons from hydrogen to Ru, which would enhance the Ru moments. The effect is, however, reversible, with depletion of hydrogen restoring the pristine characteristics of the hydrogen-unloaded sample.

19.8   Summary

In this chapter, our focus has been on ruthenates and ruthenocuprates, both exhibiting unusual superconductivity. The present status of the former has already been summarised in Section 19.3 and therefore it need not be repeated here. The most exciting aspect of the two families of ruthenocuprates is the remarkable coexistence of their superconductivity with W-FM order. The materials of the two ruthenocuprate families, especially Ru-1222, offer a rich variety of magnetic orders produced under different thermo-magnetic conditions, the origins of many of which, however, remain to be understood. Similarly, it would be natural to ask if the magnetic state of Cu carries any weight in determining superconductivity and magnetic order in ruthenocuprates. Although, by and large, the canted spin model has been accepted as an explanation of the presence of W-FM component in an otherwise AFM state, the alternative approach of phase separation advocated by the Houston group seems to have some strong points in its favour. The unusual superconductivity of ruthenocuprates is thereby accounted for, and, furthermore, the observations of superparamagnetism and a cluster spin-glass state indicate a highly inhomogeneous magnetic structure on a fine scale, which is implicit in this approach. Further experiments on better-quality samples, if not single crystals, are needed to confirm the inhomogeneity of the magnetic state and substantiate the phase-separation scenario.

CHAPTER 20

Iron-based superconductors

The advent of superconducting $LaFePO_{1-x}F_x$ with $T_c \approx 5K$, discovered in 2006 (Kamihara et al., 2006), led to an exciting new class of superconductors, the Fe-based superconductors (FBS), that surprisingly contained iron as a prime constituent. Generally, Fe is associated with ferromagnetism and considered antagonistic to superconductivity and it was because of this that the new material instantly captured attention. Two years later, the same group (Kamihara et al., 2008) created a further sensation by announcing a much higher $T_c = 26K$ for an isostructural FBS, $LaFeAsO_{1-x}F_x$ (for $x = 0.11$), formed by replacing P by As, also from the pnictogen (N, P, As, Sb, Bi) group of the periodic table. The materials of this new class are therefore also called pnictides, or oxy-pnictides if they contain oxygen. With replacement of La by other rare earths such as Ce, Nd, Pr, and Sm, the T_c was further raised, and for the Sm-containing compound $T_c = 55$ K (Ren et al., 2008a). The highest T_c reported to date for this family is 56.3 K for $Gd_{0.8}Th_{0.2}FeAsO$ (Wang et al., 2008). Within a few months of the discovery of this family, LaFeAsO was joined by a few more FBS, formed with either pnictogens (Pn) such as As, for example $Ba_xK_{1-x}Fe_2As_2$ (Rotter et al., 2008) and LiFeAs (Tapp et al., 2008), or chalcogens (Ch) such as Se or Te, for example $FeSe_{1-\delta}$ (Hsu et al., 2008), although their T_c values were lower, in the range of 8–35 K. In the hierarchy with respect to T_c values, the FBS (or HTS pnictides) are second only to ruthenocuprates and HTS cuprates, which respectively possess optimum T_c values of about 70 K and about 160 K. Some of the broad differences between cuprates and FBS are listed in Table 20.1. There are several review articles (Ishida et al., 2009; Johnston, 2010; Lumsden and Christianson, 2010), and a special issue of *Reports on Progress in Physics* (volume 74, December 2011), focusing on FBS, which the reader should find useful and informative.

20.1 Different FBS families, their crystal structures, and their general features

Primarily there exist four families of Fe-containing pnictides and chalcogenides exhibiting superconductivity, which are named according to the stoichiometries of their parental prototypes, designated as

Table 20.1 Broad differences between HTS cuprates and HTS pnictides

Property	HTS cuprate	HTS pnictide/FBS
Maximum T_c (K)	134 (164 under pressure)	56 (decreases with pressure)
Correlation effects	Strongly correlated, with $U > W$ (W = bandwidth)	Moderately correlated, with $U \approx W$
Involvement of magnetism	Parent compounds: magnetic insulators	Parent compounds magnetic poor metals or semimetallic
Band structure and order parameter	Single band, sign changing d-wave	Multiband, sine-changing s-wave ($s\pm$)
Pairing interaction	Magnetic, but still debated	Magnetic, but no consensus
Dimensionality	2D and highly anisotropic	Less 2D/3D with lesser anisotropy
Superconducting layer and substitutions	CuO_2 layer and superconductivity very sensitive to Cu-site substitutions	FeAs (FeSe) layer and superconductivity less sensitive to Fe-site substitutions

(a) RFePnO, with R = rare earth and Pn = As or P, termed 1111;
(b) BFe$_2$As$_2$, with B = alkaline earth or rare earth (Eu), termed 122;
(c) AFeAs, with A = alkali metal, termed 111;
(d) FeCh, with Ch = Se or Te, termed 11.

All these form tetragonal crystal structures, with (a), (c), and (d) having the same primitive space group $P4/nmm$, and (b) possessing a body-centred space group $I4/mmm$. Besides the above four, three less explored families are

(e) AFe$_2$Se$_2$, with A = alkali metal or Tl, which is simply the Ch analogue of (b) (Guo et al., 2010);
(f) Sr$_3$T$_2$O$_5$Fe$_2$As$_2$, with T = Sc (i.e. 32522) (Zhu, et al., 2009);
(g) Sr$_4$T$_2$O$_6$Fe$_2$(Pn)$_2$, with T = Sc or V and Pn = P or As (i.e. 42622) (Ogino et al., 2009; Zhu et al., 2009b).

The crystal structures of some the above families of superconductors are depicted in Figure 20.1. Interestingly, the presence of an identical FeAs layer is an important common feature of the displayed crystal structures. The FeAs layer is analogous to the CuO_2 layer of the cuprate superconductors in that it is where superconductivity resides in the pnictides. The analogous FeSe layer in Fe-based chalcogenides plays the same role. These Fe-containing layers are, however, not flat and are more like a slab with the Pn or Ch ions protruding above and below the plane, making the layers corrugated. Fe atoms forming the above layer are tetrahedrally coordinated with Pn or Ch atoms, so that the layers consist of FeAs$_4$ or FeSe$_4$ tetrahedra. Because of this, as a universal feature of FBS, the Fe atoms are placed closer to each other than the Cu atoms in cuprates. For instance, the Cu–Cu distances in cuprates are about 0.38 nm, while the Fe–Fe distances in FBS range from 0.285 to 0.267 nm. This is relevant to the unusual electronic structure of FBS.

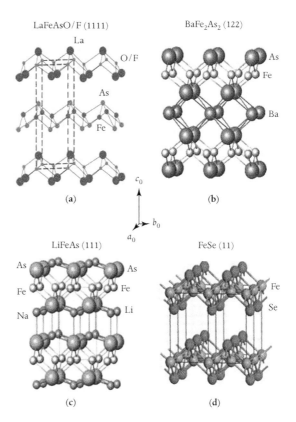

LaFeAsO/F (1111)

BaFe$_2$As$_2$ (122)

LiFeAs (111)

FeSe (11)

Figure 20.1 Crystal structures of four FBS families of superconductors.

In comparison with the CuO$_2$ planes of cuprates, the FeAs planes are less sensitive to substitutional effects and allow replacement of Fe by other cations such as Co or Ni. The ensuing T_c depression is nowhere as drastic as that due to Zn substitution at the Cu sites of HTS cuprates. Because the FeAs/FeSe layers are essentially very similar in all the pnictide/chalcogenide materials, the differences in the properties of various families arise primarily from the spacer layers and the separation between the adjoining FeAs/FeSe layers. The greater the separation (which should make the material more two-dimensional and more anisotropic), the higher is the T_c. This indicates that the other blocking or charge-reservoir layers, present as spacer layers, have a non-trivial role in superconductivity. In Sr$_2$VFeAsO$_3$, having a larger spacer of SrVO$_3$, the observed $T_c = 37.2$K (Zhu et al., 2009c). Although this is still smaller than the maximum T_c of 56 K, it is still appreciably large, considering that Sr$_2$VFeAsO$_3$ is a nominally stoichiometric material whose T_c is expected to rise with suitable doping. Indeed, the doped material Sr$_2$Sc$_{0.4}$Ti$_{0.6}$FeAsO$_3$ does possess an enhanced onset $T_c = 45$K (Chen et al., 20093). Interestingly, the 11-compound possessing only FeSe layers and no other spacer layers is a three-dimensional superconductor, which makes it of practical relevance for applications.

Table 20.2 T_S, magnetic moment of Fe atom, $T_N(\text{Fe})$, and $T_N(\text{RE})$ for compounds of different FBS families

Material	T_S (K)	$T_N(\text{Fe})$ (K)	Magnetic moment of Fe atom (μ_B)	Reference	$T_{N(RE)}$ (K)	Reference
LaFeAsO/F	155	137	0.36	de la Cruz et al. (2008)	—	—
NdFeAsO/F	150	141	0.25	Chen et al. (2008b)	2/4	Qiu et al. (2008)/Awana et al. (2011)
PrFeAsO/F	153	136	0.35	Kimber et al. (2008)	12–14	Zhao et al. (2008a)
CeFeAsO/F	155	140	0.8	Zhao et al. (2008b)	4	Zhao et al. (2008b)
SmFeAsO/F	155	135	0.40	Maeter et al. (2009)	1.6/5.4	Ryan et al. (2009)/Awana et al. (2009)
$BaFe_2As_2$	143	143	0.87	Su et al. (2009)	—	—
$SrFe_2As_2$	220	220	0.94	Zhao et al. (2008c)	—	—
$CaFe_2As_2$	173	173	0.80	Goldman et al. (2008)	—	—
$EuFe_2As_2$	190	190	0.53	Raffius et al. (1993)	19–22	Guguchia et al. (2011), Xiao et al. (2009)
$Fe_{1.068}Te$	67	67	2.25	Li et al. (2009b)	—	—
$Na_{1-\delta}FeAs$	37	37	0.09	Li et al. (2009c)	—	—

Despite the obvious differences in crystal structure, the parent (undoped) compounds of all the FBS families show a structural transformation at $T = T_S$ from a tetragonal to an orthorhombic phase. There is also a magnetic transition at $T = T_N$ from a high-temperature paramagnetic metal to a low-temperature antiferromagnetic (AFM) metal, manifesting a spin-density wave (SDW) order resulting from magnetic moments of Fe atoms. The two transitions may occur simultaneously (i.e. at $T_S = T_N$) or the temperature of the magnetic transition may be slightly (about 10–20 K) lower than that of the structural transition. This is considered unusual, since many theoretical approaches tend to regard magnetism as the cause of structural change, and therefore T_N should be greater than T_S. Typical T_S and $T_N(\text{Fe})$ values are given in Table 20.2. Both transformations in the parent compounds are suppressed by electron or hole doping, which further cooling turns the material superconducting. If the compound contains a rare-earth ion with a magnetic moment, such as Nd, Pr, Ce, or Sm, its sublattice at low temperatures may give rise to an independent AFM ordering at $T_N(\text{RE})$, coexisting with superconductivity (Table 20.2).

The structural transition shows up as a drop in resistivity at 150 K as observed for RFeAsO compounds (Chen et al., 2008a; Dong et al., 2008), as depicted in Figure 20.2 for the CeFeAsO/F compound (Chen et al., 2008a). As may be seen, for the undoped parent compound free from F (non-superconducting), there is an anomaly at $T_S \approx 150$ K, which is gradually suppressed and shifts to lower temperatures with increasing F substitution, and finally the anomaly vanishes in favour of

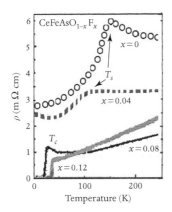

Figure 20.2 Resistivity as a function of temperature for $CeFeAsO_{1-x}F_x$. T_s (150 K for $x = 0$) is suppressed with increasing x, which favours superconductivity (after Chen et al., 2008a).

superconductivity. This structural transition was also detected by de la Cruz et al. (2008) in neutron-scattering experiments, which also showed the magnetic transition occurring at a lower temperature of $T_N = 138$ K. This was corroborated by other techniques such as ^{57}Fe Mössbauer spectroscopy and μSR (Klauss et al., 2008), and also through NMR studies (Nakai et al., 2008). Kondrat et al. (2009) detected these transitions in La1111 compounds in their measurements of specific heat, thermal conductivity, thermoelectric power, lattice parameters, and electrical resistivity. These mutually coupled structural and magnetic anomalies were absent for the superconducting samples. In the case of Ba122 compounds, Rotter et al. (2008) found the two anomalies occurring together at $T_S = T_N = 140$ K, with the structural transformation having first-order character as indicated by magnetic hysteresis and NMR studies on single-crystal samples (Kitagawa et al., 2008). As with the results on the 1111 system, the magnetic order in the 122 system was studied through neutron diffraction (Huang et al., 2008), which corroborated the first-order character of the transition at $T_S = T_N$.

20.1.1 RFeAsO: 1111 compounds

LaFeAsO$_{1-x}$F$_x$, with $T_c = 26$ K was the first Fe-containing high-temperature superconductor discovered, belonging to the 1111 pnictide family (Kamihara et al., 2008). The parent (undoped) compound is a poor metal or semimetal and a non-superconductor, and, as mentioned earlier, with cooling it undergoes a structural phase transition at $T_S \approx 150$–155 K (Table 20.2) from tetragonal to orthorhombic structure. At a temperature T_N about 10–20 K below T_S, the materials of this family become AFM, forming an SDW state. On further cooling, however, both the structural transformation and AFM are suppressed and superconductivity emerges. Superconductivity results if about 3% of O^{2-} is replaced by F^- or if the material is made slightly oxygen-deficient (Table 20.3 and 20.4). Both situations lead to electron doping and give rise to similar T_c values. The optimum T_c occurs over a broad plateau region extending from about 4.5% to about 12% of F-substitution, as

Table 20.3 Optimum T_c values of fluorinated 1111 compounds RFeAsO$_{1-x}$F$_x$ (R = rare earth) formed with x corresponding to maximum T_c

R	La	Ce	Pr	Nd	Sm	Gd	Tb	Dy	Ho
Optimum T_c (K)	26–28	41	52	52	55	36	46	45	36
x	0.11	0.16	0.11	0.11	0.1	0.17	0.1	0.1	—
Reference	Kamihara et al. (2008)	Chen et al. (2008a)	Ren et al. (2008b)	Ren et al. (2008c)	Ren et al. (2008a)	Cheng et al. (2008)	Bos et al. (2008)	Bos et al. (2008)	Rodgers et al. (2009)

Data from Ishida et al. (2009) and Johnston (2010).

Table 20.4 Optimum T_c values reported for oxygen-deficient $RFeAsO_{1-y}$

R	T_c (K)	Reference
La	26	Ren et al. (2008d), Miyazawa et al. (2009)
Pr	48	Ren et al (2008d), Miyazawa et al. (2009)
Nd	53	Ren et al. (2008d), Kito et al. (2008)
Sm	55	Ren et al (2008d)
Gd	53	Miyazawa et al. (2009), Yang et al. (2008)
Tb	52	Miyazawa et al. (2009)
Dy	52	Miyazawa et al. (2009)

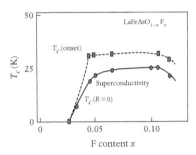

Figure 20.3 T_c as a function of F content in LaFeAsO/F, exhibiting a broad plateau from 4.5% to 12% of F content for optimum T_c (after Kamihara et al., 2008).

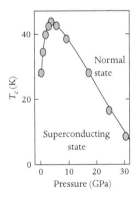

Figure 20.4 Pressure effect on T_c of LaFeAs/O (after Takahashi et al., 2008). T_c rises from 26 K to about 43 K at a pressure of 4 GPa, after which it decreases.

shown in Figure 20.3 (Kamihara et al., 2008). Under an external pressure of about 4 GPa, the T_c of La1111 was raised from 26 K to 43 K (Figure 20.4) and thereafter was suppressed, and the effect of pressure was again to suppress T_S and T_N (Takahashi et al., 2008). As with the HTS cuprates, a spectacular rise in T_c occurred with the internal chemical pressures developed by replacing larger La by smaller R-ions, such as Ce, Pr, Nd, Sm, and Gd. As may be seen from Table 20.3 and 20.4, down to R = Sm, T_c increased monotonically from 26 K for La1111 to 55 K for Sm1111, beyond which the smaller-ion substitutions caused T_c to decrease non-uniformly. T_c is maximum when FeAs$_4$ forms a regular tetrahedron, for which the Fe–Pn–Fe angle approaches 109.4°. This correlation implies that T_c is essentially controlled by the height of the Pn atoms from the Fe plane, and for optimum T_c this height is about 0.138 nm.

Superconductivity can also be realised by electron doping through partial replacement of Fe by Co, Ni, Ir, etc. Typical examples are LaFe$_{1-x}$Co$_x$AsO (Sefat et al., 2008a), LaFe$_{1-x}$Ir$_x$AsO (Qi et al., 2009b), SmFe$_{1-x}$Co$_x$AsO (Qi et al., 2008a), and SmFe$_{1-x}$Ni$_x$AsO (Li et al., 2009), with respective optimum T_c values of 14 K, 12 K, 15 K, and 10 K. The effect of partial Ir substitution for Fe in inducing superconductivity in F-free compounds is depicted in Figure 20.5 (Qi et al., 2009b). Clearly, with Fe-site substitution, the observed T_c is always significantly lower than the optimum T_c of the F-containing pristine sample, which suggests that Fe-site disorder does indeed lower T_c, although this effect is much weaker than in HTS cuprates. Interestingly, superconductivity has also been observed in the non-rare-earth containing 1111 compound AFe$_{1-x}$Co$_x$AsF (where A = Ca or Sr). Substituting La^{3+} by Ca^{2+} or Sr^{2+} essentially gives rise to hole doping. The overall effect results in T_c = 22 K for A = Ca and x = 0.1 (Matsushi et al., 2008) and T_c = 4K for A = Sr and x = 0.125 (Matsushi et al., 2008b).

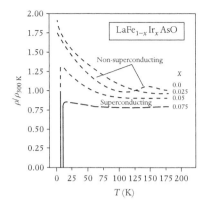

Figure 20.5 Superconductivity is induced in the F-free LaFeAsO by partial replacement of Fe by Ir (after Qi et al., 2009b).

20.1.2 AFe$_2$As$_2$ (A = alkaline earth Ba, Sr, Ca or rare earth Eu): As-based 122 compounds

These materials possess the same crystal structure (Figure 20.1(b)) as CeCu$_2$Si$_2$, CeRu$_2$Si$_2$, etc., which we discussed previously in Chapter 13 on HF superconductors. This family is, however, different from 1111 and others in that its unit cell contains two FeAs layers instead of one. In these compounds, superconductivity can be realised with either hole or electron doping or alternatively with imposed pressure (Alrieza et al., 2009). The hole doping is achieved by partly replacing divalent 'A' by monovalent 'B' (= K, Na, or Cs), that is, A$_{1-x}$B$_x$Fe$_2$As$_2$, where Fe, as in 1111 compounds, is in the 2+ valence state. For electron doping, the commonly pursued approach involves a partial replacement of Fe by Co, that is, A(Fe$_{1-x}$Co$_x$)$_2$As$_2$. The undoped parent compounds show a similar structural transformation and AFM ordering as 1111 compounds. But, unlike the 1111 system, where $T_S > T_N$, for the parent (undoped) 122 compound, $T_S = T_N = 140$ K and the transition has first-order character. Similar to 1111 compounds, 122 compounds too are not so sensitive to substitutions at the Fe sites in comparison with Cu-site substitutions in HTS cuprates. In fact, superconductivity in 122 compounds may be observed by changing each of the three constituents (Ren et al., 2009; Rotter et al., 2008). Aliovalent substitutions give rise to different charge-doping effects that influence T_c. Surprisingly, however, isovalent substitutions, such as P for As (Kasahara et al., 2010) or Ru for Fe (Rullier-Albenque et al., 2010), too are found to induce superconductivity in pnictides (Ren et al., 2009). This is generally ascribed to the chemical pressure caused by a change in the ion size. In the above instances, the observed effect is related to the possible disruption in nesting and small alterations in the Fermi surface, such as changes in the size of electron and hole pockets (Singh 2008).

Pressure, internal or external, in general, suppresses magnetism and promotes superconductivity. Kotegawa et al. (2009) found a sharp superconductive resistive transition in SrFe$_2$As$_2$, which was accompanied by suppression of the AFM state. In fact, in both SrFe$_2$As$_2$ and BaFe$_2$As$_2$, pressure alone is enough to stabilise the superconducting state (Alrieza et al., 2009). However, in 122 compounds, it is important how the pressure is applied. For instance, superconductivity of CaFe$_2$As$_2$ occurs only under non-hydrostatic pressure (Yu et al., 2009). Below T_N, these materials, as per density functional theory (DFT) calculations, order with AFM stripes along the longer orthorhombic axis and FM stripes along the shorter orthorhombic axis. Some of the prominent 122 compounds are listed in Table 20.5. The highest T_c achieved for these compounds is 38 K for Ba$_{1-x}$K$_x$Fe$_2$As$_2$ with $x = 0.4$ (Rotter et al., 2008).

Besides partial substitution of Co for Fe, which leads to electron doping, superconductivity in 122 compounds can be readily achieved also by

Table 20.5 Optimum T_c values of 122 compounds

122 compound	T_c (K)	Reference	122 compound	T_c (K)	Reference
$Ba_{1-x}K_xFe_2As_2$	38	Rotter et al. (2008)	$Sr(Fe_{1x}Ni_x)_2As_2$	10	Saha et al. (2009)
$K_{1-x}Sr_xFe_2As_2$	36	Sasmal et al. (2008)	$Ca(Fe_{1x}Co_x)_2As_2$	17	Kumar et al. (2009)
$Ba_{1-x}Rb_xFe_2As_2$	23	Bukowski et al. (2009)	$Ba(Fe_{1x}Rh_x)_2As_2$	24	Ni et al. (2009)
$Cs_{1-x}Sr_xFe_2As_2$	37	Sasmal et al. (2008)	$Ba(Fe_{1-x}Pd_x)_2As_2$	19	Ni et al. (2009)
$Ca_{1-x}Na_xFe_2As_2$	20	Wu et al. (2008)	$Sr(Fe_{1-x}Rh_x)_2As_2$	22	Han et al. (2009)
$Eu_{1-x}Na_xFe_2As_2$	35	Qi et al. (2008b)	$Sr(Fe_{1-x}Ir_x)_2As_2$	22	Han et al. (2009)
$Eu_{1-x}K_xFe_2As_2$	32	Jeevan et al. (2008)	$Sr(Fe_{1-x}Pd_x)_2As_2$	9	Han et al. (2009)
$Ba(Fe_{1-x}Co_x)_2As_2$	24	Sefat et al. (2008b)	$Ba(Fe_{1-x}Ru_x)_2As_2$	21	Ni et al. (2009)
$Ba(Fe_{1-x}Ni_x)_2As_2$	20	Li et al. (2009)	$Sr(Fe_{1-x}Ru_x)_2As_2$	14	Qi et al. (2009)

numerous other substitutions at the Fe sites, such as Ni (Li et al., 2009a), Rh (Ni et al., 2009; Han et al., 2009), Ir (Han et al., 2009), Pd (Ni et al., 2009; Han et al., 2009) and Ru (Qi et al., 2009a). But, surprisingly, electron doping through Cu (Canfield et al., 2009) or hole doping through Cr (Sefat et al., 2009) fail to induce superconductivity.

In this family, $EuFe_2As_2$ is the only member that has a rare earth at the A-site. The large localised magnetic moment of Eu^{2+} ions (about $7\mu_B$) becomes ordered below T_N (Eu) $\approx$ 19–22 K in a likely A-type AFM structure (Ren et al., 2008a) which coexists with superconductivity in the doped compound, below $T_c \approx$ 30–35 K, as with many members of the 1111 family discussed earlier. In the undoped parent compound, as it is cooled, a magneto-structural transition occurs at $T_S = T_N(Fe) \approx$ 190 K from a high-temperature paramagnetic tetragonal phase to an AFM orthorhombic phase, accompanied by an SDW ordering of Fe spins. The representative T_c and T_N values of some members of 122 family are given in Table 20.2.

20.1.3 AFe_2Se_2 (A $=$ K, Rb, Cs, Tl): Se-based 122 compounds

These compounds having T_c up to 33 K (Guo et al., 2010; Wang et al., 2011a, b) are the Se analogues of the 122-type As compounds where Fe sites are partially occupied and their ground state, unlike other systems previously mentioned, is believed to be a Mott insulator (Liu et al., 2011; Shermadini et al., 2011; Bao et al., 2011) rather than a poor metal or semimetallic. Recent μSR (Shermadini et al., 2011; Pomjakushin et al., 2011), neutron-diffraction (Bao et al., 2011; Ye et al., 2011), Raman spectroscopy, resistivity, and magnetisation (Liu et al., 2011) measurements on $A_xFe_{2-y}Se_2$(A $=$ K, Rb, Cs, orTl) have revealed coexistence of superconductivity and strong AFM. A checkerboard-type AFM order is found to occur in the Fe-deficient lattice, below $T_N(Fe) \approx$ 520 K, having an ordered magnetic moment of $3.31\mu_B$/Fe (Bao et al.,

2011; Ye et al., 2011) and superconductivity at $T_c \approx 30-40$ K (Yuan et al., 2012). Although microscopic coexistence of the two orders is not ruled out, several measurements, such as magnetisation (Shen et al., 2011), transmission electron microscopy (TEM) (Wang et al., 2011), angle-resolved photoemission spectroscopy (ARPES), and Mössbauer spectroscopy (Ryan et al., 2011), all indicate the superconducting and magnetic orders in the above materials to be phase-separated. The issue, however, remains unsettled. Interestingly, superconductivity at a relatively high temperature of 33 K found in these materials is associated with Fe vacancies (Bao et al., 2011).

20.1.4 AFeAs (A = alkali metal Li or Na): 111 compounds

The 111 compounds contain a single FeAs layer separated by double layers of A-ions (Figure 20.1(c)). The Fe–Fe ion distance in these layers is much shorter than in 1111 and 122 compounds. Superconductivity is observed in both LiFeAs and $Na_{1-x}FeAs$, in which the FeAs layers respectively carry spacer layers of Li and Na ions. LiFeAs possesses a Cu_2Sb-type tetragonal structure with an average Fe valence of 2+ as in the 1111 and 122 parent compounds. Unlike the parent 1111 and 122 compounds, which need additional doping for their superconductivity, in LiFeAs superconductivity is observed at $T_c = 18$K (Tapp et al., 2008; Wang et al., 2008; Sasmal et al., 2009) in stoichiometric composition without any extra doping, and thus this compound does not exhibit any magnetic or structural transformation at a higher temperature. NaFeAs, on the other hand, shows a transition from a high-temperature paramagnetic tetragonal phase to a low-temperature AFM orthorhombic phase at $T_S = T_N = 50$ K (Li et al., 2009c) and needs doping through Na non-stoichiometry ($Na_{1-x}FeAs$) to achieve superconductivity in the temperature range of 12–25 K (Chu et al., 2009), depending on Na content. Interestingly, T_c of stoichiometric LiFeAs decreases linearly with applied pressure while that of NaFeAs increases with pressure. Figure 20.6 shows the observed behaviour of LiFeAs (Gooch et al., 2009).

20.1.5 Fe(Se, Te): 11 compounds

This family of compounds are formed with Fe–Ch layers, but contain no spacer layers in between (Figure 20.1(d)), which makes them more three-dimensional. Similar to other pnictide compounds discussed already, non-superconducting FeCh samples, when cooled, exhibit a transition from a paramagnetic tetragonal phase to an AFM orthorhombic phase at $T_S = T_N = 70$K (Lynn and Dai, 2009). The superconducting behaviour of these compounds is complicated by the presence of excess Fe, that is, the composition $Fe_{1+y}Te$ (Bao et al., 2009), which is magnetic and destroys superconductivity (Zhang et al., 2009).

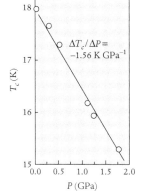

$\Delta T_c / \Delta P = -1.56$ K GPa^{-1}

Figure 20.6 Pressure effect on T_c of LiFeAs (after Gooch et al., 2009).

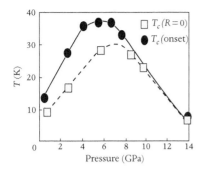

Figure 20.7 Pressure effect on T_c of FeSe (after Margadonna et al., 2009).

Nearly stoichiometric $FeSe_{0.97}$ has the PbO-type tetragonal structure with space group $P4/nmm$ and shows superconductivity at $T_c \approx 8K$ (Hsu et al., 2008). The superconducting phase is commonly represented by Se deficiency with composition $FeSe_{1-\delta}$. The T_c of 8 K was soon raised to 15 K for $FeSe_{0.5}Te_{0.5}$ (Yeh et al., 2008; Fang et al., 2008) although FeTe is a non-superconductor. Interestingly, as depicted in Figure 20.7 (Margadonna et al., 2009a), an imposed pressure of a few gigapascals has a dramatic effect of raising the onset T_c to as high as 27–37 K, which was corroborated by Medvedev et al. (2009), although such a pronounced effect has to date not been realised through chemical pressure.

20.2 Electronic structure

Because of the much closer packing of Fe ions in the Fe-based compounds, all five degenerate Fe $3d$-orbitals are able to contribute to charge carriers (Lebègue, 2007; Singh and Du, 2008), whereas in the cuprates, owing to the much larger Cu–Cu spacing, there is only one Cu $3d$-orbital responsible for this. The p-orbitals of Pn or Ch in these compounds hybridise, though weakly, with five Fe $3d$-orbitals to produce a complicated band structure (Figure 20.8) in which there are five bands crossing

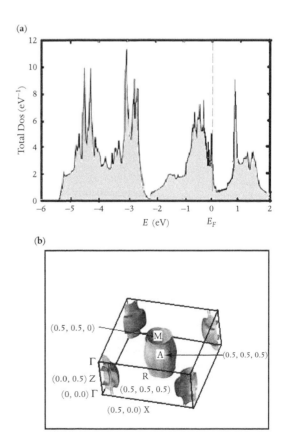

Figure 20.8 (a) Total density of states and (b) Fermi surface of LaFeAsO (after Singh and Du, 2008).

the Fermi level, giving rise to a complex multicomponent Fermi surface, as revealed through calculations (Singh and Du, 2008). The multiband electronic structure is consistent with various core-level and valence-band spectroscopic studies (Kurmaev et al., 2008; Parks Cheney et al., 2010; Bondino et al., 2010), confirming the dominant Fe character of the bands located close to E_F. As may be seen from Figure 20.8(a), there is also a dip in the density of states (DOS) at E_F, which provides a semi-metallic character to these compounds. The five d-orbitals located at E_F, in momentum space, give rise to five cylindrical sections or pockets on the Fermi surface (Singh and Du, 2008) for the room-temperature tetragonal structure. Of these, there are three hole-like cylinders at the centre of the Brillouin zone and two electron-like cylinders at the corners (Figure 20.8(b)). Such a cylindrical form of the Fermi surface with disconnected sheets is responsible for the quasi-two-dimensional nature of these materials and their unconventional superconductivity. However, in comparison with the cuprates, the ensuing anisotropy effect is much less. The members of the 1111, 122, 111, and 11(FeSe) families all have essentially similar band structures (Sefat and Singh, 2011) and, except for a few minor differences, they all have similar Fermi surfaces as depicted here.

20.3   Phase diagrams

The phase diagrams are commonly determined using various experimental tools, including μSR, Mössbauer spectroscopy, X-ray diffraction (XRD), neutron diffraction, resistivity, and magnetisation (Zhao et al., 2008b; Drew et al., 2009; Luetkens et al., 2009; Rotundu et al., 2009; Margadonna et al., 2009b). Broadly, one may identify three distinct types of phase diagrams common to most of these families, as shown in Figure 20.9. Considering the 1111 type of compounds, for example, with R = La and Pr the structural and magnetic transitions vanish like a first-order phase change upon doping, manifesting an abrupt step-like behaviour of T_N at the appearance of superconductivity (Luetkens et al., 2009, Rotundu et al., 2009). This is illustrated in Figure 20.9(a) for La1111, where the magnetic order is abruptly suppressed at the optimum doping for the onset of superconductivity.

In the case of Ce1111, there is a gradual decrease of T_N with doping (Zhao et al., 2008b), which vanishes when the doping reaches the optimal level for emergence of superconductivity (Figure 20.9(b)). Sm1111 exhibits a third type of phase diagram (Figure 20.9(c)) where the observed behaviour (Drew et al., 2009) is similar to that of Ce1111 except that, over a small doping region, there is an overlap of AFM (SDW) phase with superconductivity. This interesting situation demonstrates that a magnetic order due to Fe moments need not necessarily be totally

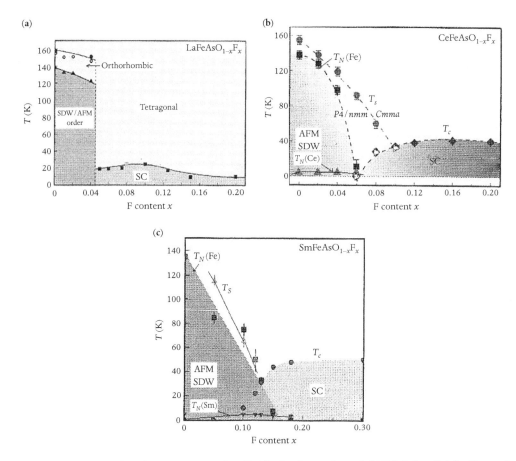

Figure 20.9 Three representative phase diagrams: (a) LaFeAsO$_{1-x}$F$_x$ (after Luetkens et al., 2009), (b) CeFeAsO$_{1-x}$F$_x$ (after Zhao et al., 2008b), and (c) SmFeAsO$_{1-x}$F$_x$ (Drew et al., 2009).

suppressed for superconductivity to emerge, although in most such cases the former eventually does get suppressed before the maximum T_c is reached with doping. In the case of Ce-, Pr-, and Sm-containing 1111 compounds, the presence of low-temperature AFM order formed below their respective T_N(RE) is found coexisting with the superconducting dome.

The third type of phase diagram is commonly manifested by the most extensively studied 122 family. Irrespective of whether the parent compound BaFe$_2$As$_2$ is hole-doped, that is, Ba$_{1-x}$K$_x$Fe$_2$As$_2$, or electron-doped, that is, BaFe$_{2-x}$Co$_x$As$_2$, their phase diagrams (Figure 20.10) contain a small doping range where the AFM/SDW state coexists with superconductivity (Ni et al., 2008; Chen et al., 2009). The main issue regarding this coexistence is whether the two orders are microscopically coexisting or are phase-separated. In the example of hole-doped Ba$_{1-x}$K$_x$Fe$_2$As$_2$, a host of experimental techniques such as NMR (Fukazawa et al., 2009), μSR, and scanning probe microscopy

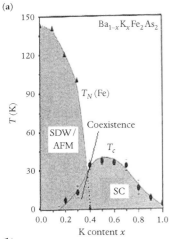

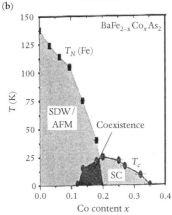

Figure 20.10 Phase diagrams of (a) $Ba_{1-x}K_x$ Fe_2As_2 (after Chen et al., 2009) and (b) $BaFe_{2-x}Co_xAs_2$ (after Wang et al., 2009).

(SPM) (Park et al., 2009), as well as neutron and X-ray diffraction (Inosov et al., 2009), all seem to corroborate the microscopic phase-separation scenario, while Mössbauer studies favour microscopic coexistence (Rotter et al., 2009). In the case of electron-doped $BaFe_{2-x}Co_xAs_2$, many of these techniques (Pratt et al., 2009; Christianson, et al., 2009; Laplace et al., 2009; Bernhard et al., 2009) surprisingly support the idea of coexistence, although some of the results could be interpreted differently in favour of phase separation. The issue still remains essentially unsettled, as with most other systems manifesting coexistence phenomenon.

20.4   Unconventional superconductivity of FBS

The large $T_c > 50$ K of FBS, which exceeds the generally accepted 30 K limit of conventional superconductors, is an indication of its unconventional superconductivity. Various studies of Raman spectroscopy (Zhang et al., 2010), resistivity (Tu et al., 2010), etc. have indicated a weak electron–phonon coupling in FBS, which is theoretically unable to explain the observed high T_c (Boeri et al., 2008; Subedi and Singh, 2008; Jishi and Alyahyaei, 2010). In terms of conventional phonon-mediated superconductivity, Mazin et al. (2008) performed an *ab initio* calculation of the electron–phonon interaction parameter for LaFeAsO/F and found a value of $\lambda \approx 0.2$, which was too small to explain its $T_c \approx 26$ K. Isotope-effect studies (Liu et al., 2009) revealed that in Sm1111 and BaK122 the oxygen isotope effect was feeble, but the Fe isotope effect was substantial, with $\alpha_0 \approx 0.35$. This suggests that phonons associated with FeAs layers may have some positive role to play in superconductivity.

In these compounds, superconductivity occurs close to AFM/SDW order and, as with the previously discussed heavy fermion, organic, cuprate, and ruthenate superconductors, such a situation strongly favours unconventional superconductivity, mediated by spin fluctuations. Despite this, early observations of the Knight shift made on single crystals of almost all categories of FBS compounds, including LaFeAsO/F (Grafe et al., 2008), PrFeAsO/F (Matano et al., 2008), BaK-122 (Matano et al., 2009; Yashima et al., 2009), $Ba(FeCo)_2As_2$ (Ning et al., 2008), LiFeAs (Jeglič et al., 2010; Li, et al., 2010a), and $BaFe_2(AsP)_2$ (Nakai et al., 2010), exhibited a vanishing Knight shift in the superconducting state in all crystallographic directions. The Knight shift measurements by Matano et al. (2008) are shown in the main part of Figure 20.11. These measurements thus strongly indicated a fully isotropic node-free gap of the conventional singlet state and simultaneously ruled out the possibility of triplet-state formation in *p*- or *g*-wave channels of FBS compounds. The spin-singlet state can occur either in the conventional *s*-wave orbital channel or in an unconventional *d*-wave channel,

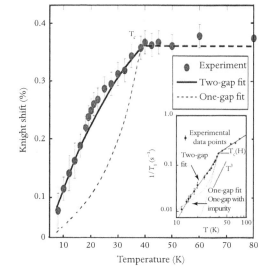

Figure 20.11 NMR studies on PrFeAsO/F showing the temperature dependence of the Knight shift; the inset on the bottom right depicts the spin–lattice relaxation rate as a function of temperature (after Matano et al., 2008).

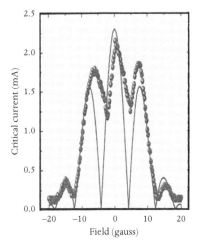

Figure 20.12 Diffraction-like Josephson current pattern along the c-direction observed in a BaK122/Pb tunnel junction (after Oh et al., 2009).

such as in HTS cuprates. The possibility of d-wave superconductivity in FBS was subsequently found to be remote because the Josephson junction formed between FBS (BaK-122) and a known s-wave superconductor (Pb) revealed a significant tunnelling current flowing along the z-direction (Oh et al., 2009) with characteristic diffraction-like pattern under a magnetic field (Figure 20.12). This would not be expected if FBS were a d-wave superconductor. Also, ARPES studies on BaK-122 showed that the gap was isotropic, in contrast to a d-wave gap with nodes (Ding et al., 2008). On the other hand, Andreev reflection data were found to be consistent with the isotropic s-wave gap (Chen et al., 2008a) without any nodes.

Despite this, the FBS compounds exhibited several experimental traits of unconventional superconductivity:

1. The unconventional superconductivity of LaFeAsO/F with highly anisotropic gap and line nodes was demonstrated by a number of researchers through the absence of a Hebel–Slichter coherence peak and a T^3 dependence of the NMR relaxation rate down to about $0.2T_c$ (Grafe et al., 2008; Matano et al., 2008; Nakai et al., 2008). This is depicted in the inset of Figure 20.11 for PrFeAsO/F (Matano et al., 2008). Similar T^3 behaviour was also observed in [75]As-NQR and ^{77}Se-NMR studies respectively of superconducting LaFeAsO$_{1-\delta}$ and FeSe$_{1-\delta}$ (Figure 20.13) (Mukuda et al., 2008).

2. Specific heat studies of 1111 compounds showed a nonlinear magnetic field dependence, with Sommerfeld coefficient $\gamma \propto H^{1/2}$ (Mu et al., 2008), characteristic of a gap with line nodes. For a fully gapped superconductor, the behaviour is expected to be linear. Surprisingly, linear and sublinear field dependences are also reported respectively

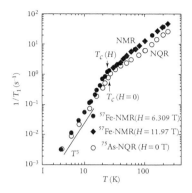

Figure 20.13 NMR/NQR studies of the temperature dependence of the spin–lattice relaxation rate, showing the absence of a coherence peak just below T_c and the T^3 behaviour at low temperatures characteristic of an unconventional gap with line nodes (Mukuda et al., 2008, see the text on p.380).

for Co-substituted (Gofryk et al., 2010) and P-substituted (Kim et al., 2010) Ba122 compounds.

3. For a fully gapped single-band s-wave superconductor, the electronic thermal conductivity is expected to follow an exponential behaviour in both temperature and field dependences, while for a superconductor with nodes the predicted variation is linear in T and an $H^{1/2}$ dependence is expected. However, the observed behaviour for 122 compounds exhibits a wide variation, ranging from conventional s-wave to d-wave, which is most surprising (Ding et al., 2009; Luo et al., 2009; Tanatar et al., 2010; Dong et al., 2010a, b; Hashimoto et al., 2010).

4. The temperature dependence of the penetration depth λ exhibited much variation and many mutual inconsistencies. PrFeAsO$_{1-\delta}$ showed a flat temperature dependence of λ_{ab} at low temperatures (Hashimoto et al., 2009a), which could not be explained in terms of the nodal gap. The observed behaviour over an extended temperature range required the presence of two isotropic gaps. Exponential behaviour of the penetration depth, characteristic of the conventional isotropic gap, was reported for BaK-122 (Hashimoto et al., 2009a), where two isotropic gaps, respectively with $\Delta_1 = 1.17k_BT_c$ and $\Delta_2 = 2.40k_BT_c$, were found necessary to explain the variation over the extended temperature range. The μSR measurements of the penetration depth also corroborated the presence of two gaps in this above material (Hiraishi et al., 2009). The T^2 dependence, characteristic of a dirty line-nodal state, was manifested by Co- and Ni-doped samples of Ba-122 (Gordon et al., 2009; Martin et al., 2010).

5. Two isotropic nodeless s-wave gaps with $\Delta_1 \approx 6$ meV and $\Delta_2 \approx 12$ meV were observed in ARPES studies of BaK-122 single crystals (Ding et al., 2008). The gaps appeared and disappeared together during a cooling and heating cycle.

6. The temperature dependence of the upper critical field (Hunt et al., 2008) (Section 20.6.1), manifesting an upward curvature (Yuan et al., 2009; Fang et al., 2010), was consistent with the presence of two gaps.

7. Muon spin relaxation has been used to measure the superfluid density and its field dependence in LaFeAsO$_{1-x}$F$_x$ (Luetkens et al., 2008). While the temperature dependence of the superfluid density closely follows the conventional BCS theory, the field dependence is found to be unconventional. The linear relation between T_c and superfluid density (the Uemura plot) for FBS falls close to that of hole-doped HTS cuprates, which are unconventional superconductors.

8. Point-contact spectroscopy showed a zero-bias conductance peak, which was interpreted as a sign change in the unconventional order parameter $s_\pm$ occurring across the Fermi surface pockets (Shan et al., 2008).

These various experiments present a rather complex situation and a wide diversity of results, with some being consistent with fully gapped behaviour and the presence of two isotropic gaps while others provide evidence for gap nodes characteristic of anisotropic gaps. Clearly, having previously ruled out p-, d-, and f-wave gap symmetries for FBS, these unusual features of these materials seem equally incompatible with the singlet s-wave order parameter of conventional superconductors. The answer to this apparent discrepancy has been provided by theoretical considerations (Mazin et al., 2008) that the pertinent symmetry, instead of being s-wave, is extended $s_{\pm}$ wave. In this, the gap function is isotropic and singlet, but, in marked contrast to the BCS model, reverses its sign between different sheets of the Fermi surface. The unconventional superconductivity and $s_{\pm}$ state of FBS are the outcomes of their disconnected Fermi surface, discussed in Section 20.2 (Mazin et al., 2008; Aoki, 2009). Superconductivity of FBS is believed to be due to a repulsive electron mechanism in which spin fluctuations mediate electron pairing. As spin fluctuations are repulsive in a singlet state, the sign change associated with an $s_{\pm}$ wave is an essential requirement to obtain Cooper pairing. In the pairing process the order parameter changes its sign under translation by a wavevector $\mathbf{Q}$ of the magnetic state, which in real space corresponds to an oscillatory potential that is attractive for electrons. $\mathbf{Q}$ is just the right vector that connects the hole and electron pockets at the Brillouin zone.

The $s_{\pm\ \text{state}}$ is fully gapped when the Fermi surface is disconnected, and although it is an unconventional order parameter, it is isotropic and free from nodes. It needs to be mentioned that without a phase-sensitive test, the $s_{\pm}$ wave gap is indistinguishable from the conventional s-wave gap. The tunnelling experiments need more complicated junction configurations, which has made reliable measurements scant. However, the pertinent sign change of the $s_{\pm}$ state has been substantiated on single crystals of $FeSe_{1-x}Te_x$ using scanning tunnelling microscopy (STM) and spectroscopy (STS) techniques to measure the magnetic field dependence of quasiparticle interference (QPI) patterns (Hanaguri et al., 2010). Further, the pertinent sign change of the order parameter gives rise to neutron spin resonance (Maier and Scalapino, 2008; Korshunov and Eremin 2008), which has also been experimentally observed in FBS materials (Christianson et al., 2008). By taking proper account of the unusual features of the multiband and multigap nature of FBS, many of the contradictory issues pertaining to the order parameter seem resolvable (Kemper et al., 2010; Maiti and Chubukov, 2010; Fernandes and Schmalian, 2010; Paglione and Greene, 2011; Basov and Chubukov, 2011). FBS thus provide the first example of multigap extended s-wave ($s_{\pm}$) superconductivity with discontinuous change of the order parameter between bands, which makes them all the more novel and interesting.

20.5   Materials synthesis

Polycrystalline bulk samples of FBS are commonly synthesised by solid state reaction of the constituent ingredients in powder form mixed together in the appropriate stoichiometry. As an illustration, to synthesise a 1111 compound, $NdFeAsO_{0.8}F_{0.2}$ for example (Awana et al., 2011), powdered As, Fe, NdF_3, Nd, and Fe_2O_3 of better than 3N purity are mixed in stoichiometric amounts, that is, $(1/15)NdF_3 + (14/15)Nd + (4/15)Fe_2O_3 + (7/15)Fe + As$ inside a glove box having less than 1 ppm level of oxygen and a humid atmosphere. The mixed powder mixture is pelletised in rectangular bar shape and further encapsulated in an evacuated (10^{-3} Torr) thick-walled quartz tube. The sealed tube is heated for 12 hours each at $550°C$, $850°C$, and $1150°C$. Finally, the sample thus synthesised, after cooling to room temperature, is black in colour and is sufficiently hard for various other characterisations. Essentially the same procedure is followed for other compounds. For $FeSe_{0.88}$, a stoichiometric powder mixture of Fe and Se in compressed pellet form, in a sealed quartz tube, is first heated for 20 hours at $680°C$, followed by 60 hours at $400°C$, before cooling to room temperature. The former heat treatment is for achieving the required stoichiometry and the latter for realising the PbO structure (Shen, 2012).

Much progress has also been made in growing thin films of these compounds. The PLD technique has also proved successful in making epitaxial thin films of all these families of pnictides and chalcogenides on LaO or STO substrates. Epitaxial thin films of Co-doped $BaFe_2As_2$ have been deposited on (La, Sr)(Al, Ta)O_3 substrates with J_c (self-field, 4.2 K) $\approx 1 \times 10^{10}$ A m^{-2}. By using buffer layers of STO or BTO on the above substrate, a fivefold increase in this value is observed (Katase et al., 2010). Similarly, thin films of 11 compounds, namely $FeSe_{0.5}Te_{0.5}$, deposited using PLD on STO and LaO substrates, have yielded $T_c^{onset} \approx$ 17 K with estimated upper critical field of about 50 T (Si et al., 2009).

Growing bulk single crystals of these pnictide compounds was initially thought to be problematic due to the toxic nature and high vapour pressure of As and of other ingredients, such as K, Na, Li, and Rb. However, the challenge has been effectively met using specially designed furnaces for growing centimetre-sized platelets of single crystals, especially of 122, 111, and 11 compounds, in either Sn or self-flux solvents (Chen et al., 2008c; Ni et al., 2008; Fang et al., 2009; Li et al., 2010; Kihou et al., 2010; Lee et al., 2010). For growing single crystals of LiFeAs of the 111 family, using the Sn-flux approach (Lee et al., 2010), a stoichiometric powder mixture of the three ingredients was added with Sn in a ratio of 1:10 in an alumina crucible, sealed inside a quartz tube under an Ar atmosphere. A glove box was used for handling the chemicals under 1 ppm level of oxygen and moisture. The sealed tube was heated for 24 hours at $250°C$ followed by 4 hours at $500°C$ to fully dissolve the ingredients in the Sn flux. To produce crystal growth, it was further heated to 850°C for 4 hours and subsequently cooled to 500°C at a rate of 3.5°C h^{-1}. At this temperature, the Sn flux was removed from the surface of the material by centrifuging. On slow cooling to room temperature, thin single-crystal platelets of 5 mm × 5 mm of LiFeAs were obtained. A similar procedure was followed by Yeh et al. (2009) for growing single crystals of FeSe$_{1-x}$Te$_x$ of the 11 family.

20.6 Upper critical field, anisotropy, and potential for applications

20.6.1 Upper critical field and its anisotropy

Like HTS cuprates, all the FBS are type II superconductors possessing a large upper critical field, albeit smaller than that of cuprates owing to their lower T_c. However, their anisotropy is much less than that of cuprates and is similar to, or even less than, that of MgB$_2$. This is a significant asset for practical utilisation. Table 20.6 depicts the upper critical field and related parameters of the different FBS families, for which Zhang et al. (2012) have pointed out some interesting features common to their H_{c2} behaviour.

Soon after the discovery of the FBS, their $\mu_0 H_{c2}^{ab}(0)$, as estimated from the initial slope of $H_{c2}^{ab}(T)$ at T_c in the Werthamer–Helfand–Hohenberg (WHH) model (Werthamer et al., 1966), turned out to be huge, ranging

Table 20.6 Measured values of upper critical fields and related parameters, including Pauli paramagnetic and orbital critical field, anisotropy factor γ, and coherence length ξ along and normal to the basal plane, for some representative Fe-based compounds of different families

Properties	Nd-1111 (F = 0.18)	Sm-1111 (O = 0.85)	Ba(K)-122 (K = 0.08)	Li-111	Fe(TeSe)-11 (Se = 0.6, Te = 0.4)
T_c (R = 0) (K)	47	50	28	18	14
Pauli paramagnetic field $H_p(0)$ (T)	86	92	52	33	26
Orbital field $\mu_0 H_{c2}^{ab(orb)}(0)$(T)	210	378	104	40	86
Orbital field $\mu_0 H_{c2}^{c(orb)}(0)$(T)	42	84	56	14	37
$\mu_0 H_{c2}^{ab}(T)$ (experimental)	57 (at 34 K)	51 (at 43 K)	57 (at 10 K)	24.2 (at 0 K)	47 (at 0 K)
$\mu_0 H_{c2}^{c}(T)$ (experimental)	43 (at 18 K)	56 (at 27 K)	55 (at 9 K)	15 (at 0 K)	47 (at 0 K)
Anisotropy γ ($\sim T_c$)	6 (at 46 K)	5 (at 49 K)	2 (at 27 K)	2.5 (at 14 K)	2 (at 12 K)
Anisotropy γ ($T < T_c$)	5.4 (at 35 K)	3.5 (at 43 K)	1.1 (at 12 K)	1.49 (at 1.4 K)	$\sim$1 (at 0.5 K)
Coherence length ξ^{ab} (nm)	2.3	17	2.1	4.8	2.7
Coherence length ξ^{c} (nm)	0.26	3.6	2.1	1.7	2.7

Data are mostly from Zhang et al. (2012).

between 100 and 400 T. The corresponding values measured along the *c*-direction were also large, around 40–100 T (Sefat et al., 2008b; Jia et al., 2008). Subsequently, these values were found to be overestimates resulting from the exclusion of the Pauli spin paramagnetic effect with its pair-breaking role. The estimated values of the Pauli paramagnetic field $H_p(0)$ and the orbital critical field along $\left(\mu_0 H_{c2}^{ab(\mathrm{orb})}\right)$ and perpendicular $\left(\mu_0 H_{c2}^{c(\mathrm{orb})}\right)$ to the *ab*-plane for different compounds are respectively shown in the second, third, and fourth rows of Table 20.6. The fifth and sixth rows depict the experimentally measured values of the upper critical field at the mentioned temperature, measured with the applied field along the above two directions. The anisotropy parameter is given by $\gamma = H_{c2}^{ab}/H_{c2}^c$ and its values at $T \approx T_c$ and at $T < T_c$ are shown in the seventh and eighth rows. As one may note, the anisotropy factor $\gamma(T_c)$ for different compounds varies from 2 to 6, and for $T < T_c$ all of them exhibit a general decrease of γ, with superconductivity of Ba(K)122 and Fe(Se, Te) becoming almost fully isotropic at sufficiently low temperatures. Decrease of γ with temperature is a universal feature of FBS, which is most unusual for layered crystals and invites practical applications for these materials.

The temperature dependences of the upper critical field of $Ba_{0.6}K_{0.4}Fe_2As_2$ (Yuan et al., 2009) and $Fe_{1.1}Se_{0.6}Te_{0.4}$ (Fang et al., 2010) are respectively depicted in Figure 20.14(a) and (b). The difference between $\mu_0 H_{c2}^{ab}$ and $\mu_0 H_{c2}^c$ is markedly reduced or even zero at a sufficiently low temperature, leading to near-isotropic behaviour. Interestingly, for these two materials, as may be seen from Table 20.6, $\xi_{ab} = \xi_c = 2.1$ and 2.7 nm. These values of ξ_c are much larger than the interlayer spacing, which makes these layered materials behave more isotropically in three dimensions. ARPES studies also find a considerable dispersion along the *z*-direction as in three-dimensional superconductors (Vilmercati et al., 2009). This suggests that for realising high T_c a two-dimensional character is not in fact a prerequisite, as has been generally believed.

The observed difference in the behaviour of $\mu_0 H_{c2}^{ab}$ and $\mu_0 H_{c2}^c$ has not been fully explained. This has been examined in terms of the WHH model and it is suggested that the upper critical field is orbitally limited for *H* parallel to *c*, while it is Pauli paramagnetically limited for *H* parallel to *ab* (Zhang et al., 2012). Interestingly, the concave upwards rise in $\mu_0 H_{c2}(T)$ is reminiscent of similar behaviour found in MgB_2, which is known to possess two superconducting energy gaps. The upwards increase is therefore attributed to the possible opening up of a second energy gap at lower temperatures. The multiband nature of FBS and their unconventional superconductivity add complexities as well as some novel features to many of the above issues, which remain puzzling.

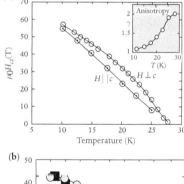

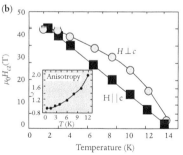

Figure 20.14 Temperature dependence of H_{c2} along and transverse to the *c*-direction for (a) (Ba, K)Fe$_2$As$_2$ (after Yuan et al., 2009) and (b) Fe$_{1.11}$Te$_{0.6}$Se$_{0.4}$ (Fang et al., 2010). The insets show the corresponding temperature variation of the anisotropy.

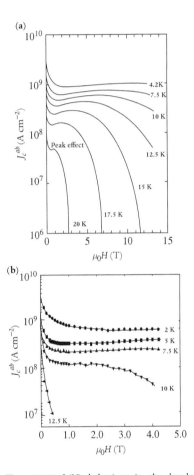

Figure 20.15 J_c(H) behaviour in the basal plane of single-crystalline (a) Co-doped Ba(Fe$_{1-x}$Co$_x$)$_2$As$_2$ (after Yamamoto et al., 2009) and (b) FeTe$_{0.61}$Se$_{0.39}$ (Taen et al., 2009).

20.6.2 Critical current in bulk single crystals, polycrystals, and thin films

Besides the features already mentioned, FBS have a few additional advantages over HTS cuprates making them more suitable for practical utilisation. For instance, as compared with cuprates, their parent compounds are metallic/semimetallic type rather than insulating and their order parameter symmetry in the superconducting state is of s-wave type, instead of d-wave; unlike the latter, the former is not detrimental to current flow across the grain boundaries. With the availability of FBS single crystals, it immediately became clear that their intragrain J_c in self-field and at about 4 K was commendably high, of the order of 10^{10} A m^{-2} (Zhigadlo et al., 2008; Yang et al., 2008; Prozorov et al., 2008; Taen et al., 2009; Yamamoto et al., 2009). The in-plane J_c of Sm1111 at 5 K remained invariant at 2×10^{10} A m^{-2} (Zhigadlo et al., 2008) in an applied magnetic field up to 7 T. Also, large peak effects were observed, with $J_c > 10^9$ A m^{-2} at 4.2 K in single crystals of (B, K)-122 (Yang et al., 2008; Yamamoto et al., 2009) and Ba(Fe, Co)$_2$As$_2$ (Figure 20.15a), whose origin was ascribed to the formation of weakly superconducting regions in a higher-T_c matrix. Similarly, $J_c > 10^9$ A m^{-2} was reported (Figure 20.15b) for single crystals of FeTe$_{0.61}$Se$_{0.39}$ by Taen et al. (2009). But, these materials also share a few unfavourable characteristics with cuprates, such as layered structure, competing phases, low carrier density, small coherence length ξ, granularity, poor connectivity of grains due to microcracks and voids, and grain-boundary weak-link effects (Durrell et al., 2011), all of which potentially hinder their practical utilisation. Nevertheless, these features are less pronounced in FBS and their adverse impact is therefore not so drastic as in cuprates. For instance, the current flow in these materials, in polycrystalline form, can withstand a larger mismatch between adjoining grains of about 9–10° (Lee et al., 2009; Katase et al., 2011) instead of 3–5° for cuprates (Hilgenkamp and Mannhart, 2002). This makes the grain-alignment criteria for achieving a high current density J_c less stringent than for wires and tapes of cuprate superconductors.

As with HTS cuprates, to achieve $J_c > 10^{10}$ A m^{-2}, FBS thin films should ideally be grown epitaxially using a biaxial substrate. The presence of the volatile ingredients of the superconductor and its structural mismatch with the substrate have made stoichiometric epitaxy a difficult problem with FBS. The subject has been reviewed by Hiramatsu et al. (2012). The compounds of the 122 family have been more extensively studied for their epitaxial growth, and J_c(self-field, 4.2 K) $\approx 2 \times 10^{10}$ A m^{-2} has been reported by several authors (Mohan et al., 2010; Lee et al., 2010; Katase et al., 2011a; Rall et al., 2011). Epitaxy has been successfully achieved on biaxial oxide substrates either through a careful control of growth conditions using ion-beam-assisted deposition (IBAD)

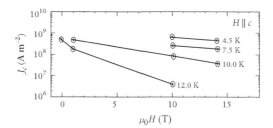

Figure 20.16 $J_c(H)$ behaviour of an epitaxial thin film of $FeSe_{0.5}Te_{0.5}$ grown on a CaF_2 substrate (after Tsukada et al., 2011).

(Iida et al., 2011; Katase et al., 2011b) or by introducing a suitable template layer between the substrate and the deposited Ba122 thin film (Lee et al., 2009, 2010). Besides 122 compounds, epitaxial thin films of other FBS have also been successfully made with high J_c. These include La1111F (Kidszun et al., 2010a, b), Nd/Sm1111F (Kawaguchi et al., 2010; Ueda et al., 2011), and $FeSe_{0.5}Te_{0.5}$ (Bellingeri et al., 2010; Eisterer et al., 2011) having self-field J_c at 4.2 K in the range of $1 \times 10^9 - 4 \times 10^{10}$ A m^{-2}. In the case of epitaxial films of $FeSe_{0.5}Te_{0.5}$ grown on CaF_2 substrates, the reported $J_c(14\ T, 4.2\ K) = 4.2 \times 10^8$ A m^{-2}, as depicted in Figure 20.16 (Tsukada et al., 2011). The success in thin-film development has paved the way towards the feasibility of coated conductors of FBS materials.

20.6.3 Progress with FBS conductor development

Fabrication of long lengths of wires and tapes of FBS carrying high transport currents in high magnetic fields has been a challenging task. Of the four different families mentioned in Table 20.6, the 1111-type superconductors possess the highest T_c and largest critical field, but also the biggest anisotropy and a much broader resistive transition. The 111 compound LiFeAs is a purer material that requires no extra doping for superconductivity, but possesses a comparatively low T_c of 18 K and a relatively small critical field showing anisotropy. All these features are unsuitable for applications. More interesting is the FeTeSe compound of the 11 family with critical field exceeding 45 T at 0 K and, at low temperature, almost free from anisotropy. But again, with $T_c = 14$ K, it belongs to the low-temperature superconductor class and, if developed as a practical superconductor, it will have to compete with the commercially long-established Nb_3Sn and Nb–Ti. More attractive is the 122 family, with $\mu_0 H_{c2}(0) > 60$ T, almost isotropic, and showing a sharp resistive T_c of 28–38 K. The compounds of this family would thus compete with MgB_2 and Nb_3Sn, the former still being in its development stage. As a result, the 122 system has attracted comparatively greater attention than its sister compounds.

The most pursued route for fabricating wires and tapes of FBS is the powder-in-tube (PIT) technique, which we described previously for various superconductors such as A-15, MgB_2, Chevrel phases, and HTS cuprates. It involves packing the precursor powder mixture of ingredients in the required stoichiometry in a metallic tube, which is mechanically drawn or rolled in a wire or tape form, followed by a high-temperature (about 1200°C) heat treatment for a few hours to form a connected length of superconductor. If the metal tube is loaded with unreacted powder mixture, the process is called *in situ*, while it is termed *ex situ* if the precursor used is already reacted in the form of FBS. Since, in the former, the reaction takes place inside a sealed tube, it is convenient when the ingredients are volatile and toxic, which is true for most FBS. If the unreacted or pre-reacted precursors are simply inserted into Nb, Ta, or Fe tubes, they come in direct contact with the tube material, which, during the several hours of the heat treatment, chemically react with the FBS ingredients to markedly deteriorate its properties. This problem is overcome by placing a thin sheath of Ag separating the precursor from the tube. Ag does not react with the precursor powders and serves as a shield for the reaction. The lower melting temperature of the Ag sheath, however, necessitates lowering of the reaction temperature to about 850–900°C. Wang et al. (2011) have found that the reaction between the Fe and superconductor is, however, much less if the annealing is carried out only for a short time of 5–15 minutes at a higher temperature of 1100°C. For (Sr, K)122, they further showed that such a heat treatment of the cold-rolled PIT tape produced fine plate-like grains with c-axis texturing, which markedly enhanced the transport J_c. In this way, they achieved a self-field transport $J_c = 3 \times 10^7$ A m^{-2} at 4.2 K, which they could raise further to 5.4×10^7 A m^{-2} by adding 10% Pb or Sn as additive. The results obtained with Sn as additive are shown in Figure 20.17.

Following this approach in the *ex situ* PIT process, Yao et al. (2012) fabricated Fe-clad (Ba, K)122 tapes containing both Ag and Pb as additives and could achieve a large transport J_c up to 1.4×10^8 A m^{-2} in self-field at 4.2 K. The effect of applying even a small magnetic field <1 T was to degrade J_c by almost two orders of magnitude, indicating pronounced weak-link effects. The Pb, however, helped in countering this drop in the low-field region and Ag did so in higher fields (Wang et al., 2010, 2011; Qi et al., 2010). Using the *ex situ* PIT process in an Ag tube and adding 10% Ag powder for better grain connectivity in (Ba, K)122, Togano et al., (2011) could achieve an improved transport J_c of 1.1×10^7 A m^{-2} in 10 T by 15–30 hours' sintering at 850°C. Although the performance of PIT-processed FBS wires and tapes indicates a rising trajectory, their weak-link problem continues to dominate. Although the weak links in FBS are comparatively less severe than in cuprates, to achieve current densities (at 4.2 K) of the order of 10^{10} A m^{-2} in self-field and 10^9 A m^{-2} at about 10 T, it would seem to be essential in FBS processing to effectively use techniques like IBAD and ISD that have

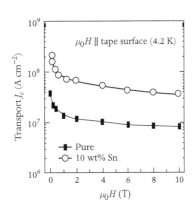

Figure 20.17 Enhanced transport current density of (Sr, K)-122 with 10% Sn as additive (after Wang et al., 2011).

been successfully followed for coated conductors of YBCO, discussed in Chapter 17. Indeed, some recent efforts to realise thin-film epitaxy using IBAD (Katase et al., 2011b; Iida et al., 2011), as already mentioned, have proved most promising in achieving biaxially textured epitaxial thin films of $Ba(Fe_{1-x}Co_x)_2As_2$ with current densities exceeding 1×10^{10} $A\ m^{-2}$. As coated conductor technology for YBCO already exists, it would be natural to expect FBS conductor development to speedily follow soon. The pertinent question, however, is whether one really needs the costly technology for these difficult materials when both MgB_2 and HTS cuprate technologies are in advanced stages of development.

20.7   Summary

In this chapter, we have seen that within a short period of four to five years from their discovery, there has been extraordinary progress in the development and understanding of FBS as a new class of high-temperature superconductors possessing T_c next only to ruthenocuprates and HTS cuprates. The discovery of FBS has revealed new facets of superconductivity and its interplay with magnetism that should improve our understanding of cuprate superconductors and the high-T_c phenomenon in general. The crystal structures of various FBS families, their unusual properties, bizarre Fermi surfaces, and interesting phase diagrams, revealing their doping-dependent interplay of magnetic and superconducting orders, have been briefly discussed. The FBS are not only unconventional in terms of their superconductivity, but are the first example of multiband, multigap superconductivity with sign-changing s_+-wave order parameter between the electron and hole bands or pockets. Progress made with materials fabrication in single crystals, epitaxial thin films, and wire and tape forms has been reviewed. Finally, FBS possess a potential for practical applications to compete with the existing conventional super conductors of Nb_3Sn and Nb–Ti, although their economic viability remains to be determined. The advent of FBS has demonstrated that high-temperature superconductivity is no longer such a rare phenomenon as once assumed, which has brought a fresh optimism that many more systems may yet be discovered, some of which may have even higher T_c or be better suited for practical applications than both cuprates and FBS.

CHAPTER 21

Miscellaneous superconductors

In this, the last chapter of the book, we discuss seven diverse superconducting systems, namely, (1) the bismuthates, (2) cobalt oxide hydrate, (3) intermetallic perovskites free from oxygen, (4) metallonitride halides, (5) pyrochlore oxides, (6) layered transition metal chalcogenides, and (7) BiS_2-based layered materials. Of these, the BiS_2-based superconductors arrived in July 2012 and are presently in the stage of development. All seven systems possess many unusual features, making them an indispensible part of the field of superconductors.

21.1 Superconducting bismuthates

21.1.1 Advent of superconducting bismuthates and their general features

Unlike HTS cuprates, which are layered perovskite oxides that make them two-dimensional and anisotropic, the bismuthates are isotropic three-dimensional perovskite oxides having the general formula ABO_3, possessing a characteristic cubic (Figure 21.1(a)) structure. Alternatively, the structure can be viewed as having the A-atom being sandwiched between four BO_6 octahedra placed above and below (Figure 21.1(b)). Any distortions and tilting in such octahedra resulting from incompatible ion size or doping give rise to structural changes in the parent cubic structure (Figure 21.1(c)).

Perovskites, known since 1925, are generally insulating or semi-metallic and more than 100 of them are ferroelectrics. It was therefore unexpected when Steglich et al. (1975) announced that the mixed perovskite compound $BaPb_{1-x}Bi_xO_3$ was metallic and superconducting with $T_c = 13$ K, which was then the highest value known for any non-transition-metal-based compound. Interestingly, the perovskites $BaBiO_3$ and $BaPbO_3$ were respectively insulating and semimetallic. $BaBiO_3$ with its half-filled band and unstable cubic perovskite structure undergoes a Peierls-like charge-density wave (CDW) transition (Mattheiss and Hamann, 1983; Hamada et al., 1989) accompanied by opening up of a semiconductor-like gap at E_F. This is caused by a strong coupling of the electron system with a breathing-mode-type lattice distortion. The

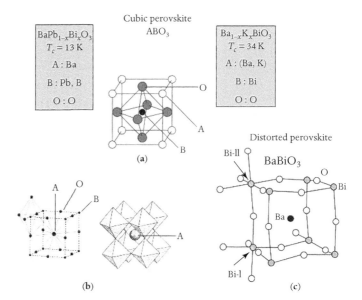

Figure 21.1 Superconducting bismuthates and perovskite structure.

distorted perovskite unit cell of $BaBiO_3$ is depicted in Figure 21.1(c). Bi generally forms compounds with valence 3+ or 5+, skipping the valence state 4+. It was therefore unusual to find Bi in the 4+ state. This was later realised as being due to disproportionation of Bi^{3+} and Bi^{5+} valence states.

In bismuthates, there is competition between the CDW and superconductivity, and the latter arises only when the former is quenched. Aliovalent substitutions on either the A (Ba) or the B (Bi) site of the ABO_3 structure, which result in electron or hole doping, displace the conduction band away from the half-filled state and suppress the insulating CDW state, and, at a certain doping level, superconductivity is induced. Partial substitution of Pb for Bi in $BaBiO_3$ made the alloyed compound $BaPb_{1-x}Bi_xO_3$, called BPBO, superconducting, with an optimum T_c of 13 K for $x = 0.25$. For $x > 0.35$, the material was semimetallic and non-superconducting.

Superconducting bismuthates and cuprates are compared in Table 21.1. While in cuprates the $Cu3d$–$O2p$ band gives rise to conduction and superconductivity, in bismuthates the $Bi(Pb)6s$–$O2p$ band is relevant. The absence of Ba (at the A-site) in the above band makes the A-site of the ABO_3 a passive site, while the B-site, where Bi is located, is an active site. The B-site substitution directly influences the conduction band and affects T_c, although the effect is not so serious as with Cu-site substitution in cuprates. However, the A-site is also important for two reasons. First, it modifies the lattice parameter and thereby influences the overlap of Bi and O orbitals. Second, the size of the A-site atom has a role in controlling the range of aliovalent substitution.

Table 21.1 Broad similarities and differences between bismuthates and HTS cuprates

Bismuthates	HTS cuprates
Three-dimensional cubic perovskite: isotropic	Layered two-dimensional perovskite: anisotropic
Low carrier density	Low carrier density
T_c sensitive to Bi–O network and affected by Bi-site substitution	T_c sensitive to CuO_2 planes and more drastically reduced by Cu-site substitution
Closer to conventional singlet s-wave superconductivity mediated by phonons; some anomalies	Unconventional singlet d-wave superconductivity mediated by spin fluctuations
Superconductivity in proximity to metal–insulator transition where the insulating phase is the CDW state	Superconductivity in proximity to metal–insulator transition where the insulating phase is the SDW / AFM state
Superconductivity competes with CDW state	Superconductivity competes with SDW / AFM state
No magnetic ions or magnetism involved	Magnetic ions involved and magnetism important
Highest T_c reached about 34 K	Highest T_c reached about 164 K

The most unexpected happening of 1988 was that when the parent compound $BaBiO_3$ was partially substituted by K at the Ba site to form $Ba_{1-x}K_xBiO_3$, called BKBO, the T_c for $x \approx 0.4$ was dramatically raised, first to 28 K and subsequently to 35 K (Cava et al., 1988, 1989). This is the highest T_c value known for any non-cuprate oxide. For $x > 0.6$, multiphases occurred (Welp et al., 1988; Hinks et al., 1988a). This substitution also suppressed the insulating CDW state of the parent compound. BKBO is considered as the only high-T_c system that was predicted theoretically (Mattheiss et al., 1988) before it was discovered. Because of the Ba-site substitution, the Bi–O network was left intact, which possibly helped in promoting the higher T_c of BKBO than of BPBO. However, as seen in Table 21.2, not all such compounds formed by A-site substitution exhibit a high T_c, which indicates the importance of other factors in controlling T_c. The discovery of BKBO led to various other members of the bismuthate family (Table 21.2) albeit of lower T_c. The listed superconductors are formed by partially substituting the A- or B- site of the corresponding parent compounds in Table 21.2 by aliovalent cations of lower or higher valence states. The former causes hole doping (e.g. Ba^{2+} by K^{1+} or Rb^{1+}) and the latter electron doping (e.g. K^{1+} by Bi^{3+} or Ca^{2+}).

The doped bismuthates possess a number of interesting features:

1. As the dopant concentration is varied, the compounds exhibit a metal–semiconductor transition (Sleight et al., 1975, 1993; Uchida et al., 1987; Mattheiss and Hamman, 1988; Mattheiss et al., 1988).

Table 21.2 Some prominent bismuthates with their T_c values

Bismuthates	Parent compound	Maximum T_c (K) (and corresponding x)	Reference
$BaPb_{1-x}Bi_xO_3$	$BaBiO_3 / BaPbO_3$	13 (0.25)	Sleight et al. (1975)
$Ba_{1-x}K_xBiO_3$	$BaBiO_3 / KBiO_3$	~34 (0.4)	Mattheis et al. (1988), Cava et al. (1988, 1989), Hinks et al. (1988a)
$Ba_{1-x}Rb_xBiO_3$	$BaBiO_3$	~29 (0.44)	Tomeno and Ando (1989)
$Sr_{1-x}K_xBiO_3$	$SrBiO_3$	12 (0.6)	Kazakov et al. (1997)
$K_{1-x}Bi_xBiO_3$	$KBiO_3$	10.2 (0.1)	Khasanova et al. (1998, 2001)
$K_{1-x}La_xBiO_3$	$KBiO_3$	12 (0.2)	Khasanova et al. (2001)
$K_{1-x}Ca_xBiO_3$	$KBiO_3$	~9 (0.25)	Khasanova et al. (2001)
$Ba_{1-x}K_xBi_{1-y}Pb_yO_3$	$BaBiO_3$	21.5 ($x = 0.4, y = 0.1$)	Marx et al. (1992)

2. The compounds are diamagnetic (Batlogg et al., 1982, 1988; Uemura et al., 1988) at all levels of doping, which excludes the Mott transition as being responsible for semiconducting behaviour.

3. Two energy gaps are exhibited by the semiconducting phases: a large optical gap and a much smaller transport gap (Uchida et al., 1987; Machida, 1988; Blanton et al., 1993).

4. Semiconducting phases manifest a CDW state (Cox and Sleight, 1976; Sugai et al., 1985; Uchida et al., 1987; Sleight et al., 1988; Cava et al., 1989).

5. Raising the dopant concentration causes structural distortions including monoclinic, tetragonal, and orthorhombic, termed pseudo-cubic (Cox and Sleight, 1976; Pei et al., 1989a, b). All such pseudo-cubic structures can be described in terms of tilting of BiO_6 octahedra along with distortions in the oxygen breathing mode (Pei et al., 1990).

6. Superconductivity is observed over a small dopant concentration range where the structure is transformed from semimetallic pseudo-cubic to metallic cubic perovskite, although superconductivity is also reported for tetragonal or orthorhombic distortions. $Ba_{1-x}K_xBiO_3$, $K_{1-x}La_xBiO_3$, $K_{1-x}Bi_xBiO_3$, and $K_{1-x}Ca_xBiO_3$ are examples of the former (Khasanova et al., 1998) and $BaPb_{1-x}Bi_xO_3$ (Sleight et al., 1975) of the latter.

21.1.1.1 BPBO and BKBO

The most extensively studied bismuthates are BPBO and BKBO. $BaPb_{1-x}Bi_xO_3$ is orthorhombic for $x < 0.1$ and $0.35 < x < 0.90$, tetragonal for $0.1 < x < 0.35$, and monoclinic for $x > 0.9$, without exhibiting the formation of the cubic perovskite phase. The phases up to $x = 0.35$ are weak metallic or semimetallic, and superconductivity

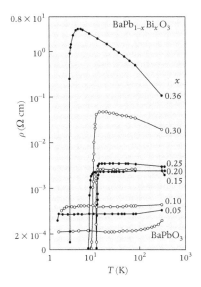

Figure 21.2 Resistivity behaviour of $BaPb_{1-x}Bi_xO_3$ (after Tranh et al., 1980).

occurs, with optimum $T_c = 13$ K at $x = 0.25$, within the tetragonally distorted composition range (Sleight et al., 1975), where Hall measurements find the carrier density to be optimum (Tranh et al., 1980). The resistivity behaviour for samples with different x is depicted in Figure 21.2 (Tranh et al., 1980). For $x = 0.25$, the normal-state behaviour is linear and temperature-independent, and with increasing x it displays a negative $d\rho/dT$ prior to the superconducting transition. For $x > 0.35$, no superconductivity is observed.

The parent compound of BKBO is $BaBiO_3$, which is a semimetallic (CDW) phase of monoclinic structure (space group $C2/m$), while $KBiO_3$, which corresponds to $x = 1$, exists in two different structural modifications. Processing under ambient conditions results in a non-superconducting and non-perovskite phase with a $KSbO_3$ structure of space group $Pn3$ and $a_0 = 1.0016(4)$ nm (Jansen, 1977). High-pressure, high-temperature (HPHT) processing, however, results in a $KBiO_3$ phase with cubic perovskite structure having $a_0 = 0.42287(2)$ nm and space group $Pm3m$, with $T_c \approx 10$ K (Khasanova et al., 1998). Between these two extremes of $x = 0$ and $x = 1$, $Ba_{1-x}K_xBiO_3$ exhibits different structures and unusual characteristics. For $0.1 < x < 0.25$, BKBO undergoes orthorhombic distortion and is non-superconducting, while for $0.3 < x < 0.0.65$, it is a cubic perovskite, manifesting superconductivity with a peak $T_c > 30$ K for $x = 0.37 - 0.41$ (Barilo et al., 2000). As x increases beyond the optimal level, T_c monotonically decreases to less than 3 K for $x = 0.76$, and with further increase of x, it rises, becoming saturated at about 10 K for $x = 1$ (Baranov et al., 2001) (Figure 21.3(a)).

With BKBO, and generally with all superconducting bismuthates, the increase in T_c with x is associated with a rise in the band-filling level as determined by the decrease in the oxidation state of Bi (Figure 21.3(b)) (Pei et al., 1990; Khasanova et al., 2001). This situation essentially promotes movement towards the half-filled band state with a higher electron density and a stronger electron–phonon interaction, both factors being responsible for the higher T_c.

Figure 21.3 (a) T_c of $Ba_{1-x}K_xBiO_3$ as a function of x (after (Baranov et al., 2001). (b) T_c of different bismuthate samples as a function of Bi valence (after Pei et al., 1990; Khasanova et al., 2001).

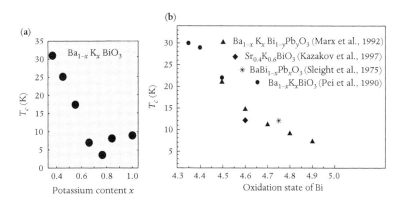

21.1.1.2 *Intriguing superconductivity of bismuthates*

Although not devoid of controversies, the superconductivity of bismuthates is closer to conventional than to unconventional. The oxygen isotope effect measured for BKBO, in contrast to HTS, does not have any unconventional features and manifests an exponent $\alpha_0 = 0.22$ (Batlogg et al., 1988), 0.41 (Hinks et al., 1988b; Zhao and Morris, 1995), and 0.35 (Kondoh et al., 1988), all of which are sufficiently large to favour conventional phonon-mediated pairing. Further, in contrast to cuprates, BKBO exhibits a gap that is isotropic, as manifested by tunnelling data (Huang et al., 1990; Sharifi et al., 1991; Sato et al., 1990; Zasadzinski et al., 1991; Kussmaul et al., 1993; Kosugi et al., 1994). Corner-junction-type tunnelling experiments corroborate this (Brawner et al., 1997). Similarly, penetration depth (Pambianchi et al., 1994; Zhao, 2007) and specific heat (Woodfield et al., 1999) measurements at $T \ll T_c$ have been consistent with a nodeless isotropic gap.

However, the linear temperature dependence of resistivity observed in the normal state just above T_c, the broadening of the resistive transition, and a sharp decrease of J_c in an applied magnetic field are features reminiscent of the unconventional HTS cuprates (Schweinfurth et al., 1992). And yet, the optical conductivity of BKBO at low frequency manifests a behaviour expected of a slightly dirty BCS superconductor, namely opening up of a gap at $T < T_c$ (Figure 21.4), given by $2\Delta \approx 4.2 k_B T_c$. The observed optical spectra of BKBO do not show all the features of a conventional BCS superconductor (Timusk, 1999). In order to explain these data, the required value of the electron–phonon coupling constant $\lambda_{e-p} = 0.2$, which is too small for a $T_c > 30$ K (Marsiglio et al., 1996). Interestingly, tunnelling measurements on BKBO have also yielded λ values that are too low (Sharifi et al., 1993). Another striking difficulty with the tunnelling measurements on both BPBO and BKBO is that, in sharp contrast with conventional low-temperature superconductors, their phonon structure as revealed in the Eliashberg spectral function $\alpha^2 F(\omega)$ fails to closely correlate with the phonon density of states $G(\omega)$ as determined from neutrons (Batlogg, 1984; Reichardt et al., 1985; Huang et al., 1990; Samuely et al., 1993). Thus, the role of phonons in their superconductivity is doubted (Batlogg et al., 1988; Jansen et al., 1995), and an indirect exchange mechanism arising from the closed-shell oxygen ions has been suggested.

21.1.2 BKBO as a practical superconductor

The bismuthates are type II superconductors, and BKBO, because of its high $T_c > 30$ K, has received more attention to determine its potential as a material for future practical applications. BKBO has been synthesised by the following four approaches: (1) solid state reaction (Hinks et al., 1988a), (2) HPHT processing (Khasanova et al., 1998),

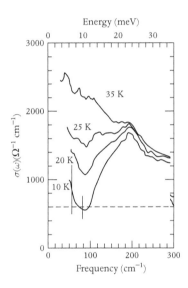

Figure 21.4 Optical conductivity of superconducting BKBO, showing superconducting gap formation below T_c (after Timusk 1999).

(3) sol–gel processing (Cui et al., 2009), and (4) electrochemical growth (Han et al., 1993; Barilo et al., 2000; Lee et al., 2000). Also, highly oriented thin films of BKBO have been deposited, using pulsed laser deposition (PLD), on STO, MgO, LAO, and alumina (Moon et al., 1991; Schweinfurth et al., 1992; Lin et al., 1993; Mijatovic et al., 2002), with transition width ranging from 0.3 K to 2 K. The presence of volatile and high-vapour-pressure ingredients is problematic in attaining the correct stoichiometry in the deposited films, and the T_c values reported are found to be a few degrees lower than those of bulk materials.

Some of the characteristic parameters of optimally doped BKBO, in different forms, are depicted in Table 21.3. The large Ginzburg–Landau parameter $\kappa > 50$ makes BKBO an extreme type II superconductor. Being a three-dimensional isotropic material gives it an edge over HTS cuprates, pnictides, and even MgB_2, all of which possess varying degrees of anisotropy. The coherence length of BKBO is much larger than of HTS cuprates, and thus it has fewer weak-link and granularity problems. The self-field J_c of about 10^{10} A m^{-2} attained to date with BKBO thin films compares well with any other material like Nb_3Sn, MgB_2, or

Table 21.3 Various parameters of optimally doped BKBO superconductor

Parameter	Material form		
	Polycrystalline bulk	Single crystal	Thin film
$\mu_0 H_{c1}(0)$ (T)	35×10^{-4} (Kwok et al., 1989) 8×10^{-4} (Jang et al., 1990)	400×10^{-4} (Barilo et al., 1994) 100×10^{-4} (Huang et al 1991)	—
$\mu_0 H_c(0)$ (T)	—	$(1800–2100) \times 10^{-4}$ (Barilo et al., 1998)	—
$\mu_0 H_{c2}(0)$ (T)	17.3 (Huang et al., 1991) 15.4 (Jang et al., 1990)	12.6 (Huang et al., 1991) 16–21 (Uchida et al., 1993)	15.2 (Schweinfurth et al., 1992)
$-\mu_0 [dH_{c2}/dT]_{T_c}$ (T K^{-1})	0.9 (Jang et al., 1990) 0.87 (Kwok et al., 1989)	0.57 (Huang et al., 1991) 0.78 (Uchida et al., 1993)	0.78 (Schweinfurth et al., 1992)
Penetration depth λ(nm)	80.3 (Jang et al., 1990) 220.0 (Kwok et al., 1989)	250 (Huang et al., 1991) 270 (Uchida et al., 1993)	330 (Pambianchi et al., 1994) 340 (Ansaldo et al., 1991)
Coherence length ξ(nm)	3.68 (Kwok et al., 1989) 3.9–4.5(Uchida et al., 1993)	3.2 (Affronte et al., 1994) 5.1 (Huang et al., 1991)	4.6 (Schweinfurth et al., 1992)
Ginzburg–Landau parameter κ	50 (Huang et al., 1991) 59 (Kwok et al., 1989)	50 (Huang et al., 1991)	72 (Schweinfurth et al., 1992; Pambianchi et al., 1994)
γ (mJ K^{-2} mol^{-1})	2.4 (Huang et al., 1991) 2.36 (Kwok et al., 1989)	2.31(Uchida et al., 1993) 2.4 (Huang et al., 1991)	—
$NE_{(F)}$ (states eV^{-1} cell^{-1})	—	0.40–0.46 (Barilo et al., 1998)	—
$2\Delta / k_B T_c$	4.4 (Zhao, 2007)	4–4.8 (Zhao and Morris, 1995) 3.5 (Gantmakher et al., 1996)	3.8 (Pambianchi et al., 1994) 3.8–4.0 (Moon et al., 1991)
Electron–phonon coupling constant λ_{e-p}	1.4 (Zhao, 2007)	0.76–1.0 (Barilo et al., 1998)	—

cuprate superconductors, which makes BKBO a promising candidate for high-field magnet applications. However, its striking drawback is that its upper critical field $\mu_0 H_{c2}(0)$ is small, at only 15–17 T, which is lower than that of Nb_3Sn, whose conductor technology has been well established since early 1970s. Except for its higher T_c, BKBO does not offer any special advantages. It will have to compete not only with the existing Nb_3Sn but also with MgB_2 and YBCO conductors, with higher T_c and critical field, whose fabrication technologies are at an advanced stage of development. Besides, presently there seems little need to develop the technology of these materials, the physics of which, however, poses many intriguing challenges.

21.2 Cobalt oxide hydrate

21.2.1 Novel form of superconductivity in cobaltate

Following the discovery of HTS cuprates, many efforts were directed to achieve superconductivity in oxides of nickel and cobalt, but it took more than 15 years to find a cobaltate that displayed superconductivity, albeit at an unexciting low temperature of about 4.5 K. Surprisingly, the discovered superconductor was a layered cobalt oxide A_xCoO_2 (where A = alkali metal) that had long been studied for its large thermoelectric power and transport properties (Fouassier et al., 1973; Mizushima et al., 1980; Terasaki et al., 1997; Terasaki, 2005) and had not been believed to be a superconductor. Takada et al. (2003) found that the γ-phase of Na_xCoO_2, when adequately hydrated for a composition $Na_{0.30}CoO_2.1.3H_2O$, turned superconducting at about 4.5 K. Analogous behaviour was observed for material deuterated using D_2O. Here, Na with concentration $x = 0.30$ provided the doping charge, while the intercalated layer of H_2O or D_2O increased the mutual separation of adjoining CoO_2 layers and thereby inducted a two-dimensionality to the material, making it superconducting (Takada et al., 2003). The less-hydrated $Na_{0.30}CoO_2.0.7H_2O$, with the same oxidation state of Co but less-separated CoO_2 layers, is not superconducting (Foo et al., 2003). The compound with larger water content has its unit cell intercalated with two layers of water molecules as against a single layer in the less-hydrated sample. The former, manifesting superconductivity, is a *bilayer hydrate (BLH)* while the latter, which is non-superconducting, is a *monolayer hydrate (MLH)*. The structures of non-hydrated, incompletely hydrated (MLH), and fully hydrated (BLH) cobaltates are shown in Figure 21.5(a), (b), and (c) respectively. Other noteworthy features are that superconductivity is induced on CoO_2 planes with Co ions placed on a triangular lattice instead of the square lattice of CuO_2 planes in cuprates. A triangular lattice tends to frustrate long-range magnetic ordering at low temperatures, and in terms of the underlying physics it is

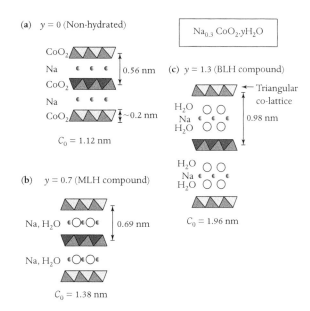

Figure 21.5 Schematic crystal structures of $Na_{0.3}CoO_2 \cdot yH_2O$: (a) non-hydrated, (b) MLH compound, and (c) BLH compound.

Table 21.4 Broad similarities and differences between cobaltates and HTS cuprates

Superconducting cobaltates	HTS cuprates
Superconducting sheets (CoO_2) contain O_2 and a transition metal (Co) with spin $1/2$	Superconducting sheets (CuO_2) contain O_2 and a transition metal (Cu) with spin $1/2$
Layered two-dimensional: anisotropic with hexagonal structure	Layered two-dimensional perovskite-related: anisotropic with tetragonal/orthorhombic structure
Structure contains triangular lattice of Co	Structure contains square lattice of Cu
$T_c < 5$ K	T_c, an order of magnitude larger, with $T_{c(\max)} = 164$ K
Optimum doping for maximum T_c, which decreases with over- and underdoping	Optimum doping for maximum T_c, which decreases with over- and underdoping
T_c shows extreme sensitivity to separation of CoO_2 layers caused by hydration, but is not so sensitive to Co-site substitution	T_c is extremely sensitive to O content and to disorder in CuO_2 planes and is drastically reduced by Cu-site substitution
Unconventional superconductivity, both singlet and triplet (p-wave) have been suggested	Unconventional singlet d-wave superconductivity mediated by spin fluctuations
Superconducting state is in proximity to charge-ordered insulator	Superconducting state is in proximity to Mott insulator
Magnetic fluctuations involved in mediating superconductivity	Magnetic fluctuations involved in mediating superconductivity

of more interest. Second, the material synthesis can be performed using soft chemical processes that are simpler than other fabrication routes. The cuprate and cobaltate superconductors are compared in Table 21.4. There are topical reviews by Takada et al. (2005a, 2007) on the progress made in cobaltate superconductors that the reader should find useful.

21.2.2 Structural aspects and synthesis of $Na_{0.30}CoO_2.1.3H_2O$; two- and three-layer compounds

The compounds A_xCoO_2 (where A = alkali metal) possess a layered hexagonal structure formed with a regular stacking of CoO_2 layer-blocks along the c-direction. Each such block, consisting of a triangular layer of Co ions sandwiched between close-packed layers of oxygen atoms, gives rise to an array of edge-shared CoO_6 octahedra. Na_xCoO_2 has two inequivalent sites for Na—Na-I and Na-II—as shown in Figure 21.6 (Lin et al., 2005). The layered structure formed may possess different polymorphs depending on the concentration x and the processing conditions followed. Different polymorphic phases of Na_xCoO_2, for example, are designated as γ-Na_xCoO_2 (P2) for $x = 0.3–0.65$, β-Na_xCoO_2 (P3) for $x = 0.55–0.6$, and α-Na_xCoO_2 (O3) for $x = 0.65–1$, according to the environment around the Na ion which can be P = trigonal prism, O = octahedron, etc. The number that follows the letters P, O, etc. indicates the number of CoO_2 layers present in their unit cell (i.e. 3 layers in α- and β-phases and 2 in the γ-phase). These polymorphs with suitable Na content x serve as parent compounds for the synthesis of superconducting phases containing two or three layers of CoO_2. Both two- and three-layer compounds (Takada et al., 2004a, 2005b; Sakurai et al., 2004; Mistry et al., 2004; Foo et al., 2005) possess nearly the same T_c, but the latter are less stable and therefore less studied.

21.2.2.1 Synthesis of a compound with two CoO_2 layers

The parent compound for synthesising superconducting $Na_{0.30}CoO_2.1.3H_2O$ containing two CoO_2 layers is γ-$Na_{0.7}CoO_2$ (P2). The compound with $x < 0.6$ does not form by conventional solid state reaction, and hence the approach pursued is first to synthesise a sample with $x > 0.6$ and then deintercalate Na by Br oxidation, followed by hydration (Lin et al., 2005). $Na_{0.61}CoO_2$ is prepared (Takada et al., 2003; Lynn et al., 2003; Lin et al., 2005) from Na_2CO_3 (taking 10%

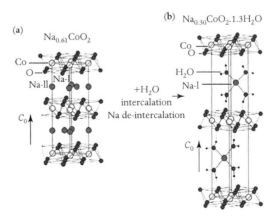

Figure 21.6 Crystal structure of Na_xCoO_2: (a) before and (b) after complete water intercalation (after Lin et al., 2005).

molar excess to compensate for volatisation loss) and Co_3O_4 heated for 12 hours in O_2 at 800°C. The neighbouring CoO_2 layers in the parent compound thus formed are attracted to each other through the Na^+ ions between them and therefore water is not able to penetrate when the compound is immersed in water. To overcome this, Br oxidation is performed before hydration. It de-intercalates Na^+ from the gallery and also electrons from the CoO_2 layers, which weakens the attraction between the CoO_2 layers and makes hydration and BLH-phase formation feasible. To de-intercalate Na from $x > 0.6$ to 0.30, $Na_{0.7}CoO_2$ as synthesised is stirred in an acetonitrile solution containing a Br_2 concentration at ambient temperature for about 120 hours. The optimum hydration corresponds to a rise in the interlayer distance of the neighbouring CoO_2 sheets from 0.56 nm for the anhydrous phase to 0.69 nm for the MLH and 0.98 nm for the BLH phase. Figure 21.5 shows that the location of the monolayer of water molecules in the unit cell of MLH is in the same plane as the Na ions, while the two water layers sandwich the Na ions in the BLH phase. This way, the H_2O molecules in the BLH phase not only make the material more two-dimensional, but are able to shield the random Coulomb potential of the Na^+ ions from CoO_2 planes, which favours a higher T_c (Takada et al., 2003). Further, water intercalation effectively compresses the CoO_2 block layers (Ihara et al., 2005) in the c-direction, making them thinner, which enhances T_c (Lynn et al., 2003, 2005).

21.2.2.2 Oxidation state of Co and the presence of oxonium ions

Purely from charge-balance considerations, the oxidation state of Co in the hydrated phase $Na_xCoO_2 \cdot yH_2O$ comes out as $s = 4 - x$. However, the value estimated from emission spectroscopy (Takada et al., 2004b) and various titration methods (Karppinen et al., 2004) is found to be smaller. Superconductivity occurs for $x < 0.5$. Initial studies (Schaak et al., 2003) had suggested the superconductivity range as $0.25 < x < 0.35$, which, however, did not take into account the formation of oxonium (H_3O^+) ions during hydration. In fact, the highest T_c of 4.9 K corresponds to $x = 0.42$, that is, closer to $x = 0.5$, the boundary of metal–insulator transition (Chen et al., 2004). The discrepancy of $s \neq 4 - x$ is resolved by the fact that some cationic species other than Na^+ are accommodated in the gallery, which compensates for the difference in $4 - x$. During hydration treatment, some of the Na^+ ions are exchanged with oxonium ions in water, as revealed by Raman spectroscopy (Fumagalli et al., 2001; Nash et al., 2001). The H_3O^+ ions, being isovalent with Na^+, share their sites in the unit cell of the BLH, whose correct chemical formula may now be written as $Na_x(H_3O)_zCoO_2 \cdot yH_2O$. Typically, $x = 0.337$, $z = 0.234$, and $y = 1.3$, giving $s = 4 - x - z \approx 3.42$–$3.43$, instead of $s = 4 - x = 3.663$. Superconductivity is particularly sensitive to the ion exchange between

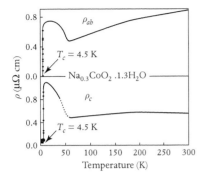

Na^+ and H_3O^+ rather than to the x or s values. The maximum T_c occurs just below $s = 3.5$, in the vicinity of the charge-ordered magnetic phase, indicating the possibility of magnetic fluctuations mediating unconventional Cooper pairing. The oxidation state of Co for optimum T_c corresponds to the non-hydrated compound $Na_{0.5}CoO_2$, where the charge-ordered phase is formed. This contention is corroborated by the resistivity manifesting a negative dR/dT (Figure 21.7) before the onset of superconductivity at 4.5 K (Jin et al., 2003). There is a strong dependence of T_c on the ratio c_0/a_0, which is greater than 6.95 for optimum T_c, and below which T_c rapidly decreases (Sakurai et al., 2004; Milne et al., 2004).

Figure 21.7 Resistivity upturn observed along c- and ab-directions for hydrated single-crystalline Na_xCoO_2 (after Jin et al., 2003).

21.2.2.3 *Characteristics of Na_xCoO_2 for different x*

The electronic properties of Na_xCoO_2 show extreme sensitivity to Na content x, as depicted in Table 21.5. It exhibits superconductivity for $x < 0.5$, with 0.25 being the lower limit of x. In this range, the material is Pauli-paramagnetic. For $x = 0.5$, the material shows a singularity manifested by the charge-ordered AFM insulating state at 52 K and two more magnetic transitions at 88 K and 24 K (Huang et al., 2004). For higher values of $x > 0.5$, Curie–Weiss behaviour is manifested and the compound also displays a rich variety of magnetic behaviour. As may be seen from the table, there are host of magnetic features interwoven with superconductivity that make the latter unconventional.

Table 21.5 Characteristic features of Na_xCoO_2 for different x

x	Properties
$0.25 < x < 0.5$	Superconductivity in hydrated sample Maximum $T_c = 4.9$ K for $x = 0.42$ (Chen et al., 2004)
0.5	Charge-ordered AFM insulator (Takada et al., 2007) at 52 K Two more magnetic transitions are observed at 88 K and 24 K (Huang et al., 2004)
$0.25 < x < 0.5$ $0.5 < x < 0.6$	Pauli paramagnetic metal (PPM) showing a weak temperature dependence of magnetic susceptibility (Sakurai et al., 2005; Wang et al., 2006)
0.55	In-plane FM ordering below 20 K with spin directions within CoO_2 plane—a situation that favours p-wave pairing (Wang et al., 2006)
$0.5 < x < 0.7$	Magnetic susceptibility showing Curie–Weiss behaviour and a negative Weiss temperature (Ray et al., 1999; Wang et al., 2003; Gavilano et al., 2004)
$0.7 < x < 0.95$	Formation of long-range antiferromagnetic order below 22 K (Mikami et al., 2003; Sugiyama et al., 2003; Bayrakci et al., 2004) For $x = 0.82$, A-type antiferromagnetism is observed in which ferromagnetic layers are coupled antiferromagnetically (Boothroyd et al., 2004)

21.2.2.4 Unconventional superconductivity of cobaltate

Various parameters of the polycrystalline BLH superconductor $Na_{0.31}CoO_2.3H_2O$, with $T_c = 4.3$ K are presented in Table 21.6 (Cao et al., 2003). However, the values reported in the literature show much scatter due to the different processing conditions used (Sakurai et al., 2003, 2006). The superconductor is extreme type II, with a Ginzburg–Landau parameter of about 140 (Table 21.6). In most cases, the upper critical field exceeds the Pauli paramagnetic limit of 7–8.5 T, which is taken as an indication of triplet pairing in the p- or f-wave channel (Cao et al., 2003; Sasaki et al., 2004; Sakurai et al., 2006). The small positive Weiss constant of 1.7 K (Table 21.6) suggests the presence of ferromagnetic fluctuations, responsible for p-wave superconductivity (Tanaka and Hu, 2003). Precise measurements of the upper critical field by Maśka et al. (2004) showed an unusual temperature dependence, namely an abrupt change in slope in the low-field regime. Moreover, the extrapolated $H_{c2}(0)$ estimated from the WHH formula (Werthamer et al., 1966) in the strong-field regime exceeded the Pauli limit, but the same quantity estimated from low-field data did not. This behaviour has been related to a field-induced transition from singlet to triplet superconductivity. However, for a single-crystal sample of this material, Sasaki et al. (2004) found that the estimated values of $H_{c2}^{ab}(T)$ were suppressed below the Pauli limit as 0 K was approached, which suggested a singlet superconducting state. But this is contradicted by the very small specific heat jump (Table 21.6), giving $\Delta C/\gamma T_c = 0.57$, which is much less than the BCS value of 1.43, which suggests unconventional pairing. Further, the triangular lattice of Co is believed to promote magnetic frustration and unconventional symmetry of the superconducting order parameter, as revealed by a host of different measurements.

The ^{16}O–^{18}O isotope effect studies of T_c performed on the superconducting cobaltate failed to produce any convincing change in T_c that did not support the conventional superconductivity (Yokoi et al., 2008).

Table 21.6 Characteristic parameters of polycrystalline BLH superconductor

$\mu_0 H_{c1}(0) = 1.3 \times 10^{-3}$ T	$\mu_0 H_c(0) = 5.05 \times 10^{-2}$ T	$\mu_0 H_{c2}(0) = 10$ T	$\mu_0 H_p = 7.8$ T
$\xi(0) = 5.7$ nm	$\lambda(0) = 790$ nm	$\kappa = 140$	$\gamma = 15.9$ mJ K^{-2} mol^{-1} per formula unit
$\theta_D = 391$ K	$N(E_F) = 6.7$ states eV^{-1} per formula unit	$\lambda_{ep} = 0.57$	Weiss constant = 1.7 K
$2\Delta(0) = 0.50$ meV	$2\Delta(0)/k_B T_c = 2.71$	$\Delta C = 6.9$ mJ K^{-1} mol^{-1} per formula unit	$\Delta C/\gamma T_c = 0.57$ (instead of the BCS value of 1.43)

Data from Cao et al. (2003).

However, the conventional mechanism is known to be suppressed by strong electron-correlation effects.

Low temperature heat capacity measurements on cobaltate superconductors were initially carried out to establish their bulk superconductivity (Jin et al., 2003). Many of them exhibited unconventional superconductivity. For $Na_{0.31}CoO_2.1.3H_2O$, instead of an exponential temperature dependence of the heat capacity below T_c, a T^3 dependence, characteristic of point nodes, was observed by Cao et al. (2003). On the other hand, a T^2 dependence was reported by Oeschler et al. (2005, 2008) and Jin et al. (2005), which could be due to the presence of line nodes or the superposition of contributions from two superconducting gaps (Oeschler et al., 2005, 2008). Jin et al. (2005) noted a sharp decrease in the heat capacity below 0.8 K, which they could not relate to the presence of two gaps. The field dependence of the Sommerfeld coefficient γ indicated $H^{1/2}$ behaviour, corroborating the presence of line nodes (Jin et al., 2005). These compounds possess large γ in zero field, corresponding to a large effective mass $m^* \approx 100m_e$ reminiscent of heavy fermion materials. The penetration depth measurements by Kanigel et al. (2004) are consistent with this contention.

^{59}Co-NQR studies of optimum-T_c samples (Ishida et al., 2003; Fujimoto et al., 2004; Ihara et al., 2005, 2007; Zheng et al., 2006a) show the absence of a coherence peak and exhibit a T^3 dependence of the nuclear spin–lattice relaxation rate $1/T_1$ below T_c, down to 1 K (Zheng et al., 2006a), which are indicative of unconventional superconductivity and the presence of line nodes in the gap structure. Fujimoto et al. (2004) argued that their observed temperature dependence could be explained in terms of chiral p-wave or d-wave pairing by considering the impurity effect. ^{59}Co-NMR studies at 7 T ($<H_{c2}$) on aBLH cobaltate sample have revealed an invariant behaviour of the Knight shift with temperature and an absence of a coherence peak below T_c (4.7 K) in the T^3 temperature dependence of the spin-lattice relaxation rate (Kato et al., 2006). Both features indicate unconventional superconductivity and the presence of a gap with line nodes. Ishida et al. (2003) found $1/T_1$ in the normal state to gradually increase as the temperature was lowered from 100 K down to T_c. This they attributed to the presence of nearly ferromagnetic fluctuations, which are believed to be responsible for the triplet pairing. Similarly, Highemoto et al. (2004) did not find any decrease of the muon Knight shift at 6 T below T_c, which gave credence to the presence of unconventional superconductivity. Some of these findings are, however, at variance with the results of Kobayashi et al. (2003, 2005, 2006) showing a decrease in the Knight shift below T_c, which favours spin-singlet superconductivity. Similarly, T_c suppression

by non-magnetic impurities was small, which conflicts with triplet pairing (Yokoi et al., 2004). On the other hand, studies of the aligned BLH cobaltate powder showed the Knight shift below T_c to remain invariant along the c-axis, but to decrease in the ab-plane (Ihara et al., 2006). This has been explained as spin-triplet superconductivity with the spin component along the c-axis. The above findings were again countered by Zheng et al. (2006b), who found the Knight shift in bulk single crystals, in both directions, to decrease below T_c, which supports singlet pairing. Clearly, the outcome of Knight-shift measurements on cobaltate samples remains confusing, as both triplet and singlet superconductivity are indicated.

21.2.2.5 Unusual phase diagrams and electronic structure

The aforementioned studies have led to unusual phase diagrams of cobaltate superconductors processed under different conditions. For the hydrated compound, with the previously mentioned general formula $Na_x(H_3O)_zCoO_2.yH_2O$, the ratio x/z, which measures the exchange between Na^+ and H_3O^+, is an important parameter controlling T_c (Sakurai et al., 2005), and T–x–z phase diagrams have been constructed, keeping the Co valence s constant. For this, soft chemical processing was suitably modified by adding an aqueous solution either of HCl (Sakurai et al., 2005) or of NaOH (Sakurai et al., 2006) as the final step of de-intercalation of Na (decrease of x) and simultaneous realisation of H_3O (increase of z). These approaches lead to samples with different x/z, with the former (HCl series of samples) resulting in a nearly constant value of $s = 3.40+$ to $3.41+$, and the latter (NaOH series of samples) $s = 3.47+$ to $3.50+$. In the case of the HCl series of samples, x decreases with increasing aqueous HCl content, while the reverse holds for the other series. The phase diagrams of both HCl and NaOH series of samples, as determined from susceptibility data obtained in low (0.001 T) and high (7.0 T) fields, are respectively depicted in Figure 21.8(a) and (b). The low-field measurements are for the superconducting phase and the high-field for the magnetic phase. As may be seen, the corresponding phase diagrams for $s \approx 3.40$ and $s \approx 3.50$ are markedly different. The HCl series manifests successive superconducting (SC1), magnetic (M), and superconducting (SC2) phases, while the NaOH series shows a single superconducting phase with a dome-like T_c behaviour. It is interesting to note that in the concentration range $0.33 < x < 0.35$ where the HCl series of samples show a magnetic phase, the NaOH series show superconductivity. These phase diagrams have also been constructed from NQR studies (Ishida et al., 2003; Ihara et al., 2005, 2007).

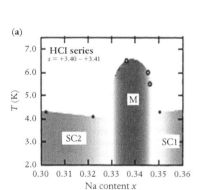

(a)

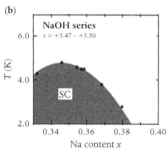

(b)

Figure 21.8 Temperature–concentration x phase diagrams of $Na_xCoO_2.H_2O$ samples of (a) HCl series and (b) NaOH series (after Sakurai et al., 2006).

21.3 Intermetallic perovskites free from oxygen: MgCNi₃ and related superconducting compounds

21.3.1 Advent of MgCNi₃

Following the unexpected discovery in 2001 of superconductivity in MgB_2 at 39 K (Nagamatsu et al., 2001), intermetallic compounds attracted fresh attention in the search for new superconductors and, interestingly, in the same year, another superconducting intermetallic of Mg, namely $MgCNi_3$, was discovered, with $T_c \approx 8$ K (He et al., 2001). $MgCNi_3$ is an ABO_3-type cubic perovskite (space group $Pm3m$) where the anion O is replaced by the cation Ni (top-right inset of Figure 21.9). Such materials formed by the interchange of an anion with a cation, or vice versa, are termed *anti-perovskites*, and $MgCNi_3$ is the first such intermetallic oxygen-free perovskite or anti-perovskite to show super-conductivity. It is amazing for a compound formed with more than 60% of ferromagnetic Ni to show superconductivity when its DOS at E_F, as revealed by energy-band calculations, is dominated by Ni d-states (Singh and Mazin, 2001; Dugdale and Jarlborg, 2001; Shim et al., 2001; Kim et al., 2002a; Rosner et al., 2002; Wang et al., 2002). The presence of a von Hove anomaly just below E_F, manifested by a large and narrow peak in the electron DOS, is shown in the main part of Figure 21.9 and its bottom-right inset. E_F is placed at an electron count of less than 0.5 electrons per formula unit above the peak maximum. The peak was experimentally corroborated by photoemission and X-ray spectroscopy studies (Shein et al., 2002; Kim et al., 2002b). Consequently, $MgCNi_3$ is electronically very close to ferromagnetic instability, and it was believed that a small amount of hole doping should transform

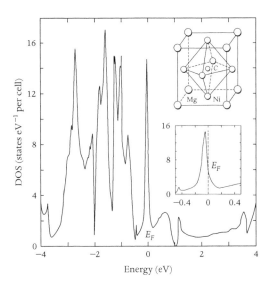

Figure 21.9 Electronic structure (main figure and lower inset) (after Singh and Mazin 2001) and crystal structure (top right inset) of $MgCNi_3$.

its superconducting state into a ferromagnetic state. In this interesting situation, it appeared natural to expect some exotic unconventional mechanism, such as p- or f-wave superconductivity, to operate via ferromagnetic fluctuations (due to Ni). Despite this, there is no experimental evidence to this effect and observations show conventional superconductivity with isotropic s-wave scenario. The role of ferromagnetism in this class of materials seems primarily to cause pair breaking in conventional superconductivity rather than pair making in unconventional superconductors, albeit with some uncommon features.

The advent of superconducting $MgCNi_3$ in 2001 led to the synthesis of many more cubic anti-perovskites, such as $ZnCNi_3$ (Park et al., 2004), $AlCNi_3$ (Dong et al., 2005), $GaCNi_3$ (Tong et al., 2006), $CdCNi_3$ (Uehara et al., 2007a, b), and $InCNi_3$ (Tong et al., 2007). Of these, the isostructural compounds formed with trivalent metals such as Al and Ga are non-superconductors, while $InCNi_3$ is magnetic and does not exhibit superconductivity. The absence of superconductivity has been attributed (Zhong et al., 2007) to their larger Debye temperature $\theta_D \approx 400$ K compared with the $\theta_D < 300$ K of $MgCNi_3$ (Dong et al., 2005; Tong et al., 2006). A larger θ_D implies a smaller electron–phonon interaction for superconductivity. Although a larger θ_D contributes to enhance T_c through the pre-exponential factor in the BCS equation, the decrease in the electron–phonon coupling constant in the exponential part of the equation dominates to suppress superconductivity. A smaller lattice parameter a_0, in general, raises θ_D, which is unfavourable for conventional superconductivity. The only superconductor in the above list is $CdCNi_3$, with $T_c \approx 3$ K. Subsequently, $ZnNNi_3$ emerged as the third superconductor ($T_c \approx 3$ K) of this class, where Zn replaced Mg and N replaced C (Uehara et al., 2009), but, interestingly, the isostructural $ZnCNi_3$ fails to exhibit superconductivity at 2 K (Park et al., 2004). The reasons for this are its smaller value of a_0, larger θ_D, and smaller electron–phonon interaction. Okoye (2009) has argued that $ZnNNi_3$ is superconducting because of its smaller DOS at E_F in comparison with that of $ZnCNi_3$. The larger DOS of the latter promotes a stronger ferromagnetic interaction, which dominates over the conventional superconductivity of $ZnCNi_3$. Some of the characteristic parameters of the three anti-perovskite superconductors are listed in Table 21.7.

21.3.2 Superconductivity of $MgCNi_3$

The typical superconducting resistive transition of $MgCNi_3$ is depicted in the inset of Figure 21.10(a) (Li et al., 2002). There are discrepancies between the experimental data and band-structure calculations. For instance, both Hall-effect and thermopower studies of $MgCNi_3$ have confirmed electrons as charge carries (Li et al., 2001, 2002; Yang et al.,

Table 21.7 Properties of $MgCNi_3$, $CdCNi_3$, and $ZnNNi_3$

Properties	$MgCNi_3$	$CdCNi_3$	$ZnNNi_3$
T_c (K)	6.4–7.6 (Lin et al., 2003; Mao et al., 2003; Uehara et al., 2007b)	2.8–3.2 (Uehara et al., 2007b)	3.0 (Uehara et al., 2009)
$\Delta C / \gamma T_c$	1.97 (Lin et al., 2003) 2.3 (Mao et al., 2003)	—	—
$2\Delta / k_B T_c$	≥ 4 (Lin et al., 2003; Mao et al., 2003)	—	—
$\mu_0 H_{c2}(0)$ (T)	10.6 (Lin et al., 2003) 14.8 (Mao et al., 2003) 14.4 (Uehara et al., 2007b)	1.8–2.2 (Uehara et al., 2007b)	0.96 (Uehara et al., 2009)
$\mu_0 H_p$ (T)	14 (Mao et al., 2003; Uehara et al., 2007b)	5.4 (Uehara et al., 2007b)	5.4 (Uehara et al., 2009)
$\mu_0 H_{c1}(0)$ (T)	0.013 (Mao et al., 2003)	0.003–0.008 (Uehara et al., 2007a)	0.0069 (Uehara et al., 2009)
$\mu_0 H_c(0)$ (T)	0.22 (Mao et al., 2003) 0.19 (Uehara et al., 2007b)	0.053 (Uehara et al., 2007b)	—
κ	46 (Mao et al., 2003) 54 (Uehara et al., 2007b)	23–32 (Uehara et al., 2007b)	17 (Uehara et al., 2009)
λ (nm)	213 (Mao et al., 2003) 248 (Uehara et al., 2007b)	276.7–427.6 (Uehara et al., 2007b)	308.9 (Uehara et al., 2009)
ξ (nm)	4.6 (Mao et al., 2003; Uehara et al., 2007b)	12.2–13.5 (Uehara et al., 2007b)	18.5 (Uehara et al., 2009)
γ (mJ K^{-2} (mol Ni)$^{-1}$)	11.2 (Lin et al., 2003) 9.2 (Mao et al., 2003) 10.03 (Uehara et al., 2007a)	6.0 (Uehara et al., 2007a)	—
θ_D (K)	287 (Lin et al., 2003) 280 (Mao et al., 2003)	352 (Uehara et al., 2007b)	—
Lattice parameter a_0 (nm)	0.3812 (Uehara et al., 2009)	0.3844 (Uehara et al., 2009)	0.3756 (Uehara et al., 2009)

2003), while band-structure studies find the hole states to dominate at E_F. The negative thermopower observed by Li et al. (2002) is illustrated in the main part of Figure 21.10. Furthermore, the pressure effects indicate a positive slope of $N(E_F)$, while the opposite is predicted theoretically. The positive dT_c/dP observed by Yang et al. (2003) (Figure 21.10(b)) is believed to be due to the weakening of spin fluctuations due to applied pressure (Kumary et al., 2002; Yang et al., 2003).

The ^{12}C–^{13}C carbon isotope effect been investigated in $MgCNi_3$, yielding an exponent $\alpha = 0.54 \pm 0.03$, indicating a prominent role of carbon-based phonons in conventional pairing (Klimczuk and Cava, 2004a). Indeed, there is a weight of evidence for conventional superconductivity in $MgCNi_3$. Its upper critical field $\mu_0 H_{c2}$ is described well by the WHH formula within the purview of conventional superconductivity. Similarly, the magnetic-field dependence of its electronic

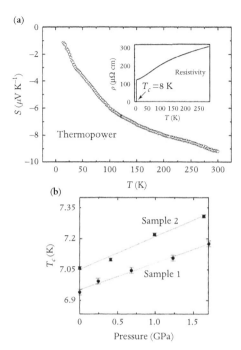

Figure 21.10 (a) Temperature dependence of the (negative) thermopower and (inset) the resistive superconducting transition (after Li et al., 2001, 2002). (b) Pressure dependence of T_c for two MgCNi$_3$ samples (after Yang et al., 2003).

specific heat is linear, which points to conventional s-wave pairing (Lin et al., 2003). Moreover, below T_c, the specific heat decreases exponentially, which is again strong evidence of BCS pairing (Lin and Yang, 2004; Mao et al., 2003; Lin et al., 2003; Shan et al., 2003, 2005; Wälte et al., 2004; Pribulová et al., 2011). As may be seen in Table 21.7, its gap coefficient $2\Delta/k_BT_c \approx 4$ and specific heat jump $\Delta C/\gamma T_c \approx$ 1.97–2.3 which indicate the superconductor to be moderately strongly electron–phonon coupled. Further, NMR studies reveal a Hebel–Slichter peak below T_c, characteristic of conventional superconductivity (Singer et al., 2001). These findings are also consistent with penetration-depth measurements (Diener et al., 2009) and point-contact tunnelling studies (Pribulová et al., 2011) on good-quality single crystals of MgCNi$_3$, although similar measurements on polycrystalline samples have shown non-BCS behaviour (Prozorov et al., 2003; Mao et al., 2003; Shan et al., 2003). These conflicting observations have motivated theorists to propose multiband models (Wälte et al., 2004), but the superconductivity of MgCNi$_3$ remains to be fully explained.

21.3.3 Substitution studies in MgCNi$_3$

The prime motivation behind substitutional studies in MgCNi$_3$ has been to examine the theoretical prediction as to whether a small change in the carrier density can indeed make the compound ferromagnetic and how superconductivity responds to such a change. Accordingly, cationic

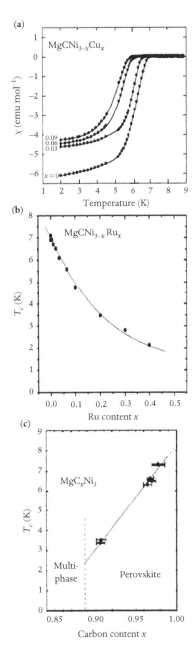

Figure 21.11 Substitution studies in MgCNi₃: (a) Cu substitution for Ni (after Hayward et al., 2001), (b) Ru for Ni (after Klimczuk et al., 2004), and (c) change in C-stoichiometry (after Amos et al., 2002).

substitutions have been carried out for all the three sites, that is, Zn, Al, and In for Mg (Park et al., 2004; Dong et al., 2005; Tong et al., 2006), B and N for C (Amos et al., 2002; Shan et al., 2003; Klimczuk et al., 2005; Uehara et al., 2009), and Co, Fe, Mn, Cu, and Ru for Ni (Hayward et al., 2001; Kumary et al., 2002; Klimczuk and Cava 2004b; Das and Kremer, 2003; Klimczuk et al., 2004) of the compound. All these substitutions that give rise to excess holes or electrons have, however, universally led to a decrease in T_c of the pristine compound resulting either from lowering of the DOS at E_F or from the magnetic cations serving as pair breakers of conventional superconductivity. There is no evidence of the theoretically speculated ferromagnetic state being induced through chemical substitutions. The results of Cu (Hayward et al., 2001) and Ru (Klimczuk et al., 2004) substitution at the Ni sites and variation in C content (Amos et al., 2002) on T_c are shown in Figure 21.11(a), (b), and (c), respectively.

Clearly, the discovery of MgCNi₃ as the first superconducting intermetallic perovskite, or so-called anti-perovskite, was a matter of considerable excitement for superconductivity researchers at the start of the present century. However, as the majority of results now seem to indicate that their superconductivity is conventional and little to do with magnetism, the interest in this class of materials, has markedly receded of late, although they still pose some challenging issues that remain to be understood. Some of the exciting speculations regarding their band-structure calculations have not been corroborated experimentally, which may either suggest a general lack of good-quality samples or call for a renewed look at the calculations.

21.4 Metallonitride halides

21.4.1 Transition metal nitride halides

We have already discussed a variety of layered superconductors in this book, such as Sr₂RuO₄, Na$_x$CoO₂, HTS cuprates, ruthenocuprates, and pnictides, all manifesting exotic superconductivity in varied forms. In the same category, we include the *metallonitride halides*, represented by the general formula MNX (M = Hf, Zr, or Ti; X = Cl, Br, or I). These were discovered by Yamanaka et al. (1996,1998) during the closing years of the last century and they captured much interest owing to their highest $T_c \approx 26$ K. Interestingly, until the advent of pnictide superconductors in 2008, this compound was the second highest in T_c among all the existing transition metal superconductors.

In the pristine state, the parent MNX compounds exist in two structural types, labelled α and β. Zr and Hf, for instance, form β-Zr(Hf)NCl (Yamanaka et al., 1996, 1998), while only recently it was found that Ti resulted in α-TiNCl (Yamanaka et al., 2009). The parent compounds, known since the 1960s (Juza and Heners, 1964), are insulators with a band gap of about 2–4 eV (Yamanaka, 2000) and have to be charge-doped to make them metallic and superconducting (Yamanaka et al., 1996, 1998). This is achieved by intercalating the compounds with alkali metals Li, Na, K, etc. Their layered structure is indeed compatible with easy intercalation of atoms and molecules in between the neighbouring Cl layers having weak van der Waals bonding. α-TiNCl possesses an orthorhombic space group *Pmmn*, which slightly changes on doping with alkali metals. In β-Zr(Hf)NCl, possessing a rhombohedral structure of space group *R-3m*, the layered structure comprises a double metal–nitrogen layer sandwiched by close-packed Cl layers. The metal–nitrogen bilayers have a graphene-like honeycomb structure, as on the left of Figure 21.12, while the main figure depicts the bulk crystal lattice. The layer sequence of the structure is represented by a stacking unit Cl–Zr(Hf)–N–N–Zr(Hf)–Cl. In the crystal lattice, the adjacent stacking units are held together by a weak van der Waals attractive force. Alkali metal atoms (Li, Na, etc.) and organic molecules, such as tetrahydrofuran (THF) or propylene carbonate (PC) can additionally be intercalated into the van der Waals gap of the parent compounds. The intercalated alkali metal atoms give rise to doping of electrons as charge carriers to Zr(Hf)–N bilayers, which serve as conducting planes, and make the compounds metallic and superconducting. Alternatively, this can be done by creating Cl vacancies by de-intercalation (Yamnaka et al., 2003) instead of by external doping. The co-intercalation of organic molecules effectively expands the interlayer separation, without changing the honeycomb layers or their doping, which enhances T_c. Co-intercalation, for instance, doubles the Hf–N layer separation d from 0.93 nm of the parent compound to 1.87 nm (Yamanaka et al., 1996, 1998). The co-intercalated compounds are thus highly two-dimensional

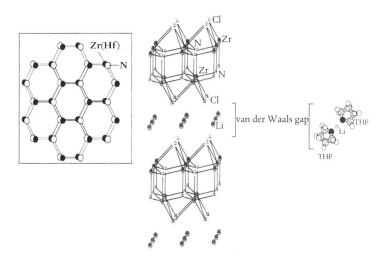

Figure 21.12 Zr-N honeycomb structure (left) and the structure of β-ZrNCl intercalated with Li, THF, etc. (after Chen, X. et al., 2002).

Table 21.8 T_c of some prominent transition metal nitride halides and, for comparison, the relevant binary nitrides with rock salt structure

Compound	T_c (K)	Reference	Compound	T_c (K)	Reference
$Li_{0.48}(THF)_{0.3}HfNCl$	25.5	Shamoto et al. (2004)	Li_xTiNCl	16.5	Yamanaka et al. (2009)
$Li_{0.18}PC_{0.15}ZrNCl$	14.6	Ito et al. (2004)	Na_xTiNCl	16.3	Yamanaka et al. (2009)
$Li_{0.18}THF_{0.08}ZrNCl$	14.4	Ito et al. (2004)	K_xTiNCl	16.3	Yamanaka et al. (2009)
$Na_{0.28}HfNCl$	22	Shamoto et al. (2004)	Rb_xTiNCl	16.3	Yamanaka et al. (2009)
$Li_{0.16}ZrNCl$	14	Chen et al. (2001c)	Py_xTiNCl	8.6	Yamanaka et al. (2009)
$Li_{0.19}ZrNBr$	13	Shamoto et al. (2001)	ZrN	10.7	Ito et al. (2004)
			HfN	8.8	Ito et al. (2004)
			TiN	5.6	Yamanaka et al. (2009)

and anisotropic. Some of the representative T_c values of the prominent compounds of this class are given in Table 21.8. Also shown are the corresponding T_c values of some of the binary nitrides, possessing a rock salt structure. As may be seen, the nitride halides possess much higher T_c.

21.4.2 Materials synthesis

Polycrystals of HfNCl and ZrNCl (Yamanaka et al., 1998; Ito et al., 2004) are respectively prepared in two steps by reaction of Hf and Zr powders or their hydrides (Shamoto et al., 1998) with vaporised ammonium chloride in the stoichiometric ratio 1:1.1 at temperatures of 740–780°C for 12 hours in evacuated and sealed silica tubes. In the second step, a temperature gradient of 100°C is introduced and the compounds are recrystallised through chemical vapour transport. Intercalation of Na is carried out by interaction of HfNCl with naphthyl sodium solutions, while Li intercalation is performed by dispersing the sample in 15% n-butyl lithium solution in hexane in an Ar-filled glove box for 24 hours. Submillimetre-size single crystals of these compounds have been grown using ammonium chloride flux at 3 GPa at 1000°C (Chen et al., 2001c). The Li intercalation in single-crystal samples is carried out as just described, but the time required is 4 months. As Li intercalates, the colour of the crystal changes from yellowish-green to black.

21.4.3 Superconducting properties

21.4.3.1 *General features*

The transition metal nitride halides are all extreme type II superconductors with Ginzburg–Landau parameter $\kappa \approx 50$–80. Because of its higher

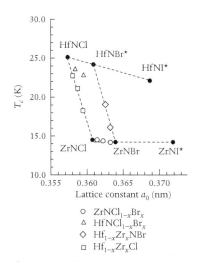

* : Sample synthesized under high pressure

Figure 21.13 Substitutions in metal nitride halides: T_c versus lattice constant (after Yamanaka et al., 2000).

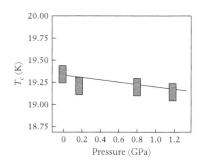

Figure 21.14 Pressure effect on T_c of $Na_{0.3}HfCl$ (after Shamoto et al., 2000)'.

T_c, HfNCl has received greater attention than its sister compounds from Table 21.8. For H parallel to ab, $[-\mu_0 dH_{c2}/dT]_{T_c} = 1.9$ T K^{-1}, while for H parallel to c, $[-dH_{c2}/dT]_{T_c} = 0.51$ T K^{-1} (Tou et al., 2001a), from which the anisotropy of the coherence length is $\xi_{ab}/\xi_c = 3.7$, where $\xi_{ab} \approx 5$ nm and $\xi_c \approx 1.35$ nm. As $\xi_c < d = 1.87$ nm, the interlayer spacing, the superconductivity along the c-axis is Josephson-coupled, which makes the material more quasi-two-dimensional. ^{35}Cl NMR studies of $Li_{0.48}(THF)_{0.3}HfNCl$ show (Tou et al., 2003a) a near-zero Knight shift, indicating the DOS at Cl-site to be practically absent, which confirms the compound to be a two-dimensional superconductor. The earlier results by Tou et al. (2001a) in this compound had shown that diamagnetic signals existed up to a temperature of $2T_c$, suggesting the presence of two-dimensional superconducting fluctuations. Extensive magnetisation and μSR studies by Ito et al. (2004) further confirmed the two-dimensional superconductivity of Li_2HfNCl and Li_2ZrNCl systems. These compounds have been subjected to substitution studies (Yamanaka et al., 2000) (Figure 21.13). Solid solutions $Hf_{1-x}Zr_xNCl$ can form over the whole composition range from $x = 0$ to $x = 1$ and T_c is found to vary linearly between the T_c values of the corresponding end members forming the compound. Partial interchange of Cl, Br, and I however did not significantly alter T_c. However, the chemical reactivity of β-HfNI is low and it does not undergo co-intercalation. Co-intercalation generally enhances T_c, but in powdered samples the overall effect is averaged out to become small.

Pressure effects have also been investigated for these compounds and found to be small. In the case of Na-doped HfNCl, $dT_c/dP = -0.13$ K GPa^{-1} (Shamoto et al., 2000) (Figure 21.14). The low value of the pressure coefficient is attributed to the competition between the increase in bandwidth and the lattice hardening taking place under imposed pressure.

21.4.3.2 Unusual superconductivity of metallonitride halides and their future prospects

Although outwardly simple, the transition metal nitride halides are found to exhibit most intriguing superconductivity with both conventional and unconventional features. In Li_xMNCl (M = Ti, Zr, or Hf) the Knight shift below T_c decreases towards zero (Akashi et al., 2012; Tou et al., 2003b, 2010), indicating spin-singlet Cooper pairing. Furthermore, tunnelling studies using break junctions and STS measurements reveal the gap function to be free from nodes (Ekino et al., 2003; Takasaki et al., 2005, 2010). These measurements, as well as those of specific heat (Taguchi et al., 2005), reveal an s-wave like gap and a large gap coefficient of $2\Delta/k_B T_c \approx 5$, characteristic of strong coupling. But, surprisingly, the doping dependences of T_c and of the gap are both most unusual. In the underdoped regime, the gap is large, but it becomes small as the doping level is increased. There is a change from strong-coupling to extremely weak-coupling superconductivity as discussed by Kuroki (2008). T_c does not exhibit the characteristic dome-like doping dependence. It is maximum at the insulator–metal transition, and for Li_xHfNCl it remains invariant in the doping range of $0.15 < x < 0.50$, beyond which it decreases with increasing x (Yamanaka et al., 1996, 1998). The spin susceptibility does not exhibit exchange enhancement and the specific heat data show the DOS at E_F to be too small to account for the observed high value of T_c. The estimated electron–phonon coupling strength λ_{e-p} is less than 0.22, which is again too small. The N isotope-effect exponent for T_c is too feeble, $\alpha \approx 0.07 << 0.5$, the BCS value, which does not support the presence of conventional phononic superconductivity (Tou et al., 2003a; Taguchi et al., 2007). The NMR spin–lattice relaxation rate does not exhibit the conventional coherence peak below T_c and μSR studies show the gap function to be highly aniso-tropic (Tou et al., 2007), in line with unconventional superconductivity. The possible candidates considered for the pairing mechanism respons-ible for the observed high T_c are spin and charge fluctuations, which have been discussed by both theorists and experimentalists, but these pos-sibilities remain much debated (Tou et al., 2001b; Takano et al., 2008). Unlike cuprates and other unconventional superconductors, metalloni-tride halides are essentially non-magnetic and are band insulators where magnetism does not appear relevant. The materials have low carrier density and a negligible mass enhancement factor, which favour charge fluctuations over spin fluctuations. However, the pairing mechanism of these interesting superconductors is yet to be explained (Yin et al., 2011).

21.5 Pyrochlore oxides

21.5.1 Superconducting pyrochlore oxides

Pyrochlore oxides belong to one of the largest structural groups of transition metal oxides with the general formula $A_2B_2O_6O'$ or $A_2B_2O_7$, where A is a large cation belonging to the rare-earth, alkali metal or post transition metal groups, while B is a transition metal. Their name is derived from the mineral *pyrochlore* (NaCa)(NbTa)O_6F/OH, having a similar structure and chemical formula. Pyrochlore oxides possess diverse properties such as metal–insulator transition, semicon-ducting behaviour, magnetism and magnetic frustration, colossal mag-netoresistance and ferroelectricity. There was much excitement when $Cd_2Re_2O_7$ was discovered (Hanawa et al., 2001) as the first supercon-ductor of the pyrochlore oxide family, albeit with a low $T_c = 1$ K, which was independently corroborated by Sakai et al. (2001) and Jin et al. (2001).

21.5.2 Crystal structure: α- and β-pyrochlore

The crystal structure of the so-called α-pyrochlores with general composition $A_2B_2O_6O'$ is fcc with space group $Fd\text{-}3m$, where there are four crystallographically distinct sites for A, B, O, and O' atoms. By keeping the origin at the B-site, the locations of the above atoms are described in terms of Wyckoff positions as $16d$ for A (i.e. the Cd atom in $Cd_2Re_2O_7$), $16c$ for B (the Re atom), $48f$ for O, and $8b$ for O'. The A- and B-sites individually form corner-sharing interpenetrating tetrahedral sublattices in which the larger A-atoms are eightfold-coordinated with six O and two O' atoms, while the smaller B-atoms are sixfold-coordinated with O atoms to form the corner-shared BO_6 octahedra. This constitutes a three-dimensional network called a *pyrochlore lattice*. In this structure, the only internal position variable is the coordinate x that characterises the $48f$ O atoms. The described α-pyrochlore structure is depicted in Figure 21.15(a). The structure is of particular interest because if either the A- or B-site is occupied by a magnetic ion having nearest-neighbour antiferromagnetic interaction, a geometrical frustration ensuing from the triangle-based lattice structure may result and give rise to unusual ground states. The situation is reminiscent of cobalt oxide hydrate, discussed in Section 21.3, which also possesses a triangular lattice geometry.

Soon after the discovery of α-pyrochlore superconductivity, another type of pyrochlore oxide, with the general formula AB_2O_6, called a β-pyrochlore, was found to be superconducting with a slightly higher T_c. This category included three osmates formed with Os as the B-atom and a large alkali metal Cs, Rb, or K as the A-atom. These pyrochlores, namely $CsOs_2O_6$, $RbOs_2O_6$, and KOs_2O_6 displayed superconductivity respectively at 3.3 K (Yonezawa et al., 2004a), 6.3 K (Yonezawa et al., 2004b; Kazakov et al., 2004; Brühwiler et al., 2004) and 9.6 K (Yonezawa et al., 2004c). Although β-pyrochlores have the same space group as the α-structure, their atom positions are slightly different. In particular, the A-atom (Cs, Rb, or K) in the new structure is shifted to the $8b$ site, which originally formed the O' site in the α-structure, leaving the $16d$ site empty (Yonezawa et al., 2004a–c; Brühwiler et al., 2004). The remaining O and B (i.e. Os) sites are left intact respectively at $48f$ and $16c$. Thus, as before, the Os atoms are coordinated with six O atoms in OsO_6 octahedra and every Os atom forms a shared corner of two Os tetrahedra in a three-dimensional pyrochlore network. Figure 21.15(b) depicts the crystal structure of the β-pyrochlore KOs_2O_6, showing the OsO_6 octahedra forming this network. There are eight formula units per unit cell. The structure shown may be compared with the α-structure of Figure 21.15(a). As may be seen, the K atoms form a diamond sublattice with K–K bonds along the $\langle 111 \rangle$ direction. The fact that the β-structure is formed by shifting a K atom from its former $16d$ site to the $8b$ site of the α-structure creates a considerable vacant space or cavity within the

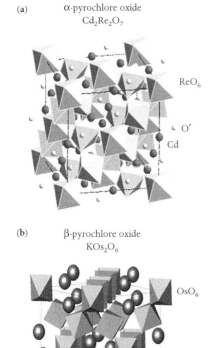

(a) α-pyrochlore oxide
$Cd_2Re_2O_7$

ReO_6

O'

Cd

(b) β-pyrochlore oxide
KOs_2O_6

OsO_6

K

Figure 21.15 Crystal structures of α- and β-pyrochlore oxides.

$Os_{12}O_{18}$ cage, inside which the K atom is located and carries 6 nearest and 12 next-nearest O neighbours. Cs- and Rb-based β-pyrochlores have similar structures. However, as the K atom is smaller than Cs and Rb, it has a larger clearance space and its bonding with the surrounding cage is weaker. This allows the K atom to *rattle* more heavily in its oversized cage than Cs and Rb (Yamaura et al., 2005; Hiroi et al., 2005), and super-conductivity of the three pyrochlores is determined by the low-energy rattling mode of the three alkali metal ions (Nagao et al., 2009).

21.5.3   Materials synthesis

Polycrystalline samples of both α- and β-pyrochlore oxides are commonly prepared by solid state reaction (Hanawa et al., 2001). For synthesising α-$Cd_2Re_2O_7$, for example, a stoichiometric mixture of CdO, ReO_3, and Re metal powder is pelletised, put inside an alumina tube, and preheated for several hours at 300°C. This is followed by several hours of heating at 1000°C, which forms the cubic α-pyrochlore oxide. To synthesise polycrystalline β-$RbOs_2O_6$, powders of Rb_2O and Os in the ratio of 1:4 are ground and pressed into a pellet, which is sealed in an evacuated silica tube and heated for 24 hours at 1050°C (Yonezawa et al., 2004b). Special care is necessary to avoid the highly toxic compound OsO_4. The oxygen partial pressure in the silica tube is controlled by adding a suitable quantity of Ag_2O to the end part of the silica tube. Ag_2O decomposes into O and Ag above 450°C and provides an oxidising atmosphere. Single crystals of these pyrochlores have also been successfully grown (Yamaura et al., 2005; Nagao et al., 2009).

21.5.4   Superconducting behaviour

21.5.4.1 *Superconductivity of α-pyrochlore oxide*

$Cd_2Re_2O_7$ with a T_c of 1 K is the only α-pyrochlore manifesting superconductivity. On cooling from ambient temperature, it shows an anomaly (Figure 21.16) at $T^* = 200$ K, confirmed through both resistivity and susceptibility studies (Hanawa et al., 2001; Sakai et al., 2001). This is associated with a second-order structural transition from one cubic structure (*Fd-3m*) to another (*F-43m*). The compound possesses a low carrier density and a relatively high normal-state resistivity above T_c. In the temperature range 1 K< T < 30 K, a T^3 rather than a T^2, behaviour of the resistivity was reported by Hiroi and Hanawa (2002), indicating that electron–electron scattering was not so dominant. Above $T^* = 200$ K, until room temperature, the behaviour is nearly temperature-independent. However, Jin et al. (2001) found a T^2 behaviour near room temperature, which is unusual.

Various superconducting parameters of $Cd_2Re_2O_7$ are listed in Table 21.9. The relatively large Sommerfeld coefficient γ indicates

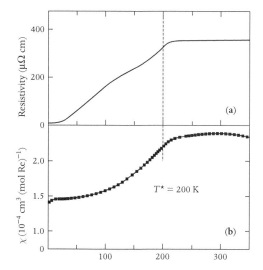

Figure 21.16 Anomaly in α-pyrochlore as observed through resistivity and magnetic susceptibility measurements (after Hanawa et al., 2001);- The x-axis of the diagram represents temperature (K).

Table 21.9 Structural and superconducting parameters of $Cd_2Re_2O_7$, $CsOs_2O_6$, $RbOs_2O_6$, and KOs_2O_6

Parameters	$Cd_2Re_2O_7$ (α-pyrochlore)	$CsOs_2O_6$ (β-pyrochlore)	$RbOs_2O_6$ (β-pyrochlore)	KOs_2O_6 (β-pyrochlore)
a_0 (nm)	1.02232	1.0149[a]	1.0114[a]	1.0099[a]
$(x, 0, 0)$ of O at $48f$	0.3137	0.3146	0.3180	0.3160
T_c (K)	1.0	3.3	6.3	9.6
$\Delta C / \gamma T_c$	1.15	~1	1.9	2.87
Gap coefficient $2\Delta / k_B T_c$	3.60[b]	3.66[c]	4.0[c]	4.57–5.0[c]
Coupling constant λ_{e-p}	0.36[c]	0.77[c]	0.80[c]–1.0	0.85[c]–1.6
γ_{exp} (mJ K^{-2} mol^{-1})	30.2	~40	~40	~71
θ_D (K)	458[c]	232[c]	186[c]	280[c]
$\mu_0 H_{c2}(0)$ (T)	0.29	~3.3	~5.5	~33
$[-\mu_0 dH_{c2}/dT]_{T_c}$ (T K^{-1})	0.42[c]	0.44[c]	0.88–1.2[c]	3.6
$\mu_0 H_{c1}(0)$ (T)	0.002[b]	0.0082[c]	0.0092	0.0116
$\mu_0 H_c(0)$ (T)	0.0148[b]	0.0589[c]	0.1249	0.2579
Pauli limit $\mu_0 H_p(0)$ (T)	1.84	6.0	11.6	17.7
Ginzburg–Landau parameter κ	14	~40	~30	~80
ξ (nm)	34	~10	7.4	3.7
λ (nm)	460	400	252	243

Data mostly from Brühwiler et al. (2006) and Hiroi et al., (2007), except: [a]Jung (2011); [b]Hiroi and Hanawa (2002); [c]Nagao et al. (2009).

the formation of heavy quasiparticles, possibly due to spin frustrations, condensing into electron pairs, and this compound therefore is the first example of coexistence of superconductivity and frustrations (Sakai et al., 2001). Both the gap coefficient and the specific heat jump (at T_c) are close to the BCS weak-coupling values. Below T_c, the specific heat decreases exponentially and its upper critical field is smaller than the Pauli limit. These observations indicate the conventional superconductivity of $Cd_2Re_2O_7$. However, the temperature dependence of the thermodynamic critical field is found to deviate from the conventional parabolic behaviour (Hiroi and Hanawa 2002), and $\lambda(T)$ measured using a μSR technique (Kadono et al., 2002) showed unconventional behaviour with an anisotropic gap. On the other hand, for a single-crystal sample, Lumsden et al. (2002), using the same technique, reported a conventional exponential dependence of $\lambda(T)$ and a fully gapped Fermi surface in the superconducting state. Thus, the conventional superconductivity of α-$Cd_2Re_2O_7$ is well supported.

21.5.4.2 *Superconductivity of β-pyrochlore oxides*

For the three superconducting osmates of β-pyrochlore structure, namely AOs_2O_6 (A = K, Rb, and Cs), T_c decreases with increasing size of the alkali metal ion A. Table 21.10 depicts T_c and various ionic-size-related parameters for the three β-pyrochlores. With increasing ion size, both the lattice parameter a_0 and the Os–O distance increase while the angle Os–O–Os decreases. T_c monotonically decreases with increasing lattice parameter a_0 (Yonezawa et al., 2004a), which corresponds to the application of negative chemical pressure on the Os-pyrochlore lattice. In BCS superconductors, T_c is generally raised under negative pressure, since this enhances the DOS at E_F. However, as may be seen from Table 21.9, the measured values of the Sommerfeld coefficient γ for Rb- and Cs-based osmates are essentially same, although their T_c values are different. This suggests that the DOS at E_F is not the deciding factor for superconductivity of these materials (Hiroi et al., 2005; Nagao et al., 2009). On the other hand, the *lattice softness index* SI (Jung 2011), as determined from Os–A interaction, is progressively enhanced when A is reduced in size (Table 21.10). A similar behaviour is manifested when considering O–A interaction. Consequently, the T_c increase from A = Cs to Rb and finally to K can be understood in terms of the progressive rise in the value of SI. Another interesting feature of these β-pyrochlores is that the gap coefficient (see Table 21.9) changes from weak coupling for $CsOs_2O_6$ to moderately strong coupling for $RbOs_2O_6$ and very strong coupling for KOs_2O_6.

Superconductivity of β-pyrochlores is believed to be of novel type. As discussed earlier, the alkali metal ion located in an oversized pyrochlore cage undergoes rattling, which is maximum for the smallest K ion. The changes in T_c and gap coefficient of the three compounds are

Table 21.10 Ion-size-related data for three β-pyrochlore osmates

β-pyrochlore AOs_2O_6	Size of ion A (nm)	Unit-cell parameter a_0 (nm)	Os–O distance (nm)	Os–O–Os angle (°)	Softness index SI
KOs_2O_6	0.133	1.0099	0.19004	139.91	46.6
$RbOs_2O_6$	0.148	1.0114	0.19211	137.08	42.4
$CsOs_2O_6$	0.169	1.0149	0.19469	134.30	37.8

From Jung (2011).

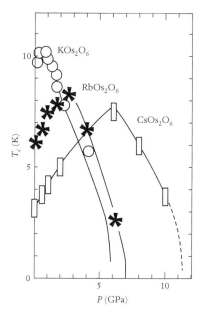

Figure 21.17 Pressure effect on T_c of three pyrochlore oxides (after Muramatsu et al., 2005).

attributed to the increase in the electron–rattle coupling with decrease in rattling frequency and the associated increase in anharmonicity (Nagao et al., 2009). Consequently, the β-pyrochlore superconductors constitute the first example of *rattling-induced superconductivity* (Nagao et al., 2009). In the presence of anharmonicity, an anomalous isotope effect, with exponent $\alpha > 0.50$, has been theoretically predicted by Oshiba and Hotta (2012), but is still to be experimentally confirmed for any of the β-pyrochlore superconductors.

The three pyrochlore oxides show interesting pressure effects. As may be seen from Figure 21.17 (Muramatsu et al., 2005), T_c first rises with pressure, following a hump respectively at 6 GPa ($T_c = 7.6$ K), 2 GPa ($T_c = 8.2$ K) and 0.6 GPa ($T_c = 10$ K) for Cs-, Rb- and K-based β-pyrochlores, which finally lose their superconductivity completely at applied pressures of about 10, 7, and 6 GPa, respectively. The unit-cell-volume dependence of T_c is depicted in Figure 21.18 (Hiroi et al., 2007). The low-temperature resistivity behaviour of the three compounds also exhibits an anomalous change. Unlike the α-pyrochlore, the normal state of these osmates does not exhibit any structural transition at T^*. Just above T_c, both Rb- and Cs-based osmates exhibit a clear T^2 dependence of resistivity below 30 K (Yonezawa et al., 2004b) and 45 K (Yonezawa et al., 2004a), which is absent in KOs_2O_6 (Yonezawa et al., 2004c). While both $RbOs_2O_6$ and $CsOs_2O_6$ manifest evidence of conventional s-wave superconductivity (Magishi et al., 2005), KOs_2O_6 behaves like an unconventional superconductor (Brühwiler et al., 2004;

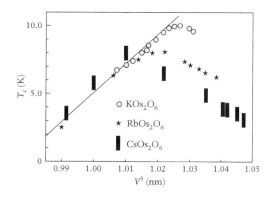

Figure 21.18 T_c as a function of cell volume for three pyrochlore oxides (after Hiroi et al., 2007).

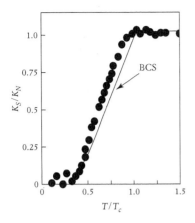

Figure 21.19 Variation of normalised Knight shift of [87]Rb-NMR in the β-pyrochlore $RbOs_2O_6$ (after Magishi et al., 2005).

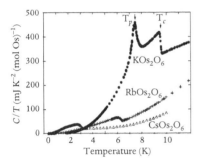

Figure 21.20 Second specific heat jump in the pyrochlore oxide KOs_2O_6 (after Hiroi et al., 2005a).

Magishi et al., 2005; Arai et al., 2006). [87]Rb-NMR studies of $RbOs_2O_6$ have revealed a coherence peak, albeit much reduced in size, in the spin–lattice relaxation rate below T_c and an exponential decrease on further cooling (Magishi et al., 2005), both of which features are in accordance with the BCS behaviour. Also, the Knight shift below T_c is found to decrease with temperature in accordance with singlet pairing (Figure 21.19). The specific heat below T_c decreases exponentially down to 1 K which supports the conventional behaviour. The upper critical fields of both $RbOs_2O_6$ and $CsOs_2O_6$ are smaller than their Pauli paramagnetic limit (see Table 21.9), which is in accord with BCS behaviour (Brühwiler et al., 2004). On the other hand, several experimental results on KOs_2O_6 manifest unconventional features, such as the absence of a coherence peak in NMR spin–lattice relaxation studies (Arai et al., 2006), a large upper critical field of about 33 T (Hiroi et al., 2004) exceeding the Pauli limit, and an anisotropic order parameter revealed by μSR experiments (Koda et al., 2005). Surprisingly, specific heat measurements on single crystalline samples of KOs_2O_6 have shown a second jump (Figure 21.20) at $T_p \approx 7.5$ K, below $T_c = 9.6$ K (Hiroi et al., 2005b), which is not due to any structural transformation of the material and whose origin is yet to be understood. Interestingly, these unconventional traits seem to have little to do with magnetism or the triangle-based geometry of the lattice. It is also not clear whether these could be attributed to the electron–rattle coupling that is a unique feature of the β-pyrochlore superconductors. Clearly, more efforts are needed to understand the unusual superconductivity of these novel oxides.

21.6 Layered transition metal chalcogenides

21.6.1 Transition metal chalcogenides

Chalcogenides (i.e. sulfides, selenides, and tellurides) in general possess more extended wavefunctions than oxides, which makes them more metallic and less magnetic than the latter. This helps them to become more favoured candidates for superconductivity. In this class, there are two main types: (i) transition metal dichalcogenides and (ii) transition metal trichalcogenides. Both are low-dimensional, the former being quasi-two-dimensional and the latter quasi-one-dimensional.

21.6.2 Dichalcogenides

The transition metal dichalcogenides (TMD) of the general formula MX_2, where M is a transition metal (Nb, Ta, or Ti) and X is a chalcogen, were perhaps the first layered superconductors to be discovered, more than four decades ago (Gamble et al., 1970). Interest in this class of materials has revived because TMDs are quasi-low-dimensional

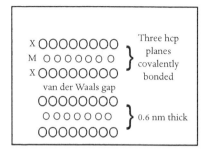

Figure 21.21 Schematic structure of an MX$_2$ compound.

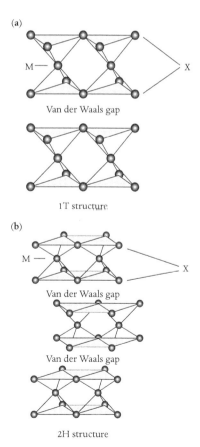

Figure 21.22 1T and 2H structures of TMD compounds.

systems that exhibit a mutual competition and possible coexistence of a charge-density wave (CDW) state and (in some cases multiband) singlet superconductivity (Wilson et al., 1975; Withers and Wilson, 1986). The CDW state occurs at a temperature $T_{CDW} > T_c$ and such materials, in analogy with magnetic superconductors, are called *CDW superconductors*. Further, they form highly anisotropic systems in which the CDW phase manifests many anomalous effects (Tidman et al., 1974; Whitney et al., 1977), which make them unusual superconductors.

The structure of TMDs comprises stacks of metallic sheets of M-layers, each sandwiched between two X-layers (Figure 21.21). The bond between M- and X-atoms is robust and of covalent type, but the neighbouring MX$_2$ sandwiches are held together by weak van der Waals-like forces. Each MX$_2$ layer block comprises three hexagonally close-packed (hcp) layers of the outer two chalcogen planes and the central metallic plane. The weak bonding between sandwiches allows them to take different relative positions along the transverse direction, which gives rise to different polymorphs called 1T, 2H (Figure 21.22), 3R, 4Ha, 4Hb, etc. Here the initia numerals indicates the number of sandwiches constituting the unit cell, while the letters T, H, and R respectively refer to trigonal, hexagonal, and rhombohedral arrangements of atoms. In 1T, the M-atoms are in octahedral coordination with X-atoms, whereas in 2H, the pertinent coordination is trigonal prismatic. In other polymorphs, the coordination is mixed, with both octahedral and trigonal prismatic being present. However, all these structures are considered hexagonal and characterised by lattice parameters a_0 and c_0. Single-crystalline samples of TMDs have been successfully grown (Prober et al., 1980; Dalrymple et al., 1986; Fang et al., 2005).

The intercalation of molecules or individual atoms (as dopants) between the adjoining sandwiches of the unit cell significantly increases the c_0-parameter and layer spacing. This vastly enhances the anisotropy of physical properties, which differ by orders of magnitude when measured along and transverse to the layers (Thomas et al., 1972). The intercalants/dopants may capture or donate charge carriers to the MX$_2$ layers and modify the electronic properties of the compound, possibly by changing the DOS at E_F. By intercalating various atomic species such as Cu, Na, or Li, and also different organic molecules such as pyridine, aniline, or collidine, superconducting critical fields of TMDs are found to be dramatically increased to 20 to 30 T (Prober et al., 1980). Interestingly, the T_c of 2H-TaS$_2$ is raised from 0.8 K to 5.5 K upon hydration (Johnston, 1982). Another way to modify the properties is by partial substitution of M- or X-atoms by other suitable cations to produce mixed TMDs such as Ta$_{1-x}$W$_x$Se$_2$, Nb$_{1-x}$Ta$_x$Se$_2$, Ta$_{0.90}$Mo$_{0.1}$S$_2$ (Dalrymple and Prober 1984; Lara et al., 2012) or TaSSe, TaS$_{0.8}$Se$_{1.2}$,etc. (Morris and Coleman, 1973; Prober et al., 1980).

Table 21.11 T_c, T_{CDW}, and upper critical field of some of the parent TMD compounds

	2H-TaS$_2$	2H-TaSe$_2$	2H-NbS$_2$	2H-NbSe$_2$	2H-TiSe$_2$
T_c (K)	0.8	0.133	6.05	7.2	—
T_{CDW} (K)	70	122 and 90	Absent	33	220
$\mu_0 H_{c2}(0)$ (T):					
Parallel to c_0-axis	2.5[a]	0.0014	4.5[b]	5.53[c]	0
Normal to c_0-axis	16[a]	0.0041	~16[b]	18.0[c]	
Reference	DiSalvo et al. (1976) [a]Na$_{0.1}$TaS$_2$ (Fang et al., 2005)	Yokota et al. (2000)	Moncton et al. (1975) [b]Kačmarčík et al. (2010)	Wilson et al. (1974) [c]Foner and McNiff (1973)	Morris et al. (1972)

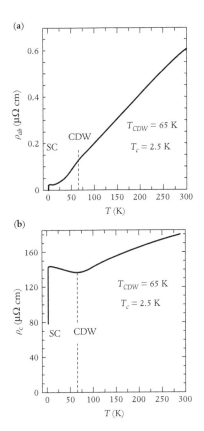

Figure 21.23 CDW and superconducting transitions in single-crystalline Na$_x$TaS$_2$ measured (a) along and (b) perpendicular to the basal plane (after Fang et al., 2005).

21.6.3 General characteristics of TMD compounds

Table 21.11 depicts T_c, T_{CDW}, and upper critical field for some of the parent compounds before intercalation. Interestingly, NbS$_2$ is an exception in that it does not undergo CDW instability (Moncton et al., 1975). In general, lower-T_c compounds exhibit the CDW transition at a higher temperature, which suggests the competing nature of the two electronic orders. Typical superconducting and CDW transitions seen in low-temperature resistivity measurements of Na$_x$TaS$_2$ single crystals are depicted in Figure 21.23 (Fand et al., 2005). Data for various intercalated compounds are presented in Table 21.12. One can see that the T_c of the various parent compounds is markedly enhanced through different intercalations and there is a simultaneous suppression of CDW. The characteristic phase diagram is depicted in Figure 21.24 (Moroson et al., 2006). It is reminiscent of the analogous behaviour seen in HTS cuprates and pnictides, with the difference that superconductivity in the present case emerges through suppression of CDW instead of SDW. Interestingly, there is also some contradictory evidence from angle-resolved photoemission spectroscopy (ARPES) studies of 2H-NbSe$_2$ indicating that charge ordering in CDW supports superconductivity (Kiss et al., 2007) and the issue therefore remains a matter of debate.

The phase diagram in Figure 21.24 summarises the effect of controlled Cu intercalation in TiSe$_2$, which was one of the first CDW-carrying materials to be found that did not exhibit superconductivity in its parent state. Cu$_x$TiSe$_2$ was, in fact, the first TMD system known where the CDW could be tuned by chemical doping, thereby allowing investigators to study its interplay with superconductivity. As may be seen in Figure 21.24, the CDW transition is continually suppressed with increasing x. Also, the semiconductor-like resistivity

Table 21.12 Effect of intercalation of parent compounds of Table 21.11

Material	T_c (K)	T_{CDW} (K)	Reference	Material	T_c (K)	T_{CDW} (K)	Reference
$Na_{0.01}TaS_2$	2.5–3.4	65	Fang et al. (2005)	$Mo_{0.1}TaS_2$	3.92	—	Lara et al., (2012)
$Na_{0.05}TaS_2$	4.0	Detected	Fang et al. (2005)	$Cu_{0.04}TiSe_2$	Superconductivity emerges	~100	Morosan et al. (2006)
$Na_{0.1}TaS_2$	4.4	Absent	Fang et al. (2005)	$Cu_{0.06}TiSe_2$	Superconductivity	<60	Morosan et al. (2006)
$Na_{0.33}TaS_2$	4.7	Absent	Sernetz et al. (1974), Lerf et al. (1979)	$Cu_{0.08}TiSe_2$	4.15 (maximum)	Absent	Morosan et al. (2006)
$Cu_{0.01}TaS_2$	2.5	—	Wagner et al. (2006)	$TaS_2(collidine)_{1/6}$	3.2	—	Prober et al. (1980)
$Cu_{0.04}TaS_2$	4.5–4.7	—	Wagner et al. (2006)	$TaS_2(pyridine)_{1/2}$	3.47	—	Prober et al. (1980)
$Cu_{0.08}TaSe_2$	2.8–3	—	Wagner et al. (2006)	$TaS_2(aniline)_{3/4}$	2.9	—	Prober et al. (1980)

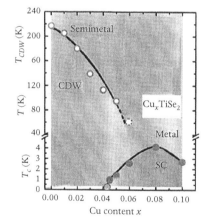

Figure 21.24 Phase diagram of Cu_xTiSe (after Morosan et al., 2006).

behaviour of the parent compound changes to metallic with Cu doping. Superconductivity emerges for $x = 0.04$ and shows a maximum $T_c = 4.15$ K for $x = 0.08$, beyond which it decreases following a dome-shaped behaviour. The CDW state is detected until $x < 0.06$, which suggests $0.04 < x < 0.06$ as the region of possible coexistence of the superconducting and CDW states. Intercalation of Cu changes the original semiconductor-like resistivity behaviour of $TiSe_2$ to metallic, and thermopower studies for $x > 0.06$ reveal the sign of the charge carries as negative over the entire temperature range of measurement. Heat capacity measurements of the optimum-T_c sample ($x = 0.08$) (Moroson et al., 2006) confirm the ratio $\Delta C / \gamma T_c = 1.49$, which is close to the BCS value of 1.43.

As an alternative to chemical tuning, T_{CDW} and T_c of TMDs can be varied by applying pressure, which generally increases T_c and decreases T_{CDW} (Morris, 1975; Nunez-Regueiro et al., 1993). The resulting pressure–temperature (P–T) phase diagrams are very similar to Figure 21.24 and exhibit dome-like behaviour of T_c (Berthier et al., 1976; Delaplace et al., 1976; Wagner et al., 2006; Bud'ko et al., 2007). The mechanism of superconductivity thus seems to arise from quantum criticality related to the fluctuations in the CDW order (Barath et al., 2008; Kusmartseva et al., 2009).

Among TMDs, 2H-NbSe$_2$ (Huang et al., 2007) and 2H-NbS$_2$ (Kačmarčík et al., 2010), with relatively high T_c values of 7.2 K and 6.05 K, respectively, have been extensively studied for their superconducting properties. Both manifest an exponential decrease of specific heat below T_c in accord with conventional superconductivity.

Interestingly, the measured data in both cases are found to fit better with the existence of two energy gaps, small and large, given by $2\Delta_S = 2.6k_BT_c$ and $2\Delta_L = 4.5k_BT_c$ for NbSe$_2$, and $2\Delta_S = 2.1k_BT_c$ and $2\Delta_L = 4.6k_BT_c$ for NbS$_2$. STM-STS studies (Guillamón et al., 2008), temperature dependence of in-plane and out-of-plane penetration depth (Fletcher et al., 2007), and ARPES measurements (Borisenko et al., 2009) have added support to the presence of a two-gap structure of NbSe$_2$.

TMDs are anisotropic type II superconductors in which the critical field measured along the metallic planes is significantly larger than in the perpendicular direction (Morris and Coleman, 1973; Foner and McNiff, 1973). For example, in TaS$_2$ samples, on intercalation with collidine, pyridine, or aniline, the in-plane upper critical fields are respectively enhanced to 32 T, 24 T, and 28 T, with the corresponding values in the transverse direction being 1.05 T, 0.97 T, and 0.75 T (Prober et al., 1980). Interestingly, although these are conventional superconductors with low T_c (Table 21.12), their in-plane critical fields exceed the respective Pauli limits. The reason for this remains to be fully understood. Despite having very high in-plane critical fields, the low T_c and large anisotropy of these materials are the major hurdles for their practical application.

21.6.4 Transition metal trichalcogenides

Unlike dichalcogenides, which are layer-based and quasi-two-dimensional, the transition metal trichalcogenides (TMTs) are chain-based and quasi-one-dimensional. They are described by the general formula MX$_3$, where M is a transition metal (Nb, Ta, or Zr) and X is a chalcogen. Compared with TMDs, TMTs have been less well studied, although it has been found that some of them again manifest a large anisotropy and an interesting competition and possible coexistence of CDW and superconductivity. The latter observed in un-intercalated compounds is of filamentary nature, not displaying the Meissner effect (Takahashi et al., 1984; Nakajima et al., 1986), and showing a low resistive $T_c \approx 2$ K when the current is passed only along specific directions. However, the phenomenon generally becomes bulk-type with a higher $T_c \approx 3$–5 K after the parent compound has been suitably doped with Cu or Ni, or when the samples are pressurised. Among TMTs, the more extensively studied compounds are NbSe$_3$, TaSe$_3$, and ZrTe$_3$, which are briefly described in the following. They possess a monoclinic space group symmetry $P2_1/m$ in which the basic structural unit is a chain of transition metal ions, surrounded by chalcogen anions, arranged in the form of distorted trigonal prisms (Figure 21.25).

21.6.4.1 TaSe$_3$

This trichalcogenide exhibits superconductivity at 2.2 K but no CDW (Sambongi et al., 1977; Yamamoto, 1978). TaSe$_3$ behaves more like a

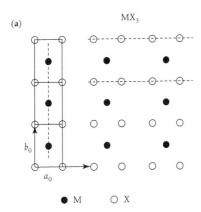

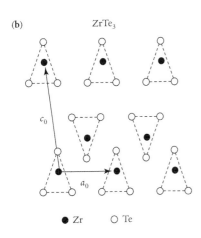

Figure 21.25 Structure of MX$_3$ compounds: (a) along the (001) plane, with dotted lines indicating M- and X-chains; (b) for ZrTe$_3$, a projected view normal to the chains; two layers are indicated through the different sizes of open and filled circles. (After an illustration from the PhD thesis, No. 3107, of S. Mitrovic (2004), EPF, Lausanne).

three-dimensional conductor, albeit one with a large anisotropy (Tajima et al., 1984). Near T_c, the Meissner effect is very small, characteristic of quasi-one-dimensional filamentary superconductivity, but it rapidly rises to 80% of its ideal value when the temperature is lowered to 0.3 K, indicating three-dimensional superconductivity. At low temperatures, superconducting filaments are believed to become Josephson-coupled to make the material three-dimensional. However, there have been reports contradicting the bulk superconductivity of $TaSe_3$ (Haen et al., 1978a).

21.6.4.2 NbSe₃

In contrast to $TaSe_3$, which does not show CDW formation, $NbSe_3$ exhibits two independent CDW transitions, associated with two different chains (Chaussy et al., 1976; Fleming et al., 1978), at $T_{CDW} = 145$ K (T_1) and 59 K (T_2), which show up as anomalies in the resistivity–temperature behaviour. The single-crystal compound, on further cooling, exhibits two more resistive drops (Haen et al., 1978b), first at 2.2 K and then at 0.4 K, but down to about 7 mK no disappearance of resistivity is observed. However, Briggs et al., (1981) found that compacted powders of the compound showed filamentary superconductivity at about 0.9 K. A moderate pressure of about 0.6 GPa (Briggs et al., 1980) is found sufficient to quench the CDW anomaly at T_2, which induces bulk superconductivity at 2.5 K. Nunez-Regueiro et al. (1992) found that the CDW temperature T_1 continually decreased with increasing pressure and the CDW anomaly was fully quenched at 3.6 GPa. This corresponded to T_c attaining its highest value of about 5 K. Further increase of pressure suppressed superconductivity. These results were corroborated by point-contact tunnelling studies by Escudero et al. (2001) carried out using $NbSe_3$ electrodes. Interestingly, the mechanical pressures exerted by the two electrodes were sufficient for the material to gain higher T_c values.

21.6.4.3 ZrTe₃

Superconductivity of $ZrTe_3$ was originally thought to be purely filamentary (Nakajima et al., 1986), but recent studies have found the phenomenon to be more complex (Yamaya et al., 2012), with specific heat measurements on a single-crystal sample revealing a tiny jump at $T < 2$ K having a broad tail extending up to about 4.5 K on the high-temperature side of the anomaly (Figure 21.26). Yamaya et al. (2012) explained these observations in terms of possible successive superconducting transitions beginning from the formation of local pairs at 5 K, responsible for the onset of filamentary superconductivity, followed by a mixture of bulk and filamentary superconductivity in the crossover region of the broad tail, and finally the opening of the

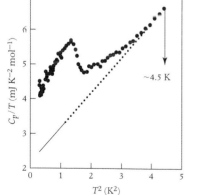

Figure 21.26 Specific heat measurements on $ZrTe_3$, showing a specific heat jump having a broad tail extending up to about 4.5 K (after Yamaya, 2010).

BCS gap of the bulk superconductivity below 2 K. Thus, superconductivity of $ZrTe_3$ is neither completely bulk nor fully filamentary, but of mixed type.

$ZrTe_3$ manifests an interesting interplay of CDW order and superconductivity under imposed pressure (Yoma et al., 2005). For pressure increasing from ambient to 2 GPa, CDW is enhanced and superconductivity suppressed, vanishing at 0.5 GPa. However, superconductivity re-emerges and rises with pressure above 5 GPa, when CDW order is quenched, with T_c reaching a maximum value of 4.7 K at 11 GPa where it is a bulk phenomenon. A similar T_c increase and occurrence of bulk superconductivity are reported for Cu- and Ni-intercalated single crystals of $Cu_{0.05}ZrTe_3$ (Zhu et al., 2011), with $T_c = 3.8$ K, and $Ni_{0.05}ZrTe_3$ (Lei et al., 2011), with $T_c = 3.6$ K. These results further showed both superconductors to be of BCS intermediate- and weak-coupling type with large anisotropy of about 7. The intercalation in both cases caused expansion of lattice parameters a_0 and c_0, partial filling of the CDW gap, and suppression of T_{CDW}. Studies by Yadav and Paulose (2012a, b) have found that polycrystalline $ZrTe_3$ intrinsically possesses a higher $T_c = 5.2$ K, albeit filamentary, which does not change on Cu or Ag intercalation, although this weakens the CDW state. The higher T_c is ascribed to internal strains of the polycrystalline samples. On the other hand, the results of Yadav and Paulose (2012b) on $Fe_{0.05}ZrTe_3$ show that Fe intercalation makes the material strongly magnetic, with $T_N = 43$ K, and quenches both CDW order and superconductivity. These results point to an exciting possibility of interplay of three different orders on magnetic ion doping of $ZrTe_3$.

21.7 BiS_2-based superconductors

21.7.1   Present status

The BiS_2 based superconductors arrived in July 2012 and during the very short time since then, commendable progress has been made, which we briefly review here. The excitement began with three novel superconductors announced in close succession by Mizuguchi and his collaborators in Japan, namely $Bi_4O_4S_3$ (Mizuguchi et al., 2012a), $LaO_{1-x}F_xBiS_2$ (Mizuguchi et al., 2012b), and $NdO_{1-x}F_xBiS_2$ (Demura et al., 2013), with T_c^{onset} of 8.6 K, 10.6 K, and 5.6 K, respectively. All three are layered compounds having common BiS_2 layers, but with different spacer layers. It is the BiS_2 layer that becomes superconducting, while the spacer layer provides the quasi-two-dimensionality and charge carriers for superconductivity. Some of the available data on these superconductors are given in Table 21.13. The close similarity of $LaO_{1-x}F_xBiS_2$ and $NdO_{1-x}F_xBiS_2$ with the high-T_c FBS pnictide $LaO_{1-x}F_xFeAs$, discussed earlier, had initially raised considerable hopes that the T_c of the new superconductors

Table 21.13 Data on three BiS$_2$-based superconductors

Material	Bi$_4$O$_4$S$_3$	LaO$_{1-x}$F$_x$BiS$_2$ ($x = 0.5$)	NdO$_{1-x}$F$_x$BiS$_2$ ($x = 0.3$)
Space group	$I4/mmm$ or $I\bar{4}2m$	$P4/nmm$	$P4/nmm$
Lattice parameters (nm)	$a_0 = 0.39592(1)$ $c_0 = 4.1241(1)$	$a_0 = 0.40527$ $c_0 = 1.33237$	$a_0 = 0.3999$ $c_0 = 1.34998$
Spacer layer	Bi$_4$O$_4$(SO$_4$)$_{1-x}$ ($x = 0.5$)	LaO	NdO
T_c (K)	$T_c^{onset} = 8.6$, $T_c^{R=0} = 4.5$ $T_c^{onset} = 4.5$, $T_c^{R=0} = 4.02^a$	10.6 (high-pressure route) 2–3 (ambient-pressure route)	5.2
$\mu_0 H_{c2}(0)$ (T)	21	10	5.2
ξ (nm)	4.3	5.8	8.4
References	Mizuguchi et al. (2012a); ^{a}Tan et al. (2012a)	Mizuguchi et al. (2012b)	Demura et al. (2013)

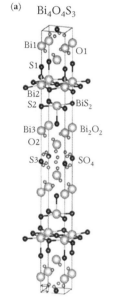

(a) Bi$_4$O$_4$S$_3$

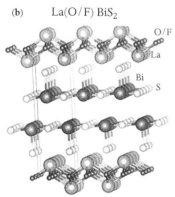

(b) La(O/F) BiS$_2$

Figure 21.27 Crystal structures of (a) Bi$_4$O$_4$S$_3$ and (b) La(O/F)BiS$_2$.

might also rise. At the same time, there was cautious speculation that the reported superconductivity of the BiS$_2$ materials ($T_c < 11$ K) might not be intrinsic but could be the outcome of Bi, present in impurity form, which in the amorphous state or under pressure possesses similar T_c values. Since the initial announcement, although new superconductors have been added to this family, these speculations have so far proved untrue. T_c has remained confined to the region of 10 K and there is growing evidence that the reported superconductivity is intrinsic and not related to Bi or any other impurities.

21.7.1.1 Bi$_4$O$_4$S$_3$

The structure of this bismuth oxysulfide comprises layer blocks of Bi$_2$S$_4$, the superconducting layers, and Bi$_4$O$_4$(SO$_4$)$_{1-x}$, serving as the spacer block. The concentration x in the latter represents defects in the form of deficiency in SO$_4^{2-}$ ions. The parent compound corresponding to $x = 0$, with composition Bi$_6$O$_8$S$_5$, is defect-free and is a band insulator, while the composition Bi$_4$O$_4$S$_3$, having $x = 0.5$, is metallic and exhibits superconductivity. The BiS$_2$ block layers have a rock-salt structure with Bi and S atoms aligned alternately on a square lattice. The spacer block comprises Bi$_2$O$_2$ (fluorite-type) and SO$_4$ layers (Figure 21.27(a)). The typical resistive transition, showing the metallic normal state with onset $T_c = 8.6$ K and zero resistance at 4.6 K (Mizuguchi et al., 2012a), is shown in Figure 21.28. Subsequent measurements have, however, brought down the onset T_c to about 5 K (Tan et al., 2012a). The thermopower behaviour has been found to be negative over the entire temperature range of measurement, indicating electrons as the majority carriers. Its large $S(T)$ and low thermal conductivity makes Bi$_4$O$_4$S$_3$ a good thermoelectric material.

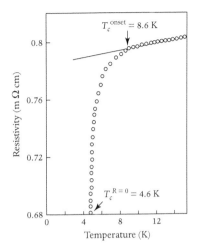

Figure 21.28 Resistive superconducting transition in $Bi_4O_4S_3$ (after Mizuguchi et al., 2012).

The material synthesis, in the double-step route (Mizuguchi et al., 2012a) involves first synthesising Bi_2S_3 by reacting Bi and S in stoichiometric proportions in an evacuated quartz tube at 500°C for 10 hours. The synthesised powder is mixed with Bi_2O_3 and S, in stoichiometric proportions, and is ground and pelletised. The pellets are annealed in a vacuum-sealed quartz tube at 510°C for 10 hours. The reacted product is ground and pelletised, and the heating schedule is repeated to achieve homogeneity. In the single-step approach (Tan et al., 2012a) Bi, S, and Bi_2O_3 are all mixed together in the required stoichiometry and heat-treated as in the double-step route to form the compound.

21.7.1.2 $LaO_{1-x}F_xBiS_2$

This ZrCuSiAs-type compound has a striking similarity (Figure 21.27(b)) with the Fe-based pnictide superconductor $LaO_{1-x}F_xFeAs$. Both these compounds possess LaO as the spacer block and their electron doping is achieved through the same mechanism, namely partial replacement of O^{2-} by F^-. Whereas in the pnictide compound the electrons are doped from LaO/F layer block to FeAs layer block, in $LaO_{1-x}F_xBiS_2$ the doping is from the same layer block to the BiS_2 layer block. There is no superconductivity for $x < 0.4$. Under ambient pressure, F has limited solubility of $x = 0.5$ at the O-site, which is found inadequate for realising the optimum $T_c = 10.6$ K (Mizuguchi et al., 2012b). The compound synthesised under ambient conditions is underdoped, with $T_c \approx 2.5$–3 K, a shielding fraction of only 10–12%, and a normal-state resistivity behaviour of activated type. The problem is overcome by following a high-pressure synthetic route (Mizuguchi et al., 2012b), which results in increased electron doping and the optimum T_c with near 100% shielding fraction.

Samples are synthesised under ambient pressure by solid state reaction of the starting materials, consisting of a stoichiometric powder mixture of La_2O_3, LaF_3, La_2S_3, Bi_2S_3, and Bi (granules), ground and pelletised as described for $Bi_4O_4S_3$ (Mizuguchi et al., 2012b). Heat treatment is carried out for 10 hours at 800°C in a sealed quartz tube. The compound has also been synthesised by using mixtures of Bi_2O_3, BiF_3, La_2S_3, Bi_2S_3, and Bi grains, heated for 10 hours at a lower temperature of 700°C (Mizuguchi et al., 2012b). In the high-pressure route (Mizuguchi et al., 2012b), a second heat treatment is carried out at 600°C under a pressure of 2 GPa using a cubic anvil press of 180 ton capacity.

21.7.1.3 $NdO_{1-x}F_xBiS_2$

This compound is formed by full replacement of the La ion by the smaller Nd ion. This substitution (Demura et al., 2013) effectively results in producing an *in situ* internal pressure, which allows a larger x and higher

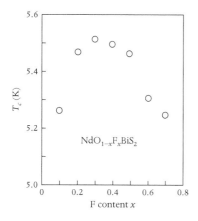

Figure 21.29 T_c as a function of x in $NdO_{1-x}F_xBiS_2$.

carrier concentration to be achieved without the need for external pressure. For $NdO_{1-x}F_xBiS_2$, the T_c versus x behaviour is dome-shaped, showing the optimum $T_c \approx 5$ K for $x = 0.3$ (Figure 21.29). The material synthesis is essentially similar to that for the La compound already described, and the material thus prepared manifests a shielding fraction greater than 90% at 2 K (Singh et al., 2012; Li, et al., 2013).

21.7.1.4 Other compounds of the BiS_2 family

Analogous to the above compounds, Pr- and Ce-containing BiS_2 compounds (i.e. $MO_{1-x}F_xBiS_2$. with M = Pr or Ce) have been synthesised, both showing an optimum $T_c \approx 3$ K for $x = 0.5$ (Xing et al., 2012; Jha et al., 2013; Wolowiec et al., 2013). Interestingly, in both cases, the normal state exhibits an activated R–T behaviour. In the presence of a magnetic field, the resistive transition in $PrO_{0.5}F_{0.5}BiS_2$ is broadened and T_c is suppressed from 3.1 K (in zero field) to 2.1 K (in 0.2 T) (Jha et al., 2013). In the case of $CeO_{1-x}F_xBiS_2$, superconductivity appears for $x > 0.33$, with the sharpest transition for $x = 0.50$, beyond which the transition broadens and T_c decreases (Figure 21.30). Analogous to the F-free $LaOBiS_2$ compound, Lei et al. (2012) have synthesised the O-free $SrFBiS_2$ compound, which has a similar structure and as a parent compound does not show superconductivity. However, subsequent studies have shown that on La doping, $Sr_{1-x}La_xFBiS_2$ (for $x = 0.5$) exhibits superconductivity at 2.8 K and its $H_{c2}(T)$ behaviour follows that of a conventional superconductor (Lin et al., 2013).

21.7.2 Pressure effects and substitution studies

These compounds exhibit interesting effects on T_c and normal-state resistivity under imposed pressure (Kotegawa et al., 2012). In the case of $Bi_4O_4S_3$, which shows metallic behaviour of resistivity above T_c, imposed pressure (up to 5 GPa) has deleterious effect on superconductivity (Figure 21.31(a)), but the original metallic behaviour remains invariant. In the case of $La_{1-x}F_xBiS_2$ (for $x = 0.5$), however, the pressure effect is more complicated. Its normal state is semiconductor-like and the effect of pressure is to rapidly change it to metallic, during which stage T_c continues rising with pressure. The rapid changeover from activated to metallic behaviour indicates that E_F is located in close proximity to some instability. However, once the compound has turned metallic, further increase in pressure suppresses superconductivity (Figure 21.31(b)), with no change in metallic behaviour. In the case of $CeO_{0.5}F_{0.5}BiS_2$, T_c first increases, under pressure, from 2.3 K to 6.7 K, and thereafter shows a decrease (Wolowiec et al., 2013). For undoped (non-superconducting) $LaOBiS_2$, pressure suppresses its semiconductor-like gap.

BiS_2 compounds have been subjected to substitution studies. We have already briefly discussed how F and La/Sr substitutions induce

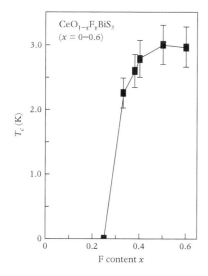

Figure 21.30 T_c as a function of F content x for $CeO_{1-x}F_xBiS_2$ (after Xing et al., 2012).

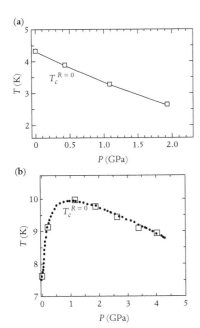

Figure 21.31 Pressure effects on T_c of (a) $Bi_4O_4S_3$ and (b) $LaO_{0.5}F_{0.5}BiS_2$ (after Kotegawa et al., 2012).

superconductivity in some of systems. In these superconductors, any cationic disorder in the BiS_2 layers is expected to seriously affect superconductivity. This is found to be the case for $Bi_{4-x}AgO_4S_3$, where $x = 0.1$ completely suppresses superconductivity (Tan et al., 2012b).

21.7.3 Superconductivity of BiS_2 compounds and unusual features

For $La_{1-x}F_xBiS_2$ ($x = 0.5$), using density functional theory with structural optimisation, Li et al. (2013) calculated an electron–phonon coupling constant $\lambda_{e-p} = 0.8$ and a corresponding $T_c = 9.1$ K, which is quite close to the observed value. This gives credence to the conventional superconductivity of BiS_2 compounds. An inherent drawback with this new family is that its superconductivity originates from hybridisation of Bi-$6p$ and S-$3p$ orbitals (Usui et al., 2012). Electrons from these orbitals are in general noted to have weak repulsive potential and a broad bandwidth that do not favour strong correlation effects. Consequently, the possibility of these materials manifesting exotic or unconventional superconductivity appears dim, although they may display some interesting new effects. For instance, the temperature dependence of resistivity of $Bi_4O_4S_3$ is found to exhibit an unusual kink at all high magnetic fields beyond $H_{c2}(T)$ (Li et al., 2013). Li et al. argue that electron pairing in this material is unusually strong and initially occurs in one-dimensional rows of Bi–S atoms. Subsequently, Josephson coupling between such linear rows establishes the bulk superconductivity of the compound. Li et al. suggest the possibility of a novel superconductivity mechanism emerging from the valence fluctuations of Bi^{2+} and Bi^{3+} to explain the presence of the kink in high magnetic fields.

Equally interesting are the recent findings of Xing et al. (2012) on $CeO_{1-x}F_xBiS_2$, which show that, contrary to band-structure calculations, the F-free ($x = 0$) parent compound is a bad metal rather than a band insulator. Further, with F-doping, the compound gradually turns from bad metallic to semiconducting, with superconductivity emerging from the semiconducting normal state, possibly connected with CDW instability. More interestingly, ferromagnetism, which may arise from the Ce magnetic moment, has been observed at low temperatures in all samples, from $x = 0$ to $x = 0.6$, suggesting the coexistence of superconductivity and magnetism. It seems difficult to reconcile these observations with conventional spin-singlet pairing for these superconductors and clearly more studies are needed to understand them. The compounds of this family are in general difficult to synthesise and widespread efforts to produce bulk single crystals and epitaxial thin films are still awaited. Undoubtedly, the newly discovered BiS_2-based system provides an interesting basis for displaying unexpected developments in the field of superconductors that may be seen in the near future.

References

Abbamonte, P., Rusydi, A., Smadici, S., Gu, G. D., Sawatzky, G. A., and Feng, D. L. (2005) Nat. Phys. 1 155.

Abrikosov, A. A. (1957) Sov. Phys. JETP 5 1174.

Abrikosov, A. A., and Gor'kov, L. P. (1960) Sov. Phys. JETP 39 1781.

Achkir, D., Poirier, M., Bourbonnais, C., Quirion, G., Lenoir, C., Batail, P., and Jérome, D. (1993) Phys. Rev. B 47 11595.

Adachi, H., Setsune, K., and Wasa, K. (1987) Phys. Rev. B 35 8824.

Adachi, S., Tokiwa-Yamamoto, A., Fukuoka, A., Usami, R., Tatsuki, T., Moriwaki, Y., and Tanabe, K. (1997) Studies of High Temperature Superconductors, Vol. 23 (Ed. Narlikar, A.). Nova Science, New York, p. 165.

Adachi, T., Ojima, E., Kato, K., Kobayashi, H., Miyazaki, T., Tokumoto, M., and Kobayashi, A. (2000) J. Am. Chem. Soc. 122 3238.

Aeppli, G., Bishop, D. J., Broholm, C., Bucher, E., Cheong, S.-W., Dai, P., Fisk, Z., Hayden, S. M., Kleiman, R., Mason, T. E., Mook, H. A., Perring, T. G., and Schroeder, A. (1999) Physica C 317–318 9.

Affronte, M., Marcus, J., Escribe Filippini, C., Sulpice, A., Rakoto, H., Broto, J. M., Ousset, J. C., Askenazy, S., and Jansen, A. G. M. (1994) Phys. Rev. B 49 3502.

Agarwal, S. K., and Narlikar, A. V. (1985a) Proc. Int. Conf. Cryogenics: INCONCRYO-85. Tata McGraw-Hill, Delhi, p. 556.

Agarwal, S. K., and Narlikar, A. V. (1985b) Sol. State. Commun. 55 563.

Agarwal, S. K., and Narlikar, A. V. (1994) Prog. Crystal Growth Charact. 28 219.

Agarwal, S. K., Samantha, S. B., Batra, V. K., and Narlikar, A. V. (1984) J. Mat. Sci. 19 2057.

Agarwal, S. K., Nagpal, K. C., and Narlikar, A. V. (1986) Sol. State. Commun. 58 89.

Agterberg, D. F. (1998) Phys. Rev. Lett. 80 5184.

Ahlers, G. A., and Walldorg, D. L. (1961) Phys. Rev. Lett. 6 (1961) 677.

Akamatu, H., Inokuchi, H., and Matsunaga, Y. (1954) Nature 173 168.

Akashi, R., Nakamura, K., Arita, R., and Imada, M. (2012) Phys. Rev. B 86 054513.

Akazawa, T., Hidaka, H., Fujiwara, T., Kobayashi, T. C., Yamamoto, E., Haga, Y., Settai, R., and Onuki, Y. (2004a) J. Phys.: Cond. Mat. 16 L29.

Akazawa, T., Hidaka, H., Kotegawa, H. Kobayashi, T. C., Fujiwara, T., Yamamoto, E., Haga, Y., Settai, R., and Onuki, Y. (2004b) J. Phys. Soc. Jpn 73 3129.

Akihama, R., Murphy, R. J., and Foner, S. (1980) Appl. Phys. Lett. 37 1107.

Aleksa, M., Pongratz, P., Eibl, O., Sauerzopf, F. M., Weber, H. W., Li, T. W., and Kes, P. H. (1998) Physica C 297 171.

Alekseevskii, N. E., Brandt, N. B., and Kostina, T. I. (1952) Izvest. Akad. Nauk. SSSR 16 233.

Alekseevskii, N. E., Garifullin, I. A., Kochelayev, B. I., and Kharakhashyan, E. G. (1973) Sov. Phys. JETP Lett. 18 189.

Alekseevskii, N. E., Mitin, A. V., Bazan, C., Dobrrovol'skii, N. M., and Raczka, B. (1978) Sov. Phys. JETP 47 199.

Alff, L., Meyer, S., Kleefisch, S., Schoop, U., Marx, A., Sato, H., Naito, M., and Gross, R. (1999) Phys. Rev. Lett. 83 2644.

Alff, L., Krockenberger, Y., Welter, B., Schonecke, M., Gross, R., Manske, R., and Naito, M. (2003) Nature 422 698.

Aliev, F. G., Prayadun, V. V., Vieira, S., Pillar, R., Levanyuk, A. P., and Yerovets, V. I. (1994) Europhys. Lett. 25 143.

Alleno, E., Neumeier, J. J., Thompson, J. D., Canfield, P. C., and Cho, B. K. (1995) Physica C 242 169.

Alloul, H., Bobbroff, J., Gabab, M., and Hershfeld, P. J. (2009) Rev. Mod. Phys. 81 45.

Alloul, H., Mahajan, A., Casalta, H., and Klein, O. (1993) Phys Rev. Lett. 70 1171.

Alrieza, P. L., Ko, Y. T. C., Gillett, J., Patrone, C. M., Cole, J. M., Lonzarich, G. G., and Sebastian, S. E. (2009) J. Phys.: Cond. Mat. 21 012208.

Amato, A., Bauer, E., and Baines, C. (2005) Phys. Rev. B 71 092501.

Amitsuka, H., Sato, M., Metoki, N., Yokoyama, M., Kuwahara, K., Sakakibara, T. S., Morimoto, H., Kawarazaki, S., Miyako, Y., and Mydosh, J. A. (1999) Phys. Rev. Lett. 83 5114.

Amitsuka, H., Tenya, K., Yokoyama, M., Schenck, A., Andreica, D., Gygax, F. N., Amato, A., Miyako, Y., Huang, Y. K., and Mydosh, J. A. (2003) Physica B 326 418.

An, J. M., and Pickett, W. E. (2001) Phys. Rev. Lett. 86 4366.

Anders, F. B. (2002) Eur. Phys. J. B 28 9.

Andersen, O. K., Klose, W., and Nohl, H. (1978) Phys. Rev. B 17 1209.

Anderson, P. W. (1962) Phys. Rev. Lett. 9 309.

Anderson, P. W. (1972) Science 177 393.

Anderson, P. W. (1984) Phys. Rev. B 30 4000.

Anderson, P. W. (1987) Science 235 1196.

Anderson, P. W. (1997) Theory of Superconductivity in High *Tc* Cuprates. Princeton University Press, Princeton, NJ.

Anderson, P. W., Baskaran, G., Zou, Z., and Hsu, T. (1987) Phys. Rev. Lett. 58 2790.

Ansaldo, T. J., Wang, Z. R., Cho, J. H., Johnston, D. C., and Riseman, T. M. (1991) Physica C 185–189 1889.

Antropov, V. P., Belashchenko, K. D., van Schilfgaarde, M., and Rashkeev, S. N. (2002) Studies of High Temperature Superconductors, Vol. 38 (Ed. Narlikar A.). Nova Science, New York, p. 91.

Aoki, D., Haga, Y., Matsuma, T. D., Tateiwa, N., Ikeda, S., Homma, Y., Sakai, H., Shiokawa, Y., Yamamoto, E., Nakamura, A., Settai, R., and Onuki, Y. (2007) J. Phys. Soc. Jpn 76 063701.

Aoki, D., Huxley, A., Ressouche, E., Braithwaite, D., Flouquet, J., Brison, J.-P., Lhotel, E., and C. Paulsen (2001) Nature 413 613.

Aoki, D., Matsuda, T. D., Taufour, V., Hassinger, E., Knebel, G., and Flouquet, J. (2009) J. Phys. Soc. Jpn 73 113709.

Aoki, H. (2009) Physica B 404 700.

Aoki, H., Uji, S., Albesaard, A. K., and Onuki, Y. (1993) Phys. Rev. Lett. 71 2110.

Aoki, Y., Namiki, T., Ohsaki, S., Saha, R., Sugawara, H., and Sato, H. (2002) J. Phys. Soc. Jpn 71 2098.

Arai, K., Kikuchi, J., Kodama, K., Takigawa, M., Yonezawa, S., Muraoka, Y., and Hiroi, Z. (2006) Phys. Rev. Lett. 96 247004.

Aronson, M. C., Thompson, J. D., Smith, J. L., Fisk, Z., and McElfresh, M. W. (1989) Phys. Rev. Lett. 63 2311.

Asano, T., Tanaka, Y., and Tachikawa, K. (1985) Cryogenics 25 503.

Asano, T., Iijima, Y., Itoh, K., and Tachikawa, K. (1986) Trans. Jpn Inst. Metals 27 204.

Asayama, K., Kitaoka, Y., Zheng, G.-Q., and Ishida, K. (1996a) Prog. Nucl. Magn. Reson. Spectrosc. 28 221.

Asayama, K., Kitaoka, Y., Zheng, G.-Q., Ishida, K., Magishi, K., Mito, T., and Tokunaga, H. (1996b) Czech. J. Phys. 46 3187.

Ashkenazi, J., and Kuper, C. G. (1989) Studies of High Temperature Superconuctors, Vol. 3 (Ed. Narlikar, A.). Nova Science, New York, p. 1.

Auban-Senzier, P., Bourbonnais, C., Jérome, D., Lenoir, C., Batail, P., Canadell, E., Buisson, J. P., Lefrant, S. (1993) J. Phys. (Paris) 3 871.

Auban-Senzier, P., Pasquier, C., Jerome, D., Carcel, C., and Fabre, J. M. (2003) Synthetic Metals, 133–134 11.

Awana, V. P. S. (2005) Frontiers in Magnetic Materials (Ed. Narlikar, A. V.). Springer, Berlin, p. 531.

Awana, V. P. S., and Narlikar, A. V. (1994) Phys. Rev. B 49 6353.

Awana, V. P. S., and Takayama-Muromachi, E. (2003) Physica C 390 101.

Awana, V. P. S., Tulapurkar, A., Malik, S. K., and Narlikar, A. V. (1994) Phys. Rev. B 50 594.

Awana, V. P. S., Kawashima, T., and Takayama-Muromachi, E. (2003) Phys. Rev. B. 67 172502.

Awana, V. P. S., Isobe, M., Singh, K. P., Shahabuddin, M., Kishan, H., and Takayama-Muromachi, E. (2006) Supercond. Sci. Technol. 19 551.

Awana, V. P. S., Pal, A., Vajpayee, A., Kishan, H., Alvarez, G. A., Yamaura, K., and Takayama-Muromachi, E. (2009) J. Appl. Phys. 105 07E316.

Awana, V. P. S., Meena, R. S., Pal, A., Vajpayee, A., Rao, K. V. R., and Kishan, H. (2011) Eur. Phys. J. B 79 139.

Aytug, T., Wu, J. Z., Kang, B. W., Verebelyi, D. T., Cantoni, C., Specht, E. D., Goyal, A., Paranthaman, M., and Christen, D. K. (2000) Physica C 340 33.

Bader, S. D., Knapp, G. S., Sinha, S. K., Schweiss, P., and Renker, B. (1976a) Phys. Rev. Lett. 37 344.

Bader, S. D., Sinha, S. K., and Shelton, R. N. (1976b) Proc. 2nd Int. Conf. on Superconductivity in *d*- and *f*-band metals (ed. Douglass, D. H.). Plenum, New York, p. 209.

Bailey, A., Russel, G. J., and Taylor, K. N. R. (1991) Studies of High Temperature Superconductors, Vol. 8 (ed. Narlikar, A.) Nova Science, New York, p. 145.

Baker, A. P., Glowacki, B. A., and Riddle, R. (1994) Adv. Cryo. Eng. 40A 201.

Baker, C. (1970) J. Mat. Sci. 5 40.

Baker, C., and Sutton, J. (1969) Phil. Mag. 19 1223.

Bakker, K., de Visser, A., Menovsky, A. A., and Franse, J. J. M. (1993) Physica B 186–188 720.

Balestrino, G., Paoletti, A., Paroli, P., and Romano, P. (1989) Appl. Phys. Lett. 54 2041.

Balicas, L., Behnia, K., Kang, W., Canadell, E., A-Senzier, P., Jérome, D., Ribaulrt, M., and Fabre, J. M. (1994) J. Phys. 4 1539.

Balicas, L., Brooks, J. S., Storr, K., Uji, S., Tokumoto, M., Tanaka, H., Kobayashi, H., Kobayashi, A., Barzykin, V., and Gor'kov, L. P. (2001) Phys. Rev. Lett. 87 067002.

Bao, W., Huang, Q., Chen, G. F., Green, M. A., Wang, D. M., He, J. B., Wang, X. Q., and Qiu, Y. (2011) Chin. Phys. Lett. 28 086104.

Bao, W., Qiu, Y., Huang, Q., Green, M. A., Zajdeb, P., Fitzsimmons, M. R., Zhernenkov, M., Chang, S., Fang, M. H., Qian, B., Vehstedt, E. K., Yang, J. H., Pham, H. M., Spinu, L., and Mao, Z. Q. (2009) Phys. Rev. Lett. 102 247001.

Bao, Z. L., Wang, R. F., Jiang, Q. D., Wang, S. Z., Ye, Z. Y., Wu, K., Li, C. Y., and Yin, D. L. (1987) Appl. Phys. Lett. 51 946.

Baranov, A. N., Kim, D. C., Kim, J. S., Kang, H. R., Park, Y. W., Pshirkov, J. S., Antipov, E. V. (2001) Physica C 357–360 414.

Barath, H., Kim, M., Karpus, J. F., Cooper, S. L., Abbamonte, P., Fradkin, E., Morosan, E., Cava, R. J. (2008) Phys. Rev. Lett. 100 106402.

Barboux, P., Tarascon, J. M., Greene, L. H., Hull, G. W., and Bagley, B. G. (1988) Appl. Phys. Lett. 63 2725.

Bardeen, J., and Stephen, M. J. (1965) Phys. Rev. 140 A1197.

Bardeen, J., Cooper, L. N., and Schrieffer, J. R. (1957) Phys. Rev. 108 1175.

Bardeen, J., Rickayzen, G., and Tewordt, L. (1959) Phys. Rev. 113 982.

Barilo, S. N., Gatalskaya, V. I., Zhigunov, D. I., Kurochkin, L. A., and Shiryaev, S. V. (1994) Supercond. Phys. Chem. Technol. 7 753.

Barilo, S. N., Gatalskaya, V. I., Shiryaev, S. V., Lynn, J. W., Baran, M., Szymczak, H., Szymczak, R., and Dew-Hughes, D. (1998) Phys. Rev. B 58 12355.

Barilo, S. N., Shiryaev, S. Y., Soldatov, A. G., Smirnova, T. V., Gatalskaya, V. I., Reichardt, W., Braden, M., Szymczak, R., and Baron, M. (2000) Supercond. Sci. Technol. 13 1145.

Barnes, L. J., and Fink, H. J. (1966) Phys. Lett. 20 583.

Baskaran, G. (1989) High Temperature Superconductors (Ed. Subramanyam, S. V., and Gopal, E. S. R.). Wiley Eastern, New Delhi, p. 167.

Baskaran, G., and Tosatti, E. (1991) Curr. Sci. 61 33.

Baskaran, G., Zou, Z., and Anderson, P. W. (1987) Sol. State. Commun. 63 973.

Basov, D. N., and Chubukov, A. V. (2011) Nature Phys. 7 272.

Basov, D. N., Timusk, T., Dabrowski, B., and Jorgensen, J. D. (1994) Phys. Rev. B 50 3511.

Batlogg, B. (1984) Physica B 126 275.

Batlogg, B., Remeika, J. P., Dynes, R. C., Barz, H., Cooper, A. S., and Garno, J. P. (1982) Superconductivity in d- and f-Band Metals (Eds. Bucker, W., and Weber, W.). KfK, Karlsruhe, p. 401.

Batlogg, B., Kourouklis, G., Weber, W., Cava, R. J., Jayaraman, A., White, A. E., Short, K. T., Rupp, L. W., and Rietmann, E. A. (1987) Phys. Rev. Lett. 59 912.

Batlogg, B., Cava, R. J., Rupp, Jr., L. W., Mujsee, A. M., Kajewski, J. J., Remeikkka, J. P., Peck, W. F. Jr., Cooper, A. S., and Espinoza, G. P. (1988) Phys. Rev. Lett. 61 1670.

Batlogg, B., Hwang, H. Y., Takagi, H., Cava, R. J., Rao, H. L., and Kwo, J. (1994) Physica C 235–240 130.

Batterman, B. W., and Barrett, C. S. (1964) Phys. Rev. Lett. 13 390.

Bauer, E., Dickey, R. P., Zapf, V. S., and Maple, M. B. (2001) J. Phys.: Cond. Mat. 13 L759.

Bauer, E. D., Frederick, N. A., Ho, P.-C., Zapf, V. S., and Maple, M. B. (2002) Phys. Rev. B 65 100506(R).

Bauer, E., Hilscher, G., Michor, H., Paul, C., Scheidt, E. W., Gribanov, A., Seropegin, Y., Noël, H., Sigrist, M., and Rogl, P. (2004) Phys. Rev. Lett. 92 027 003.

Bauer, E., Hilscher, G., Michor, H., Sieberer, M., Scheidt, E. W., Gribanov, A., Seropegin, Y., Rogl, P., Amato, A., Song, W. Y., Park, J. G., Adroja, D. T., Nicklas, M., Sparn, G., Yogi, M., and Kitaoka, Y. (2005) Physica B 359–361, 360–367.

Bauer, M., Semerad, R., and Kinder, H. (1999) IEEE. Trans. Appl. Supercond. 9 1502.

Bauernfeind, L., Widder, W., and Braun, H. F. (1995a) Physica C 254, 151.

Bauernfeind, L., Widder, W., and Braun, H. F. (1995b) High Temperature Superconductors, Vol. 6 (Eds. Barone, A., Fiorani, D., and Tampieri, A.). IV Euro Ceramics Gruppo Editoriale Faenza, Fienza.

Bauernfeind, L., Widder, W., and Braun, H. F. (1996) J. Low Temp. Phys. 105 1605.

Baumgartner, G. (1998) Doctoral thesis, EPFL, Lausanne.

Bayrakci, S. P., Bernhard, C., Chen, D. P., Keimer, B., Kremer, R. K., Lemmens, P., Lin, C. T., Niedermayer, C., and Strempfer, J. (2004) Phys. Rev. B 69 100410(R).

Beall, W. T., and Meyerhoff, R. W. (1969) J. Appl. Phys. 40 2052.

Bean, C. P. (1962) Phys. Rev. Lett. 8 250.

Bean, C. P. (1964) Rev. Mod. Phys. 36 31.

Bean, C. P., Doyle, M. V., and Pincus, A. G. (1962) Phys. Rev. Lett. 9 93.

Beauquis, S., Jimenez, C., and Weiss, F. (2004) High Temperature Superconductivity 1: Materials (Ed. Narlikar, A. V.). Springer, Heidelberg, p. 115.

Bechgaard, K., Carneiro, M., Olsen, M., and Rasmussen, F. B. (1981) Phys. Rev. Lett. 46 852.

Bednorz, J. G., and Müller, K. A. (1986) Z. Phys. B 64 187.

Beille, J., Chevalier, B., Demazeau, G., Deslandes, F., Etourneau, J., Laborde, O., Michel, C., Lejay, P., Provost, J., Raveau, B., Sulpice, A., Tholence, J. L., and Tournier, R. (1987) Physica B 146 307.

Belash, I. T., Zharikov, O. V., and Pal'nichenko, A. V. (1987a) Sol. State. Commun, 63 153.

Belash, I. T., Bronnikov, A. D., Zharikov, O. V., and Pal'nichenko, A. V. (1987b) Sol. State. Commun. 64 1445.

Belash, I. T., Zharikov, O. V., and Pal'nichenko, A. V. (1989a) Synth. Met. 34 455.

Belash, I. T., Zharikov, O. V., and Pal'nichenko, A. V. (1989b) Synth. Met. 34 47.

Belash, I. T., Bronnikov, A. D., Zharikov, O. V., and Pal'nichenko, A. V. (1989c) Sol. State. Commun. 69 921.

Belash, I. T., Bronnikov, A. D., Zharikov, O. V., and Pal'nichenko, A. V. (1990) Synth. Met. 36 283.

Belin, S., and Behria, K. (1997) Phys. Rev. Lett. 79 2125.

Bellingeri, E., Gladyshevskii, R. E., Marti, F., Dhalle, M., and Flukiger, R. (1998) Supercond. Sci. Technol. 11 810.

Bellingeri, E., Pallecchi, I., Buzio, R., Gerbi, A., Marrè, D., Cimberle, M. R., Tropeano, M., Putti, M., Palenzona, A., and Ferdeghini, C. (2010) Appl. Phys. Lett. 96 102512.

Benz, M. G. (1966) GE Report 66-C-044 (Feb.).

Benz, M. G. (1968) Trans. AIME 242 1067.

Bergeret, F. S., Volkov, A. F., and Efetov, K. B. (2004) Phys. Rev. B 69 174504.

Bergmann, G. (1976) Phys. Reports 27C, 159.

Berlincourt, T. G., and Hake, R. R. (1963) Phys. Rev. 131 140.

Bernhard, C., Tallon, J. L., Niedermayer, Ch., Blasius, Th., Golnik, A., Brücher, B., Kremer, R. K., Nokes, D. R., Stronach, C. E., and Ansaldo, E. J. (1999) Phys. Rev. B 59 14099.

Bernhard, C., Tallon, J. L., Brücher, B., and Kremer, R. K. (2000) Phys. Rev. B 61 R14960.

Bernhard, C., Drew, A. J., Schulz, L., Malik, V. K., Rossle, M., Niedermayer, C., Wolf, T., Varma, G. D., Mu, G., Wen, H. H.,

Liu, H., Wu, G., and Chen, X. H. (2009) New J. Phys. 11 055050.

Berthel, K. H., Fischer, K., Fuchs, G., Grunberger, W., Holzhauseer, W., Lange, F., Rohr, S., and Schumann, H. J. (1978) Proc. VI Int. Conf. Magnet Technology, Alpha Publishing, p. 1007.

Berthier, C., Molinié, P., and Jérome, D. (1976) Sol. State. Commun. 18 1393.

Bertman, B., Schweiitzer, D. G., and Schultz, F. P. L. (1966) Phys. Lett. 21 260.

Bhattacharya, R., and Paranthaman, M. (Eds.) (2010) High Temperature Superconductors. Wiley-VCH, Mannheim.

Bianchi, A., Movshovich, R., Jaime, M., Thompson, J. D., Pagliuso, P. G., Sarrao, J. L. (2001) Phys. Rev. B 64, 220 504.

Bianchi, A., Movshovich, R., Oeschler, N., Gegenwart, P., Steglich, F., Thompson, J. D., Pagliuso, P. G., and Sarrao, J. L. (2002) Phys. Rev. Lett. 87 057002.

Bianchi, A., Movshovich, R., Capan, C., Pagliuso, P. G., and Sarrao, J. L. (2003) Phys. Rev. Lett. 91 187004.

Bianchi, A., Movshovich, R., Vekhter, I., Pagliuso, P. G., and Sarrao, J. L. (2003a) Phys. Rev. Lett. 91 257001.

Bickers, N. E., Scalapino, D. J., and Scaletar, R. T. (1987) Int. J. Mod. Phys. B 1 687.

Biondi, M. A., and Garfunkel, M. P. (1959) Phys. Rev. 116 853

Blagoev, K. B., Engelbrecht, J. R., and Bedell, K. S. (1999) Phys. Rev. Lett. 82 133.

Blanton, S. H., Collins, R. T., Kellher, K. H., Rotter, L. D., Schlesinger, Z., Hinks, D. G., and Zheng, Y. (1993) Phys. Rev. B 47 996.

Blaugher, R. D., Heine, R. A., Cox, J. E., and Waterstrat, R. M. (1969) J. Low Temp. Phys. 1 531.

Boaknin, E., Hill, R. W., Proust, C., Lupien, C., Taillefer, L., and Canfield, P. C. (2001) Phys. Rev. Lett. 87 237001.

Bode, H. J., and Wohlleben, K. (1967) Phys. Lett. 24A 25.

Boeri, L., Dolgov, O. V. and Golubov, A. A. (2008) Phys. Rev. Lett. 101 026403.

Bohnen, K.-P., Heid, R., and Renker, B. (2001) Phys. Rev. Lett. 86 5771.

Bonalde, I., Yanoff, B. D., Salamon, M. B., Van Harlingen, D. J., Chia, E. M. E., Mao, Z. Q., and Maeno, Y. (2000) Phys. Rev. Lett. 85 4775.

Bondino, F., Magnano, E., Booth, C. H., Offi, F., Panaccione, G., Malvestuto, M., Paolicelli, G., Simonelli, L., Parmigiani, F., McGuire, M. A., Sefat, A. S., Sales, B. C., Jin, R., Vilmercati, P., Mandrus, D., Singh D. J., Mannella, N. (2010) Phys. Rev B 82 014529.

Bonney, L. A., Wills, T. A., and Larbalestier, D. C. (1995) J. Appl. Phys. 77 6377.

Boothroyd, A. T., Barratt, J. P., Bonville, P., Canfield, P. C., Murani, A., Wildes, A. R., and Bewley, R. I. (2003) Phys. Rev. B 67 104407.

Boothroyd, A. T., Coldea, R., Tennant, D. A., Prabhkaran, D., and Frost, C. D. (2004) Phys. Rev. Lett. 92 197201.

Bordet, P., LeFloch, S., Chaillout, C., Duc, F., Gorius, M. F., Perroux, M., Capponi, J. J., Toulemonde, P., and Tholence, J. L. (1997) Physica C 276 237.

Bordet, P., Mezouar, M., Núñez-Regueiro, M., Monteverde, M., Rogado, N., Regan, K. A., Hayward, M. A., He, T., Loureiro, S. M., and Cava, R. J. (2001) Phys. Rev. B 64 172502.

Borisenko, S. V., Kordyuk, A. A., Zabolotnyy, V. B., Inosov, D. S., Evtushinsky, D., Büchner, B., Yaresko, A. N., Varykhalov, A., Follath, R., Eberhardt, W., Patthey, L., and Berger, H. (2009) Phys. Rev. Lett. 102 166402.

Bormann, R., Schultz, L., Freyhardt, H. C., and Mordike, B. L. (1979) Z. Metal. 70 467.

Bormio-Nunes, C., Sandim, M. J. R., and Ghiveldor, L. (2007) J. Phys.: Cond. Mat. 19 446204.

Borsa, F., and Lecander, R. G. (1976) Sol. State. Commun. 20 389.

Bos, J.-W. G., Penny, G. B. S., Rodgers, J. A., Sokolov, D. A., Huxley, A. D., and Attfield, J. P. (2008) Chem. Commun. 31 3634.

Bouquet, F., Wang, Y., Fisher, R. A., Hinks, D. G., Jorgensen, J. D., Junod, A., and Phillips, N. E. (2001) Europhys. Lett. 56 856.

Bouquet, V., Even-Boudjada, S., Burel, L., Chevrel, R., Sergent, M., Genevey, P., Massat, H., Dubots, P., Decroux, M., Seeber, B., and Fischer, Ø. (1994) Physica C 235–240 769.

Bourne, L. C., Crommie, M. F., Zettl, A., Hans-Conrad, L., Keller, S. W., Leary, K. L., Staey, A. M., Chang, K. J., Cohen, M. L., and Morris, D. E. (1987) Phys. Rev. Lett. 58 2337.

Braccini, V., Nardelli, D., Penco, R., and Grasso, G. (2007) Physica C 456 209.

Braginski, A. I., and Roland, G. W. (1974) Appl. Phys. Lett. 25 762.

Braginski, A. I., Roland, G. W., Daniel, M. R., Santhanam, A. T., and Guardipee, K. W. (1978) J. Appl. Phys. 49 736.

Bramley, A. P., O'Connor, J. D., and Grovenor, C. R. M. (1999) Supercond. Sci. Technol. 12 R57.

Brammer, W. G. Jr., and Rhodes, C. G. (1967) Phil. Mag. 16 477.

Braun, H. F. (2005) Frontiers in Superconducting Materials (Ed. Narlikar, A. V.). Springer, Berlin, p. 365.

Brawner, D. A., Manser, C., and Ott, H. R. (1997) Phys. Rev. B 55 2788.

Briggs, A., Monceau, P., Nunez-Reguiero, M., Peyrard, J., and Ribault, M. (1980) J. Phys.: Cond. Mat. 13 2117.

Briggs, A., Monceau, P., Nunez-Reguiero, M., Ribault, M., and Richard, J. (1981) J. Phys. (Paris) 42 1453.

Brison, J. P., Flouquet, J., and Duetscher, G. (1989) J. Low Temp. Phys. 76 453.

Brison, J. P., Keller, N., Lejay, P., Huxley, A., Schmidt, L., Buzdin, A., Bernhoeft, N. R., Mineev, I., Stepanov, A. N., Flouquet, J., Jaccard, D., Julian, S. R., and Lonzarich, G. G. (1994) Physica B 199–200, 70.

Broholm, C., Kjems, J. K., Buyers, W. J. L., Matthews, P., Palstra, T. T. M., Menovsky, A. A., and Mydosh, J. A. (1987) Phys. Rev. Lett. 58 1467.

Broholm, C., Lin, H., Matthews, P., Mason, T. E., Buyers, W. J. L., Collins, M. F., Menovsky, A. A., Mydosh, J. A., and Kjems, J. K. (1991) Phys. Rev. B 43 12809.

Bromberg, L., Hashizume, H., Ito, S., Minervini, J. V., and Yanagi, N. (2010) Status of High Temperature Superconducting Magnet Development. Report PSFC/JA-10-45, US DOE Office of Fusion Energy Sciences, December.

Brown, A. R. G., Clark, D., Eastabrook, J., and Jepson, K. S. (1964) Nature 201 914.

Brown, I. D. (1989) J. Sol. State Chem. 82 122.

Brühwiler, M., Kazakov, S. M., Zhigadlo, N. D., Karpinski, J., and Batlogg, B. (2004) Phys. Rev. B 70 020503(R).

Brühwiler, M., Kazakov, S. M., Karpinski, J., and Batlogg, B. (2006) Phys. Rev. B 73 094518.

Bruls, G., Wolf, B., Finsterbusch, D., Thalmeier, P., Kouroudis, I., Sun, W., Assmus, W., Lüthi, B., Lang, M., Gloos, K., Steglich, F., and Modler, R. (1994) Phys. Rev. Lett. 72 1754.

Buchanan, M. (2001) Nature 409 11.

Bucher, B., Steiner, P., Karpinski, J., Kaldis, E., and Wachter, P. (1993) Phys. Rev. Lett. 70 2012.

Buckel, W. (1991) Superconductivity. VCH, Weinheim.

Buckel, W., and Hilsch, R. (1954) Z. Phys. 138 109.

Buckel, W., and Stritzker, B. (1973) Phys. Lett. A 43 403.

Buckett, M. I., and Labalestier, D. C. (1987) IEEE Trans. Magn. 23 1638.

Bud'ko, S. L., Canfield, P. C., Morosan, E., Cava, R. J., and Schmiedeshoff, G. M. (2007) J. Phys.: Cond. Mat. 19 176230.

Bud'ko, S. L., Demishev, G. B., Fontes, M. B., and Baggio-Saitovitch, E. M. (1996) J. Phys. Cond. Mat. 8 L159.

Bud'ko, S. L., Lapertot, G., Petrovic, C., Cunningham, C. E., Anderson, N., and Canfield, P. C. (2001) Phys. Rev. Lett. 86 1877.

Bukowski, Z., Weyeneth, S., Puzniak, R., Moll, P., Katrych, S., Zhigadlo, N. D., Karpinski, J., Keller, H., and Batlogg, B. (2009) Phys. Rev. B 79 104521.

Bulaevskii, L. N., Buzdin, A. I., Panjukov, S. V., and Rusinov, A. I. (1981) Sol. State. Commun. 40 683.

Burke, B., Crespi, V. H., Zettl, A., and Cohen, M. L. (1994) Phys. Rev. Lett. 72 3706.

Bussmann-Holder, A., Keller, H., Bishop, A. R., Simon, R., and Müller, K. A. (2005) Europhys. Lett. 72 423.

Butera, A., Fainstein, A., Winkler, E., and Tallon, J. L. (2001) Phys. Rev. B 63 054442.

Calandra, M., and Mauri, F. (2005) Phys. Rev. Lett. 95 237002.

Campbell, A. M., and Evetts, J. E. (1972) Adv. Phys. 21 199.

Campbell, A. M., Evetts, J. E., and Dew-Hughes, D. (1968) Phil. Mag. 18 313.

Canfield, P. C., Bud'ko, S. L., and Cho, B. K. (1996) Physica C 262 249.

Canfield, P. C., Bud'ko, S. L., Finnemore, D. K., and Lapertot, G. (2002) Studies of High Temperature Superconductors, Vol. 38 (Ed. Narlikar, A.). Nova Science, New York, p. 10.

Canfield, P. C., Bud'ko, S. L., Ni, N., Yan, J. Q., and Kracher, A. (2009) Phys. Rev. B 80 060501.

Cao, G., Feng, C., Xu, Y., Lu, W., Shen, J., Fang, M., and Xu, Z. (2003) J. Phys.: Cond. Mat. 15 L519.

Capan, C., Bianchi, A., Ronning, F., Lacerda, A., Thompson, J. D., Hundley, M. F., Pagliuso, P. G., Sarrao, J. L., and Movshovich, R. (2004) Phys. Rev. B 70 180502.

Capone, D. W. II, Hinks, D. G., and Brewe, D. L. (1990) J. Appl. Phys. 67 3043.

Cardosa, C. A., Araujo-Moreira, F. M., Awana, V. P. S., Takayama-Muromachi, E., De Lima, O. F., Yamayuchi, H., and Karppinen, M. (2003) Phys. Rev. B. 67 020407.

Cardwell, D. A. (1998) Mat. Sci. Eng. B 53 1.

Carlson, K. D., Williams, J. N., Geiser, U., Kini, A. M., Wang, H. H., Klemm, R. A., Kumar, S. K., Schlueter, J. A., Ferraro, J. R., Lykke, K. R., Wurz, P., Parker, D. H., and Sutin, J. D. M. (1992) Inorg. Chem. 114 10069.

Carrington, A., and Manzano, F. (2003) Physica C 385 205.

Carrington, A., Bonalde, I. J., Prozorov, R., Gianetta, R. W., Kini, A. M., Schlueter, J., Wang, H. H., Geiser, U., and Williams, J. M. (1999) Phys. Rev. Lett. 83 4172.

Carter, S. A., Batlogg, B., Cava, R. J., Krajewski, J. J., Peck, W. F. Jr., and Takagi, H. (1994) Phys. Rev. B 50 4216.

Carter, S. A., Batlogg, B., Cava, R. J., Krajewski, J. J., and Peck, W. F. Jr. (1995) Phys. Rev. B 51 12644.

Cava, R. J., and Batlogg, B. (1989) MRS Bull. 14 49.

Cava, R. J., van Dover, R. B., Batlogg, B., and Rietman, E. A. (1987) Phys. Rev. Lett. 58 408.

Cava, R. J., Batlogg, B., Krajewski, J. J., Farrow, R., Rupp, L. W. Jr., White, A. E., Short, K., Peck, W. F. Jr., and Kometani, T. (1988) Nature 332 814.

Cava, R. J., Batlogg, B., Krajewski, J. J., Farrow, R., Rupp, L. W. Jr., White, A. E., Short, K., Peck, W. F., and Kometani, T. (1988a) Nature 332 814

Cava, R. J., Batlogg, B., Krajewski, J. J., Rupp, L. W., Schneemeyer, L. F., Siegrist, T., van Dover, R. B., Marsh, P., Peck, W. F. Jr., Gallahgher, P. K., Glarum, S. H., Marshall, J. H., Farrow, R. C., Waszczak, J. V., Hull, R., and Trevor, P. (1988) Nature 336 211.

Cava, R. J., Batlogg, B., Rabe, K. M., Rietman, E. A., Gallagher, P. K., and Peck, W. F. Jr. (1988a) Physica C 156 523.

Cava, R. J., Batlogg, B., Espinosa, G. P., Ramirez, A. P., Krajewski, J. J., Peck, W. F. Jr., Rupp, L. W. Jr., and Cooper, A. S. (1989) Nature 339 L291.

Cava, R. J., Batlogg, B., van Dover, R. B., Krajewski, J. J., and Waszczak, J. V. (1990a) Nature 345 602.

Cava, R. J., Hewat, A. W., Hewat, E. A., Batlogg, B., Marezio, M., Rabe, K. M., Krajewski, J. J., Peck, W. F. Jr., Rupp, L. W. (1990b) Physica C 165 419.

Cava, R. J., Takagi, H., Batlogg, B., Zndbergen, R. W., van Dover, R. B., Peck, W. F. Jr., and Hessen, B. (1992) Physica C 191 237.

Cava, R. J., Takagi, H., Batlogg, B., Zandbergen, H. W., Krajewski, J. J., Peck, W. F. Jr., van Dover, R. B., Felder, R. J.,

Siegrist, T., Mizubashi, K., Lee, J. O., Eisaki, H., Carter, S. A., and Uchida, S. (1994a) Nature 367 146.

Cava, R. J., Takagi, H., Batlogg, B., Zandbergen, H. W., Krajewski, J. J., Peck, W. F. Jr., Van Dover, R. B., Felder, R. J., Siegrist, T., Mizubashi, K., Lee, J. O., Eisaki, H., Carter, S. A., and Uchida, S. (1994b) Nature 367 252.

Cava, R. J., Batlogg, B., Siegrist, T., Krajewski, J. J., Peck, W. F. Jr., Carter, S. A., Felder, R. J., Takagi, H., and van Dover, R. B. (1994c) Phys. Rev. B 50 12384.

Ceresara, S., Ricci, M. V., Sacchetti, N., and Sacerdoti, G. (1975) IEEE Trans. Magn. 11 263.

Cernusko, V., Frohlich, K., Machajdik, D., and Jergel, M. (1981) IEEE. Trans. Magn. 17 2051.

Chakravarty, S., and Kivelson, S. (1999) Europhys. Lett. 16 751.

Chakravarty, S., Gelfand, M. P., and Kivelson, S. (1991) Science 254 970.

Chakravarty, S., Laughlin, R. B., Morr, D. K., and Nayak, C. (2001) Phys. Rev. B 63 094503.

Chandra, P., Tripathi, V., and Coleman, P. (2005) J. Phys.: Cond. Matt. 17 5285.

Chandrasekhar, B. S. (1962) Appl. Phys. Lett. 1 7.

Chang, L. T., Tomy, C. V., McPaul, D., Anderson, N. H., Yethiraj, M. (1996a) J. Phys.: Cond. Matter. 8 2119.

Chang, L. T., Tomy, C. V., McPaul, D., and Ritter, C. (1996b) Phys. Rev. B 52 9031.

Chaudhari, P., Dimos, D., and Mannhart, J. (1990) Earlier and Recent Aspects of Superconductivity (Eds. Bednorz, J. G., and Müller, K. A.). Springer, Heidelberg, p. 201.

Chaussy, J., Haen, P., Lasjaunias, J. C., Monceau, P., Waysand, G., Waintal, A., Meerschaut, A., Molinie, P., and Rouxel, J. (1976) Sol. State. Commun. 20 759.

Cheggour, N., Decroux, M., Fischer, Ø., and Hampshire, D. P. (1998) J. Appl. Phys. 84 2181.

Chen, C.-C., and Lieber, C. M. (1992) J. Am. Chem. Soc. 114 3141.

Chen, C.-C., and Lieber, C. M. (1993) Science 259 655.

Chen, D. P., Chen, H. C., Maljuk, A., Kulakov, A., Zhang, H., Lemmens, P., and Lin, C. T. (2004) Phys. Rev. B 70 024506.

Chen, G. F., Matsubayashi, K., Ban, S., Deguchi, K., and Sato, N. K. (2006) Phys. Rev. Lett. 97 017005.

Chen, G. F., Li, Z., Wu, D., Li, G., Hu, W. Z., Dong, J., Zheng, P., Luo, J. L., and Wang, N. L. (2008a) Phys. Rev. Lett. 100 247002.

Chen, G. F., Li, Z., Dong, J., Li, G., Hu, W. Z., Zhang, X. D., Song, X. H., Zheng, P., Wang, N. L., and Luo, J. L. (2008b) Phys. Rev. B 78 224512.

Chen, G. F., Xia, T.-L., Yang, H. X., Li, J. Q., Zheng, P., Luo, J. L., and Wang, N. L. (2009) Supercond. Sci. Technol. 22 072001.

Chen, H., Ren, Y., Qiu, Y., Bao, W., Liu, R. H., Wu, G., Wu, T., Xie, Y. L., Wang, X. F., Huang, Q., and Chen, X. H. (2009) Europhys. Lett. 85 17006.

Chen, M., Soylu, B., Watson, D., Christiansen, J. K. S., Yan, Y., Glowacki, B. A., and Evetts, J. E. (1995) IEEE. Trans. Appl. Supercond. 5 1467.

Chen, S. K., Glowacki, B. A., MacManus-Driscoll, J. L., Vickers, M. E., and Majoros, M. (2004) Supercond. Sci. Technol. 17 243.

Chen, X. H., Sun, Z., Wang, K. Q., Li, S. Y., Xiong, Y. M., Yu, M., and Cao, L. Z. (2001a) Phys. Rev. B 63 64506.

Chen, X. H., Xue, Y. Y., Meng, R. L., and Chu, C. W. (2001b) Phys. Rev. B 64 172501.

Chen, X., Koiwasaki, T., and Yamanaka, S. (2001c) J. Sol. State Chem. 159 80.

Chen, X., Zhu, L., and Yamanaka, S. (2002) J. Sol. State Chem. 169 149.

Chen, Y., Lynn, J. W., Li, J., Li, G., Chen, G. F., Luo, J. L., Wang, N. L., Dai, P. C., de la Cruz, C., and Mook, H. A. (2008a) Phys. Rev. B 78 064515.

Chen, Z. Y., Sheng, Z. Z., Tang, Y. Q., Li, Y. F., Wang, L. M., and Pederson, D. O. (1993) Supercond. Sci. Technol. 6 261.

Cheng, C. H., Zhang, H., Zhao, Y., Feng, Y., Rui, X. F., Munroe, P., Zheng, H. M., Koshizuka, N., and Murakami, M. (2003) Supercond. Sci. Technol. 16 1182.

Cheng, C. H., Zhao, Y., Feng, Y., Zhang, H., Nishiyama, M., Koshizuka, N., and Murakami, M. (2005) Frontiers in Superconducting Materials (Ed. Narlikar, A. V.) Springer, Heidelberg, p. 619.

Cheng, P., Fang, L., Yang, H., Zhu, X., Mu, G., Luo, H., Wang, Z., and Wen, H.-H. (2008) Sci. Chin. G 51 719.

Chesca, B., Schulz, R. R., Goetz, B., Schneider, C. W., Hilgenkamp, H., and Mannhart, J. (2002) Phys. Rev. Lett. 88 177003.

Chester, P. F. (1967) Rep. Prog. Phys. 30 561.

Chevrel, R. (1981) Superconductor Materials Science (Eds. Foner, S., and Schwartz, B. B.). Plenum, New York, p. 685.

Chevrel, R., Sergent, M., Prigent, J. (1971) J. Sol. State Chem. 3 515.

Cho, A. (2006) Science 314 1072.

Cho, B. K., Xu, M., Canfield, P. C., and Johnston, D. C. (1996) Phys. Rev. Lett. 77 163.

Choi, H. J., Roundy, D., Sun, H., Cohen, M. L., and Louie, S. G. (2002a) Phys. Rev. B 66 020513(R).

Choi, H. J., Roundy, D., Sun, H., Cohen, M. L., and Louie, S. G. (2002b) Nature 418 758.

Christianson, A. D., Goremychkin, E. A., Osborn, R., Rosenkranz, S., Lumsden, M. D., Malliakas, C. D., Todorov, I. S., Claus, H., Chung, D. Y., Kanatzidis, M. G., Bewley, R. I., and Guidi, T. (2008) Nature 456 930.

Christianson, A. D., Lumsden, M. D., Nagler, S. E., MacDougall, G. J., McGuire, M. A., Sefat, A. S., Jin, R., Sales, B. C., and Mandrus, D. (2009) Phys. Rev. Lett. 103 087002.

Chu, C. W. (1974) Phys. Rev. Lett. 33 1283.

Chu, C. W. (1997) IEEE. Trans. Appl. Supercond. 7 80.

Chu, C. W., and Testardi, L. R. (1974) Phys. Rev. Lett. 32 766.

Chu, C. W., Hor, P. H., Meng, R. L., Gao, L., Huang, Z. J., and Wang, Y. Q. (1987a) Phys. Rev. Lett. 58 405.

Chu, C. W., Hor, P. H., Meng, R. L., Gao, L., and Huang, Z. J. (1987b) Science 235 567.

Chu, C. W., Gao, L., Chen, F., Huang, Z. J., Meng, R. L., and Xue, Y. Y. (1993) Nature 365 323.

Chu, C. W., Xue, Y. Y., Tsui, S., Cmaidalka, J., Heilman, A. K., Lorenz, B., Meng, R. L. (2000) Physica C 335 231.

Chu, C. W., Lorenz, B., Meng, R. L., and Xue, Y. Y. (2005) Frontiers in Superconducting Materials (Ed. Narlikar, A. V.). Springer, Berlin, p. 331.

Chu, C. W., Chen, F., Gooch, M., Guloy, A. M., Lorenz, B., Lv, B., Sasmal, K., Tang, Z. J., Tapp, J. H., and Xue, Y. Y. (2009) Physica C 469 326.

Cichorek, T., Mota, A. C., Steglich, F., Frederick, N. A., Yuhasz, W. M., and Maple, M. B. (2005) Phys. Rev. Lett. 94 107002.

Civale, L. (1997) Supercond. Sci. Technol. 10 A11.

Clarke, J. (1982) Advances in Superconductivity (Eds. Deaver, B., and Ruvalds, J.). Springer, New York, p. 13.

Clogston, A. M. (1962) Phys. Rev. Lett. 9 266.

Coffey, H. T. (1967) Cryogenics 7 73.

Cohen, M. L. (1964a) Phys. Rev. 134 A4511.

Cohen, M. L. (1964b) Rev. Mod. Phys. 36 240.

Coleman, P., Pepin, C., Si, Q., and Ramarzashvili, R. (2001) J. Phys.: Cond. Matter 13 R723.

Collver, M. M., and Hammond, R. H. (1973) Phys. Rev. Lett. 30 92.

Collver, M. M., and Hammond, R. H. (1977) Sol. State Commun. 22 55.

Cooper, A. S., Corenzwit, E., Longinotti, L. D., Matthias, B. T., and Zachariasen, W. H. (1970) Proc. Natl Acad. Sci. USA 67 313.

Cooper, L. N. (1956) Phys. Rev. 104 1189.

Corak, W. S., Goodman, B. B., Satterthwaite, C. B., and Wexler, A. (1956) Phys. Rev. 102 656.

Correa, V. F., Murphy, T. P., Martin, C., Purcell, K. M., Palm, E. C., Schmiedeshoff, G. M., Cooley, J. C., and Tozer, S. W. (2007) Phys. Rev. Lett. 98 087001.

Coulon, C., Delhaes, P., Amiell, J., Monceau, J. P., Fabre, J. M., and Giral, L. (1982) J. Phys. (Paris) 43 1721.

Cox, D. E., and Sleight, A. W. (1976) Sol. State. Commun. 19 969.

Cribier, D., Jacrot, B., Farnoux, B., and Madhav Rao, L. (1966) J. Appl. Phys. 37 952.

Critchlow, P. R., Gregory, E., and Zeitlin, B. (1971) Cryogenics 11 3.

Critchlow, P. R., Gregory, E., and Marancik, W. (1974) J. Appl. Phys. 45 5027.

Csányi, G., Littlewood, P. B., Nevidomskyy, A. H., Pickard, C. J., and Simons, B. D. (2005) Nature Phys. 1 42.

Cubitt, R., Eskildsen, M. R., Dewhurst, C. D., Jun, J., Kazakov, S. M., and Karpinski, J. (2003) Phys. Rev. Lett. 91 047002.

Cui, Y., Chen, Y., Wang, F., Li, J., Zhang, Y., and Zhao, Y. (2009) Rare Metal Mat. Eng. 38 0583.

Culetto, F. J., and Pobell, F. (1978) Phys. Rev. Lett. 40 1104.

Cullen, G. W. (1968) Proc. Summer Study on Superconducting Devices and Accelerators, Brookhaven National Laboratory, New York, p. 437.

Curro, N. J., Caldwell, T., Bauer, E. D., Morales, L. A., Graf, M. J., Bang, Y., Balatsky, A. V., Thompson, J. L., and Sarrao, J. L. (2005) Nature 434 622.

da Silva, R. R., Torres, J. H. S., and Kopelevich, Y. (2001) Phys. Rev. Lett. 87 147001.

Dahm, T. (2005) Frontiers in Superconducting Materials (Ed. Narlikar, A. V.). Springer, Heidelberg, p. 983.

Dahm, T., and Schopohl, N. (2003) Phys. Rev. Lett. 91 017001.

Dalrymple, B. J., and Prober, D. E. (1984) J. Low Temp. Phys. 56 545.

Dalrymple, B. J., Moroczkowski, S., and Prober, D. E. (1986) J. Crystal Growth 74 575.

Das, A., and Kremer, R. K. (2003) Phys. Rev. B 68 064503.

Dayan, M. (1981) J. Phys. F 11 L227.

de Gennes, P. G. (1966) Superconductivity of Metals and Alloys. Benjamin, New York, p. 227.

de la Cruz, C., Huang, Q., Lynn, J. W., Li, J. Y., Ratcliff, W., Zarestky, J. L., Mook, H. A., Chen, G. F., Luo, J. L., Wang, N. L., and Dai, P. C. (2008) Nature 453 899.

de Nijs, D. E., Huy, N. T., and de Visser, A. (2008) Phys. Rev. B 77 140506.

de Sorbo, W., and Healy, W. A. (1964) Cryogenics 4 257.

de Soto, S. M., Slichter, C. P., Kini, A. M., Wang, H. H., Geiser, U., Williams, J. M. (1995) Phys. Rev. B 52 10364.

de Visser, A., Puech, L., Joss, W., Menovsky, A., and Franse, J. J. M. (1987a) Jpn. J. Appl. Phys. 26 513.

de Visser, A., Menovsky, A. A., and Franse, J. J. M. (1987b) Physica B 147 81.

Deaver, B. S. Jr., and Fairbank, W. M. (1961) Phys. Rev. Lett. 7 43.

Debessai, M., Hamlin, J. J., Schilling, J. S., Rosenmann, D., Hinks, D. G., and Claus, H. (2010) Phys. Rev. B 82 132502.

Debessai, M., Matsuoka, T., Hamlin, J. J., Schilling, J. S., and Shimizu, K. (2009) Phys. Rev. Lett. 102 197002.

deBruyn, R. (1987) IEEE Trans. Magn. 23, 355.

Decroux, M., Cattani, D., Cors, J., Ritter, S., and Fischer, Ø. (1990) Physica B: Cond. Mat. 165–166 1395.

Decroux, M., Selvam, P., Cors, J., Seeber, B., Fischer, Ø., Chevrel, R., Rabiller, P., and Sergent M. (1993) IEEE. Trans. Appl. Supercond. 3 1502.

Deemyad, S., Schilling, J. S., Jorgensen, J. D., and Hinks, D. G. (2001) Physica C 361 227.

Deguchi, K., Mao, Z. Q., Yaguchi, H., and Maeno, Y. (2004) Phys. Rev. Lett. 92 047002.

Delaplace, R., Molinié, P., and Jérome, D. (1976) J. Phys. Lett. (Paris) 37 L13.

deLuca, J. A., Karas, P. L., Tkaczyk, J. E., Bednarczyk, P. J., Garbauskas, M. F., Briant, C. L., and Sorensen, D. B. (1993) Physica C 205 21.

deMarco, M., Coffey, D., Tallon, J., Haka, M., Torongian, S., and Fridmann, J. (2002) Phys. Rev. B 65 212506.

Demura, S., Mizuguchi, Y., Deguchi, K., Okazaki, H., Hara, H., Watanabe, T., Denholme, S. J., Fujioka, M., Ozaki, T., Fujihisa, H., Gotoh, Y., Miura, O., Yamaguchi, T., Takeya, H., and Takano, Y. (2013) J. Phys. Soc. Jpn 82 033708.

Dew-Hughes, D. (1971) Rep. Prog. Phys. 34 821.

Dew-Hughes, D. (1974) Phil. Mag. 30 293.

Dew-Hughes, D. (1975) Cryogenics 15 435.

Dew-Hughes, D. (1980) J. Phys. Chem. Sol. 41 851.

Dew-Hughes, D. (1987) Phil. Mag. B 55 459.

Dew-Hughes, D. (1997) Aust. J. Phys. 50 36.

Dew-Hughes, D. (1999) Proc. 12th Int. Symp. Superconductivity (ISS-99), Morioka, p. 15.

Dew-Hughes, D., and Luhman, T. S. (1978) J. Mat. Sci. 13 1868.

Dew-Hughes, D., and Suenaga, M. (1978) J. Appl. Phys. 49 357.

Dew-Hughes, D., and Witcomb, M. J. (1972) Phil. Mag. 26 73.

Dew-Hughes, D., Luhman, T. S., and Suenaga, M. (1976) Nucl. Technol. 29 268.

Dhar, S. K., Nagarajan, R., Hossain, Z., Tominez, E., Godart, C., Gupta, L. C., and Vijayaraghavan, R. (1996) Sol. State. Commun. 98 985.

Dhar, S. K., Chinchure, A. D., Alleno, E., Godart, C., Gupta, L. C., and Nagarajan, R. (2002) Pramana J. Phys. 58 885.

Diederichs, J., Schilling, J. S., Herwig, K. W., and Yelon, W. B. (1997) J. Phys. Chem. Sol. 58 123.

Dietrich, M., Wuhl, H., Fitzer, E., Brennfleck, K., Kehr, D. (1981) US Patent 4299861.

Dimos, D., Chaudhari, P., Mannhart, J., and LeGoues, F. K. (1988) Phys. Rev. Lett. 61 219.

Ding, H., Yokoya, T., Campuzano, J. C., Takahashi, T., Randeria, M., Norman, M. R., Mochiku, T., Kadowaki, K., and Giapintzakis, J. (1996) Nature 382 51.

Ding, H., Richard, P., Nakayama, K., Sugawara, K., Arakane, T., Sekiba, Y., Takayama, A., Souma, S., Sato, T., Takahashi, T., Wang, Z., Dai, X., Fang, Z., Chen, G. F., Luo, J. L., and Wang, N. L. (2008) Europhys. Lett. 83 47001.

Ding, L., Dong, J. K., Zhou, S. Y., Guan, T. Y., Qiu, X., Zhang, C., Li, L. J., Lin, X., Cao, G. H., Xu, Z. A., and Li, S. Y. (2009) New J. Phys. 11 093018.

Di Salvo, F. J., Moncton, D. E., Wilson, J. A., and Mahajan, S. (1976) J. Phys. C: Sol. State Phys. 14 1543.

Doll, R., and Nähbauer, M. (1961) Phys. Rev. Lett. 7 51.

Domb, E., and Johnson, W. L. (1978) J. Low Temp. Phys. 33 29.

Dong, A. F., Che, G. C., Huang, W. W., Jia, S. L., Chen, H., and Zhao, Z. X. (2005) Physica C 442 65.

Dong, J., Zhang, H. J., Xu, G., Li, Z., Li, G., Hu, W. Z., Wu, D., Chen, G. F., Dai, X., Luo, L. J., Fang, Z., and Wang, N. L. (2008) Europhys. Lett. 83 27006.

Dong, J. K., Zhou, S. Y., Guan, T. Y., Qiu, X., Zhang, C., Cheng, P., Fang, L., Wen, H.-H., and Li, S. Y. (2010a) Phys. Rev. B 81 094520.

Dong, J. K., Zhou, S. Y., Guan, T. Y., Zhang, H., Dai, Y. F., Qiu, X., Wang, X. F., He, Y., Chen, X. H., and Li, S. Y. (2010b) Phys. Rev. Lett. 104 087005.

Doniach, S. (1977) Physica 91B 231.

dos Sintos, D. I., Carvalho, C. L., da Silvo, R. R., Aegerter, M. A., Balachandran, U., and Poppel, R. B. (1991) Ceram. Trans. 18 263.

Dou, S. X., Soltanian, S., Horvat, J., Wang, X. L., Munroe, P., Zhou, S. H., Ionescu, M., Liu, H. K., and Tomsic, M. (2002) Appl. Phys. Lett. 81 3419.

Dou, S. X., Yeoh, W. K., Horvat, J., and Ionescu, M. (2003) Appl. Phys. Lett. 83 4996.

Dou, S. X., Pan, A. V., Qin, M. J., and Silver, T. (2005) Frontiers in Superconducting Materials (Ed. Narlikar, A. V.). Springer, Heidelberg, p. 1011.

Dressel, M. (2000) Studies of High Temperature Superconductors, Vol. 34 (Ed. Narlikar, A. V.). Nova Science, New York, p. 1.

Dressel, M., Bruder, S., Grüner, G., Carlson, K. D., Wang, H. H., and Williams, J. M. (1993) Phys. Rev. B 48 9906.

Dressel, M., Klein, O., Grüner, G., Carlson, K. D., Wang, H. H., and Williams, J. M. (1994) Phys. Rev. B 50 13603.

Drew, A. J., Pratt, F. L., Lancaster, T., Blundell, S. J., Baker, P. J., Liu, R. H., Wu, G., Chen, X. H., Watanabe, I., Malik, V. K., Dubroka, A., Kim, K. W., Rössle, M., and Bernhard, C. (2008) Phys. Rev. Lett. 101 097010.

Drew, A. J., Niedermayer, C., Baker, P. J., Pratt, F. L., Blundell, S. J., Lancaster, T., Liu, R. H., Wu, G., Chen, X. H., Watanabe, I., Malik, V. K., Dubroka, A., Rossle, M., Kim, K. W., Baines, C., and Bernhard, C. (2009) Nature Mat. 8 310.

Dubois, C., Petrović, A., Santi, G., Berthod, C., Manuel, A. A., Decroux, M., Fischer, O., Potel, M., and Chevrel, R. (2007) Phys. Rev. B. 75 104501.

Duffy, J. A., Hayden, S. M., Maeno, Y., Mao, Z., Kulda, J., and McIntyre, G. J. (2000) Phys. Rev. Lett. 85 5412.

Dugdale, S. B., and Jarlborg (2001) Phys. Rev. B 64 106508 (R).

Dunsiger, S. R., Zhao, Y., Yamani, Z., Buyers, W. J. L., Dabkowska, H., and Gaulin, B. D. (2008) Phys. Rev. B 77 224410.

Durrell, J. H., Eom, C.-B., Gurevich, A., Hellstrom, E. E., Tarantini, C., Yamamoto, A., and Larbalestier, D. C. (2011) Rep. Prog. Phys. 74 124511.

Dzyaloshinsky, J. (1958) J. Phys. Chem. Sol. 4 241.

Ebbesen, T. W., Tsai, J. S., Tanigaki, K., Tabuchi, J., Shimakawa, Y., Kubo, Y., Hirosawa, I., and Mizuki, J. (1992) Nature 355 620.

Eickemeyer, J., Selbman, D., Opitz, R., deBoer, B., Holzapfel, B., Schultz, L., and Miller, U. (2001) Supercond. Sci. Technol 14 152.

Eisaki, H., Takagi, H., Cava, R. J., Mizuhashi, K., Lee, J. O., Batlogg, B., Krajewski, J. J., Peck, W. F. Jr., and Uchida, S. (1994) Phys. Rev. B 50 647.

Eisterer, M., Raunicher, R., Weber, H. W., Bellingeri, E., Cimberle, M. R., Pallecchi, I., Putti, M., and Ferdeghini (2011) Supercond. Sci. Technol. 24 065016.

Ekimov, E. A., Sidorov, V. A., Bauer, E. D., Mel'nik, N. N., Curro, N. J., Thompson, J. D., and Stishov, S. M. (2004) Nature 428 542.

Ekin, J. (1984) Adv. Cryo. Eng. 30 823.

Ekino, T., Gabovich, A. M., and Voitenko, A. I. (2005) Low Temp. Phys. 31, 41.

Ekino, T., Takasaki, T., Muranaka, T., Fujii, H., Akimitsu, J., and Yamanaka, S. (2003) Physica B 328 23.

Eliashberg, G. M. (1960) Zh. Eksp. Teor. Fiz. 38 966.

Eliashberg, G. M. (1962) Zh. Eksp. Teor. Fiz. 43, 1005.

Elsinger, H., Wosnitza, J., Wanka, S., Hagel, J., Schweitzer, D., Strunz, W. (2000) Phys. Rev. Lett. 84 6098.

Emery, N., Hérold, C., d'Astuto, M., Barcia, V., Bellin, C., Marêché, J.-F., Lagrange, P., and Loupias, G. (2005) Phys. Rev. Lett. 95 087003.

Emery, N., Hérold, C., Marêché, J.-F., and Lagrange, P. (2008) Sci. Tech. Adv. Mat. 9 044102.

Emery, V. J., and Kivelson, S. A. (1995) Nature 374 434.

Emery, V. J., and Kivelson, S. A. (1998) J. Phys. Chem. Sol. 59 1705.

Enstrom, R. E., and Appert, J. R. (1974) J. Appl. Phys. 45 421.

Enstrom, R. E., Hanak, J. J., and Cullen G. W. (1970) RCA Rev. 31 702.

Enstrom, R. E., Hanak, J. J., Appert, J. R., and Strater, K. J. (1972) J. Electrochem. Soc. 119 743.

Eom, C. B., Lee, M. K., Choi, J. H., Belenky, L., Song, X., Cooley, L. D., Naus, M. T., Patnaik, S., Jiang, J., Rikel, M., Polyanskii, A., Gurevich, A., Cai, X. Y., Babecock, S. E., Hellstrom, E. E., Labalestier, D. C., Rogado, N., Regan, K. A., Hayward, M. A., He, T., Slusky, J. S., Inumaru, K., Haas, M. K., and Cava, R. J. (2001) Nature 411 558.

Escamilla, R., Doran, A., and Escudero, R. (2004) Physica C 403 177.

Escote, M. T., Meza, V. A., Jardim, R. F., Ben-Dor, L., Torikachvili, M. S., and Lacerda, A. H. (2002) Phys. Rev. B 66 144503.

Escudero, R., Briggs, A., and Monceau, P. (2001) J. Phys.: Cond. Mat. 13 6285.

Eskildsen, M. R., Gammel, P. L., Barber, B. P., Ramirez, A. P., Bishop, D. J., Andersen, N. H., Mortensen, K., Bolle, C. A., Lieber, C. M., and Canfield, C. P. (1997) Phys. Rev. Lett. 79 487.

Eskildsen, M. R., Gammel, P. L., Barber, B. P., Yaron, U., Ramirez, A. P., Huse, D. A., Bishop, D. J., Bolle, C. A., Lieber, C. M., Oxx, S., Sridhar, S., Andersen, N. H., Mortensen, K., and Canfield, C. P. (1997) Phys. Rev. Lett. 78 1968.

Essmann, U., and Trauble, H. (1967), Phys. Lett. 24A 526.

Everman, K., Handstein, A., Guchs, G., Cao, L., and Muller, K.-H. (1996) Physica C 266 17.

Evetts, J. E., Campbell, A. M., and Dew-Hughes, D. (1964) Phil. Mag. 10 339.

Fainstein, A., Winkler, E., Butera, A., and Tallon, J. L. (1999) Phys. Rev. B 60 R12597.

Falter, M., Hässler, H., Schobach, B., and Holzapfel, B. (2002) Physica C 372–376 46.

Fang, L., Wang, Y., Zou, P. Y., Tang, L., Xu, Z., Chen, H., Dong, C., Shan, L., and Wen, H. H. (2005) Phys. Rev. B 72 014534.

Fang, L., Luo, H., Cheng, P., Wang, Z., Jia, Y., Mu, G., Shen, B., Mazin, I. I., Shan, L., Ren, C., and Wen, H. H. (2009) Phys. Rev. B 80 140508(R).

Fang, M. H., Pham, H. M., Qian, B., Liv, T. J., Vehstedt, E. K., Liu, Y., Spinu, L., and Mao, Z. Q. (2008) Phys. Rev. B 78 224503.

Fang, M. H., Yang, J. H., Balakirev, F. F., Kohama, Y., Singleton, J., Qian, B., Mao, Z. Q., Wang, H., and Yuan, H. Q. (2010) Phys. Rev. B 81 20509.

Felner, I. (1998) Studies of High Temperature Superconductors, Vol. 26 (Ed. Narlikar, A.). Nova Science, New York, p. 27.

Felner, I. (2001) Physica C 353 11.

Felner, I. (2002) Studies of High Temperature Superconductors, Vol. 38 (Ed. Narlikar, A.). Nova Science, New York, p. 351.

Felner, I. (2003) Studies of High Temperature Superconductors, Vol. 46 (Ed. Narlikar, A. V.). Nova Science, New York, p. 41.

Felner, I. (2006) Studies of High Temperature Superconductors, Vol. 50 (Ed. Narlikar, A. V.). Nova Science, New York, p. 169.

Felner, I., and Asaf, U. (1997) Physica C 292 97.

Felner, I., Asaf, U., Levi, Y., and Millo, O. (1997) Phys. Rev. B 55 R3374.

Felner, I., Asaf, U., Goren, S., and Korn, C. (1998) Phys. Rev. B 57 550.

Felner, I., Asaf, U., Levi, Y., and Millo, O. (2000a) Physica C 334 141.

Felner, I., Sonin, E. B., Machi, T., and Koshizuka, N. (2000b) Physica C 341–345 715.

Felner, I., Galstyan, E., Lorenz, B., Cao, D., Wang, Y. S., Xue, Y. Y., and Chu, C. W. (2003) Phys. Rev. B 67 134506.

Felner, I., Galstyan, E., Awana, V. P. S., and Takayama-Moromachi, E. (2004) Physica C 408–410 161.

Felner, I., Galstyan, E., Herber, R., and Nowik, I. (2004a) Phys. Rev. B 70 094504.

Felner, I., Galstyan, E., and Nowik, I. (2005) Phys. Rev. B. 71 064510.

Felner, I., Nowik, I., Tsindlickt, M. I., Yuli, O., Asulin, I., Millo, O., Awana, V. P. S., Kishan, H., Balamurugan, S. and Takayama-Muromachi, E. (2007) Phys. Rev. B 76.

Feng, M. B., Fang, M. H., Xu, Z., and Jiao, Z. K. (2001) Chin. Phys. Lett. 18 963.

Fernandes, R. M., and Schmalian, J. (2010) Phys. Rev. B 82 014521.

Ferrel, R. A. (1964) Phys. Rev. Lett. 13 330.

Fertig, W. A., Johnston, D. C., de Long, L. E., McCallum, R. W., Maple, M. B., and Matthias, B. T. (1977) Phys. Rev. Lett. 38 987.

Fischer, C. M. (2002) MS thesis, University of Wisconsin, Madison.

Fischer, Ø. (1972) Helv. Phys. Acta 45 229.

Fischer, Ø., Odermatt, R., Bongi, G., Jones, H., Chevrel, R., and Sargent, M. (1973) Phys. Lett. A 45 87.

Fischer, Ø., Jones, H., Bongi, G., Sargent, M., and Chevrel, R. (1974) J. Phys. C 7 L450.

Fischer, Ø., Treyvaud, A., Chevrel, R., and Sergent, M. (1975) Sol. State Commun., 17 721.

Fisher, B., Chashka, K. B., Patlagan, L., and Reisner, G. M. (2003) Physica C 387 10.

Fisher, R. A., Kim, S., Woodfield, F., Phillips, N. E., Taillefer, L., Hasselbach, K., Flouquet, J., Giorgi, A. L., and Smith, J. L. (1989) Phys. Rev. Lett. 62 1411.

Fisher, R. A., Bouquet, F., Phillips, N. E., Hundley, M. F., Pagliuso, P. G., Sarrao, J. L. Fisk, Z., and Thompson, J. D. (2002a) Phys. Rev. B 65 224509.

Fisher, R. A., Bouquet, F., Phillips, N. E., Hinks, D. G., and Jorgensen, J. D. (2002b) Studies of High Temperature Superconductors, Vol. 38 (Ed. Narlikar, A.). Nova Science, New York, p. 207.

Fite, W., and Redfield, A. G. (1966) Phys. Rev. Lett. 17 381.

Fleming, R. M., Moncton, D. E., and McWhan, D. B. (1978) Phys. Rev. B 18 5560.

Fleming, R. M., Ramirez, A. P., Rosseinsky, M. J., Murphy, D. W., Haddon, R. C., Zahurak, S. M., and Makhija, A. V. (1991) Nature 352 787.

Flükiger, R. (1981) Superconductor Materials Science (Eds. Foner, S., and Schwartz, B. B.). Plenum, New York, p. 511.

Flükiger, R., Devantay, H., Jorda, J. L., and Muller, J. (1977) IEEE. Trans. Mag. 13 818.

Flükiger, R., Baillif, R., and Walker, E. (1978) Mat. Res. Bull. 13 743.

Flükiger, R., Akihama, R., Foner, S., McNiff, E. J. Jr., and Schwartz, B. B. (1979) Appl. Phys. Lett. 35 810.

Flükiger, R., Isernhagen, R., Goldacker, W., and Specking, W. (1984) Adv. Cryo. Eng. (Mat.) 30 851.

Flükiger, R., Lezza, P., Beneduce, C., Musolino, N., and Suo, H. L. (2003) Supercond. Sci. Technol. 16 264.

Foner, S., and McNiff, E. J. Jr (1973) Phys. Lett. A 45 429.

Foner, S., McNiff, E. J. Jr., and Alexander, E. J. (1974) Phys. Lett. 49A 269.

Foo, M. L., Schaak, R. E., Miller, V. L., Klimczuk, T., Rogado, N. S., Wang, Y., Lau, G. C., Craley, C., Zandbergen, H. W., Ong, N. P., and Cava, R. J. (2003) Sol. State. Commun. 127 33.

Foo, M. L., Klimczuk, T., Li, L., Ong, N. P., and Cava, R. J. (2005) Sol. State. Commun. 133 407.

Fossheim, K., Tuset, E. D., Ebbesen, T. W., Treacy, M. M. J., and Schwartz, J. (1995) Physica C 248 195.

Fouassier, C., Matejka, G., Reau, J.-M., and Haggenmuller, P. (1973) J. Sol. State Chem. 6 532.

Franse, J. J. M., de Visser, A., Menovsky, A., and Frings, P. H. (1985) 52 61.

Freericks, J. K., and Falicov, L. M. (1992) Phys. Rev. B 41(II) 874.

Friedberg, R., Lee, T. D., and Ren, H. C. (1992) Phys. Rev. B 46 14150.

Friedel, J., de Gennes, P. G., and Matricon, J. (1963) Appl. Phys. Lett. 2 119.

Friedemann, S., Westerkamp, T., Brando, M., Oeschler, N., Wirth, S., Gegenwart, P., Krellner, C., Geibel, C., and Steglich, F. (2009) Nature Phys. 5 465.

Friemel, S., Pasquier, C., and Jérome, D. (1997) Physica C 292 273.

Frings, P. H., Franse, J. J. M., de Boer, F. R., and Menovsky, A. (1983) J. Magn. Magn. Mat. 31–34 240.

Fröhlich, H. (1950) Phys. Rev. 79 845.

Fröhlich, H. (1954) Proc. R. Soc. Lond. A 223 296.

Fuchs, G., Krabbes, G., Schätzle, P., Gruss, S., Stoye, P., Staiger, T., Muller, K.-H., Fink, J., and Schultz, S. (1997) Appl. Phys. Lett. 70 117.

Fuchs, G., Schätzle, P., Krabbes, G., Gruss, S., Verges, P., Müller, K.-H., Fink, J., and Schultz, L. (2000) Appl. Phys. Lett. 76 2107.

Fujimoto, T., Zheng, Z.-Q., Kitaoka, Y., Meng, R. L., Cmaidalka, J., and Chu, C. W. (2004) Phys. Rev. Lett. 92 047004.

Fujita, M., Goka, H., Yamada, K., Tranquada, J. M., and Regnault, L. P. (2004) Phys. Rev. B 70 104517.

Fujiwara, K., Kitaoka, K., Asayama, K., Shimakawa, Y., Manako, T., Kubo, Y., and Endo, A. (1990) Physica B 165–166 1295.

Fukazawa, H., and Yamada, K. (2003) J. Phys. Soc. Jpn 72 2449.

Fukazawa, H., Yamazaki, T., Kondo, K., Kohori, Y., Takeshita, N., Shirage, P. M., Kihou, K., Miyazawa, K., Kito, H., Eisaki, H., and Iyo, H. (2009) J. Phys. Soc. Jpn 78 033704.

Fukuzumi, Y., Mizuhashi, K., Takenaka, K., and Uchida, S. (1996) Phys. Rev. Lett. 76 884.

Fulde, P., and Ferrell, R. (1964) Phys. Rev. 135 A550.

Fulde, P., and Ferrell, R. (1964) Phys. Rev. 135 A550.

Fumagalli, P., Stixrude, L., Poli, S., and Synder, D. (2001) Earth Planet. Sci. Lett. 186 125.

Funahashi, R., Matsubara, I., Ueno, K., and Ishikawa, H. (1999) Physica C 311 107.

Furrer, A. (2005) Superconductivity in Complex Systems (Eds. Müller, K. A., and Bussmann-Holder, A.). Springer, Berlin, p. 171.

Gabovich, A., Voitenko, A. I., and Ekino, T. (2004) J. Phys.: Cond. Mat. 16 3681.

Gamble, G., DiSalvo, F. J., Klemm, R. A., and Geballe, T. H. (1970) Science 168 568.

Ganguli, A. K., and Subramanian, M. A. (1991) J. Sol. State Chem. 93 250.

Gantmakher, V. F., Klinkova, L. A., Barkowskii, N. V., Tsadynzhapov, G. E., Wieger, S., and Geim, A. K. (1996) Phys. Rev. B 54 6133.

Gao, J. (2002) Studies of High Temperature Superconductors, Vol. 41 (Ed. Narlikar, A.). Nova Science, New York, p. 75.

Gao, J., Klopman, B. B. G., Aarnink, W. A. M., Reitsma, A. E., Gerritsma, G. J., and Rogalla, H. (1992) J. Appl. Phys. 71 2333.

Gao, L., Qiu, X. D., Cao, Y., Meng, R. L., Sun, Y. Y., Xue, Y. Y., and Chu, C. W. (1994b) Phys. Rev. B 50 9445.

Gao, L., Xue, Y. Y., Chen, F., Xiong, Q., Meng, R. L., Ramirez, Chu, C. W., Eggert, J. H., and Mao, H.-K. (1994a) Phys. Rev. B 50 4260.

Garoche, P., Brusetti, R., Jérome, D., and Bechgaard, K. (1982) J. Phys. Lett. 43 L147.

Garoche, P., Brusetti, R., and Bechgaard, K. (1983) J. Phys. (Paris) 44 1047.

Gasparov, V. A., Sidorov, N. S., Zver'Kova, I. I., and Kulakov, M. P. (2001) JETP Lett. 73 532.

Gavalar, J. R., Janocko, M. A., and Jones, C. K. (1974) J. Appl. Phys. 45 3009.

Gavilano, J. L., Rau, D., Pedrini, B., Hinderer, J., Ott, H. R., Kazakov, S., and Karpinski, J. (2004) Phys. Rev. B 69 100404.

Gegenwart, P., Langhammer, C., Kim, J. S., Stewart, G. R., and Steglich, F. (2004) Physica C 408–410 157.

Gegenwart, P., Si, Q., and Steglich, F. (2008) Nature Phys. 4 186.

Geibel, C., Shank, C., Theis, S., Kitazawa, H., Bredl, C. D., Bohm, A., Rau, M., Grauel, A., Caspary, R., Helfrich, R., Ahlheim, U., Weber, G., and Steglich, F. (1991) Z. Phys. B 84 1.

Genenko, Yu. A., Snezhko, A., and Fryhardt, H. C. (2000) Phys. Rev. B 62 3453.

Genoud, J.-Y., Triscone, G., Junod, A., and Muller, J. (1994) Physica C 235 437.

Ghiron, K., Salamon, M. B., Hubbard, M. A., and Veal, B. W. (1993) Phys. Rev. B 48 16188.

Giaever, I. (1960) Phys. Rev. Lett. 5 147.

Giannini, E., Clayton, N., Musolino, N., Gladyshevskii, R., and Flükiger, R. (2005) Frontiers in Superconducting Materials (Ed. Narlikar, A. V.). Springer, Heidelberg, p. 739.

Ginzburg, V. L. (1989) Phys. Scripta 27 76.

Ginzburg, V. L. (1956) Zh. Eksp. Teor. Fiz. 31 202.

Ginzburg, V. L., and Landau, L. D. (1950) Sov. Phys. JETP 20 1064.

Ginzburg, V. L., and Zharkov, G. F. (1978) Sov. Phys. Usp. 21 3811.

Giubileo, F., Bobba, F., Gombos, S., Uthayakumar, S., Veccione, A., Akimenko, A. I., and Cucolo, A. M. (2003) Int. J. Mod. Phys. B 17 3525.

Giunchi, G., Ripamonti, G., Cavallin, T., and Bassani, E. (2006) Cryogenics 46 237.

Glowacki, B. A. (2000) Supercond. Sci. Technol. 13 584.

Glowacki, B. A. (2004) High Temperature Superconductivity 1: Materials (Ed. Narlikar, A. V.). Springer, Heidelberg, p. 239.

Glowacki, B. A. (2005a) Frontiers in Superconducting Materials (Ed. Narlikar, A. V.). Springer, Berlin, p. 697.

Glowacki, B. A. (2005b) Frontiers in Superconducting Materials (Ed. Narlikar A. V.). Springer, Heidelberg, p. 833.

Glowacki, B. A., and Majoros, M. (2002) Studies of High Temperature Superconductors, Vol. 38 (Ed. Narlikar, A.). Nova Science, New York, p. 361.

Gofryk, K., Sefat, A. S., Bauer, E. D., McGuire, M. A., Sales, B. C., Mandrus, D., Thompson, J. D., and Ronning, F. (2010) New J. Phys. 12 023006.

Goldacker, W., and Schlachter, S. I. (2005) Frontiers in Superconducting Materials (Ed. Narlikar, A. V.). Springer, Heidelberg, p. 1049.

Goldacker, W., Schlachter, S. I., Reiner, H., Zimmer, S., Obst, B., Kiesel, H., and Nyilas, A. (2003) Studies of High Temperature Superconductors, Vol. 44 (Ed. Narlikar, A.). Nova Science, New York, p. 169.

Goldman, A. I., Argyriou, D. N., Ouladdiaf, B., Chatterjee, T., Kreyssig, A., Nandi, S., Ni, N., Bud'ko, S. L. Canfield, P. C., and McQueeney, R. J. (2008) Phys. Rev. B 78 100506.

Goldman, A. I., Stassis, C., Canfield, P. C., Zarestky, J., Devenagas, P., Cho, B. K., and Johnston, D. C. (1994) Phys. Rev. B 50 9668.

Goll, G. (2006) Unconventional Superconductors. Springer, Heidelberg, p. 67

Goncharov, A. F., Struzhkin, V. V., Gregoryanz, E., Hu, J., Hemley, R. J., Mao, H. K., Lapertot, G., Bud'ko, S. L., and Canfield, P. C. (2001) Phys. Rev. B 64 100509.

Goncharov, A. F., Struzhkin, V. V., Gregoryanz, E., Mao, H. K., Hemley, R. J., Lapertot, G., Bud'ko, S. L., and Canfield, P. C. (2002) Studies of High Temperature Superconductors Vol. 38 (Ed. Narlikar, A.). Nova Science, New York, p. 339.

Gonnelli, R. S., Dagfero, D., Ummarino, G. A., Stepanov, V. A., Jun, J., Kazakov, S. M., and Karpinski, J. (2002) Phys. Rev. Lett. 89 247004.

Gooch, M., Lv, B., Tapp, J. H., Tang, Z., Lorenz, B., Guloy, A. M., and Chu, C. W. (2009) Europhys. Lett. 85 27005.

Good, J. A., and Kramer, E. J. (1970) Phil. Mag. 22 329.

Goodman, B. B. (1953) Proc. R. Soc. Lond. A 66 217.

Gor'kov, L. P. (1960) Sov. Phys. JETP 10 593 998.

Gor'kov, L. P. (1959) Sov. Phys. JETP 9, 1364.

Gor'kov, L. P., and Lebed, A. G. (1984) J. Phys. Lett. 45 L433.

Gordon, R. T., Martin, C., Kim, H., Ni, N., Tanatar, M. A., Schmalian, J., Mazin, I. I., Budko, S. L., Canfield, P. C., and Prozorov, R. (2009) Phys. Rev. B 79 100506.

Gorter, C. J., and Casimir, H. B. G. (1934) Physica 1 306.

Goyal, A. (2005) Second Generation HTS Conductors (Ed. Goyal, A.) Kluwer, New York.

Goyal, A., Norton, D. P., Budai, J. D., Paranthaman, M., Specht, E. D., Kroeger, D. M., Christen, D. K., He, Q., Saffian, B., List, F. A., Lee, F., Martin, P. M., Klabunde, C. E., Hatfield, E., and Sikka, V. K. (1996) Appl. Phys. Lett. 69 1795.

Goyal, A., Norton, D. P., Kroeger, D. M., Christen, D. K., Paranthaman, M., Specht, E. D., Budai, J. D., He, Q., Saffian, B., List, F. A., Lee, D. F., Hatfield, E., Martin, P. M., Klabunde, C. E., Mathis, J., and Park, C. (1997) J. Mat. Res. 12 2924.

Graebner, J. E., Golding, B., Schultz, R. J., Hsu, F. S. L., and Chen, H. S. (1977) Phys. Rev. Lett. 39 1480.

Grafe, H.-J., Paar, D., Lang, G., Curro, N. J., Behr, G., Werner, J., Hamann-Borrero, J., Hess, C., Leps, N., Klingeler, R., and Büchner, B. (2008) Phys. Rev. Lett. 101 047003.

Gratens, X., Ferreira, L. M., Kopelevich, Y., Oliveira, Jr., N. F., Pagliuso, P. G., Movshovich, R., Urbano, R. R., Sarrao, J. L. and Thompson, J. D. (2006) ar.Xiv cond-mat/0608722v1.

Greene, R. L., and Engler, E. M. (1980) Phys. Rev. Lett. 45 1587.

Greene, R. L., Street, G. B., and Suter, L. J., (1975) Phys. Rev. Lett. 34 577.

Grier, B. H., Lawrence, J. M., Murgai, V., and Parks, R. D. (1984) Phys. Rev. B 29 2664.

Griffin, A. (1987) Phys. Rev. B 37 5943.

Grivel, J.-C., and Flukiger, R. (1996) Supercond. Sci. Technol. 9 555.

Grivel, J.-C., and Flukiger, R. (1998) Supercond. Sci. Technol. 11 288.

Grosche, F. M., Julian, S. R., Mathur, N. D., and Lonzarich, G. G. (1996) Physica B 223–224 50.

Grosche, F. M., Julian, S. R., Mathur, N. D., Carter, F. V., and Lonzarich, G. G. (1997) Physica B 237–238 197.

Grosche, F. M., Argarwa, P., Julian, S. R., Wilson, N. J., Hoselwimmer, R. K. W., Lister, S. J. S., Mathur, N. D., Carter, F. V., Saxena, S. S., and Lonzarich, G. G. (2000) J. Phys.: Cond. Mat. 12 L533.

Gross, F., Chandrasekhar, B. S., Einzel, D., Andres, K., Hirshfeld, P. K., Ott, H. R., Beurs, J., Fisk, Z., and Smith, J. L. (1986) Z. Phys. B: Cond. Mat. 64 175.

Guguchia, Z., Bosma, S., Weyeneth, S., Shengelaya, A., Puzniak, R., Bukowski, Z., Karpinski, J., and Keller, H. (2011) Phys. Rev. B 84 144506.

Guillamón, I., Suderow, H., Viera, S., Carlo, L., Diener, P., and Rodière, P. (2008) Phys. Rev. Lett. 101 166407.

Gunnarsson, O. (1997) Rev. Mod. Phys. 69 575.

Gunnarsson, O. (2004) Alkali Doped Fullerides. World Scientific, Singapore.

Gunnarsson, O., and Zwicknagl, G. (1992) Phys. Rev. Lett. 69 957.

Gunnarsson, O., Koch, E., and Martin, R. M. (1996) Phys. Rev. B 54 R11026.

Guo, J., Jin, S., Wang, G., Wang, S., Zhu, K., Zhou, T., He, M., and Chen, X. (2010) Phys. Rev. B 82 180520(R).

Gupta, A., and Narlikar, A. V. (2009) Supercond. Sci. Technol. 22 125029.

Gupta, A., Koren, G., Tsuei, C. C., Segmüller, A., and McGuire, T. R. (1989) Appl. Phys. Lett. 55 1795.

Gupta, A., Awana, V. P. S., Samanta, S. B., Kishan, H., and Narlikar, A. V. (2005) Frontiers in Superconducting Materials (Ed. Narlikar, A. V.). Springer, Heidelberg, p. 499.

Gupta, A., Deshpande, A., Awana, V. P. S., Balamurugan, S., Sood, A. K., Kishore, R., Kishan, H., Takayama-Muromachi, E., and Narlikar, A. V. (2007) Supercond. Sci. Technol. 20 1084.

Gupta, L. C. (1994) Proc. Int. Conf. on Advances in Physical Metallurgy, Bombay. Gordon & Breach, New York, p. 494.

Gupta, L. C. (1996) Physica B 223–224 56.

Gupta, L. C. (1998) Phil. Mag. 77 717.

Gurevich, A., Patnaik, S., and Braccini, V., Kim, K. H., Mielke, C., Song, X., Cooley, L. D., Bu, S. D., Kim, S. D., Choi, J. H., Belenky, L. J., Giencke, J., Lee, M. K., Tian, W., Pan, X. Q., Siri, A., Hellstroem, E. E., Eom, C. B., and Larbalestier, D. C. (2004) Supercond. Sci. Technol. 17 278.

Haddon, R. C., Hebard, A. F., Rosseinsky, M. J., Murphy, D. W., Duclos, S. J., Lyons, K. B., Miller, B., Rosamilia, J. M., Fleming, R. M., Kortan, A. R., Glarum, S. H., Makhija, A. V., Muller, A. J., Eick, R. H., Zahurak, S. M., Tyco, R., Dabbagh, G., and Thiel, F. A. (1991) Nature 350 320.

Haen, P., Lapierre, F., Monceau, P., Nunez-Regueiro, M., and Richard, J. (1978a) Sol. State. Commun. 26 725.

Haen, P., Mignot, J. M., Monceau, P., and Nunez-Regueiro, M. (1978b) J. Phys. (Paris) 39 C6–703.

Hake, R. R., Berlincourt, T. G., and Leslie, D. H. (1962) Superconductors (Eds. Tanenbaum, M., and Wright, W. V.). Interscience, New York, p. 53.

Hamada, N., Massida, S., Freeman, A. J., and Redinger, J. (1989) Phys. Rev. B 40 4442.

Hamaker, H. C., Woolf, L. D., McKay, H. B., Fisk, Z., and Maple, M. B. (1979) Sol. State. Commun. 32 289.

Hampshire, R. G., and Taylor, M. T. (1972) J. Phys. F 2 89

Han, H., Zhu, X. Y., Cheng, P., Mu, G., Jia, Y., Fang, L., Wang, Y. L., Luo, H. Q., Zeng, B., Shen, B., Shan, L., Ren, C., and Wen, H. H. (2009) Phys. Rev. B 80 024506.

Han, J. E., Gunnarsson, O., and Crespi, V. H. (2003) Phys. Rev. Lett. 90 167096.

Han, J. E., Gunnarsson, O., and Crespi, V. H. (2005) Frontiers in Superconducting Materials (Ed. Narlikar, A. V.). Springer, Heidelberg, p. 231.

Han, P. D., Chang, L., and Payne, D. A. (1993) J. Cryst. Growth 128 798.

Han, S., Ng, K. W., Wolf, E. L., Millis, A., Smith, J. L., and Fisk, Z. (1986) Phys. Rev. Lett. 57 238.

Hanaguri, T., Niitaka, S., Kuroki, K., and Takagi, S. (2010) Science 328 474.

Hanak, J. J., Strater, K., and Cullen, G. W. (1964) RCA Rev. 25 342.

Hanawa, M., Muraoka, Y., Tayama, T., Sakakibara, J., Yamaura, J., and Hiroi, Z. (2001) Phys. Rev. Lett. 87 187001.

Hanawa, M., Yamaura, J., Muraoka, Y., Sakai, F., and Hiroi, Z. (2002) J. Phys. Chem. Sol. 63 1027.

Hannay, N. B., Geballe, T. H., Matthias, B. T., Andres, K., Schmidt, P., and MacNair, D. (1965) Phys. Rev. Lett. 14 225.

Hansen, M. (1958) Constitution of Binary Alloys. McGraw-Hill, New York.

Harault, J. P. (1966) Phys. Lett. 20 587.

Hardy, W. N., Bonn, D. A., Morgan, D. C., Liang, R., and Zhang, K. (1993) Phys. Rev. Lett. 70 3999.

Harshman, D. R., Fiory, A. T., Haddon, R. C., Kaplan, M. L., Pfiz, T., Koster, E., Shinkoda, I., and Williams, D. L. (1994) Phys. Rev. B 49 12990.

Harshman, D. R., Kleiman, R. N., Haddon, R. C., Chichester-Hicks, S. V., Kaplan, M. L., Rupp, L. W. Jr., Pfiz, T., Williams, D. L., and Mitzi, D. B. (1990) Phys. Rev. Lett. 64 1293.

Hartmann, H., Ebert, F., and Bretschneider, O. (1931) Z. Anorg. Chem. 198 116.

Hasegawa, K., Yoshida, N., Fujino, K., Mukai, H., Hayashi, K., Sato, K., Ohkuma, T., Honjo, S., Ishii, H., and Hara, T. (1997) Proc. ICEC16/ICMC (Eds. Haruyama, T., Mitsui, T., and Yamafuji, K.). Elsevier, Tokyo, p. 1413.

Hasegawa, R., and Tanner, L. E. (1977) Phys. Rev. B 16 3925.

Hasegawa, T., Ohtani, N., Koizumi, T., Aoki, Y., Nagaya, S., Hirano, N., Motowidlo, L., Sokolowwski, R. S., Scanlan, R. M., Dietderich, D. R., and Hanai, S. (2001) IEEE Trans. Appl. Supercond. 11 3034.

Hashimoto, K., Shibauchi, T., Kato, T., Ikada, K., Okazaki, R., Shishido, H., Ishikado, M., Kito, H., Iyo, A., Eisaki, H., Shamoto, S., and Matsuda, Y. (2009a) Phys. Rev. Lett. 102 017002.

Hashimoto, K., Shibauchi, T., Kasahara, S., Ikada, K., Kato, T., Okazaki, R., van der Beek, C. J., Konczykowski, M., Takeya, H., Terashimas, T., and Matsuda, Y. (2009b) Phys. Rev. Lett. 102 207001.

Hashimoto, K., Yamashita, M., Kasahara, S., Senshu, Y., Nakata, N., Tonegawa, S., Ikada, K., Serafin, A., Carrington, A., Terashima, T., Ikeda, H., Shibauchi, T., and Matsuda, Y. (2010) Phys. Rev. B 81 220501(R).

Hashimoto, Y., Yoshizaki, K., and Tanetia, M. (1974) Proc. Vth Int. Cryogenic Engineering Conf. IPC Science and Technology Press, Guildford, p. 332.

Hasselbach, K., Kirtley, J. R., and Flouquet, J. (1993) Phys. Rev. B 47 509.

Hassen, A., Hemberger, J., Loidl, A., and Krimmel, A. (2003) Physica C 400 71.

Hassinger, E., Knebel, G., Izawa, K., Lejay, P., Salce, B., and Flouquet, J. (2008) Phys. Rev. B 77 115117.

Hatt, B. A., and Rivlin, V. G. (1968) J. Phys. D 1 1145.

Hauser, J. J., and Buehler, E. (1962) Phys. Rev. 125 142.

Hauser, J. J., and Helfand, E. (1962) Phys. Rev. 127 386.

Hayashi, K., Hikata, T., Kaneko, T., Ueyama, M., Mikumo, A., Ayai, N., Kobayashi, S., Takei, H., and Sato, K. (2001) IEEE Trans. Appl. Supercond. 11 3281.

Hayashi, M., Yoshioka, H., and Kanda, A. (2010) J. Phys.: Conf. Ser. 248 012002.

Hayward, M. A., Haas, M. K., Ramirez, A. P., He, T., Regan, K. A., Rogado, N., Inumaru, K., and Cava, R. J. (2001) Sol. State. Commun. 119 491.

He, T., Huang, Q., Ramirez, A. P., Wang, Y., Regan, K. A., Rogado, N., Hayward, M. A., Haas, M. K., Slusky, J. S., Inumara, K., Zandbergen, H. W., Ong, N. P., and Cava, R. J. (2001) Nature 411 54.

Heaton, J. W., and Rose-Innes, A. C. (1964) Cryogenics 4 85.

Hebard, A. F., Rosseinsky, M. J., Haddon, R. C., Murphy, D. W., Glarum, S. H., Palstra, T. T. M., Ramirez, A. P., and Kortan, A. R. (1991) Nature 350 600

Hebel, L. C., and Slichter, C. P. (1959) Phys. Rev. 113 1504.

Heeger, A. J., and Garito, A. F. (1975) Low Dimensional Cooperative Phenomena (Ed. Keller, H. J.). Plenum, New York, p. 89.

Heersche, H. B., Jarillo-Herrero, P., Oostinga, J. B., Vandersypen, L. M. K., and Morpurgo, A. F. (2007) Nature 446 56.

Heffner, R. H., Smith, J. L., Willis, J. O., Birrer, P., Baines, C., Gygax, F. N., Hitti, B., Lippelt, E., Ott, H. R., Schenck, A., Knetsch, E. A., Mydosh, J. A., and MacLaughlin, D. E. (1990) Phys. Rev. Lett. 65, 2816.

Hegger, H., Petrovic, C., Moshopoulou, E. G., Hundley, M. F., Sarrao, J. L., Fisk, Z., and Thompson, J. D. (2000) Phys. Rev. Lett. 84 4986.

Hein, R. A., Cox, J. E., Blaugher, R. D., and Waterstrat, R. M. (1971) Physica 55 523.

Hemachalam, K., and Pickus, M. R. (1975) Appl. Phys. Lett. 27 570.

Henry, J. Y., and Lapertot, G. (1991) Physica C 185–189 86.

Herrmannsdörfer, T., and Pobell, F. (2005) Frontiers in Superconducting Materials (Ed. Narlikar, A. V.). Springer, Heidelberg, p. 71.

Heussner, R. W., Marquardt, J. D., Lee, P. J., and Larbalestier, D. C. (1997) Appl. Phys. Lett. 70 901.

Hewson, A. C. (1993) The Kondo Problem to Heavy Fermions. Cambridge University Press, Cambridge, p. 1.

Higashi, I., Takahashi, Y., and Okada, S. (1986) J. Less-Common Met. 123 277.

Highemoto, W., Ohishi, K., Koda, A., Takada, K., Sakurai, H., Takayama-Muromachi, E., and Sasaki, T. (2004) Phys. Rev. B 70 134508.

Hilgenkamp, H., and Mannhart, J. (2002) Rev. Mod. Phys. 74 485.

Hinks, D. G., and Jorgensen, J. D. (2003) Physica C 385 98.

Hinks, D. G., Jorgensen, J. D., and Li, H. C. (1983) Phys. Rev. Lett. 51 1911.

Hinks, D. G., Dabrowski, B., Jorgensen, J. D., Mitchell, A. W., Richards, D. R., Pei, S., and Shi, D. (1988a) Nature 333 836.

Hinks, D. G., Richards, D. R., Dabrowski, B., Marx, D. T., and Mitchell, A. W. (1988b) Nature 335 419.

Hinks, D. G., Claus, H., and Jorgensen, J. D. (2001) Nature 411 457.

Hinks, D., Rosenmann, D., and Claus, H. (2008) Bull. Am. Phys. Soc. 53 1.

Hiraishi, M., Kadono, R., Takeshita, S., Miyazaki, M., Koda, A., Okabe, H., Akimitsu, J. (2009) J. Phys. Soc. Jpn 78 023710.

Hiramatsu, H., Katase, T., Kamiya, T., and Hosono, H. (2012) J. Phys. Soc. Jpn 81 011011.

Hiroi, M., Sera, M., Kobayashi, N., Haga, Y., Yamamoto, E., and Onuki, Y. (1997) J. Phys. Soc. Jpn 66 1595.

Hiroi, Z., and Hanawa, M. (2002) J. Phys. Chem. Sol. 63 1021.

Hiroi, Z., Yonezawa, S., and Muraoka, Y. (2004) J. Phys. Soc. Jpn 73 1651.

Hiroi, Z., Yonezawa, S., Muramatsu, T., Yamaura, J., and Muraoka, Y. (2005a) J. Phys. Soc. Jpn 74 1255.

Hiroi, Z., Yonezawa, S., Yamaura, J., Muramatsu, T., and Muraoka, Y. (2005b) J. Phys. Soc. Jpn 74 1682.

Hiroi, Z., Yamamura, J., Yonezawa, S., and Harima, H. (2007) Physica C 460–462 20.

Hitchman, M. L., and Jensen, K. F. (1993) Chemical Vapor Deposition: Principles and Applications. Academic Press, London.

Hohenberg, P. C. (1967) Phys. Rev. 158 383.

Holczer, K., Quinlivan, D., Klein, O., Grüner, G., and Wudl, F. (1990) Sol. State. Commun. 76 499.

Holm, R., and Meissner, W. (1932) Z. Phys. 74 715.

Holstein, W. L., Parisi, L. A., Wilker, C., and Flippen, R. B. (1993) IEEE Trans. Appl. Supercond. 3 1197.

Homes, C. C., Timusk, T., Liang, R., Bonn, D. A., and Hardy, W. N. (1993) Phys. Rev. Lett. 71 1645.

Hopkins, R. H., Stewart, A. M., and Daniel, M. R. (1978) Met. Trans. 9A 215.

Hor, P. H., Meng, R. L., Wang, Y. Q., Gao, L., Huang, Z. J., Bechthold, J., Forster, K., and Chu, C. W. (1987) Phys. Rev. Lett. 58, 1891.

Hornsveld, E. M. (1988) Adv. Cryo. Eng. 34 493.

Hossain, Z., Gupta, L. C., Mazumdar, C., Nagarajan, R., Dhar, S. K., Godart, C., Levy-Clement, C., Padalia, B. D., Vijayaraghavan, R. (1994) Sol. State. Commun. 92 341.

Hossain, Z., Nagarajan, R., Dhar, S. K., and Gupta, L. C. (1998) J. Magn. Magn. Mat. 184 235.

Hott, R. (2004) High Temperature Superconductivity 1 (Ed. Narlikar, A. V.). Springer, Berlin, p. 1.

Hott, R., Kleiner, R., Wolf, T., and Zwicknagl, G. (2005) Frontiers in Superconducting Materials (Ed. Narlikar, A. V.). Springer, Heidelberg, p. 1.

Hou, J. G., Lu, L., Crespi, V. H., Xiang, X.-D., Zettl, A., and Cohen, M. L. (1995) Sol. State. Commun. 93 973.

Howe, D. G., and Fancavilla, T. L. (1980) Adv. Cryo. Eng. 26 402.

Howlett, E. W. (1970) U. S. Patent 3728165.

Hsu, F.-C., Luo, J.-Y., Yeh, K.-W., Chen, T.-K., Huang, T.-W., Wu, P. M., Lee, Y.-C., Huang, Y.-L., Chu, Y.-Y., Yan, D.-C., and Wu, M.-K. (2008) Proc. Natl. Acad. Sci. USA 105 14262.

Huang, C. L., Lin, J.-Y., Chang, Y. T., Sun, C. P., Shen, H. Y., Chou, C. C., Berger, H., Lee, T. K., Yang, H. D. (2007) Phys. Rev. B 76 212504.

Huang, Q., Zasadzinski, J. F., Tralshawala, N., Gray, K. E., Hinks, D. G., Peng, J. L., and Greene, R. L. (1990) Nature 347 369.

Huang, Q., Foo, M. L., Lynn, J. W., Zandbergen, H. W., Lawes, G., Wang, Y., Toby, B. H., Ramirez, A. P., Ong, N. P., Cava, R. J. (2004) J. Phys. Cond. Mat. 16 5803.

Huang, Q., Qiu, Y., Bao, W., Lynn, J., Green, M., Chen, Y., Wu, T., Wu, G., and Chen, X. (2008) Phys. Rev. Lett. 101 257003.

Huang, S. L., Koblischka, M. R., Johansen, T. H., Bratsberg, H., and Fossheim, K. (1997) Physica C 282–287 2279.

Huang, Z. I., Fang, H. H., Hue, Y. Y., Hor, P. H., Chu, C. W., Norton, M. L., and Tang, H. Y. (1991) Physica C 180 331.

Hüfnerr, S., Hossain, M. A., Damascelli, A., and Swatzky, G. A. (2008) Rep. Prog. Phys. 71 062501.

Huhtinen, H., Awana, V. P. S, Gupta, A., Kishan, H., Laiho, R., and Narlikar, A. V. (2007) Supercond. Sci. Technol. 21 S159.

Huhtinen, H., Irjala, M., Paturi, P., Shakhov, M. A., and Laiho, R. (2010) J. Appl. Phys. 107 053906.

Hull, J. R. (2001) Supercond. Sci. Technol. 13 R1.

Hull, J. R. (2004) High Temperature Superconductivity 2: Engineering Applications (Ed. Narlikar, A. V.). Springer, Heidelberg, p. 91.

Hulm, J. K. (1950) Proc. R. Soc. Lond. A 204 98.

Hulm, J. K., and Blaugher, R. D. (1961) Phys. Rev. 123 1569.

Hunt, F., Jaroszynski, J., Gurevich, A., Larbalestier, D. C., Jin, R., Sefat, A. S., McGuire, M. A., Sales, B. C., Christen, D. K., and Mandrus, D. (2008) Nature 453 903.

Huxley, A. D., Raymond, S., and Ressouche, E. (2003) Phys. Rev. Lett. 91 207201.

Huxley, A. D., Sheikin, I., Ressouche, E., Kernavanois, N., Braithewaite, D., Calemczuk, R., and Flouquet, J. (2001) Phys. Rev. B 63 144519.

Huy, N. T., de Nijs, D. E., Huang, Y., and de Visser, A. (2008) Phys. Rev. Lett. 100 077002.

Huy, N. T., Gasparini, A., de Nijs, D. E., Huang, Y., Klaasse, J. P. S., Gortenmulder, T., de Visser, A., Hamann, A., Görlach, T., and v. Löhneysen, H. (2007) Phys. Rev. Lett. 99 067006.

Hwang, H. Y., Batlogg, B., Takagi, H., Kao, H. L., Kwo, J., Cava, R. J., Krajewski, J. J., and Peck, W. F. Jr. (1994) Phys. Rev. Lett. 72 2636.

Iavarone, M., Karapetrov, G., Koshelev, A. E., Kwok, W. K., Crabtree, G. W., Hinks, D. G., Kang, W. N., Choi, E.-M., Kim, H. J., Kim, H.-J., and Lee, S. I. (2002) Phys. Rev. Lett. 89 187002.

Ignatov, Yu., A., Tyson, T. A., and Sarasov, S. (2003) American Physical Society Meeting, March 2003, K21.010.

Iguchi, I., Wang, W., Yamazaki, M., Tanaka, Y., and Kashiwaya, S. (2000) Phys. Rev. B 62 R6131.

Iguchi, I., Yamaguchi, T., and Sugimoto, A. (2001) Nature 412 420.

Ihara, H. (1995) Adv. Supercond. 7 255.

Ihara, H. (2001a) Physica C 364–365 289.

Ihara, H. (2001b) Sol. State Phys. 35 301.

Ihara, Y., Ishida, K., Michioka, C., Kato, M., Yoshimura, K., Takada, K., Sasaki, T., Sakurai, H., and Takayama-Muromachi, E. (2005) J. Phys. Soc. Jpn 74 867.

Ihara, Y., Ishida, K., Takeya, H., Michioka, C., Kato, M., Itoh, Y., Yoshimura, K., Takada, K., Sasaki, T., Sakurai, H., and Takayama-Muromachi, E. (2006) J. Phys. Soc. Jpn 75 013708.

Ihara, Y., Takeya, H., Ishida, K., Ikeda, H., Michioka, C., Yoshimura, K., Takada, K., Sasaki, T., Sakurai, H., and Takayama-Muromachi, E. (2007) Phys. Rev. B 75 212506.

Iida, K., Hänisch, J., Trommler, S., Matias, V., Haindl, S., Kurth, F., Lucas del Pozo, I., Hühne, R., Kidszun, M., Engelmann, J., Schultz, L., and Holzapfel, B. (2011) Appl. Phys. Express 4 013103.

Iijima, K., Onabe, K., Futaki, N., Tanabe, N., Sadakata, N., Kohno, O., and Ikeno, Y. (1993) IEEE Trans. Appl. Supercond. 3 1510.

Iijima, K., Tanabe, N., Kohno, O., and Ikeno, Y. (1992) Appl. Phys. Lett., 60 769.

Iijima, Y. (1995) Presented at International Cryogenics Materials Conf.

Iijima, Y., Kosuge, Y. M., Takeuchi, T., and Inoue, K. (1997) Adv. Cryo. Eng. 42 1447.

Iijima, Y., Kikuchi, A., and Inoue, K. (2002) Physica C 372–376 1303.

Ikeda, H., and Ohashi, Y. (1998) Phys. Rev. Lett. 81 3723.

Ikeda, S., Shishido, H., Nakashima, M., Settai, R., Aoki, D., Haga, Y., Harima, H., Aoki, Y., Namiki, T., Sato, H., and Onuki, Y. (2001) J. Phys. Soc. Jpn 70 2248.

Ikuta, H. (2004) High Temperature Superconductivity 1: Materials (Ed. Narlikar, A. V.). Springer, Heidelberg, p. 79.

Ikuta, H., Mase, A., Yanagi, Y., Yoshikawa, M., Itoh, Y., Oka, T., and Mizutani, U. (1998) Supercond. Sci. Technol. 11 1345.

Ikuta, H., Mase, A., Hosokawa, T., Yanagi, Y., Yoshikawa, M., Itoh, Y., Oka, T., and Mizutani, U. (1999) Advances in Superconductivity, Vol. XI (Eds. Koshizuka, N., and Tajima, S.). Springer, Tokyo, p. 657.

Ikuta, H., Hosokawa, T., Yoshikawa, M., and Mizutani, U. (2000) Supercond. Sci. Technol. 13 1559.

Inosov, D. S., Leineweber, A., Yang, X. P., Park, J. T., Christensen, N. B., Dinnebier, R., Sun, G. L., Niedermayer, C., Haug, D., Stephens, P. W., Stahn, J., Khvostikova, O., Lin, C. T., Andersen, O. K., Keimer, B., and Hinkov, V. (2009) Phys. Rev. B 79 224503.

Inoue, I. H., Aiura, Y., Nishihara, Y., Haruyama, Y., Nishizaki, S., Maeno, Y., Fujita, T., Bednorz, J. G., and Lichtenberg, F. (1996) Physica B 223–224 516.

Inoue, N., Tsukamoto, A., Moriwaki, Y., Sugano, T., Wu, X.-J., Ogawa, A., Adachi, S., Takagi, K., and Tanabe, K. (2004) High Temperature Superconductivity 1: Materials (Ed. Narlikar, A. V.). Springer, Heidelberg, p. 213.

Ishida, K., Kitaoka, Y., Asayama, K., Ikeda, S., Nishizaki, S., Maeno, Y., Yoshida, K., and Fujita, T. (1997) Phys. Rev. B 56 R505.

Ishida, K., Mukuda, H., Kitaoka, Y., Asayama, K., Mao, Z. Q., Mori, Y., and Maeno, Y. (1998) Nature 396 658.

Ishida, K., Mukuda, H., Kitaoka, Y., Asayama, K., Mao, Z. Q., Mori, Y., and Maeno, Y. (2000a) Phys. Rev. Lett. 84 5387.

Ishida, K., Mukuda, H., Kitaoka, Y., Mao, Z. Q., Mori, Y., and Maeno, Y. (2000b) Physica B 281–282 963.

Ishida, K., Ozaki, D., Kamatsuka, T., Tou, H., Kyogaku, M., Kitaoka, Y., Tateiwa, N., Sato, N. K., Aso, N., Geibel, C., and Steglich, F. (2002) Phys. Rev. Lett. 89 037002.

Ishida, K., Ihara, Y., Maeno, Y., Michioka, M., Kato, K., Yoshimura, K., Takada, K., Sasaki, T., Sakurai, H., Takayama-Muromachi, E. (2003) J. Phys. Soc. Jpn 72 3041.

Ishida, K., Nakai, Y., and Hosono, H. (2009) J. Phys. Soc. Jpn 78 062001.

Ishiguro, T., and Tanatar, M. A. (2000) Studies of High Temperature Superconductors, Vol. 34 (Ed. Narlikar, A. V.). Nova Science, New York, p. 55.

Ishiguro, T., Yamaji, K., and Saito, G. (1998) Organic Superconductors. Springer, Berlin, p. 1.

Ishikawa, M., and Fischer, Ø. (1977) Sol. State. Commun. 24 747.

Ito, H., Ishiguro, T., Komatsu, T., Saito, G., and Anzai, H. (1994) Physica B 201 470.

Ito, T., Takenaka, K., and Uchida, S. (1993) Phys. Rev. Lett. 70 3995.

Ito, T., Fudamoto, Y., Fukaya, A., Gat-Malureanu, I. M., Larkin, M. I., Russo, P. L., Savici, A., Uemura, Y. J., Groves, K., Breslow, R., Hotehama, K., Yamanaka, S., Kyriakou, P., Rovers, M., Luke, G. M., and Kojima, K. M. (2004) Phys. Rev. B 69 134522.

Iwamoto, Y., Ueda, K., and Kohara, T. (2000) Sol. State. Commun. 113 615.

Iwasa, Y., and Montgomery, D. B. (1975) Applied Superconductivity, Vol. II (Ed. Newhouse, V. L.). Academic Press, New York, p. 387.

Iye, Y., and Tanuma, S. (1982a) Phys. Rev. B 25 4583.

Iye, Y., and Tanuma, S. (1982b) Sol. State. Commun. 44 1.

Iyo, A., Tanaka, Y., Ishiura, Y., Tokumoto, M., Tokiwa, K., Watanabe, T., and Ihara, H. (2001a) Supercond. Sci. Technol. 14 504.

Iyo, A., Aizawa, Y., Tanaka, Y., Ishiura, Y., Tokumoto, M., Tokiwa, K., Watanabe, T., and Ihara, H. (2001b) Physica C 357–360 324.

Izawa, K., Takahashi, H., Yamaguchi, H., Matsuda, Y., Suzuki, M., Sasaki, T., Fukase, T., Yoshida, Y., Settai, R., and Onuki, Y. (2001a) Phys. Rev. Lett. 86 2653.

Izawa, K., Yamaguchi, H., Matsuda, Y., Shishido, H., Settai, R., and Onuki, Y. (2001b) Phys. Rev. Lett. 87 057002.

Izawa, K., Shibata, A., Matsuda, Y., Kato, Y., Takeya, H., Hirata, K., van der Beek, C. J., and Konczykowski, M. (2001c) Phys. Rev. Lett. 86 1327.

Izumi, F., Kondo, T., Shimakawa, Y., Manako, T., Kubo, Y., Igarashi, H., and Asano, H. (1991) Physica C 185–189 615.

Jaccard, D. A., Behnia, K., and Sierro, J. (1992) Phys. Lett. 163A 475.

Jaccard, D., Wilhelm, H., Jerome, D., Moser, J., Carce, C., and Fabre, J. M. (2001) J. Phys.: Cond. Mat. 13 L89.

Jaccarino, V., and Peter, M. (1962) Phys. Rev. Lett. 9 290.

Jagadish, R., and Sinha, K. P. (1987) Curr. Sci. 56 291.

Jang, H. C., Hsieh, M. H., Lee, D. S., and Horng, H. E. (1990) Phys. Rev. B 42 2551.

Jansen, L., Block, R., and Stepankin, V. (1995) Physica A 219 327.

Jansen, M. (1977) Z. Naturforsch. B 32 1340.

Jaramillo, R, Feng, Y., Lang, J. C., Islam, Z., Srajer, G., Rønnow, H. M., Littlewood, P. B., and Rosenbaum, T. F. (2008) Phys. Rev. B 77 184418.

Jaramillo, R., Feng, Y., Lang, J. C., Islam, Z., Srajer, G., Littlewood, P. B., McWhan, D. B., and Rosenbaum, T. F. (2009) Nature 459, 405.

Jarlborg, T., and Freeman, A. J. (1980) Phys. Rev. Lett. 44 178.

Jayaram, B., Ekbote, S. N., and Narlikar, A. V. (1987) Phys. Rev. B 36 1996.

Jeevan, H. S., Hossain, Z., Kasinathan, D., Rosner, H., Geibel, C., and Gegenwart, P. (2008) Phys. Rev. B 78 092406.

Jeglič, P., Potočnik, A., Khanjšek, M., Bobnar, M., Jagodič, M., Koch, K., Rosner, H., Margadonna, S., Lv, B., Guloy, A. M., and Arčon, D. (2010) Phys. Rev. B 81 140511(R).

Jergel, M. (1995) Supercond. Sci. Technol. 8 67.

Jergel, M., Chromik, Š., Štrbík, V., Šmatko, V., Hanic, F., Plesch, G., Buchta, Š., and Valtyniova, S. (1992) Supercond. Sci. Technol. 5 225.

Jérome, D. (1991) Science 252 1509.

Jérome, D., and Pasquier, C. R. (2005) Frontiers in Superconducting Materials (Ed. Narlikar, A. V.). Springer, Heidelberg, p. 183.

Jérome, D., and Schultz, H. J. (1982) Adv. Phys. 31 299.

Jérome, D., Mazaud, A., Ribault, M., and Bechgaard, K. (1980) J. Phys. Lett. (Paris) 41 L–95.

Jha, R., Kumar, A., Sing, S. K., and Awana, V. P. S. (2013) J. Supercond. Novel Mag. 26 499.

Jha, S. S. (1989) Studies of High Temperature Superconductors, Vol. 1 (Ed. Narlikar, A.). Nova Science, New York, p. 41.

Jia, Y., Cheng, P., Fang, L., Luo, H., Yang, H., Ren, C., Shan, L., Gu, C., and Wen, H. H. (2008) Appl. Phys. Lett. 93 032503.

Jin, B. B., Dahm, T., Gubin, A. I., Choi, Eun-Mi, Kim, H. J., Lee, Sung-IK, Kang, W. N., and Klein, N. (2003) Phys. Rev. Lett. 91 127006.

Jin, C. Q., Adachi, S., Wu, X. J., Yamauchi, H., and Tanaka, S. (1994) Physica C 223 238.

Jin, R., He, J., McCall, S., Alexander, C. S., Drymiotis, F., and Mandrus, D. (2001) Phys. Rev. B 64 180503.

Jin, R., Sales, B. C., Khalifah, P., and Mandrus, D. (2003) Phys. Rev. Lett. 91 217001.

Jin, R., Sales, B. C., Li, S., and Mandrus, D. (2005) Phys. Rev. B 72 060512.

Jin, S., Tiefel, T. H., Sherwood, R. C., van Dover, R. B., Davis, M. E., Kammlott, G. W., and Fastnacht, R. A. (1988) Phys. Rev. B 37 7850.

Jin, S., Sherwood, R. C., Gyorgy, E. M., Tiefel, T. H., van Dover, R. B., Nakahara, S., Schneemeyer, L. F., Fastnacht, R. A., and Davis, M. E. (1989) Appl. Phys. Lett. 54 584.

Jin, S., Tiefel, T. H., and Graebner, J. E. (1992) Studies of High Temperature Superconductors, Vol. 10 (Ed. Narlikar, A. V.). Nova Science, New York, p. 97.

Jishi, R. A., and Alyahyaei, H. M. (2010) Adv. Cond. Mat. Phys., Article ID 804343.

Jishi, R. A., and Dresselhaus, M. S. (1992) Phys. Rev. B 45 12465.

Jishi, R. A., Dresselhaus, M. S., and Chaiken, A. (1991) Phys. Rev. B 44 10248.

Johnson, W. L., and Poon, S. J. (1975) J. Appl. Phys. 46 1787.

Johnson, W. L., Poon, S. J., and Duwez, P. (1975) Phys. Rev. B 11 150.

Johnson, W. L., Tuuei, C. C., and Chaudhari, P. (1978a) Phys. Rev. B 17 2884.

Johnson, W. L., Poon, S. J., Dusand, J., and Duwez, P. (1978b) Phys. Rev. B 18 206.

Johnston, D. C. (1982) Mat. Res. Bull. 14 797.

Johnston, D. C. (2010) Adv. Phys. 59 803.

Jorda, J.-L. (2005) Frontiers in Superconducting Materials (Ed. Narlikar, A. V.). Springer, Heidelberg, p. 833.

Jorda, J.-L., Galez, P., Phok, S., Hopfinger, T., and Jondo, T. K. (2004) High Temperature Superconductivity 1: Materials (Ed. Narlikar, A. V.). Springer, Heidelberg, p. 29.

Jorgensen, J. D., Chmaissem, O., Shakad, H., Short, S., Klamut, P. W., Dabrowski, B., and Tallon, J. L. (2001a) Phys. Rev. B 63 054440.

Jorgensen, J. D., Hinks, D. G., and Short, S. (2001b) Phys. Rev. B 63 224522.

Josephson, B. D. (1962) Phys. Lett. 1 251.

Joyce, J. J., Wills, J. M., Durakiewicz, T., Butterfield, M. T., Guziewicz, E., Sarrao, J. L., Morales, L. A., Arko, A. J., and Eriksson, O. (2003) Phys. Rev. Lett. 91 176401.

Joynt, R., and Taillefer, L. (2002) Rev. Mod. Phys. 74 235.

Julian, S. R., Tautz, F. S., McMullan, G. J., and Lonzarich, G. G. (1994) Physica B 199–200 63.

Jung, D. (2011) Bull. Korean Chem. Soc. 32 451.

Jung, J. K., Baek, S. H., Borsa, F., Bud'ko, S. L., Lapertot, G., and Canfield, P. C. (2001) Phys. Rev. B 64 012514.

Junod, A., Müller, J., Rietschel, H., and Schneider, E. (1978) J. Phys. Chem. Sol. 39 317.

Junod, A., Wang, Y., Bouquet, F., and Toulemonde, P. (2002) Studies of High Temperature Superconductors, Vol. 38 (Ed. Narlikar, A.). Nova Science, New York, p. 179.

Juza, V. R., and Heners, J. (1964) Z. Anorg. Allg. Chem. 332 173.

Kačmarčík, J., Pribulová, Z., Marcenat, C., Klein, T., Rodière, P., Cario, L., and Samuley, P. (2010) Phys. Rev. B 82, 014518.

Kadono, R., Higemoto, W., Koda, A., Kawasaki, Y., Hanawa, M., and Hiroi, Z. (2002) J. Phys. Soc. Jpn 71 709.

Kagoshima, S., Yasunaga, T., Ishiguro, T., Anzai, H., and Saito, G. (1983) Sol. State. Commun. 46 867.

Kaiser, A. B., and Uher, C. (1991) Studies of High Temperature Superconductors, Vol. 7 (Ed. Narlikar, A.), Nova Science, New York, p. 353.

Kakuyangi, K., Saitoh, M., Kumagai, K., Takashima, S., Nohara, M., Takagi, H., and Matsuda, Y. (2005) Phys. Rev. Lett., 94 047602.

Kamada, T., Setsune, K., Hirao, T., and Wasa, K. (1988) Appl. Phys. Lett. 52 1726.

Kamata, K., Tada, N., Itoh, K., and Tachikawa, K. (1983) IEEE. Trans. Magn. 19 1433.

Kamihara, Y., Hiramatsu, H., Hirano, M., Kawamura, R., Yanagi, Y., Kamiya, T., and Hosono, H. (2006) J. Am. Chem. Soc. 128 10012.

Kamihara, Y., Watanabe, T., Hirano, M., and Hosono, H. (2008) J. Am. Chem. Soc. 130 107006.

Kang, D.-J., Glowacki, B. A., Winter, P. R., and Wallach, E. R. (1996) IEEE Trans. Magn. 32 2970.

Kanigel, A., Keren, A., Patlagan, L., Chashka, K. B., Fisher, B., King, P., and Amato, A. (2004) Phys. Rev. Lett. 92 257004.

Kanoda, K., Akiba, K., Suzuki, K., and Takahashi, T. (1990) Phys. Rev. Lett. 65 1271.

Kanoda, K., Miyagawa, K., Kawamoto, A., and Nakazawa, Y. (1996) Phys. Rev. B 54 76.

Kapitulnik, A. (1994) Phys. Rev. Lett. 73 2744

Karapetrov, G., Iavarone, M., Kwok, W. K., Crabtree, G. W., and Hinks, D. G. (2001) Phys. Rev. Lett. 86 4347.

Karppinen, M., Asako, I., Motohashi, T., and Yamaguchi, H. (2004) Chem. Mat. 16 1693.

Kasahara, S., Shibauchi, T., Hashimoto, K., Ikada, K., Tonegawa, S., Okazaki, R., Ikeda, H., Takeya, H., Hirata, K., Terashima, T., and Matsuda, Y. (2010) Phys. Rev. B 81 184519.

Kasahara, Y., Nakajima, Y., Izawa, K., Matsuda, Y., Behnia, K., Shishido, H., Settai, R., and Onuki, Y. (2005) Phys. Rev. B 72 214515.

Kashiwaya, S., Tanaka, Y., Koyanagi, Y., and Kajimura, K. (1996) Phys. Rev. B 53 2667.

Kasumov, A. Yu., Klinov, D. V., Roche, P.-E., Guéron, S., and Bouchiat, H. (2004) Appl. Phys. Lett. 84 1007.

Kasumov, A. Yu., Kociak, M., Guéron, S., Reulet, B., Volkov, V. T., Klinov, D. V., and Bouchiat, H. (2001) Science 291 280.

Kasumov, A. Yu., Nakamae, S., Cazayous, M., Kawasaki, T., and Okahata, Y. (2009) Res. Lett. Nanotechnol., Article ID 540257.

Katase, T., Ishimaru, Y., Tsukamoto, A., Hiramatsu, H., Kamiya, T., Tanabe, K., and Hosono, H. (2010) Appl. Phys. Lett. 96 142507.

Katase, T., Ishimaru, Y., Tsukamoto, A., Hiramatsu, H., Kamiya, T., Tanabe, K., and Hosono, H. (2011a) Nature Commun. 2 409.

Katase, T., Hiramatsu, H., Matias, V., Sheehan, C., Ishimaru, Y., Kamiya, T., Tanabe, K., and Hosono, H. (2011b) Appl. Phys. Lett. 98 242510.

Kato, M., Michioka, C., Waki, T., Yoshimura, K., Ishida, K., Sakurai, H., Takayama-Muromachi, E., Takada, K., and Sasaki, T. (2006) J. Phys.: Cond. Mat. 18, 669.

Kaufman, A. R., and Pickett, J. J. (1970) Bull. Am. Phys. Soc. 15 838.

Kawamoto, A., Miyagawa, K., and Kanoda, K. (1997) Phys. Rev. B 55 14140.

Kawashima, T., and Takayama-Muromachi, E. (2003) Physica C 398 85.

Kazakov, S. M., Chaillollout, C., Bordet, P., Capponi, J. J., Nunez-Requero, M., Rysak, A., Tholence, J. L., Radaelli, P. G., Putilin, S. N., and Antipov, E. V. (1997) Nature 390 148.

Kazakov, S. M., Zhigadlo, N. D., Brühwiler, M., Batlogg, B., and Karpinski, J. (2004) Supercond. Sci. Technol. 17 1169.

Kealey, P. G., Riseman, T. M., Forgan, E. M., Galvin, L. M., Mackenzie, A. P., Lee, S. L., McK Paul, D., Cubitt, R., Agterberg, T. M., Heeb, R., and Maeno, Y. (2000) Phys. Rev. Lett. 84 6094.

Keizer, R. J., de Visser, A., Menovsky, A. A., Franse, J. J. M., Fak, B., and Mignot, J.-M. (1999) Phys. Rev. B 60 6668.

Keller, H. (2005) Superconductivity in Complex systems (Eds. Müller, K. A., and Bussmann-Holder, A.). Springer, Berlin, p. 143.

Keller, H., and Bussmann-Holder, A. (2010) Adv. Cond. Mat. Phys., Article ID 393526.

Kelty, S. P., Chen, C., and Lieber, M. C. (1991) Nature 352 223.

Kemper, A. F., Maier, T., Graser, S., Cheng, H.-P., Hirschfeld, P. J., and Scalapino, D. J. (2010) New J. Phys. 12 073030.

Keys, S. A., Koizumi, N., and Hampshire, D. (2002) Supercond. Sci. Technol. 15 991.

Khare, N., Gupta, A. K., Khare, S., Gupta, L. C., Nagarajan, R., Hossain, Z., and Vijayaraghavan, R. (1996) Appl. Phys. Lett. 69 1483.

Khasanov, R., Shengelaya, A., Morenzoni, E., Conder, K., Savić, I. M., and Keller, H. (2004) J. Phys. Cond. Mat. 16 S4439.

Khasanov, R., Strässle, S., Conder, K., Ponjajakushina, E., Bussmann-Holder, A., and Keller, H. (2008a) Phys. Rev. B 77 104530.

Khasanov, R., Shengelaya, A., Di Castro, D., Morenzoni, E., Maisuradze, A., Savić, I. M., Conder, K., Pomjakushina, E., Bussmann-Holder, A., and Keller, H. (2008b) Phys. Rev. Lett. 101 077001.

Khasanova, N. R., Yamamoto, A., Tajima, S., Wu, X.-J., and Tanabe, K. (1998) Physica C 305 275.

Khasanova, N. R., Yoshida, K., Yamamoto, A., and Tajima, S. (2001) Physica C 356 12.

Kidszun, M., Haindl, S., Reich, E., Hänisch, J., Iida, K., Schultz, L., and Holzapfel, B. (2010a) Supercond. Sci. Technol. 23 022002.

Kidszun, M., Haindl, S., Thersleff, T., Hänisch, J., Kauffmann, A., Iida, K., Freudenberger, J., Schultz, L., and Holzapfel, B. (2010b) Phys. Rev. Lett. 106 137001.

Kiefl, R. F., McFarlane, W. A., Chow, K. H., Dunsinger, S., Duty, D. L., Johnston, T. M. S., Schneider, J. W., Sonier, J., Brard, L., Strongin, R. M., Fischer, J. E., and Smith A. B. (1993) Phys. Rev. Lett. 70 3987.

Kihou, K., Saito, T., Ishida, S., and Nakajima, M., Tomioka, Y., Fukazawa, H., Kohori, Y., Ito, T., Uchida, S., and Iyo, A. (2010) J. Phys. Soc. Jpn 79 124713.

Kikuchi, A., Iijima, Y., Inoue, K., Kosuge, M., and Itoh, K. (2001a) IEEE. Trans. Appl. Supercond. 11 3984.

Kikuchi, A., Iijima, Y., and Inoue, K. (2001b) IEEE. Trans. Appl. Supercond. 11 3968.

Kikugawa, N., and Maeno, Y. (2002) Phys. Rev. Lett. 89 117001.

Kikugawa, N., Mackenzie, A. P., and Maeno, Y. (2002) J. Phys. Soc. Jpn 72 237.

Kim, D. G., Kim, J. S., Kim, B. H., Park, Y. W., Jung, C. U., and Lee, S. I. (2003) Physica C 387 313.

Kim, I. G., Lee, J. I., and Freeman, A. J. (2002a) Phys. Rev. B 65 064525.

Kim, J. H., Ahn, J. S., Kim, J., Park, M.-S., Lee, S. I., Choi, E. J., and Oh, S. J. (2002b) Phys. Rev. B 66 172507.

Kim, J. H., Zhou, S., Hossain, M. S. A., Pan, A. V., and Dou, S. X. (2006) Appl. Phys. Lett. 89 142505.

Kim, J. S., Hirschfeld, P. J., Stewart, G. R., Kasahara, S., Shibauchi, T., Tarashima, T., and Matsuda, Y. (2010) Phys. Rev. B 81 214507.

Kim, J. S., Kremer, R. K., Boeri, L., and Razavi, F. S. (2006) Phys. Rev. Lett. 96 217002.

Kim, M.-S., Skinta, J. A., Lemberger, T. R., Tsukada, A., Naito, M. (2003) Phys. Rev. Lett. 91 087001.

Kim, Y. J., Gu, G. D., Gog, T., and Casa, D. (2008) Phys. Rev. B 77 064520.

Kimakura, H., Kitaguchi, H., Togano, K., and Maeda, H. (1997) Studies of High Temperature Superconductors, Vol. 21 (Ed. Narlikar, A.) Nova Science, New York, p. 163.

Kimber, S. A. J., Argyriou, D. N., Yokaichiya, F., Habicht, K., Gerischer, S., Hansen, T., Chatterjee, T., Klingeler, R., Hess, C., Behr, G., Kondrat, A., and Buchner, B. (2008) Phys. Rev. B 78 140503.

Kimihara, Y., Hiramatsu, H., Hirano, M., Kawamura, R., Yanagi, H., Kamiya, T., and Hosono, H., (2006), J. Am. Chem. Soc. 128 10012.

Kini, A. M., Carlson. K. D., Dudek, J. D., Geiser, U., Wang, H. H., and Williams, J. M. (1997) Synth. Met. 85 1617.

Kini, A. M., Carlson, K. D., Wang, H. H., Schlueter, J. A., Dudek, J. D., Sirchio, S. A., Geiser, U., Lykke, K. R., and Williams, J. M. (1996) Physica C 264 81.

Kini, A. M., Schlueter, J. A., Ward, B. H., Geiser, U., and Wang, H. H. (2001) Synth. Met. 120 713.

Kishio, K., Kitazawa, K., Kanbe, S., Yasuda, T., Sugii, N., Takagi, H., Uchida, S., Fueki, K., and Tanaka, S. (1987) Chem. Lett. p. 429.

Kiss, T., Yokoya, T., Chainani, A., Shin, S., Hanaguri, T., Nohara, M., and Takagi, H. (2007) Nat. Phys. 3 720.

Kitagawa, K., Katayama, N., Ohgushi, K., Yoshida, M., and Takigawa, M. (2008) J. Phys. Soc. Jpn 77 114709.

Kitaguchi, H. (2000) Physica C 335 26.

Kitaguchi, H., Itoh, K., Takeuchi, T., Kumakura, H., Miao, H., Wada, H., Togano, K., Hasegawa, T., and Kizumi, T. (1999a) Physica C 320 253.

Kitaguchi, H., Miao, H., Kumakura, H., and Togano, K. (1999b) Physica C 320 71.

Kitaoka, Y., Hiramitsu, S., Kondo, T., and Asayama, K. (1988) J. Phys. Soc. Jpn 57 30.

Kito, H., Ikeda, S., Takekawa, S., Abe, H., and Kitazawa, H. (1997) Physica C 291 332.

Kito, H., Eisaki, H., and Iyo, A. (2008) J. Phys. Soc. Jpn 77 063707.

Kivelson, S. A., Rokhsar, D. S., and Sethna, J. P. (1987) Phys. Rev. B 35 8865.

Klamut, P. W. (2008) Supercond. Sci. Technol. 21 093001.

Klamut, P. W., Dabrowski, B., Kolesnik, S., Maxwell, M., and Mais, J. (2001a) Phys. Rev. B 63 224512.

Klamut, P. W., Dabrowski, B., Mais, J., and Maxwell, M. (2001b) Physica C 350 24.

Klamut, P. W., Dabrowski, B., Mini, S. M., Maxwell, M., Mais, J., Felner, I., Asaf, U., Ritter, F., Shengelaya, A., Khasanov, R., Savic, I. M., Keller, H., Wisniewski, A., Puzniak, R., Fita, I. M., Sulkowski, C., and Matusiak, M. (2003) Physica C 387 33.

Klauss, H.-H., Luetkens, H., Klingeler, R., Hess, C., Litterst, F. J., Kraken, M., Korshunov, M. M., Eremin, I., Drechsler, S.-L., Khasanov, R., Amato, A., Hamann-Borrero, J., Leps, N., Kondrat, A., Behr, G., Werner, J., and Büchner, B. (2008) Phys. Rev. Lett. 101 077005.

Kleefisch, S., Welter, B., Marx, A., Alff, L., Gross, R., and Naito, M. (2001) Phys. Rev. B 63 100507.

Klein, B. M., Boyer, L. L., Papaconstantopoulos, D. A., and Mattheiss, L. F. (1978) Phys. Rev. B 18 6411.

Klein, O., Holczer, K., Grüner, G., Chang, J. J., and Wudl, F. (1991) Phys. Rev. Lett. 66 655.

Klemenz, C., and Scheel, H. J. (1993) J. Cryst. Growth 129 421.

Klemenz, C., and Scheel, H. J. (1996) Physica C 265 126.

Klimczuk, T., and Cava, R. J. (2004a) Phys. Rev. B 70 212514.

Klimczuk, T., and Cava, R. J. (2004b) Sol. State. Commun. 132 379.

Klimczuk, T., Gupta, V., Lawes, G., Ramirez, A. P., and Cava, R. J. (2004) Phys. Rev. B 70 094511.

Klimczuk, T., Avdeev, M., Jorgensen, J. D., and Cava, R. J. (2005) Phys. Rev. B 71 184512.

Klimonsky, S. O., Samoilenkov, S. V., Yu, O., Gorbenko, D. A., Emelianov, A. V., Lyashenko, S. R., Lee, A. R., Kaul, Yu. D. T., Andrianov, D. G., Kalinov, A. V., and Voloshin, I. V. (2002) Physica C 383 37.

Knauf, J., Semerad, R., Prusseit, W., deBoer, B., and Eickemeyer, J. (2001) IEEE. Trans. Appl. Supercond. 11 2885.

Knebel, G., Méasson, M.-A., Salce, B., Aoki, D., Braithwaite, D., Brison, J. P., and Flouquet, J. (2004) J. Phys. Cond. Matt. 16 8905.

Knebel, G., Aoki, D., Braithwaite, D., Salce, B., and Flouquet, J. (2006) Phys. Rev. B 74 020501.

Knebel, G., Aoki, D., Brison, J.-P., and Flouquet, J. (2008) J. Phys. Soc. Jpn 77 114707.

Knee, C. S., Rainford, B. D., and Weller, M. T. (2000) J. Mat. Chem. 10 2445.

Kobayashi, A., Kato, R., Naito, T., and Kobayashi, H. (1993) Synth. Met. 56 2078.

Kobayashi, H., Akutsu, H., Arai, E., Tanaka, H., and Kobayashi, A. (1997) Phys. Rev. B 56 R8526.

Kobayashi, H., Enoki, T., Imaeda, K., Inokuchi, H., and Saito, G. (1987) Phys. Rev. B 36 1457.

Kobayashi, M., and Tsujikawa, I. (1981) Physica B 105 439.

Kobayashi, S., Kaneko, T., Ayai, N., Hayashi, K., Takei, H., and Hata, R. (2001) Physica C 357–360 1115.

Kobayashi, T. C., Miyazu, T., Takeshita, N., Shimizu, K., Amaya, K., Kitaoka, Y., and Onuki, Y. (1998) J. Phys. Soc. Jpn 67 996.

Kobayashi, Y., Yokoi, M., and Sato, M. (2003) J. Phys. Soc. Jpn 72 2453.

Kobayashi, Y., Moyoshi, T., Watanabe, H., Yokoi, M., and Sato, M. (2006) J. Phys. Soc. Jpn 75 074717.

Kobayashi, Y., Watanabe, H., Yokoi, M., Moyoshi, T., Mori, Y., and Sato, M. (2005). J. Phys. Soc. Jpn 74 1800.

Koblischka, M. R., Murlidhar, M., and Muakami, M. (2000) Studies of High Temperature Superconductors, Vol. 32 (Ed. Narlikar, A. V.). Nova Science, New York, p. 221.

Koda, A., Higemoto, W., Ohishi, K., Saha, S. R., Kadono, R., Yonezawa, S., Muraoka, Y., and Hiroi, Z. (2005) J. Phys. Soc. Jpn 74 1678.

Kodas, T., Datye, A., Lee, V., and Engler, V. (1989) J. Appl. Phys 65 2149.

Kohayashi, S., Haseyama, S., and Nagaya, S. (2000) Advances in Superconductivity, Vol. XII (Eds. Yamashita, T., and Tanabe, K.). Springer, Tokyo, p. 500.

Kohen, A., and Deutscher, D. (2001) Phys. Rev. B 64 060506(R).

Kohori, Y., Matsuda, K., and Kohara, T. (1996) J. Phys. Soc. Jpn 65 1083.

Kohori, Y., Yamato, Y., Iwamoto, Y., Kohara, T., Bauer, E. D., Maple, M. B., and Sarrao, J. L. (2001) Phys. Rev. B 64 134526.

Koike, Y., Higuchi, K., and Tanuma, S. (1978) Sol. State. Commun. 27 623.

Kolodziejczyk, A., Sarkissian, B. V., and Coles, B. R. (1980) J. Phys. F 10 L333.

Kondo, J. (1964) Prog Theor. Phys. 32 37.

Kondoh, S., Sera, M., Ando, Y., and Sato, M. (1989) Physica C 157 469.

Kondrat, A., Hamann-Borrero, J. E., Leps, N., Kosmala, M., Schumann, O., Werner, J., Behr, G., Braden, M., Klingeler, R., Büchner, B., and Hess, C. (2009) Eur. Phys. J. B 70 461.

Kopelevich, Y., Esquinazi, P., Torres, J. H. S., da Silva, R. R., Kempa, H., Mrowka, F., and Ocana, R. (2003) Studies of High Temperature Superconductors, Vol. 45 (Ed. Narlikar, A. V.). Nova Science, New York, p. 59.

Koren, G., Gupta, A., Giess, E. A., Segmüller, A., and Laibowitz, R. B. (1989) Appl. Phys. Lett. 54 1054.

Korshunov, M., and Eremin, I. (2008) Phys. Rev. B 78 140509 (R).

Kortus, J., Mazin, I. I., Belashchenko, K. D., Antropov, V. P., and Boyer, L. L. (2001) Phys. Rev. Lett. 86 4656.

Kosugi, M. (1993) IEEE Trans. Appl. Supercond. 3 1010.

Kosugi, M., Akimitsu, J., Uchida, T., Furuya, M., Nagata, Y., and Ekino, T. (1994) Physica C 229 389.

Kotegawa, H., Ishida, K., Kitaoka, Y., Muranaka, T., and Akimitsu, J. (2001) Phys. Rev. Lett. 87 127001.

Kotegawa, H., Yogi, M., Imamura, Y., Kawasaki, Y., Zheng, G-Q., Kitaoka, Y., Ohsaki, S., Sugawara, H., Aoki, Y., and Sato, H. (2003) Phys. Rev. Lett. 90 027001.

Kotegawa, H., Sugawara, H., and Tou, H. (2009) J. Phys. Soc. Jpn 78 013709.

Kotegawa, H., Tomita, Y., Tou, H., Izawa, H., Mizuguchi, Y., Miura, O., Demura, S., Deguchi, K., and Takano, Y. (2012) J. Phys. Soc. Jpn 81 103702.

Kouroudis, D., Weber, D., Yoshizawa, M., Lüthi, B., Puech, L., Bruls, G., Welp, U., Franse, J. J. M., Menovsky, A., Bucher, E., and Hufnagel, J. (1987) Phys. Rev. Lett. 58 820.

Kouznetsov, K. A., Sun, A. G., Chen, B., Katz, A. S., Bahcall, S. R., Clarke, J., Dynes, R. C., Gajewski, D. A., Han, S. H., Maple, M. B., Giapintzakis, J., Kim, J.-T., and Ginsberg, D. M. (1997) Phys. Rev. Lett. 79 3050.

Krabbes, G., Fuchs, G., Schätzle, P., Gruss, S., Park, J. W., Hardinghauss, F., Stöver, G., Hayn, R., Drechsler, S.-L., and Fahr, T. (2000) Physica C 330 181.

Krabbes, G., Fuchs, G., Verges, P., Diko, P., Stöver, G., and Gruss, S. (2002) Physica C 378–381 636.

Kramer, E. J. (1973) J. Appl. Phys. 44 1360.

Kratschmer, W., Lamb, L. D., Fostiropoulos, K., and Huffman, D. R. (1990) Nature 347 354.

Kresin, V. Z. (1987) Phys. Rev. B 35 8716.

Krimmel, A., and Loidl, A. (1997) Physica B 234–236 877.

Krimmel, A., Fischer, P., Roessli, B., Maletta, H., Geibel, C., Schank, C., Grauel, A., Loidl, A., and Steglich, F. (1992) Z. Phys. B 86 161.

Krimmel, A., Loidl, A., Knorr, K., Buschinger, B., Geibel, C., Wassilew, C., and Hanfland, M. (2000) J. Phys.: Cond. Mat. 12 8801.

Kroto, H. W., Heath, J. R., O'Brien, S. C., Curl, R. F., and Smalley, R. E. (1985) Nature 318 162.

Ku, H. C., Acker, F., and Matthias, B. T. (1980) Phys. Lett. 76A 399.

Ku, H. C., Lai, C. C., You, Y. B., Shieh, J. H., and Guan, W. Y. (1994) Phys. Rev. B 50 351.

Kubo, Y., Uchikawa, F., Utsunomiya, S., Nato, K., Katagiri, K., and Kobayashi, N. (1993) Cryogenics 33 883.

Kumagai, K., Saitoh, M., Oyaizu, T., Furukawa, Y., Takashima, S., Nohara, M., Takagi, H., and Matsuda, Y. (2006) Phys. Rev. Lett. 97 227002.

Kumakura, H. (1996) Bismuth-Based High Temperature Superconductors (Ed. Maeda, H., and Togano, K.). Marcel Decker, New York, p. 451.

Kumakura, H., Kitaguchi, H., Togano, K., and Maeda, H. (1997) Studies of High Temperature Superconductors, Vol. 21 (Ed. Narlikar, A.). Nova Science, New York, p. 163.

Kumar, A., Tandon, R. P., and Awana, V. P. S. (2011) J Appl. Phys. 110 043926.

Kumar, N., Nagalakshmi, R., Kulkarni, R., Paulose, P. L., Nigam, A. K., Dhar, S. K., and Thamizhavel, A. (2009) Phys. Rev. B 79 012504.

Kumary, T. G., Janaki, J., Mani, A., Jaya, S. M., Sastry, V. S., Hariharan, Y., Radhakrishnan, T. S., and Valsakumar, M. C. (2002) Phys. Rev. B 66 064510.

Kund, M., Müller, H., Biberacher, W., Andres, K., and Saito, G. (1993) Physica B 191 274.

Kunzler, J. E. (1961a) Rev. Mod. Phys. 33 501.

Kunzler, J. E. (1961b) Bull. Am. Phys. Soc. 6 298.

Kunzler, J. E., Beuhler, E., Hsu, F. S. L., and Wernick, J. H. (1961) Phys. Rev. Lett. 6 89.

Kuper, C. G., Revzen, M., and Ron, A. (1980) Phys. Rev. Lett. 44 1545.

Kurmaev, E. Z., Wilks, R. G., Moewes, A., Skorikov, N. A., Izyumov, Y. A., Finkelstein, L. D., Li, R. H., and Chen, X. H. (2008) Phys. Rev. B 78 220503.

Kuroki, K. (2008) Sci. Technol. Adv. Mat. 9 044202.

Kursumovic, A., Cheng, Y. S., Glowacki, B. A., Madsen, J., and Evetts, J. E. (2000) J. Cryst. Growth 218 45.

Kushch, N. D., Buravov, L. I., Khomenko, A. G., Yagubskii, E. B., Rozenberg, L. P., and Shibaeva, R. P. (1993) Synth. Met. 53 155.

Kusmartseva, A. F., Sipos, B., Berger, H., Forro, L., and Tutis, E. (2009) Phys. Rev. Lett. 103 236401.

Kussmaul, A., Hellman, E. S., Hartford, E. H. Jr., and Tedrow, P. M. (1993) Appl. Phys. Lett. 63 2824.

Kuwabara, T., and Ogata, M. (2000) Phys. Rev. Lett. 85 4586.

Kuz'min, E. V., Ovchinnikov, S. G., and Baklanov, I. O. (1999) JETP 89 349.

Kwo, J., Hong, M., Trevor, D. J., Fleming, R. M., White, A. E., Farrow, R. C., Kortan, A. R., and Short, K. T. (1988) Appl. Phys. Lett. 53 2683.

Kwok, W. K., Welp, U., Crabtree, G., Vandervoort, K. G., Hulsher, R., Zheng, Y., Dabrowski, B., and Hinks, D. G. (1989) Phys. Rev. B 40 9400.

Labbé, J., and Friedel, J. (1966a) J. Phys. (Paris) 27 153.

Labbé, J., and Friedel, J. (1966b) J. Phys. (Paris) 27 303.

Lambert, S. E., Dalichaouch, Y., Maple, M. B., Smith, J. L., and Fisk, Z. (1986) Phys. Rev. Lett. 57 1619.

Lamura, G., Aurino, M., Cifariello, G., de Gennaro, E., Andreone, A., Emery, N., Hérold, C., Marêché, J.-F., and Lagrange, P. (2006) Phys. Rev. Lett. 96 107008.

Landau, L. D., and Lifshitz, E. M. (1958) Statistical Physics. Pergamon, Oxford, p. 355.

Lang, M. (1996) Supercond. Rev. 2 1.

Lang, M., Modler, R., Ahlheim, U., Helfrich, R., Reinders, P. H. P., Steglich, F., Assmus, W., Sun, W., Bruls, G., Weber, D., and Lüthi, B. (1991) Phys. Scripta T 39, 135.

Lang, M., Toyota, N., and Sasaki, T. (1993) Physica B 186–188 1046.

Lang, M., Toyota, N., Sasaki, T., and Sato, H. (1992) Phys. Rev. Lett. 69 1443.

Lang, M., and Müller, J. (2008) Superconductivity: Conventional and Unconventional Superconductors, Vol. 1 (Eds. Bennemann, K. H., and Ketterson, J. B.). Springer, Heidelberg, p. 1155.

Langhammer, C., Helfrich, R., Bach, A., Kromer, F., Lang, M., Michels, T., Deppe, M., Steglich, F., Stewart, G. R. (1998) J. Magn. Magn. Mat. 177–181, 443.

Lao, J. Y., Wang, J. H., Wang D. Z., Tu, Y., Yang, S. X., Wu, H. L., Ren, Z. F., Verebelyi, D. T., Paranthaman, M., Aytug, T., Christen, D. K., Bhattacharya, R. N., and Blaugher, R. D. (2000) Physica C 333 221.

Laplace, Y., Bobroff, J., Rullier-Albenque, F., Colson, D., and Forget, A. (2009) Phys. Rev. B 80 140501.

Lara, N., Aranda, P., Ruiz, A. I., Manríquez, V., and Ruiz-Hitzky, E. (2012) J. Braz. Chem. Soc. 23 415.

Larbalestier, D. C. (1981) IEEE Trans. Magn. 17 1668.

Larbalestier, D. C., and West, A. W. (1984) Acta Metall. 32 1871.

Larbalestier, D. C., Madsen, P. E., Lee, J. A., Wilson, M. N., and Harlesworth, J. P. (1975) IEEE Trans. Magn. 11 247.

Larkin, A. I., and Ovchinnikov, Y. N. (1964) Zh. Eksp. Teor. Fiz. 47 1136.

Laube, F., Goll, G., Hagel, J., Löhneysen, H. V., Ernst, D., and Wolf, T. (2001) Europhys. Lett. 56 296.

Lauder, A., Wilker, C., Kountz, D. J., Holstein, W. L., and Face, D. W. (1993) IEEE Trans. Appl. Supercond. 3 1683.

Laue, von M. (1932) Phys. Z. 33 793.

Laukhin, V. N., Kostyuchenko, E. E., Sushko, Yu. V., Shchegolev, I. F., and Yagubskii, E. B. (1985) Pis'ma Zh. Eksp. Teor. Fiz. 41 68.

Le, L. P., Luke, G. M., Sternlieb, B. J., Wu, W. D., Uemura, Y. J., Brewer, H. J., Riseman, T. M., Stronach, C. E., Saito, G., Yamochi, H., Wang, H. H., Kini, A. M., Carlson, K. D., and Williams, J. M. (1992) Phys. Rev. Lett. 68 1923.

Lebedev, O. I., Van Tendeloo, G., Christiani, G., Habermeier, H.-U., and Matveev, A. T. (2005) Phys. Rev. B 71 134523.

Lebègue, S. (2007) Phys. Rev. B 75 035110.

Lee, B., Khim, S., Jung, S. K., Stewart, G. R., and Kim, K. H. (2010) Europhys. Lett. 91 67002.

Lee, D. F., Partsinevelos, C. S., Presswood, R. G. Jr., and Salama, K. (1994) J. Appl. Phys. 76 603.

Lee, I. J., Naughton, M. J., Danner, G. M., and Chaikin, P. M. (1997) Phys. Rev. Lett. 78 3555.

Lee, I. J., Chow, D. S., Clark, W. G., Strouse, J., Naughton, M. J., Chaikin, P. M., and Brown, S. E. (2003) Phys. Rev. B 68 092510.

Lee, J. H., Char, K., Park, Y. W., Zhoo, L. Z., Zhu, D. B., McIntosh, G. C., and Kaiser, A. B. (2000) Phys. Rev. B 61 14815.

Lee, P. A. (2000) Nature 406 467.

Lee, P. J., McKinnell, J. C., and Larbalestier, D. C. (1989) IEEE Trans. Magn. 25 1918.

Lee, S., Jiang, J., Weiss, J. D., Bark, C. W., Tarantini, C., Biegalski, M. D., Polyanskii, A., Zhang, Y., Nelson, C. T., Pan, X. Q., Hellstrom, E. E., Gurevich, A., Eom, C. B., and Larbalestier, D. C. (2009) Appl. Phys. Lett. 95 212505.

Lee, S., Jiang, J., Zhang, Y., Bark, C. W., Weiss, J. D., Tarantini, C., Nelson, C. T., Jang, H. W., Folkman, C. M., Baek, S. H., Polyanskii, A., Abraimov, D., Yamamoto, A., Park, J. W., Pan, X. Q., Hellstrom, E. E., Larbalestier, D. C., and Eom, C. B. (2010) Nat. Mat. 9 397.

Lehmann, M., Saemann-Ischenko, G., Adrian, H., and Nolscher, C. (1981) Physica 107B 473.

Lei, H., Wang, K., Abeykoon, M., Bozin, E. S., and Petrovic, C. (2012) arXiv:1208.3189v1 [cond-mat.supra-con].

Lei, H., Zhu, X., and Petrovic, C. (2011) Europhys. Lett. 95 1711.

Leigh, N. R., Zheng, D. N., and Hampshire, D. P. (1999) IEEE. Trans. Appl. Supercond. 9 1739.

Lerf, A., Sernetz, F., Biberacher, W., and Schöllhorn, R. (1979) Mat. Res. Bull. 14 797.

Lévy, F., Shelkin, I., Grenier, B., and Huxley, A. D. (2005) Science 309 1343.

Leyarovska, L., and Leyarovska, E. (1979) J. Less-Common Met. 67 249.

Lezza, P., Senatore, C., and Flukiger, R. (2006) Supercond. Sci. Technol. 19 1030.

Li, B., Xing, Z. W., and Huang, G. Q. (2012) Europhys. Lett. 101 47002.

Li, L. J., Luo, Y. K., Wang, Q. B., Chen, H., Ren, Z., Tao, Q., Li, Y. K., Lin, X., He, M., Zhu, Z. W., Cao, G. H., and Xu, Z. A. (2009a) New J. Phys. 11 025008.

Li, S. L., de la Crus, C., Huang, Q., Chen, Y., Lynn, J. W., Hu, J. P., Huang, Y. L., Hsu, F. C., Yeh, K. W., Wu, M. K., and Dai, P. C. (2009b) Phys. Rev. B 79 054503.

Li, S. L., de la Crus, C., Huang, Q., Chen, G. F., Xia, T. L., Luo, J. L., Wang, N. L., and Dai, P. C. (2009c) Phys. Rev. B 80 020504.

Li, S. Y., Fan, R., Chen, X. H., Wang, C. H., Mo, W. Q., Ruan, K. Q., Xiong, Y. M., Luo, X. G., Zhang, H. T., Li, L., Sun, Z., and Cao, L. Z. (2001) Phys. Rev. B 64 132505.

Li, S. Y., Mo, W. Q., Yu, M., Zheng, W. H., Wang, C. H., Xiong, Y. M., Fan, R., Yang, H. S., Wu, B. M., Cao, L. Z., and Chen, X. H. (2002) Phys. Rev. B 65 064534.

Li, S., Yang, H., Tao, J., Ding, X., and Wen, H. H. (2013) Phys. Rev. B 86 214518.

Li, W., Wang, D. Z., Lao, J. Y., Ren, Z. F., Wang, J. H., Paranthaman, M., Verebelyi, D. T., and Christen, D. K. (1999) Supercond. Sci. Technol. 12 L1.

Li, Y. K., Lin, X., Zhou, T., Shen, J. Q., Tao, Q., Cao, G. H., and Xu, Z. A. (2009d) J. Phys.: Cond. Matter 21 355702.

Li, Z.-C., Lu, W., Don, X. L., Zhou, F., and Zhou, Z.-X. (2010) Chin. Phys. B 19 026103.

Li, Z.-Z, Xuan, Y., Tao, H.-J., Ren, Z.-A., Che, G.-C., Zhao, B.-R., and Zhao, Z.-X. (2001) Supercond. Sci. Technol. 14 944.

Li, Z., Ooe, Y., Wang, X.-C., Lin, Q.-Q., Jin, C.-Q., Ichioka, M., and Zheng, G.-Q. (2010) J. Phys. Soc. Jpn 79 083702.

Liang, W. Y. (1998) J. Phys.: Cond. Mat. 10 11365.

Lieber, C. M., and Zhang, Z. (1994) Sol. State Phys. 48 349.

Lin, C. T., Chen, D. P., Lemmens, P., Zhang, X. N., Maljuk, A., and Zhang, P. X. (2005) J. Cryst. Growth 275 606.

Lin, C. T., Liang, B., Ulrich, C., and Bernhard, C. (2001) Physica C 364–365 373.

Lin, C. T., Zhou, W., Liang, W. Y., Schönherr, E., and Bender, H. (1992) Physica C 195 291.

Lin, J.-Y., and Yang, H. D. (2004) Superconductivity Research at the Leading Edge (Ed. Lewis, P. S.). Nova Science, New York, p. 111.

Lin, J.-Y., Ho, P. L., Huang, H. L., Lin, P. H., Zhang, Y.-L., Yu, R.-C., Jin, C.-Q., and Yang, H. D. (2003) Phys. Rev. B 67 52501.

Lin, W.-T., Pan, S.-M., and Chen, K. (1993) Jpn J. Appl. Phys. 32 770.

Lin, X., Ni, X., Chen, B., Xu, X., Yang, X., Dai, J., Li, Y., Yang, X., Luo, Y., Tao, Q., Cao, G., and Xu, Z. (2013) Phys. Rev. B 87 020504.

Link, P., Jaccard, D., Geibel, C., Wassilew, C., and Steglich, F. (1995) J. Phys.: Cond. Mat. 7 373.

Little, W. A. (1964) Phys. Rev. 134 A1416.

Liu, A. Y., Mazin, I. I., and Kortus, J. (2001) Phys. Rev. Lett. 87 087005.

Liu, C.-J., Shew, C.-S., Wu, T.-W., Huang, L.-C., Hsu, F. H., Yang, H. D., and Williams, G. V. M. (2005) Phys. Rev. B 71 014502.

Liu, R. H., Wu, T., Wu, G., Chen, H., Wang, X. F., Xie, Y. L., Ying, J. J., Yan, Y. J., Li, Q. J., Shi, B. C., Chu, W. S., Wu, Z. Y., and Chen, X. H. (2009) Nature 459 64.

Liu, R. H., Luo, X. G., Zhang, M., Wang, A. F., Ying, J. J., Wang, X. F., Yan, Y. J., Xiang, Z. J., Cheng, P., Ye, G. J., Li, Z. Y., and Chen, X. H. (2011) Europhys. Lett. 94 27008.

Liu, R. S., and Edwards, P. P. (1991) J. Sol. State Chem. 91 407.

Liu, R. S., Jang, L. Y., Huang, H.-H., and Tallon, J. L. (2001) Phys. Rev. B 63 212507.

Livingston, J. D. (1966) Appl. Phys. Lett. 8 319.

Livingston, J. D. (1977) J. Mat. Sci. 12 1759.

Livingston, J. D. (1978) Phys. Stat. Sol. 44A 295.

Livingston, J. D., and DeSorbo, W. (1969), Superconductivity, Vol. II (Ed. Parks, R. D.). Marcel Dekker, New York, p. 1235.

Loeser, A. G., Shen, Z.-X., Dessau, D. S., Marshall, D. S., Park, C. H., Fournier, P., and Kapitulnik, A. (1996) Science 273 325.

Lokshin, K. A., Pavlov, D. A., Putilin, S. N., Antipov, E. V., Sheptyakov, D. V., and Balagurov, A. M. (2001) Phys. Rev. B 63 064511.

London, F. (1950) Superfluids, Vol. 1. Wiley, New York.

London, F., and London, H. (1935) Proc. R. Soc. Lond. A 149 74.

Looney, C., Gangopadhyay, A. K., Klehe, A.-K., and Schilling, J. S. (1995) Physica C 252 199.

Loram, J. W., Mirza, K. A., Cooper, J. R., and Liang, W. Y. (1993) Phys. Rev. Lett. 71 1740.

Loram, J. W., Mirza, K. A., Cooper, J. R., Liang, W. Y., and Wade, J. M. (1994) J. Supercond. 7 243.

Lorenz, B., Meng, R. L., Cmaidalka, J., Wang, Y. S., Lenzi, J., Xue, Y. Y., and Chue, C. W. (2001a) Physica C 363 251.

Lorenz, B., Meng, R. L., and Chu, C. W. (2001b) Phys. Rev. B 64 012507.

Lorenz, B., Xue, Y. Y., Meng, R. L., and Chu, C. W. (2002) Phys. Rev. B 65 174503.

Lorenz, B., Xue, Y. Y., and Chue, C. W. (2003) Studies of High Temperature Superconductors, Vol. 46 (Ed. Narlikar, A. V.). Nova Science, New York. p. 1.

Lortz, R., Meingast, C., Rykov, A. I., and Tajima, S. (2003) Phys. Rev. Lett. 91 207001.

Lortz, R., Wang, Y., Demuer, A., Bottger, P. H. M., Bergk, B., Zwicknagl, G., Nakazawa, Y., and Wosnitza, J. (2007) Phys. Rev. Lett. 99 187002.

Lu, J. P. (1994) Phys. Rev. B 49 5687.

Lu, W. Q., Yamamoto, Y., Petrykin, W., Kakihana, M., Matsumoto, Y., Joshi, U. S., Koinuma, H., Hasegawa, T. (2005) Thin Sol. Films 486 79.

Luetkens, H., Klauss, H.-H., Khasanov, R., Amato, A., Klingeler, R., Hellmann, I., Leps, N., Kondrat, A., Hess, C., Köhler, A., Behr, G., Werner, J., and Büchner, B. (2008) Phys. Rev. Lett. 101 097009.

Luetkens, H., Klauss, H.-H., Ktaken, M., Litterst, F. J., Dellmann, T., Klingeler, R., Hess, C., Khasanov, R., Amato, A., Baines, C., Kosmala, M., Schumann, O. J., Braden, M., Hamann-Borrero,

J., Leps, N., Kondrat, A., Behr, G., Werner, J., and Büchner, B. (2009) Nature Mat. 8 305.

Luhman, T. S., and Suenaga, M. (1977) Adv. Cryo. Eng. 22 356.

Lumsden, M. D., Dunsiger, S. R., Sonier, J. E., Miller, R. I., Kiefl, R. F., Jin, R., He, J., Mandrus, D., Bramwell, S. T., and Gardner, J. S. (2002) Phys. Rev. Lett. 89 147002.

Lumsden, T., and Christianson, A. D. (2010) J. Phys.: Cond. Mat. 22 203203.

Luo, J. (1996) Czech. J. Phys. 46 819.

Luo, X. G., Tanatar, M. A., Reid, J.-Ph., Shakeripour, H., Doiron-Leyraud, N., Ni, S., Bud'ko, S. L., Canfield, P. C., Luo, H., Wang, Z., Wen, H.-H., Prozorov, R., and Taillefer, L. (2009) Phys. Rev. B 80 140503(R).

Lupien, C., MacFarlane, W. A., Proust, C., Taillefer, L., Mao, Z. Q., and Maeno, Y. (2001) Phys. Rev. Lett. 86 5986.

Lynn, J. W. (1981) Ternary Superconductors (Eds. Shenoy, G. K., Dunlap, B. D., and Fradin, F. Y.). North-Holland, Amsterdam, p. 55.

Lynn, J. W., and Dai, P. (2009) Physica C 469 469.

Lynn, J. W., Skanthakumar, S., Huang, Q., Sinha, S. K., Hossain, Z., Gupta, L. C., Nagarajan, R., and Godart, C. (1997) Phys. Rev. B. 55 6584.

Lynn, J. W., Keimer, B., Ulrich, C., Bernhard, C., and Tallon, J. L. (2000) Phys. Rev. B 61 R14964.

Lynn, J. W., Huang, Q., Brown, C. M., Miller, V. L., Foo, M. L., Schaak, R. E., Jones, C. Y., Mackey, E. A., and Cava, R. J. (2003) Phys. Rev. B 68 414516.

Lynn, J. W., Huang, Q., Cava, R. J., and Lee, Y. S. (2005) Proc. MRSI Symp., Vol. 840, Q4.4.1.

Lynn, J. W., Chen, Y., Huang, Q., Goh, S. K., and Williams, G. V. M. (2007) Phys. Rev. B 76 014519.

Lynton, E. A. (1962) Superconductivity. Methuen, London.

Lyubovskii, R. B., Lyubovskaya, R. N., and D'yachenko, O. A. (1996) J. Phys. (Paris) 6 1609.

Ma, Y., Kumakura, H., Matsumoto, A., Hatakeyama, H., and Togano, K. (2003) Supercond. Sci. Technol. 16 852.

Machida, K. (1988) Physica C 156 276.

Mackenzie, A. P., Julian, S. R., Diver, S. J., McMullan, G. J., Ray, M. P., Lonzarich, G. G., Maeno, Y., Nishizaki, S., and Fujita, T. (1996a) Phys. Rev. Lett. 76 3786.

Mackenzie, A. P., Julian, S. R., Diver, S. J., Lonzarich, G. G., Hussey, N. E., Maeno, Y., Nishizaki, S., and Fujita, T. (1996b) Physica C 263 510.

Mackenzie, A. P., Haselwimmer, R. K. W., Tyler, A. W., Lonzarich, G. G., Mori, Y., Nishizaki, S., and Maeno, Y. (1998) Phys. Rev. Lett. 80 161.

MacLaughlin, D. E., Lan, M. D., Tien, C., Moore, J. M., Ott, H. R., Fisk, Z., and Smith, J. L. (1987) J. Magn. Magn. Mat. 63–64 455.

Maeda, H., Tanaka, Y., Fukutomi, M., and Asano, T. (1988) Jpn J. Appl. Phys. 27 L209.

Maeno, Y., Hasimoto, H., Yoshida, K., Nissizaki, S., Fujita, T., and Lichtenberg, F. (1994) Nature 372 532.

Maeno, Y., Nishizaki, S., Yoshida, K., Ikeda, S., and Fujita, T. (1996) J. Low Temp. Phys. 105 1577.

Maeter, H., Luetkens, H., Pashkevich, Y. G., Kwadrin, A., Khasanov, R., Amato, A., Gusev, A. A., Lamonova, K. V., Chervinskii, D. A., Klingeler, R., Hess, C., Behr, G., Buchner, B., and Klauss, H. H. (2009) Phys. Rev. B 80 094524.

Magishi, K., Gavilano, J. L., Pedrini, B., Hinderer, J., Weller, M., Ott, H. R., Kazakov, S. M., and Karpinski, J. (2005) Phys. Rev. B 71 024524.

Maier, T., and Scalapino, D. J. (2008) Phys. Rev. B 78 020514 (R).

Mailfert, R., Batterman, B. W., and Hanak, J. J. (1967) Phys. Lett. A 24 315.

Mailfert, R., Batterman, B. W., and Hanak, J. J. (1969) Phys. Stat. Sol. 32 K67.

Maiti, S., and Chubukov, A. V. (2010) Phys. Rev. B 82 214515.

Maki, K. (1964) Physics 1 21; 127; 201.

Maki, K. (1966) Phys. Rev. 148 362.

Malandrino, G., and Fragalá, I. L. (2004) High Temperature Superconductivity 1: Materials (Ed. Narlikar, A. V.). Springer, Heidelberg, p. 169.

Mandal, P., Hassen H., Hemberger, J., Krimmel, A., and Loidl, A. (2002) Phys. Rev. B 65 144506.

Manzano, F., Carrington, A., Hussey, N. E., Lee, S., Yamamoto, A., and Tajima, S. (2002) Phys. Rev. Lett. 88 047002.

Mao, Z. Q., Mori, Y., and Maeno, Y. (1999) Phys. Rev. B 60 610.

Mao, Z. Q., Maeno, Y., and Fukazawa, H. (2000) Mat. Res. Bull. 35 1813.

Mao, Z. Q., Maeno, Y., Mori, Y., Sakita, S., Nimori, S., and Udagawa, M. (2001) Phys. Rev. B 63 144514.

Mao, Z. Q., Rosario, M. M., Nelson, K. D., Wu, K., Deac, I. G., Schiffer, P., Liu, Y., He, T., Regan, K. A., and Cava, R. J. (2003) Phys. Rev. B 67, 094502.

Maple, M. B. (1981) Ternary Superconductors (Ed. Shenoy, G. K., Dunlap, B. D., and Fradin, F. Y.). North-Holland, Amsterdam, p. 131.

Maple, M. B. (1985) Nature 315 95.

Maple, M. B., Fertig, W. A., Mota, A. C., DeLong, L. E., Wohlleben, D., and Fitzgerald, R. (1972) Sol. State. Commun. 11 829.

Maple, M. B., Hamaker, H. C., and Woolf, L. D. (1982) Superconductivity in Ternary Compounds II (Ed. Maple, M. B., and Fischer, O.). Springer, Berlin, p. 177.

Maple, M. B., Chen, J. W., Lambert, S. E., Fisk, Z., Smith, J. L., Ott, H. R., Brooks, J. S., and Naughton, M. J. (1985) Phys. Rev. Lett. 54 477.

Maple, M. B., Chen, J. W., Dalichaouch, Y., Kohara, T., Rossel, C., Torikachvili, M. S., McElfresh, M. W., and Thompson, J. D. (1986) Phys. Rev. Lett. 56 185.

Maple, M. B., Bauer, E. D., Zapf, V. S., Freeman, E. J., and Frederick, N. A. (2001) Acta Phys. Polon. B 32 3291.

Maple, M. B., Bauer, E. D., Zapf, V. S., and Wosnitza, J. (2008) Superconductivity, Vol. 1: Conventional and Unconventional

Superconductors (Eds. Bennemann, K. H., and Ketterson, J. B.). Springer, Heidelberg, p. 639.

Maple, M. B., Baumbach, R. E., Butch, N. P., Hamlin, J. J., and Janoschek, M. (2010) J. Low Temp. Phys. 161 4.

Margadonna, S., Takabayashi, Y., Ohishi, Y., Mizuguchi, Y., Takano, Y., Kagayama, T., Nakagawa, T., Takata, M., and Prassides, K. (2009a) Phys. Rev. B 80 064506.

Margadonna, S., Takabayashi, Y., McDonald, M. T., Brunelli, M., Wu, G., Liu, R. H., Chen, X. H., and Prassides, K. (2009b) Phys. Rev. B 79 014503.

Markert, J. T., and Maple, M. B. (1989) Sol. State. Commun. 70 145.

Markert, J. T., Seaman, C. L., Zhou, H., and Maple, M. B. (1988) Sol. State. Commun. 66 387.

Markert, J. T., Early, E. A., Bjoernholm, T., Ghamaty, S., Lee, B. W., Neumeier, J. J., Price, R. D., Seaman, C. L., and Maple, M. B. (1989) Physica B 158 178.

Marsiglio, F., Carbotte, J. P., Puchkov, A., and Timusk, T. (1996) Phys. Rev. B 53 9433.

Martin, C., Kim, H., Gordon, R. T., Ni, N., Kogan, V. G., Bud'ko, S. L., Canfield, P. C., Tanatar, M. A., and Prozorov, R. (2010) Phys. Rev. B 81 060505.

Martin, L., Turner, S. S., Day, P., Mabbs, F. E., and McInnes, E. J. L. (1997) J. Chem. Soc. Chem. Commun. p. 1367.

Martindale, J. A., Barrett, S. E., O'Hara, K. E., Schlichter, C. P., Lee, W. C., and Ginsberg, D. M. (1993) Phys. Rev. B 47 9155.

Marx, D. T., Radelli, P. G., Jorgensen, J. D., Hitterman, R. L., Hinks, D. G., Pei, S., Dabrowski, B. (1992) Phys. Rev. B 46 1144.

Maśka, M. M., Mierzejewski, M., Andrzejewski, B., Foo, M. L., Cava, R. J., and Klimczuk, T. (2004) Phys. Rev. B 70 144516.

Massalami, El., Baggio-Saitovitch, E. M., Sulpice, A. (1995) J. Alloys Compounds 228 49.

Masui, T., Yoshida, K., Lee, S., Yamamoto, A., and Tajima, S. (2002) Phys. Rev. B 65 214513.

Matano, K., Ren, Z. A., Dong, X. L., Sun, L. L., Zhao, Z. X., and Zheng, G.-Q. (2008) Europhys. Lett. 83 57001.

Matano, K., Li, Z., Sun, G. L., Sun, D. L., Lin, C.-T., Ichiko, M., and Zheng, G.-Q. (2009) Europhys. Lett. 87 27012.

Mathur, N. D., Grosche, F. M., Julian, S. R., Walker, I. R., Freye, D. M., Haselwimmer, R. K. W., and Lonzarich, G. G. (1998) Nature 394 39.

Matsuda, K., Kohori, Y., and Kohara, T. (1996) J. Phys. Soc. Jpn 65 679.

Matsuda, K., Kohori, Y., and Kohara, T. (1997) Phys. Rev. B 55 15223.

Matsuda, Y. (2006) Pramana 66 239.

Matsui, H., Goto, T., Sato, N., and Komatsubara, T. (1994) Physica B 199–200 140.

Matsumoto, A., Kumakura, H., Kitaguchi, H., and Hatakeyama, H. (2003) Supercond. Sci. Technol. 16 926.

Matsumoto, K., Kim, S. B., Wen, J. G., Hirabayashi, I., Tanaka, S., Uno, N., and Ikeda, M. (1999) IEEE Trans. Appl. Supercond. 9 1539.

Matsumoto, Y., Nakatsuji, S., Kuga, K., Karaki, Y., Horie, N., Shimura, Y., Sakakibara, T., Nevidomskyy, A. H., Coleman, P. (2011) Science 331 316.

Matsushi, S., Inoue, Y., Nomura, T., Yanagi, H., Hirano, M., and Hosono, H. (2008a) J. Am. Chem. Soc. 130 14428.

Matsushi, S., Inoue, Y., Nomura, T., Hirano, M., and Hosono, H. (2008b) J. Phys. Soc. Jpn 130 113709.

Matsushita, T., Yoshimi, D., Migita, M., and Otabe, E. S. (2001) Supercond. Sci. Technol. 14 732.

Mattheiss, L. F., and Fong, C. Y. (1977) Phys. Rev. B 15 1760.

Mattheiss, L. F., and Hamman, D. R. (1988) Phys. Rev. Lett. 60 2681.

Mattheiss, L. F., Gyorgy, E. M., and Jonston, D. W. Jr. (1988) Phys. Rev. B 37 3745.

Matthias, B. T., Suhl, H., and Corenzwit, E. (1959) J. Phys. Chem. Sol. 13 156.

Matthias, B. T. (1957) Progress in Low Temperature Physics, Vol. II (Ed. Gorter, C. J.). North-Holland, Amsterdam, p. 138.

Matthias, B. T., and Corenzwit, E. (1955) Phys. Rev. 100 626.

Matthias, B. T., Corenzwit, E., and Zachariasen, W. H. (1958) Phys. Rev. 112 89

Matthias, B. T., Geballe, T. H., and Compton, V. (1963) Rev. Mod. Phys. 35 1.

Matthias, B. T., Geballe, T. H., Wielens, R. H., Corenzewit, E., and Hull, G. W. (1965) Phys. Rev. 139 A1501.

Matthias, B. T., Zachariasen, W. H., Webb, G. W., and Engelhardt, J. J. (1967) Phys. Rev. Lett. 18 781.

Matthias, B. T., Marezio, M., Corenzwit, E., Cooper, A. S., and Barz, H. E. (1972) Science 175 1465.

Matthias, B. T., Corenzwit, E., Vandenberg, J. M., and Barz, H. E. (1977) Proc. Natl. Acad. Sci. USA 74 1334.

Maxwell, E. (1950) Phys. Rev. 78 477.

Mayaffre, H., Wzietek, P., Jérome, D., Lenoir, C., and Batail, P. (1995) Phys. Rev. Lett. 75 4122.

Mazin, I. I. (2005) Phys. Rev. Lett. 95 227001.

Mazin, I. I., and Singh, D. J. (1999) Phys. Rev. Lett. 82 4324.

Mazin, I. I., Andersen, O. K., Jepsen, O., Dolgov, O. V., Kortus, J., Golubov, A. A., Kuz'menko, A. B., and van der Marcel, D. (2002) Phys. Rev. Lett. 89 107002.

Mazin, I. I., Singh, D. J., Johannes, M. D., and Du, M. H. (2008) Phys. Rev. Lett. 101 057003.

Mazumdar, C., Nagarajan, R., Godart, C., Gupta, L. C., Latroche, M., Dhar, S. K., Levy, C., Padalia, B. D., and Vijayaraghavan, R. (1993) Sol. State. Commun. 87 413.

McDonald, W. K., Curtis, C. W., Scanlan, R. M., Smathers, D. B., Marken, K., and Larbalestier, D. C. (1983) IEEE Trans. Magn. 19 1124.

McIntyre, P. C., and Cima, M. J. (1994) J. Mat. Res. 9 2219.

McIntyre, P. C., Cima, M. J., and Fai, N. M. (1990) J. Appl. Phys. 68 4183.

McLaughlin, A. C., and Attfield, J. P. (1999) Phys. Rev. B 60 14605.

McLaughlin, A. C., Zhou, W., Attfield, J. P., Fitch, A. N., and Tallon, J. L. (1999) Phys. Rev. B. 60 7512.

McManus-Driscoll, J. L., Bravman, J. C., Savoy, R. J., Gorman, G., and Beyers, R. B. (1994) J. Am. Ceram. Soc. 77 2305.

McMillan, W. L. (1968) Phys. Rev. B 167 331.

McMillan, W. L., and Rowell, J. M. (1965) Phys. Rev. Lett. 14 108

Meddeb, D., Charfi-Kaddour, S., Bennaceur, R., and Héritier, M. (2004) Phys. State Sol. (c) 1 1817.

Medvedev, S., McQueen, T. M. Troyan, I. A., Palasyuk, T., Eremets, M. I., Cava, R. J., Naghavi, S., Casper, F., Ksenfontov, V., Worthmann, G., and Felser, C. (2009) Nat. Mat. 8 630.

Meingast, C., and Larbalestier, D. C. (1989) J. Appl. Phys. 66 5971.

Meingast, C., Blank, B., Bürkle, H., Obst, B., Wolf, T., and Wühl, H. (1990) Phys. Rev. B 41 11299.

Meingast, C., Pasler, V., Nagel, P., Rykov, A., Tajima, S., and Olsson, P. (2001) Phys. Rev. Lett. 86 1606.

Meissner, W., and Ochsenfeld, R. (1933) Naturwissenschaften 21 787.

Mellis, A. J. (2006) Science 314 1888.

Mendelssohn, K., and Moore, J. R. (1935) Proc. R. Soc. Lond. A151 334; 152 34.

Metoki, N., Haga, Y., Koike, Y., and Onuki, Y. (1998) Phys. Rev. Lett. 80 5417.

Metoki, N., Kaneko, K., Matsuda, T. D., Galatanu, A., Takeuchi, T., Hashimoto, S., Ueda, T., Settai, R., Onuki, Y., and Bernhoeft, N. (2004) J. Phys.: Cond. Mat. 16 L207.

Metskhvarishvili, I. R. (2009) J. Low Temp. Phys. 155 153.

Meul, H. W., Rossel, C., Decroux, M., Fischer, O., Remenyi, G., and Briggs, A. (1984a) Phys. Rev. Lett. 53 497.

Meul, H. W., Rossel, C., Decroux, M., Fischer, O., Remenyi, G., and Briggs, A. (1984b) Physica B 126 44.

Meyer, O. (1975), New Uses of Ion Accelerators (Ed. Ziegler, J. F.). Plenum, New York, p. 323.

Michel, C., Hervieu, M., Borel, M. M., Grandin, A., Deslandes, F., Provost, J., and Raveau, B. (1987) Z. Phys. B 68 421.

Mijatovic, D., Rijnders, G., Hilgenkamp, H., Blank, D. H. A., and Rajalla, H. (2002) Physica C 372–376 596.

Mikami, M., Yoshimura, M., Mori, Y., Sasaki, T., Funahashi, R., and Shikano, M. (2003) Jpn. J. Appl. Phys. 42 7383.

Mikheenko, P., Martinez, E., Bevan, A., Abell, J. S., and MacManus-Driscoll, J. L. (2007) Supercond. Sci. Technol. 20 S264.

Milne, C. J., Argyriou, D. N., Chemseddine, A., Aliouane, N., Veira, J., and Alber, D. (2004) Phys. Rev. Lett. 93, 247007.

Milne, I. (1972) J. Mat. Sci. 7 413.

Milne, I., and Ward, D. A. (1972) Cryogenics 12 176.

Misra, P. (2009) Heavy Fermion Systems. Elsevier Science, Oxford, p. 1.

Mistry, S., Arnold, D. C., Nuttall, C. J., Lappas, A., and Green, M. A. (2004) Chem. Commun. 21 2440.

Mito, T., Kawasaki, S., Zheng, G.-Q., Kawasaki, Y., Ishida, K., Kitaoka, Y., Aoki, D., Haga, Y., and Onuki, Y. (2001) Phys. Rev. B 63 220507.

Mitrović, V. F., Horvatić, M., Berthier, C., Knebel, G., Lapertot, G., and Flouquet, J. (2006) Phys. Rev. Lett. 97 117002.

Mityamoto, T., Katagiri, J., Nagashima, K., and Murakami, M. (1999) IEEE Trans. Appl. Supercond. 9 2066.

Miyake, A., Aoki, D., and Flouquet, J. (2008) J. Phys. Soc. Jpn 77 094709.

Miyake, K., Kohno, H., and Harima, H. (2003) J. Phys.: Cond. Mat. 15 L275.

Miyake, K., Schmitt-Rink, S., and Varma, C. M. (1986) Phys. Rev. B 34 6554.

Miyazawa, K., Kihou, K., Shirage, P. M., Lee, C. H., Kito, H., Eisaki, H., and Iyo, A. (2009) J. Phys. Soc. Jpn 78 034712.

Mizuguchi, Y., Fujihisa, H., Gotoh, Y., Suzuki, K., Usui, H., Kuroki, K., Demura, S., Takano, Y., Izawa, H., and Miura, O. (2012a) Phys. Rev. B 86 R220510.

Mizuguchi, Y., Demura, S., Deguchi, K., Takano, Y., Fujihisa, H., Gotoh, Y., Izawa, H., and Miura, O. (2012b) J. Phys. Soc. Jpn 81 115725.

Mizuno, M., Endo, H., Tsuchiya, J., Kijima, N., Sumiyama, A., and Oguri, Y. (1988) Jpn J. Appl. Phys. 27 L1225.

Mizushima, K., Jones, P. C., Wiseman, P. J., and Goodenough, J. B. (1980) Mat. Res. Bull. 15 783.

Mizutani, U., Mase, A., Ikuta, H., Yanagi, Y., Yoshikawa, M., Itoh, Y., and Oka, T. (1999) Mat. Sci. Eng. B 65 66.

Mohan, S., Taen, T., Yagyuda, H., Nakajima, Y., Tamegai, T., Katase, T., Hiramatsu, H., and Hosono, H. (2010) Supercond. Sci. Technol. 23 105016

Moler, K. A., Baar, D. J., Urbach, J. S., Liang, R., Hardy, W. N., Moncton, D. E., Axe, J. D., and DiSalvo, F. J. (1975) Phys. Rev. Lett. 34 734.

Moncton, D. E., Shirane, G., Thomlinson, W., Ishikawa, M., and Fischer, Ø. (1978) Phys. Rev. Lett. 41 1133.

Moon, B. M., Platt, C. E., Schweinfurth, R. A., and Van Harlingen, D. J. (1991) Appl. Phys. Lett. 59 1905.

Mori, N., Murayama, C., Takahashi, H., Kaneko, H., Kawabata, K., Iye, Y., Uchida, S., Takagi, H., Tokura, Y., Kubo, Y., Sasakura, H., and Yamaya, K. (1991) Physica C 185–189 40.

Morin, F. J., and Maita, J. P. (1963) Phys. Rev. 129 1115.

Moriya, T. (1960a) Phys. Rev. Lett. 4 228.

Moriya, T. (1960b) Phys. Rev. 120 91.

Morocom, W. R., Worrell, W. L., Seil, H. G., and Kaplan, H. I. (1974) Metall. Trans. 5 155.

Morosan, E., Zandbergen, H. W, Dennis, B. S, Bos, J. W. G., Onose, Y., Klimczuk, T., Ramirez, A. P., Ong, N. P., and Cava, R. J. (2006) Nat. Phys. 2 544.

Morozov, A. I. (1980) Sov. Phys. Sol. State 22 1974.

Morris, R. C. (1975) Phys. Rev. Lett. 34 1164.

Morris, R. C., and Coleman, R. V. (1973) Phys. Rev. B 7 991.

Morris, R. C., Coleman, R. V., and Bhandari, R. (1972) Phys. Rev. B 5 895.

Morse, R. R., and Bohm, H. V. (1957) Phys. Rev. 108 1094.

Mortensen, K., Tomkiewicz, Y., Schultz, T. D., and Engler, E. M. (1981) Phys. Rev. Lett. 46 1234.

Moshchalkov, V., Menghini, M., Nishio, T., Chen, Q. H., Silhanek, A. V., Dao, V. H., Chibotaru, L. F., Zigadlo, N. D., and Karpinski, J. (2009) Phys. Rev. Lett. 102 117001.

Mourachkine, A. (2008) Room-Temperature Superconductivity. Cambridge International Science Publishing, Cambridge, p. 1.

Movshovich, R., Bianchi, A., Jaime, M., Hundley, M. F., Thompson, J. D., Curro, N., Hammel, P. C., Fisk, Z., Pagliuso, P. G., and Sarrao, J. L. (2002) Physica B 312–313 7.

Movshovich, R., Graf, T., Mandrus, D., Thompson, J. D., Smith, J. L., and Fisk, Z. (1996) Phys. Rev. B 53, 8241.

Movshovich, R., Jaime, M., Thompson, J. D., Petrovic, C., Pagliuso, P. G., Fisk, Z., and Sarrao, J. L. (2001) Phys. Rev. Lett. 86 5152.

Mu, G., Zhu, X. Y., Fang, L., Shan, L., Ren, C., and Wen, H. H. (2008) Chin. Phys. Lett. 25 2221.

Mukhopadhyay, S., Sheet, G., Raychaudhuri, P., and Takagi, H. (2005) Phys. Rev. B 72 014545.

Mukuda, H., Terasaki, N., Kinouchi, H., Yashima, M., Kitaoka, Y., Suzuki, S., Miyasaka, S., Tajima, S., Miyazawa, K., Shirage, P., Kito, H., Eisaki, H., and Iyo, A. (2008) J. Phys. Soc. Jpn 77 093704.

Müller, J. (1980) Rep. Prog. Phys. 43 641.

Müller, H., Lang, M., Steglich, F., Schlueter, J. A., Kini, A. M., and Sasaki, T. (2002) Phys. Rev. B 65 144521.

Müller, J., Flukiger, R., Junod, A., Heiniger, F., and Susz, C. (1972) Proc. LT13, Vol. 3. Plenum, New York, p. 446.

Müller, K. H. (1996) Czech. J. Phys. 46 829.

Müller, K.-H., Kreyssig, A., Handstein, A., Guchs, G., Ritter, C., and Lowenhaupt, M. (1997) J. Appl. Phys. 81 4240.

Müller, T., Joss, W., and Taillefer, L. (1989) Phys. Rev. B 40 2614.

Murakami, M. (1992) Studies of High Temperature Superconductors, Vol. 9 (Ed. Narlikar, A. V.). Nova Science, New York, p. 1.

Murakami, M. (1996) Studies of High Temperature Superconductors, Vol. 22 (Ed. Narlikar, A. V.) Nova Science, New York, p. 1.

Murakami, M. (2005) Frontiers in Superconducting Materials (Ed. Narlikar, A. V.). Springer, Berlin, p. 869.

Murakami, M., Morita, M., Doi, K., and Miyamoto, K. (1989) Jpn. J. Appl. Phys. 28 1189.

Murakami, M., Oyama, T., Fujimoto, H., Gotoh, S., Yamaguchi, K., Shiohara, Y., Koshizuka, N., and Tanaka, S. (1991) IEEE Trans. Magn. 27 1479.

Muralidhar, M., Koblischka, M. R., and Murakami, M. (1998) Supercond. Sci. Technol. 12 555.

Muralidhar, M., Koblischka, M. R., and Murakami, M. (2000) Supercond. Sci. Technol. 13 693.

Muralidhar, M., Jirsa, M., Sakai, N., and Murakami, M. (2001) Appl. Phys. Lett. 79 3107.

Muralidhar, M., Nariki, S., Jirsa, M., and Murakami, M. (2002a) Physica C 372–376 1134.

Muralidhar, M., Nariki, S., Jirsa, M., Wu, Y., and Murakami, M. (2002b) Appl. Phys. Lett. 80 1016.

Muralidhar, M., Sakai, N., Jirsa, M., Koshiuka, N., and Murakami, M. (2003) Appl. Phys. Lett. 83 5005.

Muralidhar, M., Sakai, N., Jirsa, M., Murakami, M., and Hirabayashi, I. (2006) Studies of High Temperature Superconductors, Vol. 50 (Ed. Narlikar, A. V.). Nova Science, New York, p. 229.

Muralidhar, M., Sakai, N., Jirsa, M., Murakami, M., and Koshizuka, N. (2004) Physica C 412–414 575.

Muralidhar, M., Sakai, N., Jirsa, M., Murakami, M., and Koshizuka, N. (2004a) Physica C 412–414 739.

Muralidhar, M., Sakai, N., Jirsa, M., Murakami, M., Koshizuka, N., and Hirabayashi, I. (2005) Physica C 426–431 654.

Muramatsu, T., Takeshita, N., Terakura, C., Takagi, H., Tokura, Y., Yonezawa, S., Muraoka, Y., and Hiroi, Z. (2005) Phys. Rev. Lett. 95 167004.

Muranaka, T., Yokoo, T., Arai, M., Margiolaki, E., Brigatti, K., Prassides, K., Petrenko, O., and Akimitsu, J. (2002) J. Phys. Soc. Jpn 71(Suppl.) 338.

Muranaka, T., Zenitani, Y., Shimoyama, J., and Akimitsu, J. (2005) Frontiers in Superconducting Materials (Ed. Narlikar, A. V.). Springer Verlag, Heidelberg, p. 937.

Mutlu, L. H., Celik, E., Ramazanoglu, M. K., Akin, Y., and Hascicek, Y. S. (2000) IEEE Trans. Appl. Supercond. 10 1154.

Mydosh, J. A., Chandra, P., Coleman, P., and Tripathi, V. (2002) Nature 417 831.

Myers, R., Liu, Y., Fobes, D., Mao, Z. Q., Yaguchi, H., and Maeno, Y. (2009) APS Meeting, 16–20 March, Abstract V33006.

Nachtrab, T., Bernhard, C., Lin, C., Koelle, D., and Kleiner, R. (2006) C. R. Phys. 7 68.

Nagamatsu, J., Nakagawa, N., Muranaka, T., Zenitani, Y., and Akimitsu, J. (2001) Nature 410 63.

Nagao, Y., Yamaura, J., Ogusu, H., Okamoto, Y., and Hiroi, Z. (2009) J. Phys. Soc. Jpn 78 064702.

Nagarajan, R., and Gupta, L. C. (1998) Studies of High Temperature Superconductors, Vol. 26 (Ed. Narlikar, A.). Nova Science, New York, p. 1.

Nagarajan, R., Mazumdar, C., Hossain, Z., Dhar, S. K., Gopalakrishnan, K. V., Gupta, L. C., Godart, C., Padalia, B. D., and Vijayaraghavan, R. (1994) Phys. Rev. Lett. 72 274.

Nagarajan, R., Mazumdar, Chandan, Hossain, Z., and Gupta, L. C. (2005) Frontiers in Superconducting Materials (Ed. Narlikar, A. V.). Springer, Heidelberg, p. 393.

Nagata, A., Hirayama, H., Noto, K., and Izumi, O. (1977) J. Appl. Phys. 48 5175.

Naidyuk, Y. G., von Löhneysen, H., Goll, G., Yanson, I. K., and Menovsky, A. A. (1996) Europhys. Lett. 33 557.

Nakai, Y., Ishida, K., Kamihara, Y., Hirano, M., and Hosono, H. (2008) J. Phys. Soc. Jpn 77 073701.

Nakai, Y., Iye, T., Kitagawa, S., Ishida, K., Kasahara, S., Shibauchi, T., Matsuda, Y., and Terashima, T. (2010) Phys. Rev. B 81 020503(R).

Nakajima, H., Nomura, K., and Sambongi, T. (1986) Physica B+C 143 240.

Nakamura, K., and Freeman, A. J. (2002) Phys. Rev. B 66 R140405.

Nakamura, K., Park, K. T., and Freeman, A. J. (2001) Phys. Rev. B 63 024507.

Nakatsuji, S., Kuga, K., Machida, Y., Tayama, T., Sakakibara, T., Karaki, Y., Ishimoto, H., Yonezawa, S., Maeno, Y., Pearson, E., Lonzarich, G. G., Balicas, L., Lee, H., and Fisk, Z. (2008) Nat. Phys. 4 603.

Nam, M.-S., Simmington, J. A., Singleton, J., Blundell, S., Ardavan, A., Kurmoo, M., and Day, P. (1999) J. Phys.: Cond. Mat. 11 L477.

Nariki, S., and Murakami, M. (2002) Supercond. Sci. Technol. 15 786.

Nariki, S., Seo, S. J., Sakai, N., and Murakami, M. (2000) Supercond. Sci. Technol. 13 778.

Nariki, S., Sakai, N., and Murakami, M. (2001) Physica C 357–360 814.

Nariki, S., Sakai, N., Matsui, M., and Murakami, M. (2002a) Physica C 378–381 774.

Nariki, S., Sakai, N., and Murakami, M. (2002b) Physica C 378–381 631.

Nariki, S., Sakai, N., and Murakami, M. (2002c) Supercond. Sci. Technol. 15 648.

Nariki, S., Sakai, N., and Murakami, M. (2003) 4th Int. PASREG Workshop, Jena.

Narlikar, A. V. (1987) Pramana 28 529.

Narlikar, A. V. (Ed.) (2002a) Studies of High Temperature Superconductors, Vol. 38. Nova Science, New York.

Narlikar, A. V. (Ed.) (2004) High Temperature Superconductivity 2: Engineering Applications. Springer, Berlin.

Narlikar, A. V., Agarwal, S. K., and Rao, C. V. N. (1989a) Synth. Met. 33 141.

Narlikar, A. V., and Agarwal, S. K. (1988) Curr. Sci. 57 753.

Narlikar, A. V., and Dew-Hughes, D. (1964) Phys. State Sol. 6 383.

Narlikar, A. V., and Dew-Hughes, D. (1966) J. Mat. Sci. 1 317.

Narlikar, A. V., and Dew-Hughes, D. (1985) Oxford University Engineering Laboratory Report OUEL 1609/85.

Narlikar, A. V., and Dew-Hughes, D. (1987) Proc. Int. Conf. Cryogenics: INCONCRYO-85. Tata McGraw-Hill, Delhi, p. 519.

Narlikar, A. V., and Ekbote, S. N. (1983) Superconductivity and Superconducting Materials. South Asian Publishers, Delhi.

Narlikar, A. V., Rao, C. V. N, and Agarwal, S. K. (1989) Studies of High Temperature Superconductors, Vol. 1 (Ed. Narlikar, A.). Nova Science, New York, p. 341.

Narlikar, A. V., Samanta, S. B., and Dutta, P. K. (1994) Phil. Trans. R. Soc. Lond. 346 307.

Narlikar, A. V., Gupta, A., Samanta, S. B., Chen, C., Hu, Y., Wondre, F., Wanklyn, B. M., and Hodby, J. W. (1999) Phil. Mag. B 79 717.

Narlikar, A. V., Herrmann, P. S. P., Samanta, S. B., Narayan, H., Gupta, A., Kanjilal, D., Vijayaraghavan, R., Muranaka, T., and Akimitsu, J. (2002) Studies of High Temperature Superconductors, Vol. 38 (Ed. Narlikar, A.). Nova Science, New York, p. 443.

Nash, K. K., Sully, K. J., and Horn, A. B. (2001) J. Phys. Chem. A 105 9422.

Neal, D. F., Barber, A. C., Woolcock, A., and Gidley, J. A. F. (1971) Acta Metall. 19 143.

Nelson, K. D., Mao, Z. Q., Maeno, Y., and Liu, Y. (2003) Physica C 388–389 491.

Nelson, K. D., Mao, Z. Q., Maeno, Y., and Liu, Y. (2004) Science 12 1151.

Nembach, E. (1966) Phys. State Sol. 13 543.

Nembach, E. (1970) Z. Metal. 61 734.

Nemetschek, R., Prusseit, W., Holzapfel, B., Eickmeyer, J., deBore, B., Miller, U., and Maher, E. (2002) Physica C 372–376 880.

Ni, N., Tillman, M. E., Yan, J. Q., Kracher, A., Hannahs, S. T., Bud'ko, S. L., and Canfield, P. C. (2008) Phys. Rev. B 78 214515.

Ni, N., Thaler, A., Kracher, A., Yan, J. Q., Bud'ko, S. L., and Canfield, P. C. (2009) Phys. Rev. B 80 024511.

Nicol, J., Shapiro, S., and Smith, P. H. (1960) Phys. Rev. Lett. 5 461.

Niedermayer, Ch., Bernhard, C., Holden, T., Kramer, R. K., and Ahn, K. (2002) Phys. Rev. B 65 094512.

Nigam, R. (2010) PhD thesis, University of Wollongong (2010).

Nigam, R., Pan, A. V., and Dou, S. X. (2007) J. Appl. Phys. 101 09G109.

Ning, F., Ahilan, K., Imai, T., Seafat, A. S., Jin, R., McGuire, M. A., Sales, B. C., and Mandrus, D. (2008) J. Phys. Soc. Jpn 77 103705.

Nishizaki, S., Maneo, Y., and Mao, Z. Q. (1999) J. Low Temp. Phys. 117 1581.

Nishizaki, S., Maneo, Y., and Mao, Z. Q. (2000) Phys. Rev. Lett. 85 5412.

Niu, H. J., and Hampshire, D. P. (2003) Phys. Rev. Lett. 91 027002–1.

Niu, H. J., and Hampshire, D. P. (2004) Phys. Rev. B 69 174503.

Nobumasa, H., Shimizu, K., Kitano, Y., and Kawai, T. (1988) Jpn J. Appl. Phys. 27 L1669.

Noguchi, S., and Okuda, K. (1994) Physica B 194–196 1975.

Nohara, M., Suzuki, H., Mangkorntong, N., and Takagi, H. (2000) Physica C 341–348 2177.

Norman, M. R., Pines, D., and Kallin, C. (2005) Adv. Phys. 54 715.

Norton, D. P., Goyal, A., Budai, J. D., Christen, D. K., Kroeger, D. M., Specht, E. D., He, Q., Saffian, B., Paranthaman, M., Klabunde, C. E., Lee, D. F., Sales, B. C., List, F. A. (1996) Science 274 755.

Nozières, P., and Vinen, W. F. (1966) Phil. Mag. 14 667.

Nuñez-Regueiro, M., Mignot, J. M., and Castello, D. (1992) Europhys. Lett. 18 53.

Nuñez-Regueiro, M., Mignot, J. M., Jaime, M., Castello, D., and Monceau, P. (1993) Synth. Met. 56 2653.

Nuñez-Regueiro, M., Tholence, J.-L., Antipov, E. V., Capponi, J. J., and Marezio, M. (1993) Science 262 97.

Obertelli, S. D., Cooper, J. R., and Tallon, J. L. (1992) Phys. Rev. B 46 14928.

Obst, B., Pattanayak, D. R., and Hochstuhl, P. (1980) J. Low Temp. Phys. 14 595.

Odermatt, R., Fischer, Ø., Jones, H., and Bongi, G. (1974) J. Phys. C7 L13.

Oeschler, N., Fisher, R. A., Phillips, N. E., Gordon, J. E., Foo, M. L., and Cava, R. J. (2005) Chin. J. Phys. 43 574.

Oeschler, N., Fisher, R. A., Phillips, N. E., Gordon, J. E., Foo, M. L., and Cava, R. J (2008) Phys. Rev. B 78, 054528.

Oguchi, T. (1995) Phys. Rev. B 51 1385.

Oh, Y. S., Liu, Y., Yan, L. Q., Kim, K. H., Greene, R. L., and Takeuchi, I. (2009) Phys. Rev. Lett. 102 147002.

Ohara, T., Kumakura, H., and Wada, H. (2001) Physica C 13 1272.

Ohishi, K., Muranaka, T., Akimitsu, J., Koda, A., Higemoto, W., and Kadona, R. (2003) J. Phys. Soc. Jpn 72 29.

Ohmichi, E., Ishiguro, T., Sakon, T., Sasaki, T., Motokawa, M., Lyubovskii, R. B., and Lyubovskaya, R. N. (1999) J. Supercond. 12 505.

Ohnishi, T., Huh, J.-U., Hammond, R. H., and Jo, W. (2004) J. Mat. Res. 19 977

Okano, T., Sei, T., and Tsuchiya, T. (1992) J. Mat. Sci. 27 4085.

Okawa, M., Matsunami, M., Ishizaka, K., Eguchi, R., Taguchi, M., Chainani, A., Takata, Y., Yabashi, M., Tamasaku, K., Nishino, Y., Ishikawa, T., Kuga, K., Horie, N., Nakatsuji, S., and Shin, S. (2010) Phys. Rev. Lett. 104 247201–1.

Okoye, C. M. I. (2009) Proc. 2nd Int. Semin. on Theoretical Physics and National Development, 5–8 July, 2009, Abuja, Nigeria.

Okuda, S., Suenaga, M., and Sabatini, R. L. (1983) J. Appl. Phys. 54 289.

Olsen, K. M., Jack, R. F., Fuchs, E. O., and Hsu, F. S. L. (1962) Superconductors (Ed. Tanenbaum, M., and Wright, W. V.). Wiley-Interscience, New York, p. 123.

Onnes, H. K. (1908) Commun. Phys. Lab. Leiden, p. 108.

Onnes, H. K. (1911) Commun. Phys. Lab. Leiden, pp. 122b, 124c.

Oomi, G., Honda, F., Ohashi, M., Eto, T., Hai, D. P., Kamisawa, S., Watanabe, M., and Kadowaki, K. (2002) Physica B 312–313 88.

Osborn, R., Goremychkin, E. A., Kolesnikov, A. I., and Hinks, D. G. (2001) Phys. Rev. Lett. 87 017005.

Oshiba, K., and Hotta, T. (2012) J. Phys. Soc. Jpn 81 114711.

Oshima, K., Mori, T., Inokuchi, H., Urayama, H., Yamochi, H., and Saito, G. (1988a) Phys. Rev. B 38 938.

Oshima, K., Urayama, H., Yamochi, H., and Saito, G. (1988b) Synth. Met. 27 419.

Oshima, K., Urayama, H., Yamochi, H., and Saito, G. (1988c) Physica C 153–1551148.

Ott, H. R., Rudigier, H., Rice, T. M., Ueda, K., Fisk, Z., and Smith, J. L. (1984) Phys. Rev. Lett. 52, 1915.

Ott, H. R. (2008) Superconductivity: Conventional and Unconventional Superconductors, Vol. 1 (Eds. Bennemann, K. H., and Ketterson, J. B.), Springer, Heidelberg, p. 773.

Ott, H. R., Fertig, W. A., Johnston, D. C., Maple, M. B., and Matthias, B. T. (1978) J. Low Temp. Phys. 33 159.

Ott, H. R., Odoni, W., Hamaker, H. C., and Maple, M. B. (1980) Phys. Lett. 75A 243.

Ott, H. R., Rudigier, H, Fisk, Z., and Smith, J. L. (1985) Phys. Rev. B 31, 1651.

Ott, H. R., Felder, E., Bruder, C., and Rice, T. M. (1987) Europhys. Lett. 3 1123.

Otubo, J., Pourrahimi, S., Zhang, H., Thieme, C. L. H., and Foner, S. (1983) Appl. Phys. Lett. 42 469.

Ovchinnikov, S. G. (2003a) Phys. Usp. 46 21.

Ovchinnikov, S. G. (2003b) J. Magn. Magn. Mat. 258–259 210.

Paglione, J., and Greene, R. L (2010) Nat. Phys. 6 645.

Paglione, J., Tanatar, M. A., Hawthorn, D. G., Boaknin, E., Hill, R. W., Ronning, F., Sutherland, M., Taillefer, L., Petrovic, C., and Canfield, P. C. (2003) Phys. Rev. Lett. 91 246405.

Palstra, T., Menovsky, A. A., Vandenberg, J., Dirkmaat, A. J., Kes, P. H., Nieuwenhuys, G. J., and Mydosh, J. A. (1985) Phys. Rev. Lett. 55 2727.

Palstra, T. T. M., Zhou, O., Iwasa, Y., Sulewski, P. E., Fleming, R. M., and Zegarski, B. R. (1995) Sol. State. Commun. 93 327.

Pambianchi, M. S., Anlage, S. M., Hellmann, E. S., Hartford, E. H. Jr., Bruns, M., and Lee, S. Y. (1994) Appl. Phys. Lett. 64 244.

Pan, A. V., Zhou, S., Liu, H. K., and Dou, S. X. (2003) Supercond. Sci. Technol. 16 L33.

Paolasini, L., Paixao, J. A., Lander, G. H., Delapalme, A., Sato, N., and Komatsubara, T. (1993) J. Phys.: Cond. Mat. 5, 8905.

Paranthaman, M. P. (2010) High Temperature Superconductors (Ed. Bhattacharya, R., and Paranthaman, M. P.). Wiley-VCH, Weinheim, p. 93.

Paranthaman, M. P., Chirayil, T. G., List, F. A., Cui, X., Goyal, A., Lee, D. F., Specht, E. D., Martin, P. M., Williams, R. K., Kroeger, D. M., Morrell, J. S., Beach, D. B., Feenstra, R., and Christen, D. K. (2001) J. Am. Ceram. Soc. 84 273.

Park, J. T., Inosov, D. S., Niedermayer, C., Sun, G. L., Haug, D., Christensen, N. B., Dinnebier, R., Boris, A. V., Drew, A. J., Schulz, L., Shapoval, T., Wolff, U., Neu, V., Yang, X. P., Lin, C. T., Keimer, B., and Hinkov, V. (2009) Phys. Rev. Lett. 102 117006.

Park, M. S., Giim, J. S., Park, S. H., Lee, Y. W., and Choi, E. J. (2004) Supercond. Sci. Technol. 17 274.

Parkin, S. S. P., Engler, E. M., Schumaker, R. R., Lagier, R., Lee, V. Y., Scott, J. C., and Greene, R. L. (1983) Phys. Rev. Lett. 50 270.

Parks Cheney, C., Bondino, F., Callcott, T. A., Vilmercati, P., Ederer, D., Magnano, E., Malvestuto, M., Parmigiani, F., Sefat, A. S., McGuire, M. A., Jin, R., Sales, B. C., Mandrus, D., Singh, D. J., Freeland, J. W., and Mannella, N. (2010) Phys. Rev. B 81 104518.

Parrell, J. A., Field, M. B., Zhang, Y., and Hong, S. (2004) Adv. Cryo. Eng. (Mat.) 50B 369.

Paulose, P. L., Dhar, S. K., Chinchure, A. D., Alleno, E., Godart, C., Gupta, L. C., and Nagarajan, R. (2003) Physica C 399 165.

Pedron, D., Visentini, G., Bozio, R., Williams, J. M., and Schlueter, J. A. (1997) Physica C 276 1.

Pei, S., Zalusec, N. J., Hinks, D. G., Mitchell, A. W., and Richards, D. R. (1989a) Phys. Rev. B 39 811.

Pei, S., Jorgensen, J. D., Hinks, D. G., Dabrowski, B., Richards, D. R., Mitchell, A. W., Zheng, Y., Newsam, J. M., Sinha, S. K., Vaknin, D., and Jacobson, A. J. (1989b) Physica C 162–164 556.

Pei, S., Jorgensen, J. D., Dabrowski, B., Hinks, D. G., Richards, D. R., Mitchell, A. W., Newsam, J. M., Sinha, S. K., Vaknin, D., and Jacobson, A. J. (1990) Phys. Rev. B 41 4126.

Peierls, R. E. (1955) Quantum Theory of Solids. Oxford University Press, Oxford, p. 108.

Pérez-Navarro, A., Costa-Quintana, J., and López-Aguilar, F. (2000) Phys. Rev. B 61 10125.

Petrović, A. P., Lortz, R., Santi, G., Berthod, C., Dubois, C., Decroux, M., Demuer, A., Antunes, A. B., Paré, A., Salloum, D., Gougeon, P., Potel, M., and Fischer, O. (2011) Phys. Rev. Lett. 106 017003.

Petrovic, C., Movshovich, R., Jaime, M., Pagliuso, P. G., Hundley, M. F., Sarrao, J. L., Fisk, Z., and Thompson, J. D. (2001a) Europhys. Lett. 53 354.

Petrovic, C., Pagliuso, P. G., Hundley, M. F., Movshovich, R., Sarrao, J. L., Thompson, J. D., Fisk, Z., Monthoux, P. (2001b) J. Phys.: Cond. Mat. 13, L337.

Petrovic, C., Hundley, M. F., Movshovich, R., Pagliuso, P. G., Sarrao, J. L., and Fisk, Z. (2001c) J. Alloys Compounds 325 1.

Pfeiffer, I., and Hillmann, H. (1968) Acta Metall. 16 1429.

Pfleiderer, C. (2009) Rev. Mod. Phys. 81 1551.

Phillips, N. E. (1959) Phys. Rev. 114 676.

Phok, S., Galez, P., Jorda, J. L., Peroz, C., Villard, C., de Barros, D., and Weiss, F. (2003) IEEE Trans. Appl. Supercond. 13 2864.

Piano, S., Bobba, F., Giubileo, F., and Cucolo, A. M. (2005) Int. J. Mod. Phys. B 19 323.

Pickett, W. E., Weht, R., and Shick, A. D. (1999) Phys. Rev. Lett. 83 3713.

Pickus, M. R., Dariel, M. P., Holthius, J. T., Ling-Fai Wang, J., and Granda, J. (1976) Appl. Phys. Lett. 29 810.

Pickus, M. R., Holthius, J. T., and Rosen, M. (1980) Filamentary A-15 Superconductors (Eds. Suenaga, M., and Clark, A. F.). Plenum, New York, p. 331.

Pinterić, M., Prester, M., Tomić, S., Maki, K., Schweitzer, D., Heinen, I., and Strunz, W. (1999) Synth. Met. 103 1869.

Pippard, A. B. (1950) Proc. R. Soc. Lond. A 203 210.

Pippard, A. B. (1953) Proc. R. Soc. Lond. A 216 547.

Pomjakushin, V. Yu, Sheptyakov, D. V., Pomjakushina, E. V., Krzton-Maziopa, A., Conder, K., Chernyshov, D., Svitlyk, V., and Shermadini, Z. (2011) Phys. Rev. B 83 144410.

Pontius, R. B. (1937) Nature 139 1065.

Portis, A. M., Blazey, K. W., Rossel, C., and Decroux, M. (1988) Physica C 153–155633.

Powell, B. J., Annett, J. F., and Gyorffy, B. L. (2003) J. Phys. A 36 9289.

Pratt, D. K., Tian, W., Kreyssig, A., Zarestky, J. L., Nandi, S., Ni, N., Bud'ko, S. L., Canfield, P. C., Goldman, A. I., and McQueeney, R. J. (2009) Phys. Rev. Lett. 103 087001.

Pratt, F. L., Caulfield, J., Cowey, L., Singleton, J., Doporto, M., Hayes, W., Perenboom, J. A. A. J., Kurmoo, M., and Day, P. (1993) Synth. Met. 55–57 2289.

Presland, M. R., Tallon, J. L., Buckley, R. G., Liu, R. S., and Flower, N. E. (1991) Physica C 176 95.

Pribulová, Z., Kačmarčik, J., Marcenat, C., Szabó, P., Klein, T., Demuer, A., Rodiere, P., Jang, D. J., Lee, H. S., Lee, H. G., Lee, S.-I., and Samuely, P. (2011) Phys. Rev. B 83 104511.

Prober, D. E., Schwall, R. E., and Beasley, M. R. (1980) Phys. Rev. B 21 2717.

Profita, G., Continenza, A., Bernardini, F., Satta, G., and Massidda, S. (2002) Studies of High Temperature Superconductors, Vol. 38 (Ed. Narlikar, A.). Nova Science, New York, p. 117.

Prokes, K., Tahara, T., Echizen, Y., Takabatake, T., Fujita, T., Hagmusa, I. H., Klaasse, J. C. P., Brück, E., de Boer, F. R., Divis, M., and Sechosky, V. (2002) Physica B 311 220.

Promin, A. V., Pimenov, A., Loidl, A., and Krasnosvobodtsev, S. I. (2001) Phys. Rev. Lett. 87 097003.

Prozorov, R., Ni, N., Tanatar, M. A., Kogan, V. G., Gordon, R. T., Martin, C., Blomberg, E. C., Prommapan, P., Yan, J. Q., Bud'ko, S. L., and Canfield, P. C. (2008) Phys. Rev. B 78 224506.

Prozorov, R., Snezhko, A., He, T., and Cava, R. J. (2003) Phys. Rev. B 68 180502.

Puchkov A. V., Shen, Z.-X., Kimura, T., and Tokura, Y. (1998) Phys. Rev. B 58 13322.

Puech, L., Mignot, J.-M., Lejay, P., Haen, P., Flouquet, J., and Voiron, J. (1988) J. Low Temp. Phys. 70 237.

Putilin, S. N., Antipov, E. V., Abakumov, A. M., Rozova, M. G., Lokshin, K. A., Pavlov, D. A., Balagurov, A. M., Sheptyakov, D. V., and Marezio, M. (2001) Physica C 338 52.

Putilin, S. N., Antipov, E. V., Chrnaissem, O., and Marezio, M. (1993) Nature 362 226.

Putti, M., Galleani d' Agliano, E., Marrè, D., Napoli, F., Tassisto, M., Manfrinetti, P., and Palenzona, A. (2002) Studies of High Temperature Superconductors, Vol. 38 (Ed. Narlikar, A.), Nova Science, New York, p. 121.

Qi, X., and MacManus-Driscoll, J. L. (2001) Sol. State Mat. Sci. 5 291.

Qi, Y. P., Wang, L., Wang, D. L., Zhang, X. P., and Ma, Y. W. (2008a) Supercond. Sci. Technol. 21 115016.

Qi, Y. P., Gao, Z. S., Wang, L., Wang, D. L., Zhang, X. P., and Ma, Y. W. (2008b) New J. Phys. 10 123003.

Qi, Y. P., Wang, L., Gao, Z. S., Wang, D. L., Zhang, X. P., and Ma, Y. W. (2009a) Physica C 469 19212.

Qi, Y. P., Wang, L., Gao, Z. S., Wang, D. L., Zhang, X. P., Zhang, Z. Y., and Ma, Y. W. (2009b) Phys. Rev. B 80 054502.

Qi, Y. P., Wang, L., Wang, D. L., Zhang, Z. Y., Gao, Z. S., Zhang, X. P., and Ma, Y. W. (2010) Supercond. Sci. Technol. 23 055009.

Qiu, Y., Bao, W., Huang, Q., Yildrim, T., Simmons, J., Lynn, J., Gasparovic, Y., Li, J., Green, M., Wu, T., Wu, G., and Chen, X. (2008) Phys. Rev. Lett. 101 257002.

Quinn, G. C. (1977) Lawrence-Berkeley Laboratory Report LBL-6999.

Radovan, H. A., Fortune, N. A., Murphy, T. P., Hannahs, S. T., Palm, E. C., Tozer, S. W., and Hall, D. (2003) Nature 425 51.

Raffius, H., Mörsen, E., Mosel, B. D., Müller-Warmuth, W., Jeitschko, W., Terbüchte, L., and Vomhof, T. (1993) J. Phys. Chem. Sol. 54 135.

Rall, D., Il'in, K., Iida, K., Haindl, S., Kurth, F., Thersleff, T., Schultz, L., Holzapfel, B., and Siegel, M. (2011) Phys. Rev. B 83 134514.

Ramirez, A. P. (1994) Supercond. Rev. 1 1.

Ramirez, A. P., Siegrist, T., Palstra, T. T. M., Garrett, J. D., Bruck, E., Menovsky, A. A., and Mydosh, J. A. (1991) Phys. Rev. B 44 5392.

Ramirez, A. P., Kortan, A. R., Rosseinsky, M. J., Duclos, S. J., Mujsce, A. M., Haddon, R. C., Murphy, D. W., Mukhija, A. V., Zahurak, S. M., and Lyons, K. B. (1992a) Phys. Rev. Lett. 68 1058.

Ramirez, A. P., Coleman, P., Chandra, P., Bruck, E., Menovsky, A. A., Fisk, Z., and Bucher, E. (1992b) Phys. Rev. Lett. 68 2680.

Ramsbottom, H. D., and Hampshire, D. P. (1997) Physica C 274 295.

Randeria, M., Trivedi, N., Moreo, A., and Scalettar, R. T. (1992) Phys. Rev. Lett. 69 2001.

Ravindram, P., Sankaralingan, S., and Asokamani, R. (1996) Phys. Rev. B 52 12921.

Ray, R., Ghoshray, A., Ghoshray, K., and Nakamura, S. (1999) Phys. Rev. B 59 9454.

Raychaudhuri, P., Jaiswal-Nagar, D., Sheet, G., Ramakrishnan, S., and Takagi, H. (2004) Phys. Rev. Lett. 93 156802.

Reddi, B. V., Raghavan, V., Ray, S., and Narlikar A. V. (1983) J. Mat. Sci. 18 1165.

Reddi, B. V., Ray, S., Raghavan, V., and Narlikar, A. V. (1978) Phil. Mag. A 38 559.

Reichardt, W., Batlogg, B., and Remeika, J. P. (1985) Physica B 13 501.

Ren, Y., Weinstein, R., Sawh, R., and Liu, J. (1997) Physica C 282–287 2301.

Ren, Z. F., Lao, J. Y., Guo, L. P., Wang, J. H., Bidai, J. D., Christen, D. K., Goyal, A., Paranthaman, M., Specht, E. D., and Thompson, J. R. (1998) J. Supercond. 11 159.

Ren, Z.-A., Wei, L., Jie, Y., Wei, Y., Li, S. X., Cai, Z., Can, C. G., Li, D. X., Ling, S. L., Fang, Z., and Xian, Z. Z. (2008a) Chin Phys. Lett. 25 2215.

Ren, Z.-A., Yang, J., Lu, W., Yi, W., Che, G.-C., Dong, X.-L., Sun, L.-L., and Zhao, Z.-X. (2008b) Mat. Res. Innov. 12 106.

Ren, Z.-A., Yang, J., Lu, W., Yi, W., Shen, X.-L., Li, Z.-C., Chi, G.-C., Dong, X.-L., Sun, L.-L. Zhou, F., and Zhao, Z.-X. (2008c) Europhys. Lett. 82 57002.

Ren, Z.-A., Chi, G.-C., Dong, X.-L., Yang, J., Lu, W., Yi, W., Shen, X.-L., Li, Z.-C., Sun, L.-L. Zhou, F., and Zhao, Z.-X. (2008d) Europhys. Lett. 83 345.

Ren, Z.-A., Tao, Q., Jiang, S., Feng, C., Wang, C., Dai, J., Cao, G., and Xu, Z. (2009) Phys. Rev. Lett. 102 137002.

Reynolds, C. A., Serin, B., Wright, W. H., and Nesbitt, L. B. (1950) Phys. Rev. 78 487.

Riblet, G., and Winzer, K. (1971) Sol. State. Commun. 9 1663.

Rice, M. J. (1975) Sol. State. Commun. 16 1285.

Rice, T. M., and Sigrist, H. (1995) J. Phys. Cond. Mat. 7L 643.

Richards, P. L., and Tinkham, M. (1960) Phys. Rev. 119 575.

Rickel, M. O., Togonidze, T., and Tsebro, V. (1986) Sov. Phys. Sol. State 28 1496.

Riemer, D. E. (1988) Sol. State Tech. 85 107.

Rigamonti, A., Borsa, F., and Caretta, P. (1998) Rep. Prog. Phys. 61 1367.

Rijnders, G., and Blank, D. H. A. (2005) Frontiers in Superconducting Materials (Ed. Narlikar, A. V.). Springer, Heidelberg, p. 913.

Riseborough, P. S., Schmiedeshoff, G. M., and Smith, J. L. (2008) Superconductivity: Conventional and Unconventional Superconductors, Vol. 1 (Eds. Bennemann, K. H., and Ketterson, J. B.). Springer, Heidelberg, p. 1031.

Riseman, T. M., Kealey, P. G., Forgan, E. M., Mackenzie, A. P., Galvin, L. M., Tyler, A. W., Lee, S. L., Ager, C., McPaul, D., Aegerter, C. M., Cubitt, R., Mao, Z. Q., Akima, T., and Maeno, Y. (1998) Nature 396 242.

Roberge, R., and Fishey, J. L. (1977) J. Appl. Phys. 48 1327.

Roberts, B. W. (1978) NBS Tech. Note 983, US Dept of Commerce.

Rogers, B. A., and Atkins, D. F. (1955) Trans. AIME 203 1034.

Rosner, H., An, J. M., Ku, W., Johannes, M. D., Scalettar, R. T., Pickett, W. E., Shulga, S. V., Drechsler, S.-L., Eschrig, H., Weber, W., and Eguiluz, A. G. (2002) Studies of High Temperature Superconductors, Vol. 38 (Ed. Narlikar, A.). Nova Science, New York, p. 25.

Rosner, H., Weht, R., Johannes, M. D., Pickett, W. E., and Tosatti, E. (2002) Phys. Rev. Lett. 88 027001.

Rossat-Mignod, J., Regnault, L. P., Vettier, C., Bourges, P., Burlet, P., Bossy, J., Rosseinsky, M. J., Ramirez, A. P., Glarum, S. H., Murphy, D. W., Haddon, R. C., Hebard, A. F., Palstra, T. T. M., Kortan, A. R., Zahurak, S. M., and Makhija, A.V. (1991) Phys. Rev. Lett. 66 2830.

Rossel, C., Meul, H. W., Decroux, M., Fischer, O., Remenyi, G., and Briggs, A. (1985) J. Appl. Phys. 57 3099.

Rotter, M., Tegel, M., and Johrendt, D. (2008) Phys. Rev. Lett. 101 107006.

Rotter, M., Tegel, M., Schellenberg, I., Schappacher, F. M., Pottgen, R., Deisenhofer, J., Gunther, A., Schrettle, F., Loidl, A., and Johrendt, D. (2009) New J. Phys. 11 025014.

Rotundu, C. R., Keane, D. T., Freelon, B., Wilson, S. D., Kim, A., Valdivia, P. N., Bourret- Courchesne, E., and Birgeneau, R. J. (2009) Phys. Rev. B 80 144517.

Ruddlesden, S. N., and Propper, P. (1958) Acta Crystallogr. 11 54.

Rueff, J.-P., Raymond, S., Yaresko, A., Braithwaite, D., Leininger, P., Vanko, G., Huxley, A., Rebizant, J., and Sato, N. (2007) Phys. Rev. B 76 085113.

Rui, X. F., Zhao, Y., Xu, Y. Y., Zhang, L., Sun, X. F., Wang, Y. Z., and Zhang, H. (2004) Supercond. Sci. Technol. 17 689.

Rukang, Li, Chaoshui, X., Hong, Z., Bin, L., and Li, Y. (1995) J. Alloys Compounds 223 53.

Rullier-Albenque, F., Colson, D., Forget, A., Thuery, P., and Poissonne, S. (2010) Phys. Rev. B 81 224503.

Ruwalds, J. (1987) Phys. Rev. B 35 8869.

Ryan, D. H., Cadogan, J. M., Ritter, C., Canepa, F., Palenzona, A., and Putti, M. (2009) Phys. Rev. B 80 220503(R).

Ryan, D. H., Rowan-Weetaluktuk, W. N., Cadogan, J. M., Hu, R., Straszheim, W. E., Bud'ko, S. L., and Canfield, P. C. (2011) Phys. Rev. B 83 104526.

Sadewasser, S., Looney, C., Schilling, J. S., Schlueter, J. A., Williams, J. M., Nixon, P. G., Winter, R. W., and Gard, G. L. (1997) Sol. State. Commun. 104 571.

Saha, S. R., Butch, N. P., Krishenbaum, K., and Paglione, J. (2009) Phys. Rev. B 79 224519.

Saint-James, D., and de Gennes, P. J. (1963) Phys. Lett. 7 306.

Saint-James, D., Thomas, E. J, and Sarma, G., (1969) Type-II Superconductivity. Pergamon Press, Oxford.

Saito, G., and Yoshida, Y. (2000) Studies of High Temperature Superconductors, Vol. 34 (Ed. Narlikar, A. V.). Nova Science, New York, p. 133.

Saito, G., Enoki, T., Toriumi, K., and Inokuchi, H. (1982) Sol. State. Commun. 42 557.

Saito, G., Enoki, T., Inokuchi, H., and Kobayashi, H. (1983) J. Phys. (Paris) 44 C3–1215.

Saito, K., Akutsu, H., and Sorai, M. (1999) Sol. State. Commun. 111 471.

Sakai, H., Yoshimura, K., Ohno, H., Kato, H., Kambe, S., Walstedt, R. E., Matsuda, T. D., Haga, Y., and Onuki, Y. (2001) J. Phys.: Cond. Mat. 13 L785.

Sakai, H., Tokunaga, Y., Fujimoto, T., Kambe, S., Walstedt, R. E., Yasuoka, H., Aoki, D., Homma, Y., Yamamoto, E., Nakamura, A., Shiokawa, Y., Nakajima, K., Arai, Y., Matsuda, T. D., Haga, Y., and Onuki, Y. (2005) J. Phys. Soc. Jpn 74 1710.

Sakon, T., Imamura, K., Koga, N., Sato, N., and Komatsubara, T. (1994) Physica B 199–200 154.

Sakurai, H., Takada, K., Yoshii, S., Sasaki, T., Kindo, K., and Takayama-Muromachi, E. (2003) Phys. Rev. B 68 132507.

Sakurai, H., Takada, K., Sasaki, T., Izumi, F., Dilanian, R. A., and Takayama-Muromachi, E. (2004) J. Phys. Soc. Jpn 73 2590.

Sakurai, H., Takada, K., Sasaki, T., and Takayama-Muromachi, E. (2005) J. Phys. Soc. Jpn 74, 2909.

Sakurai, H., Tsujii, N., Suzuki, O., Kitazawa, H., Kido, G., Takada, K., Sasaki, T., and Takayama-Muromachi, E. (2006) Phys. Rev. B 74, 092502.

Salama, K., Selvamanickam, V., Gao, L., and Sun, K. (1989) Appl. Phys. Lett. 54 2352.

Sambongi, T., Yamamoto, M., Tsutsumi, K., Shiozaki, Y., Yamaya, K., and Abe, Y. (1977) J. Phys. Soc. Jpn 42 1421.

Samuely, P., Bobrov, N. L., Jansen, A. G. M., Wyder, P., Barilo, S. N., and Shiryaev, S. V. (1993) Phys. Rev. B 48 13904.

Sanchez, J. P., Vulliet, P., Godart, C., Gupta, L. C., Hossain, Z., and Nagarajan, R. (1996) Phys. Rev. B 54 9421.

Sano, N., Taniguchi, T., and Asayama, K. (1980) Sol. State. Commun. 33 419.

Santini, P., and Amoretti, G. (1994) Phys. Rev. Lett. 73 1027.

Sarrao, J. L., Morales, L. A., Thompson, J. D., Scott, B. L., Stewart, G. R., Wastin, F., Rebizant, J., Boulet, P., Colineau, E., and Lander, G. H. (2002) Nature 420, 297.

Sasaki, T., Badica, P., Yoneyama, N., Yamada, K., Togano, K., and Kobayashi, N. (2004) J. Phys. Soc. Jpn 73 1131.

Sasmal, K., Lv, B., Lorenz, B., Guloy, A. M., Chen, F., Xue, Y. Y., and Chu, C. W. (2008) Phys. Rev. Lett. 101 107007.

Sasmal, K., Lv, B., Tang, Z. J., Chen, F., Xue, Y. Y., Lorenz, B., Guloy, A. M., and Chu, C. W. (2009) Phys. Rev. B 79 184516.

Sato, A., Akutsu, H., Saito, K., and Sorai, M. (2001) Synth. Met. 120 1035.

Sato, H., Takagi, H., and Uchida, S. (1990) Physica C 169 391.

Sato, J., Ohata, K., Okada, M., Tanaka, K., Kitaguchi, H., Kumakura, H., Kiyoshi, T., Wada, H., and Togano, K. (2001) Physica C 357–360 1111.

Sato, M., and Kohmoto, M. (2000) J. Phys. Soc. Jpn 69 3505.

Sato, N., Imamura, K., Sakon, T., Inada, Y., Sawada, A., Komatsubara, T., Matsui, H., and Goto, T. (1994) Physica B 199–200 122.

Sauerzopf, F. M. (1991) Phys. Rev. B 43 3091.

Sauls, J. A. (1994) Adv. Phys. 43 113.

Savitskii, E. M., Baron, V. V., Efimov, Y. V., Bychkova, M. I., and Myzenkova, L. F. (1973) Superconducting Materials. Plenum, New York, p. 115.

Saxena, A. K., Das, B., Arya, S. P. S., Mandal, P., Tripathi, C. C., Singh, A. K., Tiwari, R. S., and Srivastava, O. N. (1988) High Temperature Superconductivity (Eds: Gupta, A. K., Joshi, S. K., and Rao, C. N. R.). World Scientific, Singapore, p. 397.

Saxena, S. S., Agarwal, P., Ahilan, K., Grosche, F. M., Haselwimmer, R. K. W., Steiner, M. J., Pugh, E., Walker, I. R., Julian, S. R., Monthoux, P., Lonzarich, G. G., Huxley, A., Sheikin, I., Braithwaite, D., and Flouquet, J. (2000) Nature 406 587.

Scanlan, R. M., Fietz, W. A., and Koch, E. F. (1975) J. Appl. Phys. 46 2244.

Schaak, R. E., Klimczuk, T., Foo, M. L., and Cava, R. J. (2003) Nature 424 527.

Scheel, H. J. (1994) MRS Bull. 19 26.

Scheel, H. J., Berkowski, M., and Chabot, B. (1991) J. Cryst. Growth 115 19.

Scheel, H. J., Klemenz, C., Reinhart, F. R., Lang, H. P., and Guntherodt, H. J. (1994) Appl. Phys. Lett. 65 901.

Scheenmeyer, L. F., Thomas, J. K., Siegrist, T., Batlogg, B., Rupp, L. W., Opila, R. L., Cava, R. J., and Murphy, D. W. (1988) Nature 335 421.

Schilling, A., Cantoni, M., Guo, J. D., and Ott, H. R. (1993) Nature 363 56.

Schilling, J. S., Jorgensen, J. D., Hinks, D. G., Deemyad, S., Hamlin, J., Looney, C. W., and Tomita, T. (2002) Studies of High Temperature Superconductors, Vol. 38 (Ed. Narlikar, A.). Nova Science, New York, p. 322.

Schindler, A., König, R., Herrmannsdörfer, T., Braun, H. F, Eska, G., Günther, D., Meissner, M., Mertig, M., Wahl, R., Pompe, W. (2002) Europhys. Lett. 58 885.

Schirber, J. E., Morosin, B., Merrill, R. M., Hlava, P. F., Venturini, E. L., Kwak, J. F., Nigrey, P. J., Baughman, R. J., and Ginley, D. S. (1988) Physica C 152 121.

Schirber, J. E., Overmyer, D. L., Bayless, W. R., Rosseinsky, M. J., Murphy, D. W., Zhu, Q., Zhou, O., Kniaz, K., and Fischer, J. E. (1993) J. Phys. Chem. Sol. 54 1427.

Schlabitz, W., Baumann, J., Pollit, B., Rauchschwalbe, U., Mayer, H. M., Ahlheim, U., and Bredl, C. D. (1986) Z. Phys. B 62 171.

Schlachter, S. I., Fietz, W. H., Grube, K., and Goldacker, W. (2002) Adv. Cryo. Eng. 48 809.

Schmidt, H., and Braun, H. F. (1994) Physica C 229 315.

Schmidt, H., and Braun, H. F. (1998) Studies of High Temperature Superconductors, Vol. 26 (Ed. Narlikar, A.). Nova Science, New York, p. 47.

Schmidt, H., Weber, M., and Braun, H. F. (1995) Physica C 246 177.

Schmidt, H., Weber, M., and Braun, H. F. (1996a) Physica C 256 393.

Schmidt, H., Zasadzinski, J. F., Gray, K. E., and Hinks, D. G. (2001) Phys. Rev. B 63 220504(R).

Schmidt, H., Zasadzinski, J. F., Gray, K. E., Hinks, D. G., Bud'ko, S. L., Canfield, P. C., and Finnemore, D. K. (2002) Studies of High Temperature Superconductors, Vol. 38 (Ed. Narlikar, A.). Nova Science, New York, p. 229.

Schmidt, M., Cummins, T. R., Bürk, M., Lu, D. H., Nücker, N., and Schuppler, S. (1996b) Phys. Rev. B 53 14761.

Schnell, I., Mazin, I. I., and Liu, A. Y. (2006) Phys. Rev. B 74 184503.

Schooley, J. F., Hosler, W. R., and Cohen, M. L. (1965) Phys. Rev. Lett., 12, 474.

Schrieffer, J. R. (1988) Phys. Rev. Lett. 60 944.

Schrieffer, J. R. (1994) Theory of Superconductivity. Addison-Wesley, New York.

Schrieffer, J. R., Scalapino, D. J., and Wilkins, J. W. (1963) Phys. Rev. Lett. 10 336.

Schröder, A., Lussier, J. G., Gaulin, B. D., Garrett, J. D., Buyers, W. J. L., Rebelsky, L., and Shapiro, S. M. (1994) Phys. Rev. Lett. 72 136.

Schultz, L., Freyhardt, H. C., Bormann, R., and Mordike, B. L. (1975) Proc Int. Conf. Low Temperature Physics: LT14, p. 59.

Schwartz, B. B., and Gruenberg, L. W. (1969) Phys. Rev. 177 747.

Schwartz, J., Effio, T., Liu, X., Le, Q. V., Mbaruku, A. L., Schneider-Muntau, H. J., Shen, T., Song, H., Trociewitz, U. P., Wang, X., and Weijers, W. (2008) IEEE Trans. Appl. Supercond. 18 70.

Schweinfurth, R. A., Platt, C. E., Teepa, M. R., and Van Harlingen, D. J. (1992) Appl. Phys. Lett. 61 480.

Schwenk, H., Andres, K., Wudl, F., and Aharon-Shalom, E. (1983) Sol. State. Commun. 45 767.

Schwenk, H., Hess, E., Andres, K., Wudl, F., and Aharon-Shalom, E. (1984) Phys. Lett. 102A 57.

Seaman, C. L., Neumeier, J. J., Maple, M. B., Le, L. P., Luke, G. M., Sternlieb, B. J., Uemura, Y. J., Brewer, J. H., Kadono, R., Kiefl, R. F., Krietzman, S. R., and Riseman, T. M. (1990) Phys. Rev. B 42 6801.

Seeber, B. (1998) Handbook of Applied Superconductivity, Vol. I (Ed. Seeber, B.). IOP Publishing, Bristol, p. 429.

Seeber, B., Decroux, M., and Fischer, Ø. (1989) Physica B 155 129.

Sefat, A. S., and Singh, D. J. (2011) MRS Bull. 36 614.

Sefat, A. S., Huq, A., McGuire, M. A., Jin, R. Y., Sales, B. C., Mandrus, D., Cranswick, L. M. D., Stephens, P. W., and Stone, K. H. (2008a) Phys. Rev. B 78 104505.

Sefat, A. S., Jin, R. Y., McGuire, M. A., Sales, B. C., Singh, D. J., and Mandrus, D. (2008b) Phys. Rev. Lett. 101 117004.

Sefat, A. S., McGuire, M. A., Sales, B. C., Jin, R. Y., Howe, J. Y., and Mandrus, D. (2008c) Phys. Rev. B 77 174503.

Sefat, A. S., Singh, D. J., Van Bebber, L. H., Mozharivskyj, Y., McGuire, M. A., Jin, R. Y., Sales, B. C., Keppens, V., and Mandrus, D. (2009) Phys. Rev. B 79 224524.

Sekine, H., Iijima, Y., Itoh, K., Tachikawa, K., Tanaka, Y., and Furuto, Y. (1983) IEEE. Trans. Magn. 19 1429.

Selvam, P., Cors, J., Cattani, D., Decroux, M., Fischer, Ø., and Seibt, E. W. (1995) Appl. Phys. A 61 615.

Serin, B. (1965) Phys. Lett. 16 217.

Sernetz, F., Lerf, A., and Schöllhorn, R. (1974) Mat. Res. Bull. 9 1597

Shamoto, S., Kato, T., Ono, Y., Miyazaki, Y., Ohoyama, K., Ohashi, M., Yamaguchi, Y., and Kajitani, T. (1998) Physica C 306 7.

Shamoto, S., Lizawa, K., Koiwasaki, T., Yasukawa, M., Yamanaka, S., Petrenko, O., Bennington, S. M., Yoshida, H., Ohoyama, K., Yamaguchi, Y., Ono, Y., Miyazaki, Y., and Kajitani, T. (2000) Physica C 341–348 747.

Shamoto, S., Kato, Y., Oikawa, K., and Kajitani, T. (2001) J. Phys. Soc. Jpn 70 283.

Shamoto, S., Takeuchi, K., Yamanaka, S., and Kajitani, T. (2004) Physica C 402 283.

Shan, L., Liu, Z. Y., Wen, H. H., Ren, Z. A., Che, G. C., and Zhao, Z. X. (2003) Phys. Rev. B 68 024523.

Shan, L., Liu, Z. Y., Ren, Z. A., Che, G. C., and Wen, H. H. (2005) Phys. Rev. B 71 144516.

Shan, L., Wang, Y., Zhu, X., Mu, G., Fang, L., Ren, C., Wen, H. H. (2008) Europhys. Lett. 83 57004.

Sharifi, F., Pargellis, A., Dynes, R. C., Miller, B., Hellman, E. S., Rosamilia, J., and Hartford, E. H. Jr. (1991) Phys. Rev. B 44 12521.

Sharoni, A., Felner, I., and Millo, O. (2001) Phys. Rev. B 63 220508.

Sheikin, I., Steep, E., Braithwaite, D., Brison, J. P., Raymond, S., Jaccard, D., and Flouquet, J. (2001) J. Low. Temp. Phys. 122 591.

Shein, I. R., Ivanovskii, A. L., Kurmaev, E. Z., Moewes, A., Chiuzbian, S., Finkelstein, L. D., Neumann, M., Ren, Z. A., and Che, G. C. (2002) Phys. Rev. B 66 024520.

Shelton, R. N., Lawson, H. C., and Johnston, D. C. (1975) Mat. Res. Bull. 10 297.

Shen, B., Zeng, B., Chen, G. F., He, J. B., Wang, D. M., Yang, H., and Wen, H. H. (2011) Europhys. Lett. 96 37010.

Shen, X.-L. (2012) J. At. Mol. Sci. 3 89.

Sheng, Z. Z., and Hermann, A. M. (1988) Nature 332 55, 138

Shengelaya, A., Khasanov, R., Eshchenko, D. G., Felner, I., Asaf, U., Keller, H., and Müller, K. H. (2004) Phys. Rev. B 69 024517.

Shermadini, Z., Krzton-Maziopa, A., Bendele, M., Khasanov, R., Luetkens, H., Conder, K., Pomjakushina, E., Weyeneth, S., Pomjakushin, V., Bossen, O., and Amato, A. (2011) Phys. Rev. Lett. 106 117602.

Shi, L., Li, G., Zhang, X. D., Feng, S. J., and Li, G. (2003) Physica C 383 450.

Shibaeva, R. P., Rozenberg, L. P., Yagubskii, E. B., and Kushch, N. D. (1993) Kristallografiya 38 91.

Shibata, H. (2002) Phys. Rev. B 65 180507.

Shim, J. H., Kwon, S. K., and Min, B. I. (2001) Phys. Rev. B 64 180510(R).

Shimahara, H. (1997) J. Phys. Soc. Jpn 66 541.

Shimahara, H. (2003) J. Phys. Soc. Jpn 72 (2003) 1851.

Shimahara, H. (2004) J. Phys. Soc. Jpn 73 2635.

Shimizu, K., Suhara, K., Ikumo, M., Eremets, M. I., and Amaya, K. (1998) Nature 393 767.

Shimizu, K., Kimura, T., Furomoto, S., Takada, K., Kotani, K., Onuki, Y., and Amaya, K. (2001) Nature 412 316.

Shimojo, Y., Ito, H., Ishiguro, T., Kondo, T., and Saito, G. (1999) J. Supercond. 12 501.

Shimojo, Y., Kamiya, S., Tanatar, M. A., Ohmichi, E., Kovalev, A. E., Ishiguro, T., Yamochi, H., Saito, G., Yamada, J., Anzai, H., Kushch, N. D., Lyubovskii, R. B., Lyubovskaya, R. N., and Yagubskii, E. B. (2003) Synth. Met. 133–134 197.

Shimoyama, J., Kadowaki, K., Kitaguchi, H., Kumakura, H., Togano, K., Maeda, H., Nomura, K., and Seido, M. (1993) Appl. Supercond. 1 43.

Shoenberg, D. (1938) Nature 142 874.

Shoenberg, D. (1952) Superconductivity. Cambridge University Press, Cambridge.

Shubnikov, L. W., Khotkevich, W. I., Shepelev, J. D., and Riabinin, J. N. (1937) Zh. Eksp. Teor. Fiz. 7, 221.

Si, Q., and Steglich, F. (2010) Nature 329 1161.

Si, Q., Rabello, S., Ingersent, K., and Smith, L. (2001) Nature 413 804.

Si, W., Lin, Z.-W., Jie, Q., Yin, W.-G., Zhou, J., Gu, G., Johnson, P. D., and Li, Q. (2009) Appl. Phys. Lett., 95 052504.

Sidis, Y., Braden, M., Bourges, P., Hennion, B., Nishizaki, S., Maeno, Y., and Mori, Y. (1999) Phys. Rev. Lett. 83 3320.

Sidorov, V. A., Nicklas, M., Pagliuso, P. G., Sarrao, J., Bang, Y., Balatsky, A. V., and Thompson J. D. (2002) Phys. Rev. Lett. 89 157004.

Siegrist, T., Zandbergen, H. W., Cava, R. J., Krajewski, J. J., and Peck, W. F. Jr. (1994) Nature 367 254.

Sigrist, M., and Rice, T. M. (1994) J. Low Temp. Phys. 95, 389.

Silcox, J., and Rollins, R. W. (1964) Rev. Mod. Phys. 36 52.

Silsbee, F. B. (1916) J. Wash. Acad. Sci. 6 597.

Silver, R. M., Talvacchio, J., and de Lozanne, A. L. (1987) Appl. Phys. Lett. 51 2149.

Singer, P. M., Imai, T., He, T., Hayward, M. A., and Cava, R. J. (2001) Phys. Rev. Lett. 87 257601

Singh, D. J. (1995) Phys. Rev. B 52 1358.

Singh, D. J. (2008) Phys. Rev. B 78 094511.

Singh, D. J., and Du, M. H. (2008) Phys. Rev. Lett. 100 237003.

Singh, D. J., and Mazin, I. I. (2001) Phys. Rev. B 64 140507.

Singh, S. K., Kumar, A., Gahtori, B., Kirtan, S., Sharma, G., Patnaik, S., and Awana, V. P. S. (2012) J. Am. Chem. Soc. 134 16504.

Singleton, J., Symington, J. A., Nam, M.-S., Ardavan, A., Kurmco, M., and Day, P. (2000) J. Phys.: Cond. Mat. 12 L641.

Sinha, K. P. (1990) Studies of High Temperature Superconductors, Vol. 5 (Ed. Narlikar, A.). Nova Science, New York, p. 119.

Sinha, S. K., Lynn, J. W., Grigereit, T. E., Hossain, Z., Gupta, L. C., Nagarajan, R., and Godart, C. (1995) Phys. Rev. B 51 861.

Skinta, J. A., Lemberger, T. R., Greibe, T., and Naito, M. (2002) Phys. Rev. Lett. 88 207003.

Skolnick, M. S., Saker, M. K., Robertson, D. S., Satehell, J. S., Reed, L. J., Singleton, J., and Bagguley, D. M. S. (1987) J. Phys. C: Sol State Phys. 20 L435.

Skoskiewicz, T. (1972) Phys. Stat. Sol. (a) 4 K132

Sleight, A. W. (1988) Chemistry of Oxide Superconductors (Ed. Rao, C. N. R.). Blackwell, Oxford, p. 27.

Sleight, A. W., Gillson, J. L., and Bierstedt, P. E. (1975) Sol. State. Commun. 17 27.

Sleight, A. W., Gillson, J. L., and Bierstedt, P. E. (1993) Sol. State. Commun. 88 841.

Slusky, J. S., Rogado, N., Regan, K. A., Hayward, M. A., Khalifah, P., He, T., Inumaru, K., Loureiro, S. M., Haas, M. K., Zandbergen, H. W., and Cava, R. J. (2001) Nature 410 343.

Smathers, D. B., Marken, K. R., Lee, P. J., Larbalestier, D. C., McDonald, W. K., and O'Larey, P. M. (1985) IEEE. Trans. Magn. 21 1133.

Smith, J. A., Cima, M. J., and Sonnenberg, N. (1999) IEEE. Trans. Appl. Supercond. 9 1531.

Smith, J. L., Fisk, Z., Willis, J. O., Batlogg, B., and Ott, H. R. (1984) J. Appl. Phys. 55, 1996.

Solovyov, V. F., Wiesmann, H. J., Wu, L. J., Zhu, Y., Suenaga, M., and Freenstra, R. (1997) Physica C 309 269.

Soltanian, S., Wang, X. L., Horvat, J., Qin, M. J., Liu, H. K., Munroe, P. R., and Dou, S. X. (2003) IEEE Trans. Appl. Supercond. 13 3273.

Sonin, E. B. (1998) J. Low Temp. Phys. 110 411.

Sonin, E. B., and Felner, I. (1998) Phys. Rev. B. 57 14000.

Sparn, G., Thompson, J. D., Huang, S.-M., Kaner, R. B., Diederichs, F., Wetten, R. L., Grüner, G., and Holczer, K. (1991) Science 252 1829.

Sparn, G., Borth, R., Lengyel, E., Pagliuso, P. G., Sarrao, J., Steglich, F., and Thompson, J. D. (2002) Physica B 319 262.

Steglich, F. (2005) Physica B 359–361, 326.

Steglich, F., Aarts, J., Bredl, C. J., Lieke, W., Meschede, D., Franz, W., and Schäfer, H. (1979) Phys. Rev. Lett. 43 1892.

Steglich, F., Gegenwart, P., Geibel, C., Hinze, P., Lang, M., Langhammer, C., Sparn, G., Tayama, T., Trovarelli, O., Sato, N., Dahm, T., and Varelogiannis, G. (2001) More is Different—Fifty Years of Condensed Matter Physics. Princeton University Press, Princeton, NJ, p. 191.

Steglich, F., Arndt, J., Friedemann, S., Krellner, C., Tokiwa, Y., Westerkamp, T., Brando, M., Gegenwart, P., Geibel, C., Wirth, S., and Stockert, O. (2010) J. Phys.: Cond. Mat. 22 164202.

Stenger, V. A., Pennington, C. H., Buffinger, D. R., and Ziebarth, R. P. (1995) Phys. Rev. Lett. 74 1649.

Stephens, P. W., Mihaly, L., Lee, P. L., Whetten, R. L., Huang, S. M., Kanier, R., Diederichs, F., and Holczer, K. (1991) Nature 351 632.

Stewart, G. R. (1984) Rev. Mod. Phys. 56 755.

Stewart, G. R., Fisk, Z., Willis, J. O., and Smith, J. L. (1984) Phys. Rev. Lett. 52 679.

Stojkovic, B. P., and Pines, D. (1996) Phys. Rev. Lett. 76, 811.

Strand, J. D., Bahr, D. J., Van Harlingen, D. J., Davis, J. P., Gannon, W. J., and Halperin, W. P. (2010) Science 328 1368.

Stritzker, B. (1979) Phys. Rev. Lett., 42 1769

Stritzker, B., and Buckel, W. (1972) Z. Phys. 257 1.

Strobel, P., Korezak, W., and Fournier, T. (1991) Physica C 172 435.

Su, X., Zuo, F., Schlueter, J. A., Williams, J. M., Nixon P. G., Winter, R. W., and Gard, G. L. (1999) Phys. Rev. B 59 4376.

Su, X. D., Witz, G., Kwasnitza, K., and Flukiger, R. (2004) High Temperature Superconductivity 1: Materials (Ed. Narlikar, A. V.). Springer, Heidelberg, p. 281.

Su, Y., Link, P., Schneidewind, A., Wolf, T., Adelmann, P., Xiao, Y., Meven, M., Mittal, R., Rotter, M., Johrendt, D., Buueckel, T., and Loewenhaupt, M. (2009) Phys. Rev. B 79 064504.

Subba Rao, G. V., and Shenoy, G. K. (1981) Current Trends in Magnetism (Eds. Satyamurthy, N. S., and Madhav Rao, L.). IPA Publication, Bombay, p. 338.

Subedi, A., and Singh, D. J. (2008) Phys. Rev. B 78 132511.

Subramanian, M. A., Gopalakrishnan, J., Torardi, C. C., Askew, T. R., Flippen, R. B., Sleight, A. W., Lin, J. J., and Poon, S. J. (1988) Science 240 495.

Suenaga, M. (1981) Superconductor Materials Science (Eds. Foner, S., and Schwartz, B. B.). Plenum, New York, p. 201.

Suenaga, M., Horigami, O., and Luhman, T. S. (1974) Appl. Phys. Lett. 25 624.

Suenaga, M., Klamut, C. J., Higuchi, N., and Kuroda, T. (1985) IEEE. Trans. Magn. 21 305.

Sugai, S., Uchida, S., Kitazawa, K., Tanaka, S., and Katsui, A. (1985) Phys. Rev. Lett. 55 426.

Sugawara, H., Kobayashi, M., Osaka, S., Saha, S. R., Namiki, T., Aoki, Y., and Sato, H. (2005) Phys. Rev. B 72 014519.

Sugiyama, J., Itahara, H., Brewer, J. H., Ansaldo, E., Motohashi, T., Karppinen, M., and Yamauchi, H. (2003) Phys. Rev. B 67 214420.

Suhl, H. (2001) Phys. Rev. Lett. 87 167007.

Sumption, W. D., Bhatia, M., Dou, S. X., Rindfleisch, M., Tomsic, M., Arda, L., Ozdemir, M., Hascicek, Y., and Collings, E. W. (2004) Supercond. Sci. Technol. 17 1180.

Sumption, W. D., Bhatia, M., Rindfleisch, M., Tomsic, M., Soltanian, S., Dou, S. X., and Collings, E. W. (2005) Appl. Phys. Lett. 86 092507.

Sun, A. G., Gajewski, D. A., Maple, M. B., and Dynes, R. C. (1994) Phys. Rev. Lett. 72 2267.

Sun, G. F., Wong, K. W., Xu, B. R., Xin, Y., and Lu, D. F. (1994) Phys. Lett. 192A 122.

Sunshine, S. A., Siegrist, T., Schneemeyer, L. F., Murphy, D. W., Cava, R. J., Batlogg, B., van Dover, R. B., Fleming, R. M., Glarum, S. H., Nakahara, S., Krajewski, J. J., Zahurak, S. M., Waszczak, J. V., Marshall, J. H., Rupp, L. W. Jr., and Peck, W. F. (1988) Phys. Rev. B 38 893.

Sutherland, M., Doiron-Leyraud, N., Taillefer, L., Weller, T., Ellerby, M., and Saxena, S. S. (2007) Phys. Rev. Lett. 98 067003.

Sweedler, A. R., Cox, D. E., Schweitzer, D. G., and Webb, G. W. (1974) Proc. Applied Superconductivity Conf.

Sweedler, A. R., Snead, C. L. Jr., and Cox, D. E. (1979) Treatise on Materials Science and Technology, Vol. 14 (Eds. Luhman, T., and Dew-Hughes, D.). Academic Press, New York, p. 490.

Swift, R. M., and White, D. (1957) J. Am. Chem. Soc. 79 3641.

Szabó, P., Samuely, P., Ka, J., Klein, T., Marcus, J., Fruchart, D., Miraglia, S., Marcenat, C., and Jansen, A. G. M. (2001) Phys. Rev. Lett. 87 137005.

Tachikawa, K. (1970) Int. Cryogenic Engineering Conf., Berlin.

Tachikawa, K., Burt, R. J., and Hartwig, K. T. (1977) J. Appl. Phys. 48 3623.

Tachikawa, K., Fukuda, S., and Tanaka, Y. (1967) Proc. 1st Int. Conf. Cryogenic Engineering, p. 154.

Tachikawa, K., Sakinada, T., and Kobayashi, M. (1993) Cryogenics 33 1091

Tachikawa, K., Tanaka, Y., Yoshida, Y., Asano, T., and Iwasa, Y. (1979) IEEE. Trans. Magn. 15 391.

Tachikawa, K., Yoshida, Y., and Rinderer, L. (1972) J. Mat. Sci. 7 1154.

Taen, T., Tsuchiya, Y., and Tamegai, T. (2009) Phys. Rev. B 80 092502.

Taguchi, Y., Kawabata, T., Takano, T., Kitora, A., Kato, K., Takata, M., and Iwasa, Y. (2007) Phys. Rev. B 76 064508.

Tajima, Y., Yamaya, K., and Abe, Y. (1984) Int. Symp. Nonlinear Transport and Related Phenomena in Inorganic Quasi One-Dimensional Conductors.

Takada, K., Sakurai, H., Takayama-Muromachi, E., Izumi, F., Dilanian, R. A., and Sasaki, T. (2003) Nature 422 53.

Takada, K., Sakurai, H., Takayama-Muromachi, E., Izumi, F., Dilanian, R. A., and Sasaki, T. (2004a) Adv. Mat. 16 1901.

Takada, K., Fukuda, K., Osada, M., Nakai, I., Izumi, F., Dilanian, R. A., Kato, K., Takata, M., Sakurai, H., Takayama-Muromachi, E., and Sasaki, T. (2004b) J. Mat. Chem. 14 1448.

Takada, K., Sakurai, H., and Takayama-Muromachi, E. (2005a) Frontiers in Superconducting Materials (Ed. Narlikar, A. V.). Springer, Heidelberg, p. 651.

Takada, K., Osada, M., Izumi, F., Sakurai, H., Takayama-Muromachi, E., and Sasaki, T. (2005b) Chem. Mat. 17 2034.

Takada, K., Sakurai, H., Takayama-Muromachi, E., Izumi, F., Dilanian, R. A., Sasaki, T. (2007) New Research on Superconductivity and Magnetism (Ed. Tran, L. K.). Nova Science, New York, p. 143.

Takagi, H., Uchida, S., Kitazawa, K., and Tanaka, S. (1987) Jpn. J. Appl. Phys. 26 L123.

Takahashi, H., Igawa, K., Arii, K., Kamihara, Y., Hirano, M., and Hosono, H. (2008) Nature 453 376.

Takahashi, S., Sambongi, T., Brill, J. W., and Roark, W. (1984) Sol. State. Commun. 49 1031.

Takahashi, T., Jérome, D., and Bechgaard, K. (1982) J. Phys. Lett. (Paris) 43 L565.

Takahashi, T., Kanoda, K., Akiba, K., Sakao, K., Watabe, M., Suzuki, K., and Saito, G. (1991) Synth. Met. 41–43 2005.

Takahashi, T., Sato, T., Souma, S., Muranaka, T., and Akimitsu, J. (2001) Phys. Rev. Lett. 86 4915.

Takano, T., Kishiume, T., Taguchi, Y., and Iwasa, Y. (2008) Phys. Rev. Lett. 100 247005.

Takasaki, T., Ekino, T., Muranaka, T., Fujii, H., and Akimitsu, J. (2002) Physica C 378–381 229.

Takasaki, T., Ekino, T., Fujii, H., and Yamanaka, S. (2005) J. Phys. Soc. Jpn 74 2586.

Takasaki, T., Ekino, T., Sugimoto, A., Shohara, K., Yamanaka, S., and Gahovich, A. M. (2010) Euro Phys J. B 73 471.

Takeuchi, T. (2000) Supercond. Sci. Technol. 13 R101.

Takeuchi, T., Togano, K., and Tachikawa, K. (1987) IEEE. Trans. Magn. 23 956.

Takeya, H., and Takei, H. (1989) Jpn. J. Appl. Phys. 28 L229.

Takigawa, M., Yasuoka, H., and Saito, G. (1987) J. Phys. Soc. Jpn 56 873.

Takigawa, M., Reyes, A. P., Hammel, P. C., Thompson, J. D., Heffner, R. H., Fisk, Z., and Ott, H. (1991) Phys. Rev. B 43 247.

Tallon, J. L. (2005) Frontiers in Superconducting Materials (Ed. Narlikar, A. V.). Springer, Heidelberg, p. 295.

Tallon, J. L., and Loram, J. W. (2000) Physica C 349, 53.

Tallon, J. L., Pooke, D. M., Buckley, R. G., Presland, M. R., and Blunt, F. J. (1990) Phys. Rev. B 41 7220.

Tallon, J. L., Cooper, J. R., deSilva, P., Williams, G. V. M., and Loram, J. W. (1995) Phys. Rev. Lett. 75 4114.

Tallon, J. L., Bernhard, C., Bowden, M., Gilberd, P., Stoto, T., and Pringle, D. (1999) IEEE Trans. Appl. Supercond. 9 1696.

Tallon, J. L., Loram, J. W., Williams, G. V. M., Cooper, J. R., Fisher, I. R., Johnson, J. D., Staines, M. P., and Bernhard, C. (1999) Phys. Stat. Sol. (b) 215 531.

Tallon, J. L., Loram, J. W., Williams, G. V. M., and Bernhard, C. (2000) Phys. Rev. B 61 6471.

Tamura, T., Adachi, S., Wu, X.-J., Tatsuki, T., and Tanabe, K. (1997) Physica C 277 1.

Tan, S. G., Li, L. J., Liu, Y., Tong, P., Zhao, B. C., Lu, W. J., and Sun, Y. P. (2012a) Physica C 483 94.

Tan, S. G., Tong, P., Liu, Y., Lu, W. J., Li, L. J., Zhao, B. C., and Sun, Y. P. (2012b) Preprint.

Tanaka, A., and Hu, X. (2003) Phys. Rev. Lett. 91 257006.

Tanaka, S. (2003) Physica C 392–396 1.

Tanaka, Y., Ishizuka, M., He, L. L., Horiuchi, S., and Maeda, H. (1997) Studies of High Temperature Superconductors, Vol. 21 (Ed. Narlikar, A.). Nova Science, New York, p. 145.

Tanatar, M. A., Laukhin, V. N., Ishiguro, T., Ito, H., Kondo, T., and Saito, G. (1999a) Phys. Rev. B 60 7536.

Tanatar, M. A., Ishiguro, T., Tanaka, H., Kobayashi, A., and Kobayashi, H. (1999b) J. Supercond. 12 511.

Tanatar, M. A., Nagai, S., Mao, Z. Q., Maeno, Y., and Ishiguro, T. (2001) Phys. Rev. B 63 064505.

Tanatar, M. A., Ishiguro, T., Tanaka, H., and Kobayashi, H. (2002) Phys. Rev. B 66 134503.

Tanatar, M. A., Reid, J.-Ph., Shakeripour, H., Luo, X. G., Doiron-Leyraud, N., Ni, N., Bud'ko, S. L., Canfield, P. C., Prozorov, R., and Taillerfer, L. (2010) Phys. Rev. Lett. 104 067002.

Tang, Z. K., Lingyun, Z., Wang, N., Zhang, X. X., Wen, G. H., Li, G. D., Wang, J. N., Chan, C. T., and Sheng, P. (2001) Science 292 2462.

Tanigaki, K., Ebbesen, T. W., Saito, S., Mizuki, J., Tsai, J. S., Kubo, Y., and Kuroshima, S. (1991) Nature 352 222.

Tanigaki, K., Hirosawa, I., Ebbesen, T. W., Mizuki, J., Shimakawa, Y., Kubo, Y., Tsai, J. S., and Kuroshima, S. (1992) Nature 356 419.

Tanigaki, K., Hirosawa, I., Manako, T., Tsai, J. S., Mizuki, J., and Ebbesen, T. W. (1994) Phys. Rev. B 49 12307.

Tapp, J. H., Tang, Z., Lv, B., Sasmal, K., Lorenz, B., Chu, P. C. W., and Guloy, A. M. (2008) Phys. Rev. B 78 060505(R).

Tayama, T., Lang, M., Luhmann, T., Steglich, F., and Assmus, W. (2003) Phys. Rev. B 67, 214504.

ten Haken, B., Godeke, A., and ten Kate, H. H. J. (1996) IEEE. Trans. Magn. 32 2739.

Terasaki, I. (2005) Frontiers in Magnetic Materials (Ed. Narlikar, A. V.). Springer, Heidelberg, p. 327.

Terasaki, I., Sassago, Y., and Uchinokura, K. (1997) Phys. Rev. B 56 12685(R).

Terasaki, I., Tsukada, I., and Iguchi, Y. (2002) Phys. Rev. B 65 195106.

Testardi, L. R. (1973) Physical Acoustics, Vol. 10 (Ed. Mason, W. P., and Thurston, R. N.). Academic Press, New York, p. 1.

Testardi, L. R. (1975) Rev. Mod. Phys. 47 637.

Testardi, L. R., Wernick, J. H., and Royer, W. A. (1974) Sol. State. Commun. 15 1.

Thalmeier, P., Zwicknagl, G., Stockert, O., Sparn, G., and Steglich, F. (2005) Frontiers in Superconducting Materials (Ed. Narlikar, A. V.). Springer, Heidelberg, p. 109.

Thomas, A. H., Gamble, F. R., and Koehler, E. F. (1972) Phys. Rev. B. 5 2811.

Thomas, F., Thomasson, J., Ayache, C., Geibel, C., and Steglich, F. (1993) Physica B 186–188 303.

Thomlinson, W., Shirane, G., Moncton, D. E., Ishikawa, M., and Fischer, O. (1981) Phys. Rev. B. 23 4455.

Thompson, J. D., Parks, R. D., and Borges, H. (1986) J. Magn. Magn Mat. 54–57 377.

Thompson, J. R., Krusin-Elbaum, L., Christen, D. K., Song, K. J., Paranthaman, M., Ulmann, J. L., Wu, J. Z., Ren, Z. F., Wang, J. H., Tkaczyk, J. E., and DeLuca, J. A. (1997) Appl. Phys. Lett. 71 536.

Tidman, J. P., Singh, O., and Curzon, A. E. (1974) Phil. Mag. 30 1191.

Timusk, T. (1999) Physica C 317–318 18.

Timusk, T., and Statt, B. (1999) Rep. Prog. Phys. 62 61.

Tissen, V. G., Nefedova, M. V., Kolesnikov, N. N., and Kulakov, M. P. (2001) Physica C 363 194.

Togano, K., and Tachikawa, K. (1975) Phys. Lett. 54A 205.

Togano, K., Kumakura, H., Kadowaki, K., Kitaguchi, H., Maeda, H., Kase, J., Shimoyama, J., and Nomura, J. (1992) Adv. Cryo. Eng. 38 1081.

Togano, K., Matsumoto, A., and Kumakura, H. (2011) Appl. Phys. Express 4 043101.

Tokumoto, M., Anzai, H., Murata, K., Kajimura, K., and Ishiguro, T. (1987) J. Phys. Soc. Jpn 26 1977.

Tokunaga, Y., Kotegawa, H., Ishida, K., Kitaoka, Y., Takagiwa, H., and Akimitsu, J. (2001) Phys. Rev. Lett. 86 5767.

Tokura, Y., Takagi, H., and Uchida, S. (1989) Nature 337 345.

Tollis, S., Daumens, M., and Buzdin, A. (2005) Phys. Rev. B 71 024510.

Tomeno, I., and Ando, K. (1989) Phys. Rev. B 40 2690.

Tomić, S., Jérome, D., Mailly, D., Ribault, M., and Bechgaard, K. (1983) J. Phys. (Paris) 44 1075.

Tomita, M., and Murakami, M. (2000) Supercond. Sci. Technol. 13 722.

Tomita, M., and Murakami, M. (2001) Physica C 354 358.

Tomita, M., and Murakami, M. (2003) Nature 421 517.

Tomita, T., Hamlin, J. J., Schilling, J. S., Hinks, D. G., and Jorgensen, J. D. (2001) Phys. Rev. B. 64 092505.

Tong, P., Sun, Y. P., Zhu, X. B., and Song, W. H. (2006) Phys. Rev. B 73 245106.

Tong, P., Sun, Y. P., Zhu, X. B., and Song, W. H. (2007) Sol. State. Commun. 141 336.

Torrance, J. B., Pedersen, H. J., and Bechgaard, K. (1982) Phys. Rev. Lett. 49 881.

Tou, H., Maniwa, Y., Koiwasaki, T., and Yamanaka, S. (2001a) J. Mag. Mag. Mat. 226–230 330.

Tou, H., Maniwa, Y., Koiwasaki, T., and Yamanaka, S. (2001b) Phys. Rev. Lett. 86 5775.

Tou, H., Maniwa, Y., Yamanaka, S., and Sera, M. (2003a) Physica B 329 1323.

Tou, H., Maniwa, Y., and Yamanaka, S. (2003b) Phys. Rev. B 67 100509(R).

Tou, H., Oshiro, S., Kotegawa, H., Taguchi, Y., Kasahara, Y., and Iwasa, Y. (2010) Physica C 470 S658.

Tranh, T. D., Koma, A., and Tanaka, S. (1980) Appl. Phys. 22 205.

Tranquada, J. M., Sternlieb, B. J., Axe, J. D., Nakamura, Y., and Uchida, S. (1995) Nature 375 561.

Träuble H., and Essman, U. (1968) J. Appl. Phys. 39, 4052.

Träuble H., and Essman, U. (1969) Phys. Stat. Sol. 32 337.

Tristan Jover, D., Wijngaarden, R. J., Griessen, R., Haines, E. M., Tallon, J. L., and Liu, R. S. (1996) Phys. Rev. B. 54 10175.

Troyanovski, A. M., v. Hecke, M., Saha, N., Aarts, J., and Kes, P. H. (2001) Phys. Rev. Lett. 89, 147006.

Tsaur, B. Y., Dilorio, M. S., and Strauss, A. J. (1987) Appl. Phys. Lett. 51 858.

Tse, P. K., Aldred, A. T., and Fradin, F. Y. (1979) Phys. Rev. Lett. 43 1825.

Tsebro, V. I., Omel'yanovskii, O. E., and Moravskii, A. P. (1999) JETP Lett. 70 462.

Tsuda, S., Yokoya, T., Takano, Y., Kito, H., Matsushita, A., Yin, F., Itoh, J., Harima, H., and Shin, S. (2003) Phys. Rev. Lett. 91 127001.

Tsuei, C. C. (1973) Science 180 57.

Tsuei, C. C., and Kirtley, J. R. (2000a) Rev. Mod. Phys. 72 969.

Tsuei, C. C., and Kirtley, J. R. (2000b) Phys. Rev. Lett. 85 182.

Tsuei, C. C., and Kirtley, J. R. (2002) Physica C 367 1.

Tsuei, C. C., Kirtley, J. R., Chi, C. C., Yu-Jahnes, L.-S., Gupta, A., Shaw, T., Sun, J. Z., and Ketchen, M. B. (1994) Phys. Rev. Lett. 72 93.

Tsukada, I., Hanawa, M., Akiike, T., Nabeshima, F., Imai, Y., Ichinose, A., Komiya, S., Hikage, T., Kawaguchi, T., Ikuta, H., and Maeda A. (2011) Appl. Phys. Express 4 053101.

Tu, J. J., Li, J., Liu, W., Punnoose, A., Gong, Y., Ren, Y. H. M, Li, L. J., Cao, G. H., Xu, Z. A., and Homes C. C. (2010) Phys. Rev. B 82 1745091.

Tyco, R., Dabbagh, G., Rosseinsky, M. J., Murphy, D. W., Ramirez, A. P., and Fleming, R. M. (1992) Phys. Rev. Lett. 68 1912.

Uchida, S., Kitazawa, K., and Tanaka, S. (1987) Phase Trans. 8 95.

Uchida, T., Nakamura, S., Suzuki, N., Nagata, Y., Mosley, W. D., Lan, M. D., Klavins, P., and Shelton, R. N. (1993) Physica C 215, 350.

Ueda, S., Yamagishi, T., Takeda, S., Agatsuma, S., Takano, S., Mitsuda, A., and Naito, M. (2011) Physica C 471 1167.

Uehara, M., Amano, T., Takano, S., Kori, T., Yamazaki, Y., and Kimishama, Y. (2007a) Physica C 440 6.

Uehara, M., Yamazaki, Y., Kori, T., Kashida, T., Kimishama, Y., and Hase, I. (2007b) J. Phys. Soc. Jpn 76 034714.

Uehara, M., Uehara, A., Kozawa, K., and Kimishama, Y. (2009) J. Phys. Soc. Jpn 78 033702.

Uemura, Y. J. (1991) Physica B 169 99.

Uemura, Y. J., and Luke, G. M. (1993) Physica B 186–188 223.

Uemura, Y. J., Sternlib, B. J., Cox, D. E., Brewer, J. W., Kadano, R., Kempton, J. R., Keifl, R. E., Kreitzman, S. R., Luke, G. M., Mulhern, P., Riseman, T., Williams, D. L., Kossler, W. J., Yu, X. H., Stronach, C. E., Subramanian, M. A., Gopalakrishnan, J., and Sleight, A. W. (1988) Nature 335 151.

Uji, S., Shinagawa, H., Terashima, T., Yakabe, T., Teral, Y., Tokumoto, M., Kobayashi, A., Tanaka, H., and Kobayashi, H. (2001) Nature 410 908.

Uji, S., Terakura, C., Terashima, T., Yakabe, T., Teral, Y., Tokumoto, M., Kobayashi, A., Sakai, F., Tanaka, H., and Kobayashi, H. (2002) Phys. Rev. B 65 113101.

Uji, S., Terashima, T., Nishimura, M., Takahide, Y., Konoike, T., Enomoto, K., Cui, H., Kobayashi, H., Kobayashi, A., Tanaka, H., Tokumoto, M., Choi, E.-S., Tokumoto, T., Graf, D., and Brooks, J. S. (2006) Phys. Rev. Lett, 97 157001.

Ullrich, M., Muller, D., Heinemann, K., Niel, L., and Freyhardt, H. C. (1993) Appl. Phys. Lett. 63 406.

Ummarino, G., Calzolari, A., Daghero, D., Gonnelli, R., Stepanov, V., Masini, R., Cimberle, M. (2003) Proc. 6th EUCAS Conf., 14–18 Sept. 2003, Sorrento.

Upton, M. H., Forrest, T. R., Walters, A. C., Howard, C. A., Ellerby, M., Said, A. H., and McMarrow, D. F. (2010) Phys. Rev. B 82 134515.

Upton, M. H., Walters, A. C., Howard, C. A., Rahnejet, K. C., Ellerby, M., Hill, J. P., McMarrow, D. F., Alatas, A., Leu, B. M., and Ku, W. (2007) Phys. Rev. B 76 220501.

Usui, H., Suzuki, K., and Kuroki, K. (2012) Phys. Rev. B 22 220510.

Usui, N., Ogasawara, T., and Yasukochi, K. (1968) Phys. Lett. 27A 529.

Uwatoko, Y., Oomi, G., Canfield, P. C., and Cho, B. K. (1996) Physica B 216 329.

Vajpayee, A., Huhtinen, H., Awana, V. P. S., Gupta, A., Rawat, R., Lalla, N. P., Kishan, H., Lahio, R., Felner, I., and Narlikar, A. V. (2007) Supercond. Sci. Technol. 20 S155.

van der Meulen, H. P., Tarnawski, Z., de Visser, A., Franse, J. J. M., Perenboom, J. A. A. J., Althof, D., and van Kempen, H. (1990) Phys. Rev. B 41 9352.

van Dijk, N. H., Fak, B., Charvolin, T., Lejay, P., and Mignot, J. M. (2000) Phys. Rev. B 61 8922.

van Sprang, M., Boer, R. A., Riemersma, A. J., Roeland, L. W., Menovsky, A., Franse, J. J. M., and Schoenes, J. (1988) J. Magn. Magn. Mat. 76–77 229.

van Vijfeijken, A. G., and Niessen, A. K. (1965) Phys. Lett. 16 23.

Vargoz, E., and Jaccard, D. (1998) J. Magn. Magn. Mat. 177–181, 294.

Varma, C. M. (1997) Phys. Rev. B 55 14554.

Varma, C. M., and Zhu, L. (2006) Phys. Rev. Lett., 96 036405.

Varma, C. M., Schmitt-Rink, S., and Abrahams, E. (1987) Sol. State. Commun. 62 681.

Veringa, H., Hoogendam, P., and van Wees, A. C. A. (1983) IEEE Trans. Magn. 19 773.

Veringa, H., Hornsveld, E. M., and Hoogendam, P. (1984) Adv. Cryo. Eng. 30 813.

Vieland, L. J., Cohen, R. W., and Rehwald, W. (1971) Phys. Rev. Lett. 26 373.

Vieland, L. T., and Wickland, A. W. (1974) Phys. Lett. 49A 407.

Vilmercati, P., Fedorov, A., Vobornik, I., Manju, U., Panaccione, G., Goldoni, A., Sefat, A. S., McGuire, M. A., Sales, B. C., Jin, R., Mandrus, D., Singh, D. J., and Mannella, N. (2009) Phys. Rev. B 79 220503.

Vollmer, R., Faisst, A., Pfleiderer, C., v. Löhneysen, H., Bauer, E. D., Ho, P.-C., Zapf, V., and Maple, M. B. (2003) Phys. Rev. Lett. 90 5700.

Volovik, G. E. (1988) JETP 67 1804.

Vostner, A., Sun, Y. F., Weber, H. W., Cheng, Y. S., Kursumovic, A., and Evetts, J. E. (2003) Physica C 399 120.

Vuletic, T., Auban-Senzier, P., Pasquier, C., Tomi, S., Jérome, D., Héritier, M., and Bechgaard, K. (2002) Eur. Phys. J. B 25 319.

Wacenovsky, M., Miletich, R., Weber, H. W., and Murakami, M. (1993) Cryogenics 33 70.

Wagner, K. E., Morosan, E., Hor, Y. S., Tao, J., Zhu, Y., Sanders, T., McQueen, T. M., Zandbergen, H. W., Williams, A. J., West, D. V., and Cava, R. J. (2006) Phys. Rev. B 78 104520.

Waldram, J. (1964) Adv. Phys. 13 1.

Walsh, W. M. Jr., Wudl, F., Aharon-Shalom, E., Rupp, L. W. Jr., Vandenberg, J. M., Andres, K., and Torrance, J. B. (1982) Phys. Rev. Lett. 49 885.

Wälte, A., Fuchs, G., Müller, K. H., Handstein, A., Nenkov, K., Narozhnyi, V. N., Drechsler, S.-L., Shulga, S., Schultz, L., and Rosner, H. (2004) Phys. Rev. B 70 174503.

Wälti, C., Ott, H. R., Fisk, Z., and Smith, J. L. (2000) Phys. Rev. Lett. 84 5616.

Wälti, C., Felder, E., Degen, C., Wigger, G., Monnier, R., Delley, T., and Ott, H. R. (2001) Phys. Rev. B 64 172515.

Wang, A. F., Ying, J. J., Yau, Y. J., Liu, R. H., Luo, X. G., Li, Z. Y., Wang, X. F., Zhang, M., Ye, G. J., Chang, P., Xiang, Z. J., and Chen, X. H. (2011a) Phys. Rev. B 83 060512.

Wang, C. H., Chen, X. H., Wu, T., Luo, X. G., Wang, G. Y., and Luo, J. L. (2006) Phys. Rev. Lett. 96 216401.

Wang, C., Li, L., Chi, S., Zhu, Z., Ren, Z., Li, Y., Wang, X. L., Luo, Y., Jiang, S., Xu, X., Cao, G., and Xu, Z. (2008) Europhys. Lett. 83 67006.

Wang, H. D., Dong, C. H., Li, Z. J., Mao, Q. H., Zhu, S. S., Feng, C. M., Yuan, H. Q., and Fang, M. H. (2011b) Europhys. Lett. 93 47004.

Wang, J. L., Xu, Y., Zeng, Z., Zeng, Q. Q., and Lin, H. Q. (2002) J. Appl. Phys. 91 8504.

Wang, J. S., and Wang, S. Y. (2005) Frontiers in Superconducting Materials (Ed. Narlikar, A. V.). Springer, Berlin, p. 885.

Wang, J., Bugoslavsky, Y., Berenov, A., Cowey, L., Caplin, A. D., Cohen, L. F., and MacManus Driscol, J. L. (2002) Appl. Phys. Lett. 81 2026.

Wang, L., Qi, Y. P., Wang, D. L., Zhang, X. P., Gao, Z. S., Zhang, Z. Y., Ma, Y. W., Awaji, S., Nishijima, G., and Watanabe, K. (2010) Physica C 470 183.

Wang, L., Qi, Y. P., Zhang, X. P., Wang, D. L., Gao, Z. S., Wang, C. L., Yao, C., and Ma, Y. W. (2011c) Physica C 471 1689.

Wang, X. C., Liu, Q., Lv, V., Gao, W., Yang, X. L., Yu, R. C., Li, F. Y., and Jin, C. (2008) Sol. State. Commun. 148 538.

Wang, X. F., Wu, T., Wu, G., Liu, R. H., Chen, H., Xie, Y. L., and Chen, X. H. (2009) New J. Phys. 11 045003.

Wang, X. L., Soltanian, S., Horvat, J., Li, A. H., Qin, M. J., Liu, H. K., and Dou, S. X. (2001) Physica C 361 149.

Wang, Y., Rogado, N. S., Cava, R. J., and Ong, N. P. (2003) Nature 423 425.

Wang, Z., Song, Y. J., Shi, H. L., Wang, Z. W., Chen, Z., Tian, H. F., Chen, G. F., Guo, J. G., Yang, H. X., and Li, J. Q. (2011d) Phys. Rev B 83 140505.

Warren, W. W. Jr., Walstedt, R. E., Brennert, G. F., Cava, R. J., Tycko, R., Bell, R. F., and Dabbagh, G. (1989) Phys. Rev. Lett. 62 1193.

Wastin, F., Boulet, P., Rebizant, J., Colineau, E., and Lander, G. H. (2003) J. Phys.: Cond. Mat. 15 82279.

Watanabe, S., and Miyake, K. (2002) J. Phys. Soc. Jpn 71 2489.

Watanabe, K., Yamane, H., Kurosawa, H., Hirai, T., Kobayashi, N., Iwasaki, H., Noto, K., and Muto, Y. (1989) Appl. Phys. Lett. 54 575.

Watanabe, M., Hai, D. P., and Kadowaki, K. (2004a) 4th Int. Symp. Intrinsic Josephson Effect and Plasma Oscillations in High T_c Superconductors, Tsukuba.

Watanabe, T., Wada, K., Ohashi, Y., Ozaki, M., Yamamoto, K., Maeda, T., and Hirabayashi, I. (2002) Physica C 378–381 911.

Watanabe, T., Izawa, K., Kasahara, Y., Haga, Y., Onuki, Y., Thalmeier, P., Maki, K., and Matsuda, Y. (2004b) Phys. Rev. B 70 184502.

Watanabe, T., Fujii, T., and Matsuda, A. (2004c) arXiv:cond-mat/0401448v3 [cond-mat.supr-con] 30 Jan. 2004.

Webb, G. W., Vieland, L. T., Miller, R. E., and Wickland, A. W. (1971) Sol. State. Commun. 9 1769.

Weger, M. (1964) Rev. Mod. Phys. 36 175.

Weger, M., and Goldberg, I. B. (1973) Solid State Physics, Vol. 28 (Ed. Seitz, F., and Turnbull, D.). Academic Press, New York, p. 1.

Weinstein, R., Liu, J., Ren, Y., Sawh, R.-P., Parks, D., Foster, C., and Obot, V. (1996) Proc. 10th Anniversary HTS Workshop on Physics, Materials and Applications (Eds. Batlogg, B., Chu, C. W., Chu, W. K., Gubser, D. U., and Müller, K. A.). World Scientific, Singapore, p. 625.

Weller, T. E., Ellerby, M., Saxena, S. S., Smith, R. P., and Skipper, N. T. (2005) Nature Phys. 1 39.

Welp, U., Kwok, W. K., Crabtree, G. W., Claus, H., Vandervoort, K. G., Dabrowski, B., Mitchell, A. W., Richards, D. R., Marx, D. T., and Hinks, D. G. (1988) Physica C 156 27.

Werthamer, N. R., Helfand, E., and Hohenberg, H. C. (1966) Phys. Rev. 147 295.

Wertheim, G. K., Dislavo, F., and Chiang, S. (1976) Phys. Rev. B 13 5476.

West, A. W., and Rawlings, R. D. (1977) J. Mat. Sci. 12 1962.

White, G. K. (1962) Phil. Mag. 7 271.

White, G. K. (1962) Cryogenics 2 292.

White, G. K. (1992) Studies of High Temperature Superconductors, Vol. 9 (Ed. Narlikar, A.). Nova Science, New York, p. 121.

Whitney, D. A., Fleming, R. M., and Coleman, R. V. (1977) Phys. Rev. B 15 3405.

Williams, G. V. M., and Krämer, S. (2000) Phys. Rev. B 62 4132.

Williams, G. V. M., and Ryan, M. (2001) Phys. Rev. B. 64 094515.

Williams, J. (1990) Organic Superconductivity (Eds. Kresin V. Z., and Little, W. A.). Plenum Press, New York.

Williams, J. M., Kini, A. M., Wang, H. H., Carlson, K. D., Geiser, U., Montgomery, L. K., Pyrka, G. J., Watkins, D. M., Kommers, J. M., Boryschuk, S. J., Strieby-Crouch, A. V., Kwok, W. K., Schirber, J. E., Overmeyer, D. L., Jung, D., and Whangbo, M.-H. (1990) Inorg. Chem. 29 3272.

Wilson, J. A., DiSalvo, F. J., and Mahajan, S. (1974) Phys. Rev. Lett. 32 882.

Wilson, J. A., DiSalvo, F. J., and Mahajan, S. (1975) Adv. Phys. 24 117.

Wilson, M. N., Walter, C. R., Lewin, J. D., and Smith, P. F. (1970) J. Phys. D 3 1518.

Winkel, A. V., and Bakker, H. (1985) J. Phys. F 15 1556.

Witanachchi, S., Kwok, H. S., Wang, X. M., and Shaw, T. W. (1988) Appl. Phys. Lett. 53 234.

Witcomb, M. J., and Narlikar, A. V. (1972) Phys. Stat. Sol. (a) 11 311.

Witcomb, M. J., Echarri, A., Dew-Hughes, D., and Narlikar, A. V. (1968) J. Mat. Sci. 3 191.

Withers, R. L., and Wilson, J. A. (1986) J. Phys. C: Sol. State Phys. 19 4809.

Wollman, D. A., Van Harlingen, D. J., Giapintzakis, J., and Ginsberg, D. M. (1995) Phys. Rev. Lett. 74 797.

Wolowiec, C. T., Yazici, D., White, B. D., Huang, K., and Maple, M. B. (2013) arXiv:1307. 4157v1[cond-mat.supr-con].

Wong, J. (1962) Superconductors (Eds. Tanenbaum, M., and Wright, W. V.). Wiley-Interscience, New York, p. 83.

Woodfield, B. F., Wright, D. A., Fisher, R. A., Phillips, N. E., and Tang, H. Y. (1999) Phys. Rev. Lett. 83 4622.

Wosnitza, J. (2000) Studies of High Temperature Superconductors, Vol. 34 (Ed. Narlikar, A. V.). Nova Science, New York, p. 97.

Wosnitza, J., Wanka, S., Hagel, J., Reibelt, M., Schweitzer, D., and Schlueter, J. A. (2003) Synth. Met. 133–134 201.

Wu, G., Chen, H., Wu, T., Xie, Y. L., Yan, Y. J., Liu, R. H., Wang, X. F., Ying, J. J., and Chen, X. H. (2008) J. Phys.: Cond. Mat. 20 422201.

Wu, J. Z., Xie, Y. Y., Xing, Z. W., and Aga, R. S. (2002) Physica C 382 62.

Wu, M. K., Ashburn, J. R., Torng, C. J., Hor, P. H., Meng, R. L., Gao, L., Huang, Z. J., Wang, Y. Q., and Chu, C. W. (1987) Phys. Rev. Lett. 58 908.

Wu, X. D., Dijjkkamp, D., Ogale, S. B., Inam, A., Chase, E. W., Miceli, P. E., Chang, C. C., Tarascon, J. M., and Venkatesan, T. (1989) Appl. Phys. Lett. 55 2271.

Wu, X.-D., Xu, K.-X., Fang, H., Jiao, Y.-L., Xiao, L., and Zheng, M.-H. (2009) Supercond. Sci. Technol. 22 125003.

Wu, X. J., Jin, C. Q., Adachi, S., and Yamauchi, H. (1994) Physica C 224 175.

Wzietek, P., Mayaffre, H., Jérome, D., and Brazovskii, S. (1996) J. Phys. (Paris) 6 2011.

Xi, X. X. (2005) Frontiers in Superconducting Materials (Ed. Narlikar, A. V.). Springer, Heidelberg, p. 1073.

Xiao, Y., Su, Y., Meven, M., Mittal, R., Kumar, C. M. N., Chatterjee, T., Price, S., Persson, J., Kumar, N., Dhar, S. K., Thamizhavel, A., and Brueckel, Th. (2009) Phys. Rev. B 80 174424.

Xing, J., Li, S., Ding, X., Yang, H., and Wen, H. H. (2012) Phys. Rev. B 86 214518.

Xu, G. J., Grivel, J.-C., Abrahmsen, A. B., and Andersen, N. H. (2004) Physica C 406 95.

Xu, M., Kitazawa, H., Takano, Y., Ye, J., Nishida, K., Abe, H., Matsushita, A., Tsujii, N., and Kido, J. (2001) Appl. Phys. Lett. 79 2779.

Xue, Y. Y., Tsui, S., Cmaidalka, J., Meng, R. L., Lorenz, B., and Chu, C. W. (2000) Physica C 341 483.

Xue, Y. Y., Lorenz, B., Bailakov, A., Cao, D. H., Li, Z. G., and Chu, C. W. (2002) Phys. Rev. B 66 014503.

Xue, Y. Y., Lorenz, B., Cao, D. H., and Chu, C. W. (2003) Phys. Rev. B 67 184507.

Yadav, C. S., and Paulose, P. L. (2012) J. Phys.: Cond. Mat. 24 235702.

Yadav, C. S., and Paulose, P. L. (2012a) AIP Conf Proc. 1447, 895.

Yagubskii, E. B., Shchegolev, I. F., Laukhin, V. N., Kononovich, P. A., Kartsovnik, M. V., Zvarikina, A. V., and Buravov, L. I. (1984) JETP Lett. 39 12.

Yamada, K., Lee, C. H., Kurahashi, K., Wada, J., Wakimoto, S., Ueki, S., Kimura, H., Endoh, Y., Hosoya, S., Shirane, G., Birgeneau, R. J., Greven, M., Kastner, M. A., and Kim, Y. J. (1998) Phys. Rev. B 57 6165.

Yamada, Y. (1992) Sumitomo Electr. Tech. Rev. 33 83.

Yamada, Y. (2000) Supercond. Sci. Technol. 13 82.

Yamada, Y., and Hirabayashi, I. (2001) J. Cryst. Growth 229 343.

Yamada, Y., and Shiohara, Y. (2004) High Temperature Superconductivity 1: Materials (Ed. Narlikar, A. V.). Springer, Heidelberg, p. 291.

Yamamoto, A., Takao, C., Masui, T., Izumi, M., and Tajima, S. (2002) Physica C 383 197.

Yamamoto, A., Shimoyama, J., Ueda, S., Iwayama, I., Horii, S., and Kishio, K. (2005) Supercond. Sci. Technol. 18 1323.

Yamamoto, A., Jaroszynski, J., Tarantini, C., Balicao, L., Jiang, J., Gurevich, A., Larbalestier, D. C., Jin, R., Sefat, A. S., McGuire, M. A., Sales, B. C., Christien, D. K., and Mandrus, D. (2009) Appl. Phys. Lett. 94 062511.

Yamamoto, M. (1978) J. Phys. Soc. Jpn 45 431.

Yamanaka, S. (2000) Annu. Rev. Mat. Sci. 30 53.

Yamanaka, S., Hotehama, K., and Kawaji, H. (1998) Nature 392 580.

Yamanaka, S., Kawaji, H., Hotehama, K., and Ohashi, M. (1996) Adv. Mat. 8 771.

Yamanaka, S., Hotehama, K., Koiwasaki, T., Kawaji, H., Fukuoka, H., Shamoto, S., and Kajitani, T. (2000) Physica C 341–348 699.

Yamanaka, S., Zhu, L., Chen, X., and Tou, H. (2003) Physica B 328 6.

Yamanaka, S., Yasunaga, T., Yamaguchi, K., and Tagawa, M. (2009) J. Mat. Chem. 19 2573.

Yamane, H., Kurosawa, H., Iwasaki, H., Hirai, T., Kobayashi, N., and Muto, Y. (1989) J. Appl. Phys. 28 L827.

Yamaura, J., Yonezawa, S., Muraoka, Y., and Hiroi, Z. (2005) J. Sol. State Chem. 179 336.

Yamaya, K., Takayangi, S., and Tanda, S. (2012) Phys. Rev. B. 85 184513.

Yang, H. D., Mollah, S., Huang, H. L., Ho, P. L., Huang, W. L., Liu, C.-J., Lin, J.-Y., Shang, Y.-L., Yu, R. C., and Jin, C.-Q. (2003) Phys. Rev. B 68 92507.

Yang, H., Luo, H., Wang, Z., and Wen, H. H. (2008) Appl. Phys. Lett. 93 142506.

Yang, J., Li, Z. C., Lu, W., Yi, W., Shen, X. L., Ren, Z. A., Che, G. C., Dong, X. L., Sun, L. L., Zhou, F., and Zhao, Z. X. (2008) Supercond. Sci. Technol. 21 082001.

Yang, P., and Lieber, Ch. M. (1996) Science 273 1836.

Yao, C., Wang, C., Zhang, X., Wang, L., Gao, Z. S., Wang, D. L., Wang, C., Qi, Y. P., Ma, Y. W., Awaji, S., and Watanabe, K. (2012) Supercond. Sci. Technol. 25 035020.

Yao, Q. W., Wang, X. L., Soltanian, S., Li, A. H., Horvat, J., and Dou, S. X. (2004) Ceram. Int. 30 1603.

Yaron, U., Gammel, P. L., Ramirez, A. P., Huse, D. A., Bishop, D. J., Goldman, A. I., Stassis, C., Canfield, P. C., Mortensen, K., and Eskildsen (1996) Nature 382 236.

Yashima, M., Nishimura, H., Mukuda, H., Kitaoka, Y., Miyazawa, K., Shirage, P. M., Kihou, K., Kito, H., Eisaki, H., and Iyo, A. (2009) J. Phys. Soc. Jpn 78 103702.

Yatskar, A., Budraa, N. K., Bayerman, W. P., Canfield, P. C., and Bud'ko, S. L. (1996) Phys. Rev. B 54 R3772.

Ye, F., Chi, S., Bao, W., Wang, X. F., Ying, J. J., Chen X. H., Wang, H. D., Dong, C. H., and Fang, M. H. (2011) Phys. Rev. Lett. 107 137003.

Yeh, K.-W., Huang, T.-W., Lin Huang, Y., Chen, T.-K., Hsu, F.-C., Wu, P. M., Li, Y.-C., Chu, Y.-Y., Chen, C.-L., Luo, J.-Y., Yan, D.-C., and Wu, M. K. (2008) Europhys. Lett. 84 37002.

Yeh, K.-W., Ke, C. T., Huang, T.-W., Chen, T. K., Huang, Y. L., Wu, P. M., and Wu, M. K. (2009) Cryst. Growth Des. 30 900675e.

Yeoh, W. K., Kim, J. H., Horvat, J., Xu, X., Qin, M. J., Dou, S. X., Jiang, C. H., Nakane, T., Kumakaru, H., and Munroe, P. (2006a) Supercond. Sci. Technol. 19 596.

Yeoh, W. K., Kim, J. H., Horvat, J., Dou, S. X., and Munroe, P. (2006b) Supercond. Sci. Technol. 19 L5.

Yethiraj, M., McPaul, D., Tomy, C. V., and Forgan, E. M. (1997) Phys. Rev. Lett. 78 4849.

Yildirim, T., Barbedette, L., Kniaz, K., Fischer, J. E., Lin, C. L., Bykovetz, N., Stephens, P. W., Sulewski, P. E., and Erwin, S. C. (1995) MRS Proc. 359 273.

Yildirim, T., Barbedette, L., Fischer, J. E., Lin, C. L., Robert, J., Petit P. and Palstra, T. T. M. (1996) Phys. Rev. Lett. 77 167.

Yin, Q., Ylvisaker, E. R., and Pickett, W. E. (2011) Phys. Rev. B 83 014509.

Yogi, M., Kitaoka, Y., Hashimoto, S., Yasuda, T., Settai, R., Matsuda, T. D., Haga, Y., Onuki, Y., Rogl, P., and Bauer, E. (2004) Phys. Rev. Lett. 93 027 003.

Yokoi, M., Watanabe, H., Mori, Y., Moyoshi, T., Koboyashi, Y., and Sato, M. (2004) J. Phys. Soc. Jpn 73 1297.

Yokoi, M., Koboyashi, Y., Sato, M., and Sugai, S. (2008) J. Phys. Soc. Jpn 77 094713.

Yokota, K., Kurata, G., Matsui, T., and Fukuyama, H. (2000) Physica B 284–288 551.

Yokoya, T., Chainani, A., and Takahashi, T. (1996) Phys. Rev. B 54 13311.

Yokoyama, T., Sawa, Y., Tanaka, Y., and Golubov, A. A. (2007) Phys. Rev. B 75 020502(R).

Yoma, R., Yamaya, K., Abliz, M., Hedo, M., and Uwatoko, Y. (2005) Phys. Rev. B 71 132508.

Yonezawa, S., Muraoka, Y., and Hiroi, Z. (2004a) J. Phys. Soc. Jpn 73 1655.

Yonezawa, S., Muraoka, Y., Matsushita, Y., Hiroi, Z. (2004b) J. Phys. Soc. Jpn 73 819.

Yonezawa, S., Muraoka, Y., Matsushita, Y., Hiroi, Z. (2004c) J. Phys.: Cond. Mat. 16 L9.

Yoshida, M., Nakamoto, T., Kitamura, T., Hyun, O. B., Hirabayashi, I., and Tanaka, S. (1994) Appl. Phys. Lett. 65 1714.

Yoshida, Y., Ito, Y., Hirabayashi, I., Nagai, H., and Takai, Y. (1996) Appl. Phys. Lett. 69 845.

Yoshida, Y., Mukai, A., Settai, R., Miyake, K., Inada, Y., Onuki, Y., Betsuyaku, K., Harima, H., Matsuda, T. D., Aoki, Y., and Sato, H. (1999) J. Phys. Soc. Jpn 68 3041.

Yoshizaki, K., Wakata, M., Miyashita, S., Fujiwara, F., Taguchi, O., Imaijumi, M., and Hashimoto, Y. (1985) IEEE. Trans. Magn. 21 301.

Yoshizaki, R., Sawada, H., Iwazumi, T., and Ikeda, H. (1988) Sol. State. Commun. 65 1539.

Young, D. P., Goodrich, R. G., Adams, P. W., Chan, J. Y., Fronczek, F. R., Drymiotis, F., and Henry, L. L. (2002) Phys. Rev. B 65 180518(R).

Yu, Y., Aczel, A. A., Williams, T. J., Budko, S. L., Ni, N., Canfield, P. C., and Luke, G. M. (2009) Phys. Rev. B. 79 020511.

Yuan, H. Q., Grosche, F. M., Deppe, M., Geibel, C., Sparn, G., Steglich, F. (2003) Science 302 2104.

Yuan, H. Q., Singleton, J., Balakirev, F. F., Baily, S. A., Chen, G. F., Luo, J. L., and Wang, N. L. (2009) Nature 457 565.

Yuan, R. H., Dong, T., Song, Y. J., Zheng, P., Chen, G. F., Hu, J. P., Li, J. Q., and Wang, N. L. (2012) Sci. Rep. (China) 2:221 1.

Yvon, K. (1979) Curr. Top. Mat. Sci. 3 53.

Zakhidov, A. A., Imaeda, K., Petty, D. M., Yakushi, K., Inokuchi, H., Kikuchi, K., Ikemoto, I., Suzuki, S., and Achiba, Y. (1992) Phys. Lett. 164A 355.

Zandbergen, H. W., Huang, Y. K., Menken, M. J. V., Li, J. N., Kadowaki, K., Menovsky, A. A., van Tendeloo, G., and Amelinckx, S. (1988) Nature 332 620.

Zandbergen, H. W., Cava, R. J., Krajewski, J. J., and Peck, W. F. Jr. (1994) J. Sol. State Chem. 110 196.

Zapf, V. S., Freeman, E. J., Bauer, E. D., Petricka, J., Sirvent, C., Frederick, N. A., Dickey, R. P., and Maple, M. B. (2001) Phys. Rev. B 65 014506.

Zareapour, P., Hayat, A., Zhao, S. Y. F., Kreshchuk, M., Jain, A., Kwok, D. C., Lee, N., Cheong, S.-W., Xu, Z., Yang, A., Gu, G. D., Jia, S., Cava, R. J., and Burch, K. S. (2012) Nature Commun. 3 Article 1056.

Zarestky, J., Stassis, C., Goldman, A. I., Canfield, P. C., Dervenagas, P., Cho, B. K., and Johnston, D. C. (1995) Phys. Rev. B 51 678.

Zasadzinski, J. F., Tralshawala, N., Huang, Q., Gray, K. E., and Hinks, D. G. (1991) IEEE Trans. Magn. 27 833.

Zeitlin, B. A., Ozeryansky, G. M., and Hemachalam, K. (1985) IEEE Trans. Magn. 21 293.

Zeritis, D., Iwasa, Y., Ando, T., Takahashi, Y., Nishi, M., Nakajima, H., and Shimamoto, S. (1991) IEEE Trans. Magn. 27 1829.

Zhang, J., Zhao, J., Marcy, H. O., Tonge, L. M., Wessels, B. W., Marks, T. J., and Kannewurf, C. R. (1989) Appl. Phys. Lett. 54 1166.

Zhang, J. L., Jiao, L., Chen, Y., Yuan, H. Q. (2012) arXiv:1201.2548v1[cond-mat.supr-con].

Zhang, K., Kwak, B. S., Boyd, E. P., Wright, A. C., and Erbil, A. (1989) Appl. Phys. Lett. 54 380.

Zhang, L. J., Singh, D. J., and Du, M. H. (2009) Phys. Rev. B 79 012506.

Zhang, L., Guan, P. F., Feng, D. L., Chen, X. H., Xie, S. S., and Chen, M. W. (2010) J. Am. Chem. Soc. 132 15223.

Zhang, S.-C. (1997) Science 275 1089.

Zhao, G.-M. (2007) Phys. Rev. B 76 020501(R).

Zhao, G.-M., and Morris, D. E. (1995a) Phys. Rev. B 51 12848.

Zhao, G., and Morris, D. E. (1995b) Phys. Rev. B 51 16487.

Zhao, G. M., and Wang, Y. S. (2003) Unpublished.

Zhao, G., Conder, K., Keller, H., and Müller, K. A. (1998) J. Phys.: Cond. Mat. 10 9055.

Zhao, G., Keller, H., Conder, K. (2001) J. Phys.: Cond. Mat. 13 R569.

Zhao, J., Huang, Q., de la Cruz, C., Lynn, J. W., Lumsden, M. D., Ren, Z. A., Yang, J., Shen, X., Dong, X., Zhao, Z., and Dai, P. (2008a) Phys. Rev. B 78 132504.

Zhao, J., Huang, Q., de la Cruz, C., Li, S., Lynn, J. W., Chen, Y., Green, M. A., Chen, G. F., Li, G., Li, Z., Luo, J. L., Wang, N. L., and Dai, P. (2008b) Nature Mat. 7 953.

Zhao, J., Ratcliff, W., Lynn, J. W., Chen, G. F., Luo, J. L., Wang, N. L., Hu, J. P., and Dai, P. C. (2008c) Phys. Rev. B. 78 140504.

Zhao, Y., Cheng, C. H., Xu, M., Choi, C. H., and Yao, X. (2002a) Studies of High Temperature Superconductors, Vol. 41 (Ed. Narlikar, A.). Nova Science, New York, p. 45.

Zhao, Y., Cheng, C. H., Xu, M., Choi, C. H., and Yao, X. (2002b) Studies of High Temperature Superconductors, Vol. 41 (Ed. Narlikar, A.). Nova Science, New York, p. 77.

Zhao, Y., Cheng, C. H., Feng, Y., Zhou, L., Koshizuka, N., and Murakami, M. (2003) Studies of High Temperature Superconductors, Vol. 44 (Ed. Narlikar, A.). Nova Science, New York, p. 141.

Zheng, D. N., Ramsbottom, H. D., and Hampshire, D. P. (1995) Phys. Rev. B 52 12931.

Zheng, D. N., Ali, S., Hammid, H. A., Eastell, C., Goringe, M., and Hampshire, D. P. (1997) Physica C 291 49.

Zheng, G., Tanabe, T., Mito, K., Kawasaki, S., Kitaoka, Y., Aoki, D., Haga, Y., and Onuki, Y. (2001) Phys. Rev. Lett. 86 4664.

Zheng, G., Matano, K., Meng, R. L., Cmaidalka, J., and Chu, C. W. (2006a) J. Phys.: Cond. Mat. 18 L63.

Zheng, G., Matano, K., Chen, D. P., and Lin, C. T. (2006b) Phys. Rev. B. 73 180503(R).

Zhigadlo, N. D., Katrych, S., Bukowski, Z., Weyneth, S., Puzniak, R., and Karpinski, J. (2008) J. Phys.: Cond. Mat. 20 342202.

Zhitomirsky, M. E., and Rice, T. M. (2001) Phys. Rev. Lett. 87 057001.

Zhong, G. H., Wang, J. L., Zeng, Z., Zheng, X. H., and Lin, H. Q. (2007) J. Appl. Phys. 101 09G520.

Zhu, X. Y., Han, F., Mu, G., Zheng, B., Cheng, P., Shen, B., and Wen, H. H. (2009a) Phys. Rev. B. 79 024516.

Zhu, X. Y., Han, F., Mu, G., Cheng, B., Shen, B., Zheng, B., and Wen, H. H. (2009b) Phys. Rev. B 79 020512.

Zhu, X. Y., Han, F., Mu, G., Cheng, B., Shen, B., Zheng, B., and Wen, H. H. (2009c) Phys. Rev. B 79 220512.

Zhu, X., Leh, H., and Petrovic, C. (2011) Phys. Rev. Lett. 106 246404.

Zhu, Y., Li, Q., Wu, L., Volkov, V., Gu, G., and Moodenbaugh, A. R. (2002) Studies of High Temperature Superconductors, Vol. 38 (Ed. Narlikar, A.). Nova Science, New York, p. 421.

Zou, F., Brooks, J. S., McKenzie, R. H. (2000) Phys. Rev. B 61 750.

Zou, Z., Ye, J., Oka, K., and Nishihara, Y. (1998) Phys. Rev. Lett. 80 1074

Index